Routledge Handbook of World-Systems Analysis

World-systems analysis has developed rapidly over the past thirty years. Today's students and junior scholars come to world-systems analysis as a well-established approach spanning all of the social sciences. The best world-systems scholarship, however, is spread across multiple methodologies and more than half a dozen academic disciplines. Aiming to crystallize forty years of progress and lay the groundwork for the continued development of the field, the *Routledge Handbook of World-Systems Analysis* is a comprehensive review of the state of the field of world-systems analysis since its origins almost forty years ago.

The *Handbook* includes contributions from a global, interdisciplinary group of more than 80 world-systems scholars. The authors include founders of the field, mid-career scholars, and newly emerging voices. Each one presents a snapshot of an area of world-systems analysis as it exists today and presents a vision for the future.

The clear style and broad scope of the *Handbook* will make it essential reading for students and scholars of sociology, anthropology, archaeology, geography, political science, history, and development economics.

Salvatore J. Babones is a senior lecturer in Sociology and Social Policy at The University of Sydney and the author of the *Benchmarking America* blog. He is the author or editor of three books focusing on development, social change, and global health inequalities.

Christopher Chase-Dunn is the Director of the Institute for Research on World-Systems at the University of California at Riverside and was the founding editor of the *Journal of World-Systems Research*. He is the author or editor of fifteen books on topics ranging from pre-modern world-systems to the future of capitalism.

Routledge Handbook of World-Systems Analysis

Edited by Salvatore J. Babones and Christopher Chase-Dunn

Routledge
Taylor & Francis Group
LONDON AND NEW YORK

First published 2012 by Routledge

2 Park Square, Milton Park, Abingdon, Oxon OX14 4RN
711 Third Avenue, New York, NY10017, USA

Routledge is an imprint of the Taylor & Francis Group, an informa business

First issued in paperback 2017

British Library Cataloguing in Publication Data
A catalogue record for this book is available from the British Library

Library of Congress Cataloging-in-Publication Data
Routledge handbook of world-systems analysis / edited by
Salvatore J. Babones and Christopher Chase-Dunn.
p. cm.
Includes bibliographical references and index.
1. Social systems. 2. Economics. 3. Social sciences. 4. System theory.
5. System analysis. I. Babones, Salvatore J. II. Chase-Dunn, Christopher K.
HM701.R68 2012
300.1'1- -dc23
2011032740

ISBN: 978-0-415-56364-2 (hbk)
ISBN: 978-1-138-10719-9 (pbk)

Typeset in Bembo
by Integra Software Services Pvt. Ltd, Pondicherry, India

Contents

Illustrations

Figures

Tables

Contributors

Younes Abouyoub Visiting scholar, Columbia University

E. N. Anderson Department of Anthropology, University of California–Riverside

Kelly F. Austin Department of Sociology & Anthropology, North Carolina State University

Philipp Babcicky Department of Sociology, University of Graz

Salvatore J. Babones Department of Sociology and Social Policy, University of Sydney

Mikhail Balaev Department of Sociology and Social Work, Northern Michigan University

John Barnshaw Department of Sociology, University of South Florida

Albert J. Bergesen Department of Sociology, University of Arizona

Richard E. Blanton Department of Anthropology, Purdue University

Elson E. Boles Department of Sociology, Saginaw Valley State University

Volker Bornschier Department of Sociology, Zurich University

John Borrego Department of Latin American & Latino Studies, University of California–Santa Cruz

Nicole Bousquet Department of Sociology, Université Laval

Ian Breckenridge-Jackson Department of Sociology, University of California–Riverside

Paul D. Buell Horst-Goertz-Institute, Charité – Medical University Berlin

Thomas J. Burns Department of Sociology, University of Oklahoma

Jon D. Carlson Department of Political Science, University of California–Merced

Christopher Chase-Dunn Department of Sociology, University of California–Riverside

Jenny Chesters Research School of Economics, Australian National University

Paul S. Ciccantell Department of Sociology, Western Michigan University

Rebecca Clausen Department of Sociology, Fort Lewis College

Donald A. Clelland Department of Sociology, University of Tennessee–Knoxville

Samuel Cohn Department of Sociology, Texas A&M University

Sandra Curtis Comstock Department of Sociology, Reed College

Gary Coyne Department of Sociology, University of California–Riverside

Robert A. Denemark Department of Political Science and International Relations, University of Delaware

Raymond J. Dezzani Department of Geography, University of Idaho

Kevin Doran Department of Sociology, Indiana University

Marilyn Dudley-Flores School of Social Sciences, Atenisi University

Wilma A. Dunaway Program in Government and International Affairs, Virginia Polytechnic Institute

Edwin Elias Department of Sociology, University of California–Riverside

Elena Ermolaeva Department of Sociology and Anthropology, Marshall University

Robbie Ethridge Department of Sociology and Anthropology, University of Mississippi

David W. Everson Department of Sociology, University of Notre Dame

Lane F. Fargher Department of Anthropology, Purdue University

James V. Fenelon Department of Sociology, California State University

R. Scott Frey Department of Sociology, University of Tennessee–Knoxville

Zbigniew Galor Department of Social Sciences, Poznan University of Life Sciences

Thomas Gangale School of Social Sciences, Atenisi University

Brian J. Gareau Department of Sociology, Boston College

Paul K. Gellert Department of Sociology, University of Tennessee

Barry K. Gills School of Geography, Politics and Sociology, Newcastle University

W. L. Goldfrank Department of Sociology, Latin American & Latino Studies, University of California–Santa Cruz

Peter E. Grimes Department of Sociology, Anthropology, and Social Work, Central Michigan University

Leonid Grinin Volgograd Center for Social Research

John L. Gulick Department of Sociology, Brooklyn College

Thomas D. Hall Department of Sociology and Anthropology, DePauw University

Alf Hornborg Human Ecology Division, Lund University

Shahadat Hossain Department of Sociology, University of Dhaka

Hiroko Inoue Institute of Research on World-Systems, University of California–Riverside

Andrew K. Jorgenson Department of Sociology, University of Utah

Golam Sarwar Khan Department of Sociology, University of Sydney

Edward L. Kick Department of Sociology & Anthropology, North Carolina State University

Piotr Konieczny Department of Sociology, University of Pittsburgh

Andrey Korotayev Program in Anthropology of the East, Russian State University for the Humanities

Kirk S. Lawrence Department of Sociology, University of California–Riverside

Frederick William Lee Department of Sociology, University of Sydney

Richard E. Lee Department of Sociology, Binghamton University

Lynn Letukas Department of Sociology and Criminal Justice, University of Delaware

Ligaya Lindio-McGovern Department of Sociology, History, and Political Science, Indiana University Kokomo

Delario Lindsey Department of Africana-World Studies, William Paterson University

Stefano B. Longo Department of Environmental Studies, University of Illinois-Springfield

James Love Department of Sociology, University of California–Riverside

Sergey Malkov Department of Applied Mathematics, Russian State Social University

Laura A. McKinney Department of Sociology & Anthropology, North Carolina State University

Eric Mielants College of Arts and Sciences, Fairfield University

Valentine M. Moghadam Department of Sociology, Purdue University

Timothy Patrick Moran Department of Sociology, Stony Brook University

Melsome Nelson-Richards Department of Sociology, State University of New York at Oswego

J. B. Owens Department of History, Idaho State University

Christof Parnreiter Department of Geography, Hamburg University

Aaron Z. Pitluck Political Economy Research Group, Central European University

Ellen Reese Department of Sociology, University of California–Riverside

Thomas Ehrlich Reifer Department of Sociology, University of San Diego

James Rice Department of Sociology, New Mexico State University

Matthew R. Sanderson Department of Sociology and Anthropology, Lehigh University

Stephen J. Scanlan Department of Sociology & Anthropology, Ohio University

Jon Shefner Department of Sociology, University of Tennessee

Kyoung-Ho Shin Department of Psychology, Sociology and Counseling, Northwest Missouri State University

Kristen Shorette Department of Sociology, University of California–Irvine

Jeremy Simpson Department of Sociology and Social Policy, University of Sydney

Leslie Sklair Department of Sociology, London School of Economics and Political Science

David A. Smith Department of Sociology, University of California–Irvine

Alvin Y. So Division of Social Science, Hong Kong University of Science and Technology

Roksolana Suchowerska School of Social and Political Sciences, University of Melbourne

Christian Suter Department of Sociology, University of Neuchâtel

Michael Timberlake Department of Sociology, University of Utah

Eric Vanhaute Department of Sociology, Ghent University

Claudia von Werlhof Department of Political Science, University of Innsbruck

Immanuel Wallerstein Department of Sociology, Yale University

David Wilkinson Department of Political Science, University of California–Los Angeles

Jessie Winitzky Department of Sociology, University of Utah

Nobuyuki Yamada Department of Sociology, Komazawa University

Introduction

Salvatore J. Babones and Christopher Chase-Dunn

Most academic handbooks have imprinted on them their editors' distinctive visions of the fields they cover. The editors set the boundaries of the field; the editors map out the most important areas of inquiry; the editors select specific topics to be covered; the editors solicit the authors who will cover them. The editors impose an internal section and chapter order on what is externally an anarchic intellectual space. Editing a handbook with 80 contributions is a major organizational challenge, and one might forgive editors for organizing it in accordance with their own wills. In the words of Robert Michels, "who says organization, says oligarchy."

This handbook is different.

This handbook has been compiled from the bottom up. Our goal was to reflect the state of world-systems scholarship as it is, rather than as we would like to see it. We put out a call for papers to the community of world-systems scholars. We actively encouraged shy junior scholars to submit their work. We twisted the arms of busy senior scholars to make sure they contributed. We strictly limited the length of contributions in order to make sure that we could include as many different voices as possible. Having found our contributors, we steadfastly refused to tell them what to contribute. We insisted that we wanted their best work, whatever that might be. We then organized the handbook around the submissions we received. We provided the necessary organization, minus the oligarchy.

This approach was firmly in keeping with the character of the field. As Immanuel Wallerstein explains in his conclusion to this volume, world-systems analysis is neither a perspective nor a theory; it is a knowledge movement. It is a holistic approach to understanding the human world using knowledges that range from formal econometric modeling to traditional indigenous learning. Increasingly it extends to interpreting the interaction between the human world and the natural world as well.

World-systems analysts interpret human worlds as whole systems, including cultural, economic, and political subsystems. World-systems analysts do this with a sense of historical depth. While today there clearly exists a single, global world-system, in the past there were many world-systems spread across both historical and geographical space. A key feature of world-systems analysis is that, at its best, it marries formal-analytical social-scientific rigor to interpretive historical-geographical understanding. Though its academic origins lie in sociology, its prominent practitioners include anthropologists, archaeologists, ecologists, economists, geographers, historians, political scientists, and (not coincidentally) many social and political activists.

The close link between world-systems analysis and social and political activism is a testament to the applicability of world-systems analysis beyond the academy. Though many academic world-systems analysts are socially and politically active, the main flow is in the opposite direction: activists use world-systems analysis as a tool for understanding, interpreting, and changing the world. This is a second key feature of world-systems analysis. Like other knowledge movements—environmentalism, feminism, indigenous knowledges—it is engaged in and with the world.

A foundational work of world-systems analysis, cited dozens of times in this handbook, is Immanuel Wallerstein's (1974) *The Modern World-System, Vol. I: Capitalist Agriculture and the Origins of the European World-Economy in the Sixteenth Century*. In this work, Wallerstein presented an analytical history of the emergence of a capitalist world-economy in the long sixteenth century, from 1450 to 1640 CE. He described a crisis of European feudalism in which a squeeze on the seigniorial revenues garnered by feudal lords spurred the emergence of agrarian capitalism and a new hierarchical division of labor between Western and Eastern Europe.

A key advance in Volume 1 of *The Modern World-System* was the discovery of an early core/periphery hierarchy that was based on a dominance/dependence structure that did not involve formal empire. The emergence of serfdom in Eastern Europe, especially Poland, was understood as a form of peripheral capitalism using coerced labor analogous to the expansion of slavery in the plantation complex that was emerging in the New World. This insight formed the basis for Wallerstein's contention that forms of labor relationships that had previously been termed "pre-capitalist" were in fact capitalist—that is, embedded in networks of production for market exchange.

The Modern World-System also includes an eleven-page "Theoretical Reprise" that is the standard sourcebook for Wallerstein's early views on the new field of world-systems analysis. In it he lays out the now-familiar vocabularies of mini-systems, world-empires, and world-economies; core, semiperiphery, and periphery; semiperipheries as agents of change and hegemons as agents of stability. Students looking for an overview of the world-systems analysis as originally conceived would be well-advised to read it for themselves. It remains one of the best short introductions to world-systems analysis.

Less known is the fact that Wallerstein published an article-length summary of the world-systems approach in the same year (1974), "The Rise and Future Demise of the World Capitalist System: Concepts for Comparative Analysis" (*Comparative Studies in Society and History*, Vol. 16, pp. 387–415). Though it covered much the same ground as the more famous "Theoretical Appendix," it is interesting for two reasons. First, it deals with (then and still) current issues, rather than with historical subject matter that some readers might (incorrectly) dismiss as irrelevant to contemporary debates. People with no interest in the sixteenth century can read it, understand it, learn from it, and enjoy it.

Second, and more importantly, it presents a nascent case for understanding world-systems analysis as an integrative knowledge movement. Those words are not used in the 1974 essay, but Wallerstein draws on and exposes the limitations of modernization theory, dependency theory, orthodox Marxism, and Braudelian historiography as precursors to world-systems analysis. Also in this paper, Wallerstein explicitly emphasizes the need to understand distinct cultural, political, and economic bases of social stratification systems and how they interact. In short, the main points of Wallerstein's conclusion to this handbook are prefigured in this 1974 essay. This early essay is at once seminal and valedictory: in short, a classic.

Though some think of Wallerstein as the sole creator of the world-systems perspective, very similar approaches were emerging in the works of several of his colleagues and collaborators in the 1970s, especially Samir Amin, Giovanni Arrighi, Andre Gunder Frank, and Terence Hopkins. Together these seminal thinkers discovered, or rediscovered, and reinterpreted the modern system of national states and the capitalist world-economy that emerged with the rise of Europe. Their vocabularies were slightly different, but this was clearly a single interactive intellectual project and all of these authors should be understood as the founders of world-systems analysis.

The founders realized that core/periphery relations have been characteristic of the rise of capitalism in Europe for centuries. This was not just a phase of "primitive accumulation" that set the motor of capitalism going. It was an evolving system in which new institutional structures of exploitation emerged as earlier structures (like formal colonialism) came to an end. This insight forms the basis for nearly the entire broad array of views on the genesis, operation, and future of the modern world-system presented in this handbook. World-systems analysis, however, has since taken on a life of its own well beyond the work of its founding generation. Numerous scholars from many different social science disciplines have produced important work utilizing a world-systems perspective.

There are now many institutional bases of support for world-systems scholarship. The first of these to be founded was the Fernand Braudel Center for the Study of Economies, Historical Systems and Civilizations at what is now Binghamton University, established under Wallerstein's leadership in 1976 to "engage in the analysis of large-scale social change over long periods of historical time." Giovanni Arrighi, Terrence Hopkins, and many world-systems scholars who are still active today have been associated with the Braudel Center for substantial portions of their careers. Since 1977, the Braudel Center has published the influential journal *Review*. It also regularly sponsors conferences related to world-systems analysis and has published many volumes of proceedings from these conferences.

A few years later, in 1980, a group of sociologists got together to form the Political Economy of the World-System (PEWS) section of the American Sociological Association (ASA). The PEWS section has grown to become one of the ASA's most vibrant, with over 400 members, regular mini-conferences held before or after the ASA meetings, and an annual spring conference held independently of the ASA. A selection of papers presented at the PEWS spring conference is published every year as an edited volume. World-systems analysis regularly features on the ASA's main program, and world-systems sociologists are very active in the association. Since 2008 the PEWS section has sponsored the *Journal of World-Systems Research*, which is now an official ASA journal.

The *Journal of World-Systems Research* (*JWSR*) was founded in 1995. It was one of the first fully peer-reviewed academic journals to be published online—almost certainly the first in the social sciences. To put this in context, the first web browser, Mosaic 1.0, was released only two years earlier. *JWSR* is still one of the very few peer-reviewed journals that is freely available without a subscription. Its stated editorial goal is to "develop and disseminate scholarly research on topics that are relevant to the analysis of world-systems" but its influence extends much further. *JWSR* is widely read in the activist community as well as in academia. Its free status makes it especially popular among independent scholars and in countries where heavy subscription fees put most academic journals out of reach.

Web hosting for *JWSR* is provided by the Institute for Research on World-Systems (IROWS) at the University of California–Riverside. Founded in 2000, IROWS publishes a major working paper series, hosts the World-Systems Archives, and sponsors *Cliodynamics: The Journal of Theoretical and Mathematical History*. The IROWS working paper series publishes roughly six to eight titles a year, most of which are later published in peer-reviewed journals. The World-Systems Archive includes online historical materials, such as datasets, syllabi, and the complete records of the World-Systems Network discussion board from 1995–2004. These are a potential treasure trove for researchers interested in the history of world-systems analysis. The World Society Foundation at the University of Zurich, founded at the behest of Swiss sociologist Peter Heinz, has strongly supported the research of world-systems scholars.

World-systems analysis is also prominently represented in professional associations around the world. An interdisciplinary group of world-systems scholars whose work is focused on world historical systems is active in the International Political Economy section of the International Studies Association. World historians who take a world-systems approach in their research are also active in the World History Association. World-systems sociologists are active in Research Committee 02 (Economy and Society) of the International Sociological Association. Finally, many world-systems analysts are active in the World Social Forum movement.

Though it arose (institutionally) in sociology, world-systems analysis is not an exclusively sociological approach to understanding the world. It is fundamentally transdisciplinary. The resulting diversity of viewpoints is reflected in the diversity of topics covered in this handbook.

The handbook is organized into six major parts on the origins of the modern world-system, theory in world-systems analysis, the contemporary world-economy, development and underdevelopment, sustainability, and society broadly construed. Each part contains two or three sets of related essays. The essays are of two types: "contributions" essays that represent chapter-length treatments of a particular topic, and "perspectives" essays that briefly present additional viewpoints within the same general area. The organization of the essays has emerged organically, based on the universe of proposals we received from an open call for papers. Together, the contributions to the handbook represent the full breadth of the research being done today by people who identify themselves as world-systems analysts.

Part I of the handbook is concerned with origins. We open in the pre-history of the modern world-system with a set of essays on historical world-systems before the sixteenth century. Before the world became integrated into "the" world-system there were many world-systems, and world-systems analysts have been at the forefront of studying these for clues about the general characteristics of world-systems that can be distinguished from the specific characteristics of the world-system in which we happen to live today. The second set of essays in Part I focus on the increasing integration of the entire world into the modern world-system.

After the world-historical bang of the opening, we quiet down in Part II to a theoretical whisper. A series of carefully argued theoretical essays delve into the role of theory in world-systems analysis, the prospects for the application of explicit modeling approaches, and the weaknesses of world-systems analysis. The essays on weaknesses in some ways go further to propose new directions for world-systems analysis. We consider these critiques to be some of the most valuable essays in the collection.

The essays in Part III explore different aspects of the contemporary world-economy: markets, networks and distribution. The modern world-system is specifically a capitalist world-economy, meaning that the economic system (the world market) encompasses all of the political systems (mainly countries) and cultural systems (language groups, ethnic groups, families and kinship networks) that exist in the world today. While world-systems analysts would argue that all aspects of social life are ultimately embedded in the contemporary world-economy, global markets themselves are institutions that can only really be understood at the world-systemic level of analysis.

World-systems analysis has both its biographical and academic roots in the study of development. The three sets of essays collected in Part IV of the handbook cover different aspects of development and underdevelopment. First, a continuing inspiration for world-systems analysts—and activists in particular—has been the resilience of indigenous nations in the face of (almost always) violent forced incorporation into the world-system. The set of essays on indigeneity is followed by sets on understanding economic growth (and lack of growth) in poor countries and the related problems of food and agriculture. Indigeneity, poverty and agriculture are all closely linked in historical processes of underdevelopment.

Several of the essays on indigeneity, development, and agriculture might just as easily have been included in Part V on sustainability. World-systems analysis has become a major node of research on sustainability at all levels, from local farmers' cooperatives to multinational agribusiness and from forest preservation to global warming. Part V includes sets of essays on natural resources and the natural environment, but many more essays on sustainability are spread throughout the handbook. We estimate that around a quarter of all the papers currently being published that use world-systems analysis are in some way concerned with sustainability. Within a few years, this literature will be large enough to warrant its own handbook. Watch this space.

The final part of the handbook, Part VI, deals with society, broadly construed. Though world-systems analysis is pitched at the very highest level at which human societies can be understood, that of super-systems of societies, the character of the world-system in which people live has implications right down to the level of the family and the individual. In particular, the household sector of the world-economy is still very large, and in many places dominant. Market exchange cannot be properly understood without understanding how it affects and is affected by the household economy. We close the handbook with a set of essays that pull out from the household to focus on super-national and global society.

All in all, this handbook arises out of a good-faith attempt to represent world-systems analysis as it is practiced in the field, not as it was laid out by the founders or defined by the editors. At first glance, it might not be obvious how pre-Hispanic Mesoamerica is related to twenty-first century global civil society via natural resources and unpaid household labor, but it is. The people who study each of these topics all read and make use of each others' work. They attend conferences together and share space on panels. They understand that the problems of today's world are not unique to today, nor are they emblematic of some kind of tragic, immutable "human condition." The problems of today's world can be solved, and world-systems analysts are dedicated to using all the intellectual tools at our disposal—historical knowledge, analytical modeling, theory construction, and empirical data—to solving them.

So what then is world-systems analysis? World-systems analysis is the fuzzy overlap among the research, writing, and practice of those people who identify themselves as world-systems analysts. There is not a single body of theory that defines world-systems analysis, nor is there a single object of study. World-systems analysts, however, all share several qualities. They are globally aware and particularly sensitive to global hierarchies that span particular countries and cultures. They read and cite the work of scholars from multiple disciplines. They take history seriously. All of us in the world-systems world have idiosyncratic outside interests, but the space where our interests overlap is what we would call world-systems analysis. In this collective space we study the world in order to improve it. That is the hallmark of our knowledge movement, or to call it what it really is, our community.

★ ★ ★

It remains for us to thank those who have helped us make this handbook a reality. Nelda Thomas at Riverside, Francisco Morales at Sydney, and Jennifer Dodd at Routledge all provided invaluable editorial and administrative support. Gerhard Boomgaarden at Routledge conceived the book, commissioning us to write it. Most of all, however, we must thank our authors. They all wrote without exception to the length and dates we set for them, cooperated with our requests for revisions, and delivered first-rate essays—all for free. We have shamelessly exploited their uncompensated intellectual labor. We hope they will graciously accept our thanks in return.

Part I
Origins

1. Before the long sixteenth century

1.1

Market cooperation and the evolution of the pre-Hispanic Mesoamerican world-system

Richard E. Blanton and Lane F. Fargher

The emergence of the early modern European world-system, according to Wallerstein (1974), represented a sea change in human socioeconomic organization inasmuch as there was a vast increase in the interregional exchange of staple goods, unlike world-empires and mini-systems that principally featured the long-distance exchange of preciosities. The latter, Wallerstein argued, were only elite symbols of conspicuous consumption and hence lacked the system-shaping potential of staple goods that account "more of men's economic thrusts than luxuries" (ibid.: 42). This argument has been criticized (e.g., in Abu Lughod 1989; Blanton and Feinman 1984; Schneider 1977), and here we elaborate these critiques by developing a goods-centered approach to expand on Wallerstein's structural opposition of preciosities and staples, before showing how the approach can be applied in the case of pre-Hispanic Mesoamerica. We trace the long-term history of Mesoamerican goods systems, then identify what we consider to be the central driving force in the emergence of the Mesoamerican world-system, namely, a changing consumer propensity to consume exotic and costly goods that we identify as bulk luxury goods. Lastly, we contextualize bulk luxury consumption in relation to a growing core-zone market economy.

Outline of a theory of goods

Research on the role of goods in culture and social life found in sources including Appadurai (1986), Douglas and Isherwood (1979), Gregory (1982), McCracken (1988), Mintz (1985), Sahlins (1994), and Schneider (1987) informed our application of a goods perspective to world-systems analysis (Blanton et al 2005, with our colleague Verenice Heredia Espinoza). In our formulation, both preciosities and staples are viewed as potentially system-shaping and will figure into any explanation of cultural, social, demographic, technological, and agroecological change, including world-system change. However, how, and the degree to which, any particular good is implicated in change varies across space and time. To characterize this variation, we identify three broad categories of goods: Prestige Goods, Regional Goods, and Bulk Luxury Goods.

Prestige goods

Prestige goods are similar to those identified by Wallerstein as preciosities in that they serve, in part, as symbols of elite status. However, anthropologists have come to better understand how such goods figure into processes of social change, principally in relation to the emergence of a governing elite in many chiefdoms and pre-modern states (Brumfiel and Earle 1987; Hayden 1995; Peregrine 1992). Exclusive goods manufactured from rare materials and using technologically sophisticated methods are an ideal medium for calculated gifting aimed at political faction-building and for purposes of competitive feasting between members of an elite (Blanton et al 1996). However, typically, they have little role in the economic activities of the great majority of people and play only a limited direct role in world-system evolution.

Regional goods

Regional goods facilitate the transition to increased production and the attendant revisions in work scheduling and foodways as commoner households respond to both growing tax obligations and growing commercial opportunities (e.g., Blanton and Fargher 2010; Blanton et al 1999: Ch. 4; Fargher 2009). In general, regional goods were produced and exchanged at the geographical scale of regional tribute flows and periodic markets rather than the world-system scale.

Bulk luxury goods

A bulk luxury goods category (the phrase is from Kepecs 2003: 130) is a necessary amendment to Wallerstein's concept of a preciosity as they occupy a middle ground between the rare and exclusive prestige goods and the regional goods that were available to many consumers (Blanton et al 2005: 274). Although costly and often exotic, they are consumed by commoner as well as elite households, for example in rites of household social reproduction such as feasts, weddings and funerals (e.g., Mintz 1985). There are profound system-shaping consequences when commoner households in an emerging core region increase bulk luxury consumption. Supplier regions undergo periphery incorporation as substantial numbers of households respond to external demand, potentially bringing change in agroecology, demography, labor mobilization, production methods, work intensification and regional market systems (Blanton et al 2005). We argue that bulk luxuries are a key causal factor promoting the development and integration of the core-periphery structures of world-systems, whether in early modern Europe or elsewhere.

The world of goods and world-system evolution in pre-Hispanic Mesoamerica

We contextualize the emergence of a Mesoamerican world-system by outlining the ways in which goods have featured in sociocultural change over a span of approximately 3,000 years. Mesoamerican long-distance exchange of prestige goods such as jade, stingray spines, marine shell and magnetite mirrors intensified during the Early and Middle Formative Periods (1400–500 BCE), but these goods were produced and consumed only in small quantities, principally by high status households (e.g., Marcus and Flannery 1996: Chapters 8 and 9). The Late Formative through Classic Periods (500 BCE–CE 950) marked the emergence of states, cities, market systems and regional goods, as indicated by the intensification and standardization of utilitarian pottery (e.g., Blanton et al 1999: Ch. 4; Feinman et al 1984). For example, during this time in the Valley of Oaxaca, Mexico (Figure 1.1.1), we detect the emergence of a regional division of labor including intensification of food production in some localities. In other areas, there was an increased production of handicrafts such as pottery, palm frond products, salt, chert tools and fiber (Fargher 2004; Feinman and Nicholas 2010). World-system growth is first evident after about CE 300 with the advent of bulk luxuries such as "Thin Orange" pottery, which was used widely at

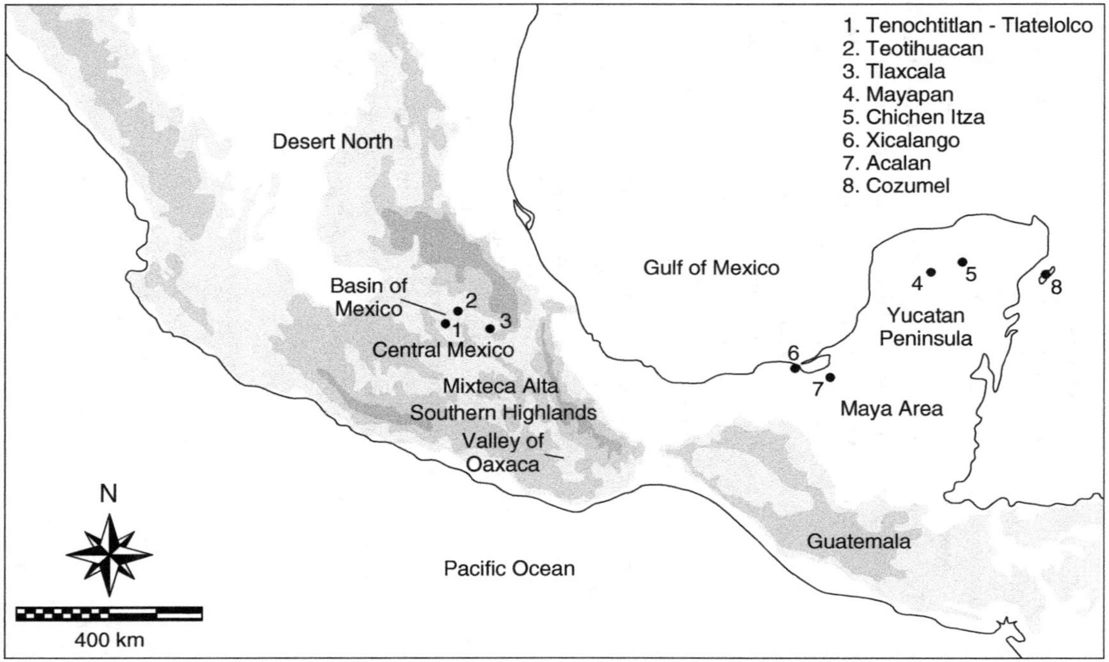

Figure 1.1.1 Places mentioned in the text

Teotihuacan and elsewhere (Kolb 1986). At this time we first detect a process of periphery incorporation, for example in tropical regions, where cotton production was increased to satisfy growing consumer demand in arid and high-elevation cities such as Teotihuacan, where cotton cannot be cultivated (e.g., Stark et al 1998).

During the Postclassic Period (CE 950–1521), important bulk luxury goods that moved between regions at the scale of Mesoamerica and were consumed across social sectors included polychrome pottery, green obsidian blades, cacao (chocolate), fine sea salt and cotton (Blanton et al 2005). From a world-systems perspective, these are analogs of the wool, cotton, sugar, spices and tea that featured in the growth of the early modern European world-system (cf. Mintz 1985; Sahlins 1994; Wallerstein 1974).

The growing consumption of these bulk luxury goods was powerfully system-shaping, bringing unprecedented levels of economic growth as well as social, demographic, agroecological and technological change. For example, obsidian mining became more sophisticated, involving shaft and tunnel techniques surpassing earlier shallow pit mining (Charlton and Spence 1982: 24; Cobean 2002) and specialist producers developed more efficient methods for blade tool production (e.g., Clark 1985) that made obsidian (especially green obsidian) a commodity more widely available across the length and breadth of Mesoamerica. Cotton production also underwent a technological revolution during the Postclassic as it became a highly sought after and widely consumed commodity, especially in Central Mexico (Anawalt 2001; Berdan 1987; Hicks 1994). Growing cotton consumption also spurred the development of new markets for secondary industries, especially cochineal, a brilliant and fast red dye produced in regions including Tlaxcala, the Mixteca Alta and the Valley of Oaxaca (e.g., Donkin 1977). The production of low-grade salt for use as a dye-fixing mordant was another important secondary industry in the Basin of Mexico (Parsons 2001: 248). The Postclassic saw a major expansion in the production and consumption of sea salt, especially in the Maya lowlands (Kepecs 1999: 315, 516, 2003), which required new forms of labor mobilization, including seasonal labor migration to work coastal saltpans in marginal locations (Kepecs 1999: 422). In Guatemala, late pre-Hispanic cacao production probably exceeded six million pounds per year (Bergmann 1969: 94) from large monocrop plantations or estates that had been carved out of tropical forest (Bergmann 1969, Millon 1955: 130–33). Cacao production involved major changes in agroecology, production scheduling (as cacao production would displace local subsistence production) and labor mobilization.

Discussion

Wallerstein (1974: Ch. 2) saw the decline of the feudal mode of production of medieval European and the corresponding rise of the capitalist yeoman farmer as an important social transition underlying a new European, and, eventually, world-wide division of labor. Here we address the analogous social and cultural foundations of the emerging Mesoamerican world-system, but we take a direction unlike Wallerstein's mode of production analysis by focusing on the evolution of marketplace institutions in the emerging Central Mexican core zone. As Central Mexico emerged as Mesoamerica's preeminent core zone, its trade and political interactions with other Mesoamerican regions intensified, especially involving the flow of commodities exchanged with tropical lowland regions. Expansion of core zone influence stimulated the growth of secondary foci of economic activity in semiperipheral polities linking the highlands to tropical regions, for example, in trade emporia in the Xicalango and Acalan regions. Increased macro-regional integration altered regional networks of cities across a wide zone from Central Mexico to the Maya area, including the growth of new commercial and political centers in Yucatán at

Chichen Itzá and Mayapán (Kepecs 2007; Smith and Berdan 2003) and trade emporia such as Cozumel (Sabloff and Rathje 1975).

The Central Mexican region emerged as Mesoamerica's major Postclassic core zone, especially in the Basin of Mexico where a Late Postclassic population that exceeded one million persons was integrated politically by the Aztec Triple Alliance empire (Berdan et al 1996) and economically through a complex division of labor and market system (Berdan 1985; Blanton 1996; Hassig 1985; Hodge and Smith 1994; Nichols et al 2002; Smith 2010; Smith and Berdan 2003). Commercial growth in Central Mexico extended to commoners who attended distant markets outside their own polity and ethnic boundaries (e.g., Durán 1967: Vol. 1, 177) and who had a wide range of goods choices extending to bulk luxury goods. The region's, and Mesoamerica's, most important marketplace at Tlatelolco was a vast walled concourse that served up to an estimated 50,000 people on major market days (Cortés 1986: 103).

We attribute the relative elaboration of the Central Mexican commercial system to social and cultural changes that allowed for an expansion of market cooperation across polities and across social sectors. We think this was a key ingredient in world-system growth because, in pre-Hispanic Mesoamerica, as in all complex societies, market cooperation presents problems, often restricting commoner commercial participation to only local market venues where participants are socially familiar with commercial transaction partners (Blanton forthcoming). The lack of trust that market transactions can take place honestly will be a disincentive to market participation, but other disincentives to participation abound, for example when marketers must travel through dangerous foreign territory to reach distant markets and when market transactions are between persons of different ethnicities or polities. Additionally, commoners will not likely participate in markets managed by biased local officials whose adjudicative decisions may favor persons of their own ethnicity, political group, or social standing. Geographically and socially broadly based market participation is based on trust that markets will be safe, that most market participants will observe an ethic of fairness, and that there will be adjudicative certainty so that when market disputes occur they will be resolved in a neutral and fair manner.

Uniquely in Mesoamerica, in Central Mexico market cooperation problems were resolved through the development and wide acceptance of novel institutional and cultural changes accompanying the rise of the Postclassic polities. In one of these changes, three of the major antagonists of the militarily conflictive Middle Postclassic Period (the Mexica, Acolhua, and Tepaneca) agreed to form a political alliance to consolidate imperial power in Central Mexico and beyond (Berdan et al 1996), thus enhancing possibilities for safe interregional travel. Of equal importance is the fact that during the Postclassic Period, most market management, including marketplace adjudication of disputes, came to be vested in an organized group of commoner merchants called the Pochteca. The Pochteca was a vast Central Mexican organization that operated as a semiautonomous paragovernmental system within the larger authority structure of the Aztec empire (van Zantwijk 1985: Ch. 7). Although the Pochteca were commoners operating largely outside the official structure of the state, they maintained a high-level autonomous judicial authority within marketplace venues (Offner 1983: 156), were known for their judicial neutrality in commercial matters (Sahagún 1950–69, Book 8: 69; cf. van Zantwijk 1985: Ch. 7), and upheld an ethical code that emphasized fairness in matters of trade (van Zantwijk 1985: 171). The most important commercial venue in all of Mesoamerica by the end of the pre-Hispanic sequence was the massive market concourse at Tlatelolco, which was designated as a distinct social domain in which the judicial authority of the Aztec state was suspended, leaving market management entirely in the hands of the Pochteca (Sahagun 1950–69, Book 9: 24). No comparable institutional arrangement for market management was present in other Mesoamerican regions (Blanton forthcoming).

In addition to safer travel provided by imperial integration and effective market management, institutional and cultural changes during the Central Mexican Postclassic Period weakened notions of ethnic identity while serving to minimize the divide between nobility and commoners to a degree that was unusual in Mesoamerica. The cultural foundations of these two important social outcomes, we suggest, are to be found in the promulgation of a slightly variant but largely uniform mythic history shared among the Postclassic Nahua (Central Mexican) peoples (Fargher et al 2010). According to mythic history, the different ethnic groups making up the Nahua peoples could trace their origins to "primitive" Chichimecs from northern arid regions who migrated into Central Mexico from a shared origination point (Boone 1991). While different migrating groups eventually colonized distinct regions of Central Mexico to establish polities and adopt "civilization" ("neo-Toltec" Nahua culture and language), a mythic scheme emphasizing a shared originating point and shared ethnic origins resulted in a weak and contingent sense of ethnicity that largely dissolved ethnic boundaries in Central Mexico, producing what Stark (2008: 44) refers to as an "ethnic mosaic" (cf. Brumfiel 1994; van Zantwijk 1985). We think this would have facilitated market participation beyond local systems.

Mythic history documented how "primitive" desert peoples were transformed from nomadic bands into neo-Toltec societies governed by hereditary nobility (Boone 1991: 148; cf. Gillespie 1989: 20), but Nahua cultural logic defined authority and moral capacity in a dualistic sense by setting up a logical opposition between two principal deities, who, working side-by-side, created what the Nahua peoples thought of as the current world: Quetzalcoatl (associated with a prior but collapsed Toltec civilization and its noble descendants who held most positions of authority) (Nicholson 2001), on the one hand, and Tezcatlipoca (often associated with the arid north and the Chichimec peoples), on the other (e.g., van Zantwijk 1985: 96–97). In this dual scheme, while the neo-Toltec person is one who enjoys a civilized way of life (sedentary, farming, with great material wealth), mythic history elaborates on a more fundamental sense of self in which the culturally "primitive" Chichimec peoples, although culturally and technologically deprived, are endowed with a well-developed sense of moral purpose and understanding (Fargher et al 2010). This can be seen in the symbolism of Tezcatlipoca, who is thought to have the capacity to perceive the potential for virtue in any person, regardless of social standing, including the idea that success can be the product of achievement, not just noble birth (Heyden 1991: 189; Sahagun 1950–82, II: 5). This egalitarian cultural code provided the foundations for new legal systems that were socially enacted as judicial systems and were expected to provide fair treatment for both commoners and the nobility (e.g., Offner 1983). Hence, persons of different ethnic groups, as well as commoners and elite, were freer to engage equitably in market transactions and to stand in front of market arbiters with the certainty that all categories of persons—commoner, nobleman and those from different ethnic backgrounds—have the potential for moral capacity and can be judged accordingly. Further, because commoners were viewed as potentially virtuous and morally equal to the nobility, they possessed the right to consume luxury items, although we should note that this more extroverted consumer behavior was met with some opposition by moralists who argued that the growing luxury paved the way to decadence, moral promiscuity, and self-destruction (summarized in Blanton et al 2005: 279–80).

Conclusions

The idea that a modified world-system theory might contribute to understanding the Mesoamerican past has only a short history, beginning in the 1980s (Blanton and Feinman 1984; Whitecotton and Pailes 1986). Since these modest beginnings, world-system and related multiregional and multiscalar approaches have gained many converts (e.g., Peregrine and Feinman

1996; Schortman and Urban 1992; cf. Feinman and Nicholas 1991, 2010), culminating in a substantial collection of papers edited by Michael E. Smith and Frances F. Berdan titled "The Postclassic Mesoamerican World" (2003). While we applaud this rapid growth of research literature, we suggest that, of the many possible agendas that can contribute to further understanding of world-systems, researchers should, as we have done here in a preliminary way, investigate the role of cultural codes that favored commercial development. The growth of the Postclassic Mesoamerican world-system was made possible when people envisioned more egalitarian forms of political economy that widened possibilities for economic and political participation. We see this generally across Mesoamerican regions to some degree, for example, in the Maya area with the Postclassic emergence of council-based governance that replaced rule by divine kings (e.g., Chase and Chase 1988; Cobos 2007). But innovative cultural and institutional changes that facilitated market cooperation and that were widely adopted in Central Mexico had the most profound world-system outcomes.

We propose that researchers would benefit from understanding how the evolution of market cooperation impacted world-system evolution in other world areas, including early modern Europe. According to Bridbury (1986: 117), in medieval European history, market development was only achieved with great difficulty owing to persistent elements of the feudal cultural code. These include notions of property rights that had to be relaxed to allow property to be "private to individuals and their families, corporations, and other voluntary associations" (Kerridge 1986: 121). This change allowed for mutually profitable trade across social sectors with no fear that profits, or goods bought and sold, could be confiscated according to feudal concepts of fealty. Bridbury (1986: 83) also alludes to the importance of changes in moral concepts surrounding how wealth can be used over the course of medieval European history. Commoners were not likely to trust the market until new norms of behavior that placed moral restrictions on the uses of wealth were developed and widely accepted, because in the Middle Ages, "anything could be bought," including justice (ibid.).

References

Abu Lughod, J. L. (1989) *Before European Hegemony: The World System, A.D. 1250–1350*, Austin: University of Texas Press.

Anawalt, P. R. (2001) "Clothing", in S. T. Evans and D. L. Webster (eds) *Archaeology of Ancient Mexico and Central America: An Encyclopedia*, New York: Garland Publishing.

Appadurai, A. (ed.) (1986) *The Social Life of Things: Commodities in Cultural Perspective*, Cambridge: Cambridge University Press.

Berdan, F. F. (1985) *Trade, Tribute, and Market in the Aztec Empire*, Unpublished thesis, The University of Texas.

Berdan, F. F. (1987) "Cotton in Aztec Mexico: production, distribution, and uses", *Mexican Studies/Estudios Mexicanos*, 3: 235–62.

Berdan, F. F., R. E. Blanton, E. H. Boone, M. G. Hodge, M. E. Smith, and E. Umberger (1996) *Aztec Imperial Strategies*, Washington, DC: Dumbarton Oaks Research Library and Collection.

Bergmann, J. F. (1969) "The distribution of cacao cultivation in pre-Columbian America", *Annals of the American Association of Geographers*, 59: 85–96.

Blanton, R. E. (forthcoming) "The moral economy of the marketplace", in K. Hirth and J. Pillsbury (eds) *Merchants, Trade and Exchange in the Pre-Columbian World*, Washington, DC: Dumbarton Oaks Research Library and Collection.

Blanton, R. E., and L. F. Fargher (2010) "Evaluating causal factors in market development in premodern states: A comparative study, with comments on the history of ideas about markets", in C. Garraty and B. Stark (eds) *The Evolution of Markets and Market Systems in the Premodern World*, Boulder: University Press of Colorado.

Blanton, R. E., L. F. Fargher, and V. Y. Heredia Espinoza (2005) "The Mesoamerican world of goods and its transformations", in R. E. Blanton (ed.) *Settlement, Subsistence, and Social Complexity: Essays Honoring the Legacy of Jeffrey R. Parsons*, Los Angeles: Cotsen Institute of Archaeology, the University of California.

Blanton, R. E., and G. M. Feinman (1984) "The Mesoamerican world-system", *American Anthropologist*, 86: 673–82.
Blanton, R. E., G. M. Feinman, S. A. Kowalewski, and L. M. Nicholas (1999) *Ancient Oaxaca: The Monte Albán State*, Cambridge: Cambridge University Press.
Blanton, R. E., G. M. Feinman, S. A. Kowalewski, and P. N. Peregrine (1996) "A dual-processual theory for the evolution of Mesoamerican civilization", *Current Anthropology*, 37: 1–14.
Boone, E. H. (1991) "Migration histories as ritual performance", In D. Carrasco (ed.) *Aztec Ceremonial Landscapes*, Boulder: University Press of Colorado.
Bridbury, A. R. (1986) "Markets and freedom in the Middle Ages", in B. L. Anderson and A. J. H. Latham (eds) *The Market in History*, London: Croom Helm.
Brumfiel, E. M. (1994) "Ethnic groups and political development in ancient Mexico", in E. M. Brumfiel and J. W. Fox (eds) *Factional Competition and Political Development in the New World*, Cambridge: Cambridge University Press.
Brumfiel, E. M., and T. K. Earle (1987) "Specialization, exchange, and complex societies: An introduction", in E. M. Brumfiel and T. K. Earle (eds) *Specialization, Exchange, and Complex Societies*, Cambridge: Cambridge University Press.
Charlton, T. H., and M. W. Spence (1982) "Obsidian exploitation and civilization in the Basin of Mexico", in P. C. Weigand and G. Gwynne (eds) *Mining and Mining Techniques in Ancient Mesoamerica, special issue of Anthropology*, 6: 7–86.
Chase, D. Z., and A. F. Chase (1988) *A Postclassic Perspective: Excavations at the Maya Site of Santa Rita Corozal, Belize*, San Francisco, CA: Pre-Columbian Art Research Institute, Monograph 4.
Clark, J. E. (1985) "Platforms, bits, punches, and vices: A potpourri of Mesoamerican blade technology", *Lithic Technology*, 14: 1–15.
Cobean, R. H. (2002) *A World of Obsidian: The Mining and Trade of a Volcanic Glass in Ancient Mexico*, Pittsburgh, PA Mexico, D.F.: Instituto Nacional de Antropología e Historia.
Cobos, R. (2007) "Multepal or centralized kingship? New evidence on governmental organization at Chichén-Itzá", in J. K. Kowalski and C. Kristan-Graham (eds) *Twin Tollans: Chichén Itzá, Tula, and the Epiclassic to Early Postclassic Mesoamerican World*, Washington, DC: Dumbarton Oaks Library and Collection.
Cortés, H. (1986) *Hernán Cortés: Letters From Mexico*, translated by A. Pagden, New Haven, NC: Yale University Press.
Donkin, R. A. (1977) *Spanish Red: An Ethnogeographical Study of Cochineal and the Opuntia Cactus*, Philadelphia, PA: Transactions of the American Philosophical Society 67.
Douglas, M., and B. Isherwood (1979) *The World of Goods: Towards and Anthropology of Consumption*, New York: W. W. Norton.
Durán, D. (1967) *Historia de las Indias de Nueva España*, in A. M. Garibay K. (ed.) Mexico, D. F.: Porrua.
Fargher, L. F. (2004) *A Diachronic Analysis of the Valley of Oaxaca Economy from the Classic to the Postclassic*, Unpublished thesis, the University of Wisconsin.
Fargher, L. F. (2009) "A comparison of the spatial distribution of agriculture and craft specialization in five state-level societies", *Journal of Anthropological Research*, 65: 353–87.
Fargher, L. F., R. E. Blanton, and V. Y. Heredia Espinoza (2010) "Egalitarian ideology and political power in pre-Hispanic Central Mexico: The case of Tlaxcallan", *Latin American Antiquity*, 21: 227–51.
Feinman, G. M., S. A. Kowaleski, and R. E. Blanton (1984) "Modeling ceramic production and organizational change in the pre-Hispanic Valley of Oaxaca", in S. van der Leeuw and A. C. Pritchard (eds) *The Many Dimensions of Pottery: Ceramics in Archaeology and Anthropology*, Amsterdam: Universiteit van Amsterdam.
Feinman, G. M., and L. M. Nicholas (1991) "New perspectives on preHispanic highland Mesoamerica: A macroregional approach", *Comparative Civilizations Review*, 24: 13–33.
Feinman, G. M., and L. M. Nicholas (2010) "A multiscalar perspective on market exchange in the Classic-Period Valley of Oaxaca", in P. Garraty and B. L. Stark (eds) *Archaeological Approaches to Market Exchange in Ancient Societies*, Boulder: University Press of Colorado.
Gillespie, S. D. (1989) *The Aztec Kings: The Construction of Rulership in Mexica History*, Tucson: University of Arizona Press.
Gregory, C. A. (1982) *Gifts and Commodities*, Cambridge: Cambridge University Press.
Hassig, R. (1985) *Trade, Tribute, and Transportation: The Sixteenth-Century Political Economy of the Valley of Mexico*, Norman: University of Oklahoma Press.
Hayden, B. (1995) "Pathways to power: Principles for creating socioeconomic inequalities", in T. D. Price and G. M. Feinman (eds) *Foundations of Social Inequality*, New York: Plenum.

Heyden, D. (1991) "Dryness before the rain: Toxcatl and Tezcatlipoca", in D. Carrasco (ed.) *Aztec Ceremonial Landscapes*, Boulder: University Press of Colorado.

Hicks, F. (1994) "Cloth in the political economy of the Aztec state", in M. Hodge and M. E. Smith (eds) *Economies and Polities in the Aztec Realm*, Albany: Institute for Mesoamerican Studies, the State University of New York.

Hodge, M. G., and M. E. Smith (eds) (1994) *Economies and Polities in the Aztec Realm*, Albany: Institute for Mesoamerican Studies, the State University of New York.

Kepecs, S. (1999) *The political economy of Chikinchel, Yucatán, Mexico*, Unpublished thesis, the University of Wisconsin.

Kepecs, S. (2003) "Salt sources and production", in M. E. Smith and F. F. Berdan (eds) *The Postclassic Mesoamerican World*, Salt Lake City: The University of Utah Press.

Kepecs, S. (2007) "Chichén-Itzá, Tula, and the Epiclassic/Early Postclassic world-system", in J. K. Kowalski and C. Kristan-Graham (eds) *Twin Tollans: Chichén Itzá, Tula, and the Epiclassic to Early Postclassic Mesoamerican World*, Washington, DC: Dumbarton Oaks Library and Collection.

Kerridge, E. (1986) "Early modern English markets", in B. L. Anderson and A. J. H. Latham (eds) *The Market in History*, London: Croom Helm.

Kolb, C. C. (1986) "Commercial aspects of Classic Teotihuacan Period 'thin orange' wares", in B. L. Isaac (ed.) *Research in Economic Anthropology, Supplement 2*, Greenwich: JAI Press.

Marcus, J., and K. V. Flannery (1996) *Zapotec Civilization: How Urban Society Evolved in Mexico's Oaxaca Valley*, New York: Thames and Hudson.

McCracken, G. (1988) *Culture and Consumption: New Approaches to the Symbolic Character of Consumer Goods and Activities*, Bloomington: Indiana University Press.

Millon, R. (1955) *When money grew on trees*, Unpublished thesis, Columbia University.

Mintz, S. (1985) *Sweetness and Power: The Place of Sugar in Modern History*, New York: Penguin Books.

Nichols, D. L., E. M. Brumfiel, H. Neff, M. G. Hodge, T. Charlton, and M. Glascock (2002) "Neutrons, markets, cities, and empires: A 1000-year perspective on ceramic production and distribution in the Postclassic Basin of Mexico", *Journal of Anthropological Archaeology*, 21: 25–82.

Nicholson, H. B. (2001) *Topiltzin Quetzalcoatl: The Once and Future Lord of the Toltecs*, Boulder: University Press of Colorado.

Offner, J. A. (1983) *Law and Politics in Aztec Mexico*, Cambridge: Cambridge University Press.

Parsons, J. R. (2001) *The Last Saltmakers of Nexquipayac, Mexico: An Archaeological Ethnography*, Ann Arbor: Anthropological Papers 82, the Museum of Anthropology, the University of Michigan.

Peregrine, P. N. (1992) *Mississippian Evolution: A World-System Perspective*. Madison, NJ: Prehistory Press.

Peregrine, P. N. and G. M. Feinman (eds) (1996) *Pre-Columbian World Systems*, Madison, WI: Prehistory Press.

Sabloff, Jeremy and William Rathje (eds) (1975) *A Study of Changing Pre-Columbian Commercial Systems*, Cambridge: Peabody Museum of Archaeology and Ethnology, Harvard University.

Sahagún, Fray B. de (1950–69) *Florentine Codex: General History of the Things of New Spain*, translated by C. E. Dibble and A. J. O. Anderson, Austin: University of Texas Press.

Sahlins, M. (1994) "Cosmologies of capitalism: The trans-Pacific sector of 'the world system'", in N. B. Dirks, G. Eley, and S. B. Ortner (eds) *Culture/Power/History: A Reader in Contemporary Social Theory*, Princeton, NJ: Princeton University Press.

Schneider, J. (1977) "Was there a pre-capitalist world-system?", *Peasant Studies*, 6: 20–29.

Schneider, J. (1987) "The anthropology of cloth", *Annual Review of Anthropology*, 16: 409–48.

Schortman, E. M., and P. A. Urban (eds) (1992) *Resources, Power, and Interregional Interaction*, New York: Plenum.

Smith, M. E. (2010) "Regional and local market systems in Aztec-Period Morelos", in C. Garraty and B. L. Stark (eds) *The Evolution of Markets and Market Systems in the Premodern World*, Boulder: University Press of Colorado.

Smith, M. E., and F. F. Berdan (2003) "Spatial structure of the Mesoamerican world-system", in M. E. Smith and F. F. Berdan (eds) *The Postclassic Mesoamerican World*, Salt Lake City: The University of Utah Press.

Smith, M. E. and F. F. Berdan (eds) (2003) *The Postclassic Mesoamerican World*, Salt Lake City: The University of Utah Press.

Stark, B. L. (2008) "Archaeology and ethnicity in Postclassic Mesoamerica", in F. F. Berdan, J. K. Chance, A. R. Sandstrom, B. L. Stark, J. M. Taggart, and E. Umberger (eds) *Ethnic Identity in Nahua Mesoamerica: The View from Archaeology, Art History, Ethnohistory, and Contemporary Ethnography*, Salt Lake City: University of Utah Press.

Stark, B. L., L. Heller, and M. Ohnersorgen (1998) "People with cloth: Mesoamerican economic change viewed from the perspective of cotton in south-central Veracruz", *Latin American Antiquity*, 9: 7–36.

Van Zantwijk R. (1985) *The Aztec Arrangement: The Social History of Pre-Spanish Mexico*, Norman: University of Oklahoma Press.

Wallerstein, I. (1974) *The Modern World-System: Capitalist Agriculture and the Origins of the European World-Economy in the Sixteenth Century*, New York: Academic Press.

Whitecotton, J., and R. Pailes (1986) "New world pre-Columbian world systems", in F. Mathien and R. McGuire (eds) *Ripples in the Chichimec Sea*, Carbondale: Southern Illinois University Press.

1.2

Assessing the debate between Abu-Lughod and Wallerstein over the thirteenth-century origins of the modern world-system

Elson E. Boles

Janet Abu-Lughod's *Before European Hegemony, The World System A. D. 1250–1350* (1989) caught the attention of world-systems scholars and instigated debates concerning the origins of the modern world-system. Most of her book details the vast trading circuits among "subsystems" or "world-economies and world-empires" spanning the Eurasian continent. Abu-Lughod's two main theses are firstly, that this network was a "world-system" (or "world system"—she uses the hyphenated and unhyphenated terms without distinction) and secondly, that its rise and demise explains the formation of the West and the modern world-system in the sixteenth century. She argues, "… the Eurocentered 'modern' world system … was built on the ruins of the thirteenth-century world system" and that the "pathways and routes developed by the thirteenth century were later 'conquered' and adapted by a succession of European powers. Europe did not need to *invent* the system, since the basic groundwork was already in place … " (1989: 361, 369). The decline of the East, in particular China's withdrawal, opened a "vacuum of power" in Asia that enabled Europeans to take over established routes (Abu-Lughod 1989: 361). Abu-Lughod thus criticizes Eurocentric explanations of the West's rise to global dominance for overlooking the existence and role of the thirteenth-century "world-system" (Abu-Lughod 1993: 22). She "rejects the facile answer that Europe had unique qualities that allowed her to. Europe pulled ahead because the 'Orient' was temporarily in disarray" (1989: 18). In this regard, she sees her study as a corrective "to Immanuel Wallerstein's work on the sixteenth century *et seq.* world-system" (Abu-Lughod 1996: 278) because his explanation begins too late: "The failure to begin the story early enough has resulted, therefore, in a truncated and distorted causal explanation for the rise of the west" (1989: 20; 1996: 278). Causality is indeed central to the debate. It is key to the 500-year gap in historical timing in Abu-Lughod's argument between, on the one hand, starting "early enough" to take into account the apogee of the network in the thirteenth century and its decline in the fourteenth, and on the other hand, the fact that Europeans conquered the Americas in the sixteenth century and only several centuries *later* came to dominate Asian societies in the eighteenth and nineteenth centuries.

The problem with Abu-Lughod's first thesis begins with her assumption that the criteria by which Wallerstein defines a world-system are the same as those she herself employs.

Abu-Lughod's principal criterion turns out to be long-distance trade, though she never makes this explicitly clear in her work, whereas Wallerstein's main criterion is a division of labor. Both entail trade, but very different intensities of causality. The consequence of her failure to address this fundamental difference is that Abu-Lughod's argument is unrewarding. The debate could have been productive if Abu-Lughod addressed or attacked, rather than entirely overlooked, Wallerstein's emphasis in *The Modern World-System I* (1974) on the difference of trade and causality associated with a division of labor in relation to defining a world-system. Abu-Lughod's neglect of Wallerstein's criteria, and her assumption that there is no difference in trade and causal relations in a division of labor and in long-distance or "global" trade, gives rise to her puzzlement as to why he "refuses to 'dignify' any pre-sixteenth century patterns of global trade by applying the term 'world-system' to them. Indeed, he defends reserving *that* term only for the *modern* world-system, with its capitalist structure" (Abu-Lughod 1993: 278). This claim is unfounded. Abu-Lughod repeats the mistaken idea that Wallerstein sees the modern world-system as the first world-system in at least two more distinct places in her text (Abu-Lughod 1989: 9; 1993: 279). In so doing, she misrepresents his argument, which is that many past world-systems existed, including within the area of the thirteenth-century trading network. However, the long-distance trade networks among them did not constitute a world-system and the modern world-system of the sixteenth century is merely the first *capitalist* world-system.

Abu-Lughod's claims that the thirteenth-century trading network and the modern world-system are both world-systems rests on certain differences and similarities. One of the differences she notes (in agreement with Wallerstein) is the distinct capitalist nature of the modern world-system and that the thirteenth-century long-distance trading network was not a capitalist system, though it had nascent capitalist areas (Abu-Lughod 1996: 278). Unlike the modern world-system, the thirteenth-century network was, "organized on very different principles. Rather than a single hegemon, there were a number of coexisting 'core' powers ... " (Abu-Lughod 1989: 364–65). By "'core' powers" Abu-Lughod uses the term " 'core' powers" to refers to world-empires and world-economies. The absence of hierarchy among them is why, "this earlier world-system was organized in a very different way from the one over which Europe would ultimately exercise hegemony" (Abu-Lughod 1996: 278). The terms "world-empires and world-economies" are precisely those that Wallerstein uses in reference to two types of world-systems, not subsystems and core zones. For Wallerstein the lack of hierarchy shows a lack of integration, which logically follows from the long-distance trade relations among them. The language Abu-Lughod uses to describe the thirteenth-century trading network tends to support Wallerstein's view. Further, Abu-Lughod acknowledges that the thirteenth-century system did not develop a division of labor as the sixteenth-century world-system did, at least eventually. According to Abu-Lughod: "Nor was the world [in the thirteenth century] the global village of today, sharing common consumer goals and assembly-line work in a vast international division of labor ... " (Abu-Lughod 1989: 33). To be sure, she goes on to argue that the modern world-system in the sixteenth century was about as integrated as the thirteenth-century network. That is true only in the sense that by using long-distance trade to define the modern world-system in the sixteenth century, she includes Asia. Wallerstein, using the criteria of a division of labor, includes only Europe and parts of the New World. Wallerstein refers to smaller and tightly integrated systems in both the thirteenth and sixteenth century; Abu-Lughod does not.

In describing the similarities, Abu-Lughod refers to similar scales and technologies of trade (1989: 10–11, 353–54) and how in her view, both systems are also uneven and socially similar (1989: 354). These similarities, she argues, indicate that, "Wallerstein tends to overemphasize the discontinuity between the new Eurocentered capitalist world-economy that began to come into being and *the system of* world-empires and world-economies that preceded it" (Abu-Lughod

1996: 278, italics added). But note the wording I have emphasized. As noted previously, Abu-Lughod describes world-empires and world-economies as contained within a larger system, as subsystems of a world-system. This is nonsensical in Wallerstein's perspective since world-empires and world-economies are themselves world-systems. Abu-Lughod's failure to debate Wallerstein on his own terms is both puzzling and misleading.

Abu-Lughod never counters Wallerstein's claim that the thirteenth-century trade network did not entail a division of labor. Nor does she ever attack the most fundamental argument of his career, that from its beginning, the modern world-system did feature a division of labor. On the contrary, she agrees with this claim. She observes that in the earlier network, "the proportion of population and even production involved in international exchange constituted only a very small fraction of the total productivity of the societies" (Abu-Lughod 1993: 3). Regarding the European part of it, she notes that, "long-distance trade was not making a significant contribution to the internal economies [there]" (Abu-Lughod 1989: 13). Overall, "the subsystems in the thirteenth century were more self-sufficient than they are today and, therefore, less vitally dependent upon one another for common survival" (1989: 33). And yet, for Abu-Lughod, the *lack* of tight integration becomes the main reason to argue that the thirteenth-century network should be considered a "world-system":

> Since such production and exchange were relatively unimportant to the subsistence economies of all participating regions, I did not have to defend an unrealistic vision of a tightly entailed international system of interdependence. Clearly, this was not the case but it was also true in the sixteenth century. Thus, if it is possible to argue [as Wallerstein does] that a world-system began in that later century, it is equally plausible to acknowledge that it existed three hundred years earlier.
>
> *(Abu-Lughod 1993: 3)*

Predictably, for Wallerstein, Abu-Lughod's premise that the thirteenth-century network did not entail integrated interdependence is the very reason to reach the conclusion that it was not a world-system. This is why, based on Abu-Lughod's own facts, Wallerstein's description of the thirteenth-century trading network is more accurate than her own: "Abu-Lughod's 'world system' represents a long-distance trading linkup, on a nonhierarchical basis, of a series of temporally coexistent world-systems, each of which was internally hierarchal" (Wallerstein, 1992: 130).

The essence of this debate is the importance in the different degrees of systemic causality associated with a division of labor and that associated with long-distance trade. Here too, Abu-Lughod's descriptions support Wallerstein's view. In order to explain the West's rise, she firstly stresses how Europeans introduced in Asia new "rules of the game" of "short-term plunder rather than long-term exchange" in the sixteenth century, and how "Arab and Indian vessels proved no match for the Portuguese 'men-of-war' that appeared in their waters in the early 1500s" (1989: 19). However, she acknowledges that this new European approach in Asia did not result in European hegemony in Asia in the sixteenth century. She begins to concede that Portuguese and Spanish advances into Asian regions in the sixteenth century may not be so relevant to the "rise of the West" or the formation of the modern world-system. She goes on to recognize that the new European takeover of long-distance trade routes in Asia were not of comparable significance to the European conquest of entire societies in the New World (accomplished through the establishment of entirely new shipping routes). She explains:

> What *decisively transformed* the shape of the 'modern' world-system was not so much the Portuguese takeover of the 'old world' but the Spanish incorporation of the 'new

> world' ... This geographic reorientation *displaced the center of world gravity in a decisive manner* ... .
>
> *(Abu-Lughod 1989: 363, emphasis added)*

Europe's conquest of the New World raises the fundamental question: what is the relevance of the thirteenth-century long-distance trading network if it had little to do with Europe's conquest of the New World, which marked Europe's rise? Abu-Lughod responds by changing her thesis: that Europe's conquest of the Americas enabled Europe to become hegemonic over Asia *several centuries later* and not in the sixteenth century:

> With respect to European development, it is important to recognize that it was not until several centuries later that the new globalized system incorporating the Americas yielded its full return and catapulted Europe to world hegemony. By that time, Eastern trade had revived, but under a *new* hegemonic system that was at first exercised by proxies (the state-supported European trading companies) but eventually by strong states themselves. But that is an entirely different story.
>
> *(Abu-Lughod 1989: 364)*

It was an entirely different story indeed. According to Abu-Lughod herself, Europe's power over Asia in later centuries derived not from Asia's decline (a resulting power vacuum), nor from reconstructing long-distance trade routes dating to the thirteenth century. Given that Europe's conquest of Asia occurred some five hundred years after the decline of the thirteenth-century network and Asian societies' withdrawal from it, the relevance of the network and the "Fall of the East" for Europe's rise to power in the sixteenth century is quite unconvincing.

The difference in Europe's relations to the New World and to Asia in the sixteenth century relate directly to the differences in Wallerstein and Abu-Lughod's definitions of a world-system. Because Abu-Lughod does not distinguish between the interactions and goods involved in long-distance trade and those involved in a division of labor, she has no explanation for the significant difference in Europe's trade and power relations with the New World and with Asia. In Abu-Lughod's view, both areas are equally part of the modern world-system in the thirteenth and sixteenth centuries because the type of trade and associated relations are undistinguished. Unfortunately, Wallerstein does not rehash this theoretical-empirical issue with Abu-Lughod and the debate suffers for it. However, he does mention it in response to Frank, explaining that "the distinction between trade in 'luxury' goods and trade in 'bulk' goods or 'necessities,' ... permits us to distinguish trade *within* a historical system (primarily in 'necessities') and trade *between* separate systems (primarily in 'luxuries')" (Wallerstein 1996: 294). Contrary to Abu-Lughod's assumption regarding the size of the modern world-system in the sixteenth century, Wallerstein argues that:

> we cannot simply include in it any part of the world with which 'Europe' traded ... it would be *prima face* hard to argue that either Monomotapa or Japan were part of the European world-economy at that time ... [or] that Asian primary production was an integral part ... of the European division of labor ... [Thus,] Asia, or even Indian Ocean border regions, did not become part of the European world-economy in the sixteenth century. Asia was an external arena with which Europe traded, on somewhat unequal terms to be sure.
>
> *(Wallerstein 1974: 301, 332)*

Abu-Lughod's view that European dominance in Asia began well after the sixteenth century is quite consistent with Wallerstein's own argument in *The Modern World-System III*, except that his argument explains *why* this was the case, whereas hers does not. He describes how the Indian subcontinent remained part of the modern world-system's external arena through long-distance trade until incorporation began in the mid-eighteenth century, at which point it became part of the division of labor at the same time that Europeans brought China into the system's external arena and, beginning in the mid-nineteenth century, into its division of labor (along with Japan it could be added) (Wallerstein 1989: 129, 168).

In other words, the basis of inclusion and exclusion of regions, of European incorporation vs. mere dominance in long-distance trade, is one of intense causality associated with the total conquest and destruction of societies that permits the establishment of a division of labor. This is in contrast to the much weaker causality associated with mere domination of long-distance trade that does not involve the conquest of societies. As Wallerstein explains:

> To appreciate why we do not consider the Indian Ocean trading area to be part of the European world-economy despite the fact that it was so completely dominated by a European power, we must look successively at the meaning of this dominance for the Asian countries affected, its meaning for Europe and how it compares with those parts of the Americas under Iberian rule.
>
> *(Wallerstein 1974: 328)*

How does it compare? On the one hand, in the sixteenth century, "Europe did not conquer Asia because she could not … and this military balance would only change with the Industrial Revolution" (1974: 332). By contrast, in the sixteenth century, "The Americas were easily conquered. Even the structured states, like the Aztecs and the Incas, were no match for European arms [and germs]. Asia was another matter altogether" (1974: 338). The difference in the power of the societies in the New World and in Asia, and secondarily their different distances from Europe, is what explains, as Wallerstein insightfully observes, the different capabilities of European powers in the two regions, and why Europe could incorporate the New World into its division of labor but not Asia, where it engaged primarily in long-distance trade: "It is striking how *Iberian* policy was roughly similar in both areas. For in the sixteenth century, Iberia establishes *colonies* in the Americas, but *trading-posts* in Asia" (1974: 335).

Yet both Wallerstein and Abu-Lughod trace, if inadequately, the dynamics of the sixteenth-century conquest of the Americas to the thirteenth century. As noted, Abu-Lughod points to Europe's geographic reorientation toward the Atlantic and emphasizes that the dynamic behind it had little to do with the long-distance trading network: "Genoa's move into the Atlantic was primarily motivated by changes that were occurring within the local European subsystem, rather than in the 'world system,' and was at first directed only toward Europe" (1989: 363). She also suggests that the "rules of the game" that she says decisively transformed the modern world-system in the sixteenth century and displaced the center of world gravity also originated in the thirteenth-century *European* world-economy: "In the Mediterranean, a perpetual state of naval warfare existed from the ninth century onward, and commercial ships, therefore, always traveled in military convoys. That was not true in the Indian Ocean" (1989: 275). Wallerstein had already pointed this out. Preceding the Spanish and Portuguese advances in the sixteenth century, "the technological superiority of the gunned ship … had been developed in Atlantic Europe in the *two prior centuries*" (Wallerstein 1974: 326, my emphasis).

Wallerstein's arguments about those prior centuries indicate that Abu-Lughod's claim that his analysis does not start early enough is only partially true. Like Abu-Lughod, he observes that, "From about 1150 to 1300, there was an expansion in Europe … an expansion at once geographic, commercial, and demographic" (Wallerstein 1974: 37). This expansion was part cause and part consequence of the rise of the modern states in Europe, which he argues,

> dates not from the sixteenth century but from the thirteenth century in Western Europe ... It was at this period that not only were the boundary lines decided but, even more important, it was decided that there would be boundary lines.
>
> *(Wallerstein 1974: 31–32)*

Wallerstein also explains that the centralized bureaucracy, which formed the basis of the "great restorers of internal order in western Europe" in the fifteenth century following the Black Plague, was a process that:

> had started already in the twelfth and thirteenth centuries. With the cessation of the invasions, which had previously preoccupied and exhausted the princes, the growth of population, the rival of trade and hence the more abundant circulation of money, there was a basis for the taxation which could pay for salaried officials and troops. This was true not only in France, England, and Spain but in the principalities of Germany as well.
>
> *(Wallerstein 1974: 29)*

That is, the rise of the interstate system was tied to the conversion of domains and estates to capitalist agriculture and the rise of yeoman farmers. Both increasingly marketed their surpluses which provided taxes and fed into the creation of the town-country division of labor and the creation of two European world-economies. As Wallerstein explains: "In Europe in the late Middle Ages ... there were at least two smaller world-economies, a medium-sized one based on the city-states of northern Italy and a smaller one based on the city-states of Flanders and northern Germany" (Wallerstein 1974: 37). Whereas the new states arose within the northern world-economy, it was in the South that the dynamics of overseas expansion originated. Well before Abu-Lughod, Wallerstein traced the Genoese reorientation to Iberia and the Atlantic—and Iberia to the Americas—to the dynamics of the southern world-economy:

> The Genoese, the great rivals of the Venetians, decided early on to invest in Iberian commercial enterprise and to encourage their efforts at overseas expansion ... K.M. Panikkar points to Genoa's desire to capture the India trade from the thirteenth century on ... once there, the Italians would proceed to play an initiating role in the Iberian colonization efforts because, by having come so early, 'they were able to conquer the key positions of the Iberian Peninsula itself'.
>
> *(Wallerstein 1974: 49, n. 125, 50)*

In fact, Wallerstein considers the rise of the modern world-system as an outcome of the linkage of the two thirteenth-century European world-economies:

> I think it most fruitful to think of the sixteenth-century European world as being constructed out of the linkage of two formerly more separate systems, the Christian Mediterranean system centering on the Northern Italian cities and the Flanders-Hanseatic trade network of north and northwest Europe, and the attachment to this complex on the one hand of East Elbia, Poland, and some other areas of Eastern Europe, and on the other hand of the Atlantic islands and parts of the New World.
>
> *(Wallerstein 1974: 68)*

If the strength of Wallerstein's analysis (1974) is his reconstruction of the sixteenth-century division of labor that included Europe and parts of the Americas, the weakness lies in not detailing or following through with these thirteenth-century trajectories and instead focusing on the so-called "crisis of feudalism," 1350–1450 as the turning point. That period now appears to be the collapse of the emergent capitalist world-economy *and* the long-distance trade network, while the crisis of feudalism is presented as an expansion-crisis rather than a collapse-crisis: the commercial revolution of the 950–1350 period, or what Wallerstein calls the new "classic feudalism." In fact, he notes that, "For most previously 'free' laborers, this *new system* in fact involved a considerable increase in their exploitation, combined with a relative opportunity for some to improve their situation" (Wallerstein 1992: 583, italics added). In other words, the expansion temporarily strengthened some feudal nobility who made cash profits from selling surpluses generated with coerced labor. This is why Hibbert contends that, "Both fact and theory suggest that in earlier medieval times trade was by no means [an immediate] solvent of feudal society, but that it was a natural product of that society and that feudal rulers up to a point favoured its growth" (Hibbert 1953: 17). It is no paradox or contradiction that the new "classic feudalism" of the 950–1350 period was part of the commercial revolution, and vice-versa, and that it brought to an end, as Wallerstein explains, the older "manorial system, with its combination of slave labor on the demesne and 'free' laborers [that] collapsed by the end of the tenth century" (1992: 583).

In light of such developments, one can say that if the weakness of Abu-Lughod's analysis is attributing a causality to the thirteenth-century long-distance trading network for the rise of "the West" that it did not have in the sixteenth century, the strength of her study is in revealing that the trading network factored significantly into the thirteenth-century *European* world-economy's formation and expansion, and more importantly, that European world-economies seem to have already fused by 1350, if weakly. Her description of the Champagne Fairs, for example, reveals significant integration of northern and southern European capitalist activities and the conversion of aristocrats to regulators, protectors and direct participants (Abu-Lughod 1989: 58, 63–64). She sums up by citing Braudel:

> Thus it was that the two major economic zones—the Low Countries and Italy—slowly and *simultaneously* came into being. And it was between these poles, these two potential 'core zones,' that the Champagne fairs had their day. Neither North nor South triumphed … in this early form of the European world-economy … [However,] all the international and above all most of the modern aspects of the Champagne fairs were controlled, on the spot or at a distance, by Italian merchants whose firms were often huge concerns.
>
> *(cited in Abu-Lughod 1989: 57, 69, italics added)*

Regarding southern Europe, she explains how the leading Italian city-states dominated the most profitable activities across Europe:

> [D]uring the latter part of the thirteenth century and, even more, in the early part of the fourteenth century, the port cities of Genoa and Venice expanded their reach to incorporate virtually every part of the developing European world-economy … Not only did they trade in all parts of Western Europe but they had made significant inroads into the Central Asian and Middle East-North African world-economies that continued to mediate the far Eastern trade.
>
> *(Abu-Lughod 1989: 122)*

But taking all this into consideration suggests that if there was continuity between two systems, it was not, *contra* Abu-Lughod, between the thirteenth-century long-distance Eurasian trading network and the sixteenth-century world-system, but between the thirteenth-century *European* world-economy and the sixteenth-century modern world-system. Although Wallerstein argues that nowhere prior to the sixteenth century did capitalist entrepreneurs survive countermeasures that would have permitted the transformation of a world-economy into a fully capitalist system, he acknowledges that "such a transformation seemed to be beginning … in the Mediterranean basin between 1150–1300" (1979: 142). It would seem that these entrepreneurs did survive and, following the 1350–1450 collapse, expanded that system on the basis of pre-existing dynamics. However, if Wallerstein doesn't fully trace back the trajectories of the sixteenth century to the thirteenth, Abu-Lughod stops at 1350 without ever explaining exactly how Iberia's military superiority and successful colonization of the New World in the sixteenth century developed out of the thirteenth-century commercial revolution.

Still, Abu-Lughod's work poses a subtle challenge to the theoretical conception of world-systems as *the* largest scale unit of analysis in two ways. She demonstrates that the thirteenth-century European world-economy cannot be explained as entirely internal to itself, at least insofar as European involvement in long-distance trade both stimulated its formation *and* its collapse (with germs imported along with goods). In other words, if the Eurasian long-distance network reconstructed by Abu-Lughod was not a world-system, it nevertheless is a systemic network that is not accounted for in Wallerstein's perspective. It emerged out of relatively separate world-systems. The interactions among the world-empires and world-economies did not give rise to the same degree of causal impact or integrative influence compared to the internal interactions defining them as world-empires and world-economies. This is why the world-empires and world-economies did not become, and ought not to be conceptualized as, "subsystems" *contra* Abu-Lughod, yet alone as a 5,000 year system as argued by Frank and Gills (1996), among others. If anything, this long-distance trade network constituted a "subsystem" of those world-empires and world-economies, for it had an epiphenomenal existence. It exhibited degrees of causality sufficient to be theoretically conceptualized as a type of systemic pattern, but not a world-system.

Beyond this, the rise of the network and those world-systems presupposed longer-run, larger-scale evolutions in the form of human social systems (see Diamond 1999; Johnson and Earle 2000; Sanderson 1995). In fact, Wallerstein's response to the long-distance trade network is to locate it historically in the very long run, rather than deal with it theoretically. He expresses agreement that,

> … world-empires entered into long-distance trade (often? regularly? this is still to be demonstrated) with each other. I agree too that these world-empires included in the trading network of the oikumene various zones that were not organized as 'world-empires.' I even agree that, as a consequence, there may have been some common economic rhythms between them.
>
> *(Wallerstein 1996: 293–94)*

By situating such long-distance trading systems within the historical evolution and crowding of world-systems, he suggests that they also have an epiphenomenal nature. However, if the thirteenth-century long-distance network was the expression and medium of the "'crowding together' [that] accounts in part for the outbreak of the malady that is capitalism" (Wallerstein 1996: 295), this idea conflicts with his earlier argument, that the governing 'logics' of mini-systems and world-systems " 'determine' the largest part of sequential reality" (Wallerstein 1991: 244). It is less true when crowding reduces the degree of *relative* autonomy of such systems, or when a long-distance trade network becomes a regularized pattern that contributes to common rhythms among world-systems. Moreover, historical social systems are historical not only in having a limited timespace existence, but also in the sense that their autocatalytic developments are part of the very long run, arrow-of-time, developments that date to the spread of humanity across the planet, and the rapid evolutions (plural) of historical social systems in different regions, the diffusions of cultures, and interactions among systems during the past 11,000 years (Diamond 1999; Johnson and Earle 2000; Sanderson 1999). In these senses, world-systems analysis becomes more historical by taking new historical evidence into account theoretically while revealing the empirical weaknesses of arguments that lump thousands of years of historical social system evolutions into a single systemic process.

References

Abu-Lughod, Janet (1989) *Before European Hegemony: The World System A.D. 1250–1350*, New York: Oxford University Press.

Abu-Lughod, Janet (1993) *The World System in the Thirteenth Century: Dead-end or Precursor?* Washington, DC: American Historical Association.

Abu-Lughod, Janet (1996) "Discontinuities and Persistence, One World System or a Succession of Systems?" in Frank, A. G. and Gills, B. K. (eds), *The World System: Five Hundred Years or Five Thousand?* London: Routledge.

Diamond, Jared M. (1999) *Guns, Germs, and Steel: The Fates of Human Societies*, New York: W. W. Norton and Company

Frank, A. G. and Gills, B. K. (eds) (1996) *The World System: Five Hundred Years or Five Thousand?* London: Routledge.

Hibbert, A. B. (1953) "The origins of the medieval town patriciate", *Past and Present*, 3(1): 15–27.

Johnson, Allen W. and Earle, Timothy (2000) *The Evolution of Human Societies, From Foraging to Agrarian State*, Stanford: Stanford University Press

Sanderson, Stephen K. (ed.) (1995) *Civilizations and World Systems*, Walnut Creek, CA: AltaMira Press.

Sanderson, Stephen K. (ed.) (1999) *Social Transformations*, New York: Rowman & Littlefield.

Wallerstein, Immanuel (1974) *The Modern World-System I*, San Diego, CA: Academic Press.

Wallerstein, Immanuel (1979) *The Capitalist World-Economy*, New York: Cambridge University Press.

Wallerstein, Immanuel (1980) *The Modern World-System II*, San Diego, CA: Academic Press.

Wallerstein, Immanuel (1984) *The Politics of the World-Economy*, Cambridge, MA: Cambridge University Press.

Wallerstein, Immanuel (1989) *The Modern World-System III*, San Diego, CA: Academic Press.

Wallerstein, Immanuel (1991) "Call for a debate about the paradigm", in *Unthinking Social Science*. Cambridge: Polity Press. Original title of essay, "World systems analysis" in Giddens, A. and Turner, J. (eds), (1987) *Social Theory Today*, Cambridge: Polity Press.

Wallerstein, Immanuel (1992) "The West, capitalism, and the modern world-system" *Review*, XV (4): 561–619.

Wallerstein, Immanuel (1995) "Hold the tiller firm: on method and the unit of analysis" in Stephen K. Sanderson (ed.), *Civilizations and World Systems*, Walnut Creek, CA: AltaMira Press.

Wallerstein, Immanuel (1996) "World system versus world-systems: a critique" in Frank, A. G. and Gills, B. K. (eds), *The World System: Five Hundred Years or Five Thousand?* London: Routledge.

Wallerstein, Immanuel (2005) *World-Systems Analysis, An Introduction*, Durham, NC: Duke University Press.

1.3

The Afroeurasian world-system

Genesis, transformations, characteristics

Leonid Grinin and Andrey Korotayev

The Afroeurasian world-system (AEWS throughout) is the largest historical world-system. In the period of its largest expansion in the thirteenth to fifteenth centuries it encompassed almost all the societies of Europe and Asia and a substantial part of African societies. In the sixteenth to nineteenth centuries the AEWS incorporated all the rest of the world-systems, as a result of which it was transformed into the planetary world-system. Thus, the AEWS was the predecessor of the modern world-system, which is why the roots of many of its relationships and phenomena go deep into the history of the emergence and transformations of the AEWS (Chase-Dunn and Babones 2006; Chase-Dunn and Hall 1997; Grinin and Korotayev 2009; Korotayev 2005, 2008; Korotayev et al 2006a).

In addition to the AEWS, there were several world-systems in existence (in the Americas, Oceania, Australia) prior to the transformation of the AEWS into the modern planetary world-system. However, from the time of its formation and during the course of the subsequent millennia, the global supremacy of the AEWS was constant. It had the greatest tendency toward expansion, growth of complexity and growth rates. It is significant that by the early first millennium CE, it encompassed more than 90 percent of the world's population (Durand 1977: 256).

Outline of Afroeurasian world-system history

In the tenth to eighth millennia BCE the transition from foraging to food production took place in the Fertile Crescent area, as a result of which one could observe a significant growth in the complexity of the respective social systems.

In the eighth and fifth millennia BCE the AEWS expanded and the formation of effective informational, cultural, and even trade links between its parts was apparent.

In the fourth and third millennia, a large number of cities formed, first in Southern Mesopotamia, and then in other parts of the AEWS. Writing systems, large-scale irrigation agriculture, and new technologies of tillage developed. The first early states and civilizations formed on this basis.

In the late third and the second millennia BCE in Mesopotamia, the succession of such large-scale political entities as the Kingdom of Akkad, the third Dynasty of Ur, the Old Babylonian and Assyrian Kingdoms took place. The hegemony struggle in the core of the AEWS moved to a new level with the clash between the New Kingdom of Egypt and the Hittite Empire. In the second millennium BCE, a new AEWS center emerged in the Far East with the formation of the first Chinese state of Shang/Yin.

In the late second and first millennia BCE, iron metallurgy was diffused throughout the AEWS, which led to a significant growth of agricultural production in the areas of non-irrigated agriculture of Europe, North Africa, the Middle East, South Asia, and the Far East. This also led to the rise of crafts, trade, urbanization, and military capabilities. In the first millennium BCE, the hegemony struggles moved far beyond the Near East and the enormous Persian Empire emerged. In the second half of the fourth century BCE, Alexander the Great's campaign created (albeit for a short period of time) a truly Afroeurasian empire encompassing vast territories in all three parts of the world.

In the late first millennium BCE, two new empires formed: the Roman Republic and the Chinese Empire (Qin, and later Han). An unusually long network of trade routes (the Silk Route) then developed between the western and eastern centers of the AEWS.

In the first millennium BCE and the early first millennium CE, in connection with climate change and some important technological innovations (saddle, stirrup, etc.), a new type of nomadic society emerged. These new nomads were able to cover enormous distances and to transform themselves very quickly into a sort of mobile army. As a result, the entire enormous landmass of the Eurasian steppe belt became the nomadic periphery of the AEWS.

In the first centuries CE, as a result of mass migrations and military invasions of peoples from the barbarian periphery, the ethnic and cultural landscape of the AEWS experienced very significant changes. The Western Roman Empire disappeared as a result of the barbarians' onslaught. The Han Empire in China collapsed earlier. As a result of stormy events within the AEWS, a considerable number of new states emerged; it is noteworthy that some of them (such as Turkic khaganates) played roles as trade links between the East and the West. The first millennium CE saw the emergence and wide diffusion of new world religions.

In the first half of the second millennium CE, the Crusades (the eleventh to thirteenth centuries CE) were one of the most important world-system events; among other things they opened a channel of spice trade with Europe. An enormous role was played by the Mongolian conquests of the thirteenth century that brought unprecedented destruction and political perturbation. However, the later emergence of an exceptionally large Mongolian empire secured the diffusion of a number of extremely important technologies throughout AEWS; it also established a network of trade routes connecting East Asia with Europe that was unparalleled as regards its scale and efficiency.

New qualitative changes within AEWS were connected with the start of the Great Geographic Discoveries and the AEWS' transformation into the capitalist world-system.

Some special features of the Afroeurasian world-system and its main world-system processes

The AEWS featured some important peculiarities that stemmed from its scale, its extreme age and certain specific geographic conditions:

1. **The special complexity (supercomplexity) of its structure** that was determined by the size of its territory and the population concentration patterns. A very large world-system such as the AEWS is a *supersystem that integrates numerous subsystems*, including states, stateless polities and various spatial-cultural and cultural-political entities, such as civilizations, alliances, and cultural areas.
2. **The primary/autochthonous character of the major part of social and technological innovations**. Numerous borrowings and technological diffusion currents occurred almost

exclusively within the AEWS due to the enormous diversity of the available sociopolitical and economic conditions. Sea communications and landscape features allowed major flows of information, technologies and commodities to reach all the major AEWS centers sooner or later. This secured a certain (albeit imperfect) synchronization of processes in the different parts of the AEWS and raised the general speed of its development, as well as its stability.

3 **An exceptionally high speed of change**. The larger and more diverse the world-system, the higher the speed of its development. As a result, within the AEWS (as the largest world-system), growth rates were extremely high as the contacts became increasingly dense and the evolution of individual social systems was influenced ever more by macroevolutionary innovations diffusing throughout the system. This resulted in a significantly higher speed of development within the AEWS than in smaller world-systems (Diamond 1999).

4 **A succession of qualitative transformations** changed the structure of the AEWS due to the high speed and substantial continuity of its development. The Near Eastern center emerged first, with the South Asian and Far Eastern centers forming later. After this came the emergence of the European center that eventually became the leading center.

5 **The particularly important role of the barbarian (especially nomadic) periphery** was connected with certain peculiarities of climate and landscape, notably within the Eurasian Steppe Belt. For quite a long time, the development of the AEWS proceeded to a very considerable extent through the integration of its periphery, the transformation of a number of peripheral societies into semiperipheral ones, and the transformation of some of the semiperipheral societies into core ones (Hall et al 2009). As a result, the structure of the AEWS constantly changed; the flow of information and merchandise grew in complexity, as did the military-political interactions.

6 **The vital role of water communications**. A number of communication networks with particularly high levels of contact density were able to emerge due to the presence of water communication systems (Mediterranean, Baltic Sea, Indian Ocean networks, etc.). The growth of the AEWS proceeded to a considerable extent through the incorporation of coastal areas suitable for colonization and trade and their hinterlands (Phoenician, or Greek colonization, Sawahili cities along the East African coast, etc.).

World-system processes and transformations can be understood much better if these systemic properties are taken into account. Such systemic properties account for the synchronicity or asynchronicity of certain processes and the presence of positive and negative feedbacks that can be traced for very long periods of time. We believe that special attention should be paid to the idea suggested by Chase-Dunn and Hall (1997: XI–XII) that a world-system is constituted not just by intersocietal interactions but by the whole set of such interactions. Thus, the level of analysis that is most important for our understanding of social development is not the level of societies and states, but that of the world-system as a whole. In this way, a fundamental system property (the whole is more than just a sum of its parts) is realized within world-systems. Changes and transformations in certain parts of a world-system can produce changes in other parts via what may be called an *impulse transformation*. Such change may be manifested in various forms (producing sometimes rather unexpected consequences). Thus, the blocking of spice deliveries to Europe due to the Turkish conquests in the fifteenth century stimulated the search for the sea route to India, which finally changed the whole set of relationships within the AEWS. Due to systemic properties, processes that started in certain parts of the AEWS were able to diffuse rather rapidly to most other parts of it (e.g., the rapid diffusion of the Black Death pandemic in the fourteenth century).

A particularly interesting type of systemic property manifestation is represented by **synchronized processes** that took place in various parts of the AEWS. One example of this is the East/West synchrony in the growth and decline of the population sizes of the largest cities from 500 BCE to 1500 CE in West Asia and East Asia (Chase-Dunn and Manning 2002). There is a similar synchrony in the territorial sizes of the largest empires (Hall et al 2009). Barfield (1989) argues that large steppe confederacies usually cycle synchronously with the rise and fall of the large sedentary agrarian states that they raid. These cycles are one hypothesized mechanism of the systemic linkages between East and West Asia (Hall et al 2009). Such synchronized processes within the AEWS have also been detected by students of the Bronze Age and earlier periods (Chernykh 1992; Frank 1993; Frank and Thompson 2005). Other examples of synchronized processes are the Axial Age transformations of the first millennium BCE (Jaspers 1953) and the military revolution and formation of a new type of statehood in Europe and Asia in the late fifteenth and sixteenth centuries CE (Grinin 2008).

Whilst considering the general trends of AEWS development, it is necessary to note that: (a) the AEWS (phase) transition to a new phase produced a diffusion effect (through borrowing, modernization, incorporation, etc.) of the respective innovations throughout territories that turned out to be unprepared for the respective independent transformation; and (b) the development of the AEWS was frequently accompanied by the decline/underdevelopment of some of its parts, which meant that the flourishing of some societies led to a temporary decrease in the overall level of development/complexity of the AEWS. In addition, all of the major processes (especially the communication development) within the AEWS were greatly affected by migrations that frequently caused chain reactions involving the movement of peoples and warfare, which created the necessary conditions for large-scale transformations (Frank 1993).

The development of world-system links

The movement of the AEWS to every new level of development was inevitably connected with the expansion and strengthening of communication links and networks. Chase-Dunn and Hall (1997: 59) singled out the following main types of spatial links in world-systems: bulk goods exchange, prestige goods exchange, political-military interaction, and information exchange. They also note that the religions of the world constituted major innovations in the information networks and technologies of ideological power (Chase-Dunn and Hall 1997: 185). For this reason, it might make sense to single out cultural-ideological (civilization) interactions as a special type of world-system link, as they differ substantially from the usual information flows. Cultural-ideological interaction played a very important role within the AEWS, especially in its maturity. Since the eighth century CE all the civilized parts of the AEWS (with a partial exception of South Asia) consisted of actively interacting world religion areas (for more detail on the influence of the world religions on the evolution of AEWS see Korotayev 2004).

Initially, world-system analysis paid most attention to bulk goods trade (Wallerstein 1974), but for the period of the AEWS' formation, the most important role was played by information links, particularly the diffusion of innovations (Grinin and Korotayev 2009; Korotayev 2005, 2008). The presence of a pan-AEWS information network secured the diffusion of innovations throughout the AEWS. In general, the processes of innovation generation and diffusion played an immensely important role throughout the whole history of the AEWS.

A large-scale trade in strategic economically important items was already present in the framework of the emerging AEWS in West Asia. In particular, obsidian was transported from the Anatolian Plateau throughout the AEWS as early as the seventh millennium BCE. This is likely to have been accompanied by trade in foodstuffs, leather and textiles (Lamberg-Karlovsky and Sabloff 1979). In the fifth and fourth millennia BCE, we have evidence of large-scale trade in metals (Chernykh 1992; Frank 1993), and there is even more evidence on large-scale trade in the third and second millennia BCE (Frank 1993; Wilkinson 1987). In the first millennium BCE, long-distance trade (including sea trade) became even more developed (Chase-Dunn and Hall 1997).

In the second half of the first millennium CE the Indian Ocean Basin saw the formation of a prototype of an oceanically connected world-system. In this enormous network of international trade an important role was played by merchants from countries which include Persia, Arabia and India (see Bentley 1996 for more detail). It is important to note that trade in this region was not restricted to luxury items but included a considerable number of bulk goods (dates, timber, construction materials, etc.).

Later on, the thirteenth and fourteenth centuries saw the emergence of a vigorous transcontinental trade network through the territories of the Mongolian states that connected, in a very tangible way, all the main zones of the AEWS. According to Abu-Lughod (1989), this world-system trade network had a greater complexity of organization and a larger volume than any previously existing network.

Main phases of genesis and transformation

There are many of points of view regarding possible dates of the formation of the AEWS. For example, Frank and Thompson date its origins to the fourth and third millennia BCE (Frank 1993; Frank and Thompson 2005), while Wilkinson (1987) and Berezkin (Березкин 2007: 92–93) consider the second millennium as its beginning. We date the emergence of AEWS to a considerably earlier period: the tenth to eighth millennia BCE (Grinin and Korotayev 2009; Korotayev and Grinin 2006). However, some other world-system students believe that it only came into real existence in the late first millennium BCE (Chase-Dunn and Hall 1997, 2008; Hall et al 2009).

The approaches to this issue differ considerably depending on the world-system criteria employed. The criteria are as follows: the bulk good criterion (a more rigid one), the prestige good, or the information network (both softer criteria). The more rigid the approach, the more recent is the dating that it produces. However, dating also depends upon the general approaches to the emergence of the AEWS. For example, if, together with Chase-Dunn and Hall (1997: 150) we believe that by the moment of the initiation of the Silk Route there were three main independent world-systems (West Asian, Chinese, and South Asian) that merged later into a single world-system (AEWS), then it appears very logical to date the emergence of the single AEWS to the late first millennium BCE. However, if we begin from the basis that the West Asian world-system was, from the very beginning, the leader in respect of innovations, technologies, and social relations; that it greatly influenced the development of South Asia and the Far East; and that by the late first millennium BCE, the influence in the opposite direction was negligible (and hence we should speak about the incorporation of South and East Asia into the AEWS, rather than a merger of three equally important world-systems); then the origins of the AEWS must have been significantly more ancient.

In any case it is quite clear that the emergence of the AEWS was a rather prolonged process. It should also be noted that it was the Near East that experienced the earliest transition to food production (in particular the cultivation of cereals), large-scale irrigated agriculture, urban settlement patterns, metallurgy, writing, statehood, and empires. Hence, however we date the initiation of the AEWS, it is perfectly clear that the roots of its formation reside deep in time and certainly up to the beginnings of the agrarian ("Neolithic") revolution in West Asia in the tenth to eighth millennia BCE.

Within this prolonged process of the genesis and transformation of the AEWS, it is possible to single out a few major phases:

The first phase: The formation of contours and structure of the Middle Eastern core of AEWS (the eighth to fourth millennia)

This period saw the finalization of the first stage of the agrarian revolution in the Near East. Following this, the second stage of the Agrarian Revolution was connected with the formation of large-scale irrigation and later intensive plow agriculture in the fourth to first millennia BCE (Grinin 2007; Korotayev and Grinin 2006). This period evidenced the beginning of the formation of long-distance and permanent information/exchange contacts. Those processes were accompanied by the formation of medium-complexity early agrarian societies, relatively complex polities, and settlements that distantly resembled cities.

In the fifth millennium BCE the Ubaid culture emerged in Southern Mesopotamia, within which the material and social basis of the Sumerian civilization were developed. The Uruk culture that succeeded the Ubeid was characterized by the presence of a considerable number of reasonably large settlements. Thus, by the end of the first phase, the Urban Revolution took place within AEWS; this revolution can be regarded as a phase transition of AEWS to a qualitatively new level of social, political, cultural, demographic, and technological complexity (Березкин 2007). By the end of the first phase, urbanized societies emerged (Pollock and Bernbeck 2005: 17), as well as the first early states, their analogues (Grinin 2003, 2008; Grinin and Korotayev 2006), and civilizations.

The second phase: Development of AEWS centers in the Bronze Age (the third and second millennia BCE)

This period saw fast growth in agricultural intensiveness and population in the AEWS. A relatively rapid process of urbanization in the AEWS was observed in the second half of the fourth millennium and the first half of the third millennium BCE, which then slowed down significantly until the first millennium BCE. One of the most important results of this period was the growth of political integration in the AEWS core societies, which was a consequence of complex military-political and other interactions. Primarily, in the AEWS, core political complexity grew very significantly from cities and small polities to large and developed early states (Grinin 2008; Grinin and Korotayev 2006). Secondly, the first empires emerged. Thirdly, from the third millennium BCE, there were cycles of political hegemony involving upswings and downswings (Frank and Gills 1993; see also Chase-Dunn et al 2010).

Parts of Europe were firmly included in the AEWS communication network, while new world-system centers emerged in South Asia and the Far East. However, none of them could compete with the West Asian center, whose territory and complexity grew rapidly. In this framework, the prestige goods' trade system reached a high level and was often supported by states. Trade links with South Asia were established through the Persian Gulf.

Key West Asian technologies (cultivation of West Asian cereals, breeding of cattle and sheep, some important metallurgy, transportation and military technologies) penetrated to East Asia (possibly through the Andronovo intermediaries), which is marked archaeologically by the transition from the Yangshao culture to the Longshan (see, e.g., Березкин 2007). In this way the formation of the main AEWS centers took place; these centers developed throughout the subsequent history of the AEWS, yet during this period such development was marked by the technological (and other) leadership of the West Asian center and the strengthening of (still rather weak) communication links between the various centers.

The third phase: The Afroeurasian world-system as a belt of expanding empires and new civilizations (the first millennium BCE till 200 BCE)

This is the time of the early Iron Age. In the first part of this period, the agrarian revolution within the AEWS was already finalized through the diffusion of the technology of plow non-irrigation agriculture based on the use of cultivation tools with iron working parts (see Korotayev and Grinin 2006). From this production base, enormous changes in the trade and military-political spheres took place, accompanied by a new upswing in urbanization and state development, including the emergence of a group of developed states (see Grinin 2008; Grinin and Korotayev 2006).

This is the Axial Age period, the period of the emergence of the second generation civilizations. The development of all the AEWS centers proceeded vigorously. The West Asian center was finally integrated with the Mediterranean world, whereas the European areas of the barbarian periphery were linked more and more actively with the AEWS centers through military, trade and cultural links. In South Asia, a new civilization formed and the first world religion—Buddhism—emerged. Trade links were established in the space stretching from Egypt to Afghanistan and the Indus Valley (Bentley 1996), and in general, all this territory developed military and political connections. The East Asian center of the AEWS also developed very rapidly; this period evidenced the emergence of its own super-ethnic quasi-religion, Confucianism.

During this period the belt of empires expanded, with New Babylonian, Median, Persian, Macedonian (and its heirs) empires in the center, the Mauryan Empire in South Asia, and Carthage in the West. The end of the period is marked by the formation of major empires at both extremes of the AEWS: the Roman Republic in the extreme West and the Chinese Empire in the extreme East.

The fourth phase: The AEWS is integrated by the steppe periphery (200 BCE—the early seventh century CE)

Around the second century BCE, relatively stable trade links (albeit involving preciosities rather than bulk goods) were established between the "marcher empires" of the AEWS through the so-called Silk Route, a significant part of which went through the territories of the nomadic periphery and semiperiphery. Thus, in this period the periphery closed the circuit of AEWS trade links. The AEWS expansion proceeded for a long period to a considerable extent through the expanding interaction between civilizations and their barbarian peripheries. The larger and more organized civilizations grew, the more active and organized their peripheries became. During the fourth phase, this process was sharply amplified and the Great Migration epoch evidenced the acquisition of a world-system scale and synchronicity of influence by the barbarian periphery itself. The disintegration of the Western Roman Empire, the weakening of the Eastern Roman Empire, the fast diffusion of Christianity in the western part of the AEWS and a new rise of the Chinese Empire in its eastern part prepared the AEWS for major geopolitical changes involving a new level of complexity. In contrast, the growth of the AEWS population by the end of the first millennium BCE up to 9-digit numbers led to an increased level of threat from pathogens. Thus, the Antonine and Justinian pandemics led to catastrophic depopulations throughout the AEWS in the second and sixth centuries, contributing (in addition to the onslaught of the barbarian peripheries) in a very substantial way to the significant slowdown of demographic and economic growth for the AEWS in the first millennium CE (Korotayev et al 2006b).

The fifth phase: AEWS apogee—world religions and world trade (the seventh to fourteenth centuries)

In this period, the level of development of world-system links reached the upper limits of what could be achieved on an agrarian basis. In this respect one should particularly note the formation and development of all the major world religions. In certain aspects, during this phase the AEWS developed as a supersystem of contacting and competing third generation civilizations, which created firm cultural-information links between all the AEWS centers, including South Asia (which had remained in relative isolation during the preceding period). Note also an unprecedented sweep of military-political contacts and the growth of the level of development of state structures.

Important preconditions for the transformation of the AEWS into the capitalist world-system began to form. In this respect one should note especially: (a) the formation of particularly dense oceanic trade links in the second half of the first millennium in the Indian Ocean Basin; (b) the creation of vigorous major transcontinental land routes that directly connected the main AEWS centers; and (c) the beginnings of the formation of an urbanized zone stretching from northern Italy up to the Netherlands, where commodity production became the dominant form of economic activity (Wallerstein 1974).

The sixth phase: Transformation of the Afroeurasian world-system (AEWS) into the world-system (the fifteenth to eighteenth centuries)

This phase was connected with the start of the industrial revolution that determined the transformation of AEWS into the capitalist world-system (now corresponding rather well to Wallerstein's 1974 notion of the world-system, as from this point its development involved mass movements of bulk goods throughout). However, as the agrarian production principle still prevailed absolutely, the extreme development of some previous trends are evident, especially in the non-European centers of the world-system. In particular, East Asia still continued to develop along its own trajectory, demonstrating indubitable achievements in the development of state and cultural structures and outstanding demographic growth.

The Great Geographic Discoveries sharply extended the contact zone of the AEWS. As a result of this, together with Europe's technological breakthrough, the world-system began to take on a new structure. The trade-capitalist core emerged in Europe, while previous world-system centers (in particular, the one in South Asia) were transformed into exploited periphery (this process became even more active at the subsequent phase of world-system evolution). Thus the phenomenon of the world-system periphery experienced a significant transformation.

References

Abu-Lughod, J. (1989) *Before European Hegemony: The World System A. D. 1250–1350*, New York: Oxford University Press.

Barfield, T. (1989) *The Perilous Frontier: Nomadic Empires and China*, Cambridge: Blackwell.

Bentley, J. (1996) "Cross-Cultural Interaction and Periodization in World History", *American Historical Review*, 101(3): 749–70.

Березкин, Ю. (2007) "О структуре истории: временные и пространственные составляющие" в П. Турчин и др. (ред.) *История и Математика: Концептуальное пространство и направления поиска*, Москва: URSS.

Chase-Dunn, C. and Babones, S. (2006) *Global Social Change: Comparative and Historical Perspectives*, Baltimore, MD: Johns Hopkins University Press.

Chase-Dunn, C. and Hall, T. (1997) *Rise and Demise: Comparing World-Systems*, Boulder, CO: Westview.

Chase-Dunn, C. and Hall, T. (2008) "East and West in World-Systems Evolution". IROWS Working Paper #42.

Chase-Dunn, C. and Manning, S. (2002) "City Systems and World-Systems", *Cross-Cultural Research*, 36(4): 379–98.

Chase-Dunn, C., Niemeyer, R., Alvarez, A., Inoue, H. and Love, J. (2010) "Cycles of Rise and Fall, Upsweeps and Collapses", in L. Grinin et al (eds) *Evolution*, Volgograd: Uchitel.

Chernykh, E. (1992) *Ancient Metallurgy in the USSR: The Early Metal Age*, Cambridge: Cambridge University Press.

Diamond, J. (1999) *Guns, Germs, and Steel: The Fates of Human Societies*, New York: Norton.

Durand, J. (1977) "Historical Estimates of World Population. An Evaluation", *Population and Development Review*, 3(3): 253–96.

Frank, A. (1993) "Bronze Age World System Cycles", *Current Anthropology* 34(4): 383–413.

Frank, A. and Gills, B. (eds) (1993) *The World System: Five Hundred Years of Five Thousand?* London: Routledge.

Frank, A. and Thompson, W. (2005) "Afro-Eurasian Bronze Age Economic Expansion and Contraction Revisited", *Journal of World History*, 16: 115–72.

Grinin, L. (2003) "The Early State and its Analogues", *Social Evolution & History*, 1(1): 131–76.

Grinin, L. (2007) "Production Revolutions and Periodization of History: A Comparative and Theoretic-mathematical Approach", *Social Evolution & History*, 6(2): 11–55.

Grinin, L. (2008) "Early State, Developed State, Mature State: The Statehood Evolutionary Sequence", *Social Evolution & History*, 7(1): 69–83.

Grinin, L. and Korotayev A. (2006) "Political Development of the World System: A Formal Quantitative Analysis", in P. Turchin et al (eds) *History and Mathematics. Historical Dynamics and Development of Complex Societies*, Moscow: URSS.

Grinin, L. and Korotayev, A. (2009) "Social Macroevolution: Growth of the World System Integrity and a System of Phase Transitions", *World Futures* 65/7: 477–506.

Hall, T., Chase-Dunn, C. and Niemeyer, R. (2009) "The Roles of Central Asian Middlemen and Marcher States in Afroeurasian World-System Synchrony", in G. Trinchur (ed) *The Rise of Asia and the Transformation of the World-System*, Boulder, CO: Paradigm Press.

Jaspers, K. (1953) *The Origin and Goal of History*, New Haven, CT: Yale University Press.

Korotayev, A. (2004) *World Religions and Social Evolution of the Old World Oikumene Civilizations*, Lewiston, NY: Mellen.

Korotayev, A. (2005) "A Compact Macromodel of World System Evolution", *Journal of World-System Research*, 11: 79–93.

Korotayev, A. (2008) "Compact Mathematical Models of World-System Development", in G. Modelski (ed.) *Globalization as Evolutionary Process*, London: Routledge.

Korotayev, A. and Grinin, L. (2006) "Urbanization and Political Development of the World System: A Comparative Quantitative Analysis", in P. Turchin et al (eds) *History and Mathematics. Historical Dynamics and Development of Complex Societies*, Moscow: URSS.

Korotayev, A., Malkov, A. and Khaltourina, D. (2006a) *Introduction to Social Macrodynamics: Compact Macromodels of the World System Growth*, Moscow: URSS.

Korotayev, A., Malkov, A. and Khaltourina, D. (2006b) *Introduction to Social Macrodynamics: Secular Cycles and Millennial Trends*, Moscow: URSS.

Lamberg-Karlovsky, C. and Sabloff, J. (1979) *Ancient Civilizations. The Near East and Mesoamerica*, Menlo Park, CA: Benjamin/Cummings.

Pollock, S. and Bernbeck, R. (eds) (2005) *Archaeologies of the Middle East: Critical Perspectives*, Oxford: Blackwell.

Wallerstein, I. (1974) *The Modern World-System. Vol. I. Capitalist Agriculture and the Origins of the European World-Economy in the Sixteenth Century*, New York: Academic Press.

Wilkinson, D. (1987) "Central Civilization", *Comparative Civilizations Review*, 17: 31–59.

1.4
Agricultural origins and early development

E. N. Anderson

Countless theories of agricultural origin exist. Graeme Barker (2006: 383) gives an alphabetical list, with dozens of theories, mostly improbable or vague (see also Bayliss-Smith 2007: 133–34; Cohen 2009; Denham and White 2007; Denham et al 2007.)

Agriculture arose in approximately seven places. The earliest was in the Near East, around 9000–10000 BCE. Agriculture in China was almost as early, where millets (*Setaria italica, Panicum miliaceum*) in the north and rice in the center-south appear to have been domesticated separately by 8000 BCE. The New World had agriculture in Mexico, Peru and the lowland tropics by 5000 BCE or earlier. These appear to be three separate agricultural origins, although they may be connected. New Guinea invented agriculture independently, perhaps by 6000 BCE. The earliest North American agriculture north of Mexico appears in the Mississippi Valley around 2500 BCE; it involves native plants of the area and has no discernable connection with Mexican domestication. It may or may not be a separate event.

The earliest cultivated plants were usually grains. In Mexico, however, early cultigens included squash, gourds and chillies. In South America, tubers were early—potatoes in the mountains, manioc and others in the lowlands. New Guinea's earliest agriculture was probably based on local yams and taro.

As R. S. MacNeish pointed out long ago (MacNeish 1991), the early centers of agriculture show common traits. They are mountainous or hilly, with complex interweaving of ecological zones. They are warm-temperate or tropical-alpine in climate. They have an enormous variety of plants and usually, fertile soil. Most are in areas with sharply seasonal rain; this puts a premium on producing much food in the rainy season and storing for the dry. All are at the centers of huge land masses, at or near points where great trade routes meet, thus maximizing opportunities for trade and information. Typically, different forms of early agriculture arose in these areas as part of wide regional development. Areas that were otherwise similar but existed as isolated cul-de-sacs never developed agriculture.

Many theories of agriculture are based on the assumption that people needed more food due to population increase and the death of Pleistocene megafauna (Cohen 1977), drought (Childe 1954), or some other reason. However, Carl Sauer (1952) pointed out that agriculture required time and effort to develop. Thousands of years were required to develop really productive agriculture (Barker 2006; Willcox 2007; Zeder 2008; Zeder et al 2007). People were generally well nourished at the time agriculture was invented, but became less and less so as they became more dependent on staple crops (mostly mere starch) (Cohen and Armelagos 1984). Agriculture also led to diseases caught from domestic animals and to water pollution—though it also permitted civilization, sciences and the flowering of the arts.

Agriculture came well after the extinction of the megafauna, and developed in areas that were not dependent on hunting. Agriculture developed and flourished with the rapid rebound of plant resources after the Ice Ages. Agriculture was perhaps a response to more plant availability. It developed only in areas where highly diversified foraging and plant use already existed, and where some degree of sedentary life had developed.

The only theories that fit the facts involve trade, communication or spatial translocation of crops. Perhaps crops were moved from drying areas to moister ones (Childe 1954). Perhaps they were brought nearer the house to protect them from raiders and to make them more available for trade. In New Guinea, for instance, it was dangerous to go far from the village, so intensive agriculture developed (see Brown 1972; Meggitt 1977). Trade with nonagricultural neighbors was clearly important everywhere.

The regions of initial invention dominated intensification and innovation for millennia. In some cases, speakers of languages from those regions expanded their populations and moved outward, carrying agriculture with them and thus spreading the practice (Bellwood and Renfrew 2002). Other groups picked up agriculture from their neighbors. How much of agricultural spread was due to migration, and how much to borrowing, remains controversial.

New crops accumulated: protein sources, flavorings, medicines, spices, perfumes, and industrial and fiber crops. Animals (starting with dogs) were domesticated. The evidence fits a model of development more often for trade and luxury than necessity. Drug plants were soon domesticated; by far the most widely cultivated crop in the pre-Columbian Americas was tobacco. Ester Boserup's theory (1965) of intensification driven by necessity is not widely applicable to such developments.

References

Barker, Graeme (2006), *The Agricultural Revolution in Prehistory: Why Did Foragers Become Farmers?* Oxford: Oxford University Press.

Bayliss-Smith, Tim (2007), "The Meaning of Ditches: Interpreting the Archaeological Record from New Guinea Using Insights from Ethnography", pp. 126–48 in Tim Denham, José Iriarte and Luc Vrydaghs (eds), *Rethinking Agriculture: Archaeological and Ethnoarchaeological Perspectives*, Walnut Creek, CA: Left Coast Press.

Bellwood, Peter, and Colin Renfrew (eds) (2002), *Examining the Farming/Language Dispersal Hypothesis*, Cambridge: McDonald Institute Monographs.

Boserup, Ester (1965), *The Conditions of Agricultural Growth*, Chicago, IL: Aldine.

Brown, Paula (1972), *The Chimbu: A Study of Change in the New Guinea Highlands*, Cambridge, MA: Schenkman.

Childe, V. Gordon (1954), *What Happened in History*, Harmondsworth: Pelican.

Cohen, Mark Nathan (2009), "Rethinking the Origins of Agriculture," *Current Anthropology*, 50(5): 585–712.

Cohen, Mark Raymond (1977), *The Food Crisis in Prehistory: Overpopulation and the Origins of Agriculture*, New Haven: Yale University Press.

Cohen, Mark, and George Armelagos (eds) (1984), *Paleopathology and the Origins of Agriculture*, New York: Academic Press.

Denham, Tim, José Iriarte, and Luc Vrydaghs (eds) (2007), *Rethinking Agriculture: Archaeological and Ethnoarchaeological Perspective*, Walnut Creek, CA: Left Coast Press.

Denham, Tim, and Peter White (eds) (2007), *The Emergence of Agriculture: A Global View*, 2nd edn., New York: Routledge.

MacNeish, R. S. (1991), *The Origins of Agriculture and Settled Life*, Norman: University of Oklahoma Press.

Meggitt, Mervyn (1977), *Blood Is Their Argument: Warfare among the Mae Enga Tribesmen of the New Guinea Highlands*, Palo Alto, CA: Mayfield.

Sauer, Carl O. (1952), *Agricultural Origins and Dispersals*, Berkeley: University of California Press.

Willcox, George (2007), "Agrarian Change and the Beginnings of Cultivation in th Near East: Evidence from Wild Progenitors, Experimental Cultivation and Archaeobotanical Data", pp. 217–41 in Tim Denham and Peter White (eds), *The Emergence of Agriculture*, 2nd edn., New York: Routledge.

Zeder, Melinda (2008), "The Neolithic Macro-(R)evolution: Macroevolutionary Theory and the Study of Culture Change," *Journal of Archaeological Research*, 17: 1–64.

Zeder, Melinda A., Daniel G. Bradley, Eve Emshwiller, and Bruce D. Smith (eds) (2007), *Documenting Domestication: New Genetic and Archaeological Paradigms*, Berkeley: University of California Press.

1.5

Qubilai and the Indian Ocean

A new era?

Paul D. Buell

The Mongols, creators of the greatest empire in history, are best known for their advances over land, through the deserts of Central Asia, into Iran, Syria, and Palestine, into Russia and even Europe. To be sure, thanks to Marco Polo, many are aware that Qubilai-qan (r. 1260–94) conquered and ruled China but knowledge of Mongol expansion almost always stops there. In this paper, I will argue that the advance into China was the beginning of a much larger enterprise, which coupled with the great maritime expansion of Ming times, ushered in our age. It was Westerners such as the Portuguese and later, the Dutch and British, who gained the most from what took place. The Mongol advance changed history by altering fundamental relationships between East and West, West and East. I will also argue that conventional wisdom claiming that Qubilai-qan's military defeats in Vietnam and in Sumatra ended his plans for a takeover in the Indian Ocean is incorrect. These defeats simply forced the Mongols to realize their goals in new ways.

In the history of the Mongol Empire, the critical event leading to its collapse was the 1251 election of Möngke-qan (r. 1251–59) from the house of Tolui-noyan, the youngest son of Cinggis-qan (r. 1206–27). The problem was that his election involved putting aside the house of Ögödei (r. 1229–41; Güyük, r. 1246–48), third son of Cinggis-qan, by force via a coup, which resulted in fissures in the Mongolian world that never closed again. Qubilai tried to gain recognition as Great Khan of the old Empire but was never able to reunite it under his banner. He even had to face a pretender to the old throne of Ögödei, Qaydu (1236–1301), who fought him unsuccessfully for decades, and eventually outlasted him.

In reality, Qubilai and his house were confined to Mongol China and adjacent parts of Central Asia. His enemies controlled the rest of Central Asia, Turkistan (the Ca'adai qanate, held by descendants of the second son of Cinggis-qan), and Russia, ruled by the house of Joci and descended from the eldest son of Cinggis-qan. Only in Iran did there exist a regime friendly to Qubilai and ruled by descendants of his younger brother Hüle'ü (died 1265), but how to establish contact? In fact, Qubilai did establish contact and formed a close relationship with Mongol Iran, the Ilqanate. The desire to establish this contact was a major driver in his push south into Southeast Asia and the Indian Ocean and beyond. In so doing, Qubilai sought to establish a southern axis—China on the one side, the Ilqanate on the other—to parallel the north-south (and east-west, via the Mediterranean) economic alliance of the Golden Horde of Russia with Mamluk Egypt.

The Mongol advance into China began before Qubilai. The north was conquered in 1235. Möngke-qan died trying to continue the march south. Qubilai, once secure in China, resumed this advance in 1267. He finally conquered all of Song China in 1279, defeating the last Song forces in a great naval battle at Yaishan, near modern Macau. At the time of its collapse, Song had a flourishing trade with the South Seas, with Song ships going as far as Africa. Qubilai inherited its networks and its sailors, its ships and ship yards and began to build on these foundations, initially through military action, even before the final stages of conflict in China. These were the well-known attacks on Japan (1274, 1281), on northern Vietnam (1285–88), and Champa (1281–85), located near modern Hue, and, farthest of all, against Java (1292). All were mounted with relatively large fleets and armies, particularly the campaigns into Indochina. All were unsuccessful militarily. If this had been the final outcome the net result would have been failure for Qubilai's policies, but this was clearly not the case. The great expeditions were visible and expensive and ultimate failure manifest, but side-by-side with military actions was a diplomatic offensive of even greater scope. It took Mongol representatives all over the Indian Ocean. Qubilai-qan's successor, Temür-öljeitü (r. 1294–1307) was able to discontinue the war in Vietnam and abandon another attempt on Japan, because there was no longer any need for such operations. Chinese ships were freely sailing the Indian Ocean and the link with Iran was by then secure. In addition, Mongol Iran itself had by then veered away from being purely a land power. As its coinage shows, it had also become an activist in ocean trade.

This trade was of massive proportions. From the one side came high value commodities such as blue and white porcelain, which was in high demand in much of the Old World. Tea was likely to have been another commodity. Porcelain was moved in bulk; this was no luxury trade. From the other side came other bulk goods, including cobalt to make the coveted glaze of China's blue and white porcelain, and—a truly large commodity—horses. In addition, and also moving in large numbers were people, ambassadors from China to Iran and from Iran to China. They included Marco Polo, who came to Iran with a princess. Alongside bulk goods came spices and medicinals (Mongol China was flooded with Indian Ocean medicines). Essentially, all that no longer moved via the old Silk Route, in the hands of the enemies of Mongol China and Iran, together with new products were put on ships and became part of the trade.

And it did not stop there. The Mongol dynasty in China collapsed but its Ming successor began a new age of maritime expansion and the voyages of Zheng. He followed the Mongol lead. A new age had dawned. Ming was the world's greatest economy, providing what the Portuguese and the rest went looking for. What followed was a change in scale, not essence, of a trading system that persists.

2. Historical processes of incorporation and development

2.1

Incorporation into and merger of world-systems

Thomas D. Hall

World-systems typically expand. Thus, they incorporate or absorb new peoples, new territories, or both. In situations where more than one world-system exists, sufficient expansion might lead to the merger or combination of one or more into a larger system. Alternatively, a merger might be seen as the incorporation of one world-system by another. The point of this article is to summarize and expand upon discussions of the concepts of incorporation into world-systems and the merger of world-systems. I also provide a brief account of how frontier zones fit into these discussions and conclude with some suggestions of possible directions for future research.

Wallerstein's concept of incorporation

Immanuel Wallerstein developed the concept of incorporation over several publications (1974, 1989; Hopkins and Wallerstein 1987; Hopkins et al 1987). His notion of incorporation was developed primarily to map the expansion of the modern Europe-centered system and to explain uneven development and what Andre Gunder Frank (1966) had called "development of underdevelopment" in the Global South. Thus, Wallerstein focused on the process of peripheralization. He also distinguished between incorporated peripheries and regions that were outside of the modern system, which he called "external arenas."

Wallerstein's key points regarding incorporation are:

> The question we are dealing with now is the nature of the process by which a zone which was at one point in time in the external arena of the world-economy came to be, at a later point in time, in the periphery of that same world-economy. We think of this transition as a period of medium duration and we denominate it the period of "incorporation." Hence, the model we are using involves three successive moments for a "zone": being in the external arena, being incorporated and being peripheralized. None of these moments is static; all of them involve processes.
>
> *(Wallerstein 1989: 129–30)*

Despite continuing trade and some modifications of the local mode of production:

> ... the trade zone in which these particular processes are located must be considered to remain in the external arena of the world-economy, despite the existence of the trade links, and no matter how extensive or profitable the ongoing "trade" seems to be.
>
> *(Wallerstein 1989: 130)*

In short:

> … incorporation involves "hooking" the zone in the orbit of the world-economy in such a way that it can virtually no longer escape, while peripheralization involves a continuing transformation of the ministructures of the area in ways that are sometimes referred to as the deepening of capitalist development.
>
> *(Wallerstein 1989: 130)*

Critiques of the initial formulations of "incorporation"

In the mid 1980s, Hall (1986, 1987, 1989a) developed a critique of Wallerstein's concept of incorporation. In Hall's studies of what is now the US southwest (approximately New Mexico, Arizona, Texas, California, and contiguous areas), he found that Wallerstein's analysis gave too little attention to how incorporated peoples resist incorporation and tended to ignore indigenous peoples. Hall argued that incorporation:

1. begins almost with initial contact, much earlier than originally conceived;
2. includes the "plundering" of some regions that have been termed "external arenas";
3. is actually a continuum from very weak to very strong;
4. has "fuzzy" or vague limits at the weaker pole;
5. has generally asymmetrical impacts, but not always to the benefit of core regions;
6. engenders resistance by those absorbed, albeit not always successful; and
7. is, especially at weaker levels, to some extent reversible, yet is "sticky" and usually tends eventually toward stronger incorporation.

Hall's studies of European intrusions into the southwestern United States showed that even very weak incorporation can have drastic effects on incorporated regions and peoples (Hall 1989a; see Hollis 2005 for the southeastern United States). Thus incorporation is a continuum from weak to strong. In both theoretical and methodological aspects, it is significant that the "fuzziness" of weak incorporation is not a result of poor measurement. Rather, it reflects an empirical condition that is inherently vague. While there are impacts on both incorporated and incorporating areas and peoples, the impacts are generally asymmetric, to the net benefit of the incorporating agent. All levels of incorporation, typically more so the weaker the degree, engender resistance and attempts to control or negotiate the process. In short, incorporation is always an interactive process. While incorporation is to some extent reversible, over time it tends to strengthen. Changes in the degree of incorporation are seldom explained by only one or two factors.

It is no surprise that the merger of world-systems was not part of the original formulation of world-systems analysis (WSA), since there was only one world-system. Chase-Dunn and Hall (1991, 1997) later argued that there were many earlier world-systems whose mergers eventually gave rise to the modern world-system. Here I summarize only those aspects of their argument relevant to mergers.

If there were many world-systems then merger, and separation, were both possible. An underlying assumption, backed by empirical evidence, is that a key factor in change in societies was their context amid other societies. When this interacting group was virtually self-contained (almost no societies are completely isolated), it constituted its own "world," which is how Wallerstein (1993: 294) defines the "world" in world-system. Chase-Dunn and Hall also argued that all world-systems pulsate, that is, expand and contract, or expand more before their expansion proceeds less rapidly. Pulsation is a manifestation of various cycles within the system and is one drive to incorporate new areas and/or peoples. Occasionally, it results in mergers of formerly unconnected world-systems.

They also argued that there have been different types of world-systems driven by different modes of accumulation—kin-ordered, tributary and capitalist—each with its own internal dynamics. Thus, they reject Andre Gunder Frank's (Frank and Gills 1993) claim that there has been only one expanding world-system for five millennia. More importantly, they argued that world-systems have four types of boundaries, demarcated by relative densities of interaction in different networks: bulk goods networks (BGNs), political-military networks (PMNs), prestige (or luxury) goods networks (PGNs) and information networks (INs). The PMNs are networks on polities that ally and/or make war with one another. This corresponds with what international relations theorists call "international systems." These four sets of boundaries seldom coincide and are progressively larger in the order listed, though PGNs and INs are usually of the same order of magnitude. Thus, merger and incorporation can occur at any of these boundaries. For these reasons, studies of incorporation should specify which kinds of interaction are becoming linked. The strength of systemness increases as interaction comes to include political-military contact and the exchange of bulk goods.

While mergers clearly have occurred, the processes have not been discussed widely. In Afroeurasia the major world-systems were the eastern, centered on what is now China, southwestern Asia and the Mediterranean, centered on various empires, and the southern, centered on what is now the South Asian subcontinent. The emergence of the Silk Roads in the mid-to-late first millennium BCE merged the eastern and western systems at the prestige goods level, together with the southern system (Yang 2009). Continuing research suggests that Central Asian states and pastoral confederations played important roles in this and later mergers (Chase-Dunn et al 2010). The Mongol Conquest in the thirteenth century CE merged the eastern and western PMNs, but these broke apart when the Mongol Empire disintegrated. Finally, the modern Europe-centered system eventually brought all the other systems into a single globe-wide PMN by the end of the late nineteenth century.

Unsurprisingly, as world-systems began to merge, areas caught in the middle tried to resist, giving rise to contested semiperipheries and contested peripheries (Allen 1997, 2005). Indeed, it may well be that such contested arenas easily became semiperipheral by playing one system against another in order to gain a modicum of autonomy. This seems to have happened in Central Asia (Chase-Dunn et al 2010). Mergers are a topic needing further research, of which more has been done on incorporation.

Incorporation: Recent developments

Through the 1990s, Wilma A. Dunaway (1994, 1996a, b, c, 1997, 2000) pursued the analysis of incorporation processes in studies of the southern Appalachian region in what is the now the United States, especially for the Cherokee Nation. She explicitly recognized the interplay of incorporation processes with both gender and ecological concerns. She furthered the emphasis on incorporation as interactive, "a dialectical historical process" (1996b: 467). She also noted that incorporation processes that attempt to impose cultural changes usually generate indigenous resistance. Dunaway examined how involvement in the fur trade had served to disempower Cherokee women, because it increased the relative power of men and developed a dependency on the goods acquired via the fur trade. These processes differed significantly from those in northeastern North America.

Jon Carlson (2001, 2002, 2011) has widened the continuum of incorporation even further to what he calls a "zone of ignorance" and modifies the concept of "external arena" with respect to weak forms of incorporation. In this range there is a process of "grooming" in which an area and/or people are prepared for incorporation, and at times held somewhat at bay. By "zone of ignorance" Carlson means a "mythic domain, largely unexplored and unknown to the current members of the world-system" (2001: 241). This is an area where "the *grooming* process of conditioning the area toward capitalist exchange and production is initiated" (241). This is followed by another phase of "*incorporation*, whereby an external arena's contact and involvement with the world-system is developed, ultimately producing 'nominal incorporation,' building toward 'effective incorporation'" (242). During this process the zone of ignorance is transformed into an external arena: an area seen as still outside the system but interacting weakly with it.

This analysis "stretches out" the weaker pole of incorporation, encompassing the areas Hall had labeled "external arena" and "contact" peripheries. Carlson builds this analysis from an amalgamation of Wallerstein's approach and the analyses of Hall and Chase-Dunn. Carlson's close study of the history of the otter hide trade in the Pacific Northwest in the seventeenth and eighteenth centuries and of the Red Sea and the Indian Ocean over the last millennium produced his conceptual refinements.

Carlson's reconceptualization calls to mind Brian Ferguson and Neil Whitehead's concept of "war in the tribal zone" (1992a, b). Ferguson and Whitehead use the term "tribal" in an ironic way, explicitly recognizing that "tribe" is a vague and often misleading term. The "tribal zone" is a transition zone, a frontier, beyond the boundary of state expansion into nonstate territories. They make several key points. First, when states expand, the effects of their contact ripple well beyond the area of direct contact. This spreading effect is driven by the efforts of indigenous leaders to gain access to state-produced goods which they can use to gain followers. Often peoples further from the contact zone try to bypass such middlemen and thereby come into conflict with them. An example is seen in the hoards of Roman coins found hundreds of miles beyond the limit of Roman expansion into Europe (Wells 1999a, b). In the US southwest, war could be generated by slave or captive trade (Hall 1989a). Similarly, diseases could spread far beyond the zone of immediate contact and new goods, such as horses, spread from the southwest far into the northwest. *War in the Tribal Zone* contains many accounts, both ancient and modern, of such effects. These accounts serve as a useful corrective to the mistaken assumption that capitalism is the only system that does this. Ancient systems also expanded, but not as efficiently.

A second consequence of the "tribal zone" effect is that the first literate observers did not see pristine conditions. Rather, they saw conditions already significantly altered by state contact. For instance, Lewis and Clark encountered indigenous peoples in the northwest who had had horses for generations, though they were the first Europeans to travel through that area (Fenelon and Defender-Wilson 2004). There were no horses in the Americas before the Spaniards brought them in the sixteenth century. Thus, many first-hand accounts of the "tribal zone" do not provide good evidence about the true nature of precontact societies.

Early phases of incorporation are very murky and fuzzy. First-hand accounts need to be used with a great deal of caution. Still, it is clear that those contacted often do succeed in shaping the interaction to a considerable degree.

Shirley Hollis (2004, 2005) examines what happened during the sixteenth century in the region that became the southeastern United States. She argues that incorporation began well before the early 1600s. This is an example of the "grooming" of a "zone of ignorance" discussed by Carlson. Hollis argues that these early contacts "not only contributed to the establishment of trade relations with other parts of the emerging world-system, but also led to the subsequent planting of European settlements on the North American continent" (2005: 122). Further, New World plantations were replacing declining plantations in the Mediterranean and the Canaries, where Europeans had already begun to develop the institutions of imperial exploitation.

She concludes that these refined concepts of incorporation are more useful for understanding the interactions with indigenous peoples than Wallerstein's original concepts.

Caleb Bush extends the critiques of incorporation to the twentieth and twenty-first centuries (2005a, b), thereby filling a gap that Hall (1987) recognized, but did not address. Bush argues that incorporation is a process that continues as more and more goods and relations are commodified, and more pressure is exerted on indigenous peoples, using the example of the Diné (Navajo), who assimilated further into middle-class US culture. In short, incorporation has not ended. Further, it does not end only with "dependent" peripheralization, but goes beyond material exchanges to include cultural assimilation.

Nick Kardulias (2007) develops a concept of "negotiated peripherality" wherein incorporated peoples succeed in negotiating the terms of their incorporation. He does this through a comparison of how ancient Cyprus adopted features from Near Eastern and Greek cultures with analogous processes between indigenous fur harvesters and European traders in northern North America. He uses this nuanced comparison to argue that negotiation processes are almost universal, but have features that are unique to each setting. Kardulias further argues that "cultural changes occur at an accelerated rate in contact zones" (55). His work underscores the importance of studying these "marginal" places in order to discover the nuances of processes of cultural contact and interaction. Kardulias transforms the problematic "tribal zone" into a zone of opportunity for a deeper understanding of world-systems processes.

Frontiers are where the action is

Frontiers are where incorporation takes place. Indeed, one could argue that areas where incorporation is highly active are one definition of a frontier. Wallerstein says very little about frontiers (1974). He uses frontier in the loose way it was used by many writers around that time. Yet he offers an interesting, teasing comment about frontiers. In commenting that the development of underdevelopment is more general than Frank originally argued, Wallerstein cites Owen Lattimore, who wrote "Civilization gave birth to barbarism" (Lattimore 1962: 93 in Wallerstein 1974: 98). Wallerstein goes on to note that societies on either side of frontiers shape each other. These comments prompted Hall to elaborate on his concept and to connect it to frontiers.

Hall has written extensively about frontiers (1989a, b, 1998, 2000, 2006, 2009). To summarize this work would require a separate article (see 2009); however, a few points are germane to this discussion. First, incorporation typically creates frontier zones that have some width, especially when studied in detail. Second, they are differentially permeable both with respect to what passes through them and to the direction of flow. Third, both sides—and many frontiers are actually multi-sided—influence and shape each other's historical trajectories of change. Fourth, frontier zones are places where ethnicity and/or race are created, shaped, reshaped and occasionally eliminated. Fifth, and most significant with respect to incorporation, frontier zones are often the best, and some times the only places where these processes can be observed in any detail (1989b). Finally, much of what Hall (2009) argues about comparative strategies for studying frontiers, applies *mutatis mutandis* to the study of incorporation. The two topics—incorporation and frontiers—are closely connected, but not identical. Frontiers are concerned with the creation, transformation and elimination of boundary zones. Incorporation is about how new areas and/or peoples are absorbed into and resist absorption into a world-system.

Future research problems

There are many opportunities for comparative studies of incorporation and mergers. There can be multi-tiered comparisons, organized by world-systemic processes and contexts. Contexts might include the following (e.g., Chase-Dunn and Babones 2006; Chase-Dunn and Hall 1997; Hall 2009):

- types of world-systems (kin-based, tributary, capitalist);
- position in the world-system: core vs. semiperiphery vs. periphery;
- incorporation along the edge of each bounding network: BGN, PMN, PG, N, or IN;
- condition of the world-systems that are coming into contact (expanding or contracting);
- cyclical phases:
    - a- or b-phase in Kondratieff cycles (about 50 years)
    - rising or falling in hegemonic cycles (about 100 years)
    - rising or falling "long waves" (about 200 years)
    - rising or falling Ibn Kaldhun cycles (about 300 years) (Turchin and Hall 2003)
    - dark age cycles (about 600 years) (Chew 2001, 2007, 2008)
- reason for incorporation: land, people, trade, access, etc.;
- types of people(s) or society/ies being incorporated: band, chiefdom state, and sub types;
- level of core (or even semiperipheral) states competition for these peoples or territories.

One could add further conditions or qualifications. The point is to make intelligent comparisons that take conditions and properties into account. With a sufficient number of cases, a Boolean strategy (Ragin 1987) might be employed, similar to that used by John Foran (2005) in his comparative study of Third World revolutions. For cyclical processes, attention to the phase of a cycle is an important factor. There are significant differences between a rising (a-phase) and a falling (b-phase) of the Kondratieff cycle (e.g., Boswell and Chase-Dunn 2000). Whether the evidence for the existence these cycles is solid or controversial, lack of awareness of cyclical processes can lead to flawed conclusions, especially for comparisons of processes occurring at different phases of a cycle (see Hall 2009: Fig. 2, p. 31 for an illustration).

Similarly, researchers could address the inherent fuzziness of incorporation, especially in its early or weak phases, in the formation of contact peripheries, or Carlson's grooming of zones of ignorance for incorporation. Here, fuzzy set analysis (Ragin 2000) might be a useful tool. This might even be effective in addressing an issue raised in studies of globalization that claim that contemporary state boundaries and frontiers are becoming less important, and that transnational process are becoming more important.

These strategies of comparison might be fruitfully used in combination with what Philip McMichael calls incorporated comparisons, which entail three processes (1990, 2000). First, comparison is "internal" to historical inquiry, where process-instances are comparable because they are historically connected and mutually conditioning (2000: 671). Second, structure and context of units derives from their relations to the whole and each other. Third, comparisons can be made over time or space, or both.

All three of these strategies—Boolean, fuzzy set, or incorporated—facilitate comparisons that at first glance might seem inappropriate. One might reasonably argue that a comparison of Chinese incorporation of Central Asia steppe territory might be compared to Spanish incorporation of territory occupied by nomadic indigenous peoples either on far northern New Spain (now the US southwest) or far southern La Plata (see Guy and Sheridan 1998). While these instances occur at widely different times, on separate continents, involving different types of agrarian states and different kinds of nomadic peoples, they share slow expansion into territory occupied by nonstate peoples, with the goal of increasing land holdings. One might also note another shared condition: both seldom worked effectively, and when they did it was at great cost in military power and treasure. Comparison between modern and ancient instances of incorporation will probably underscore what Ferguson and Whitehead (1992a, b) found: that states are a fundamental source of incorporation. It is not only capitalism; Chase-Dunn and Mann's study of the Wintu (1998) show that even nonstate societies expand, albeit in different ways.

Through many such studies, a fuller understanding of incorporation can be achieved. Such comparisons should include closer attention to roles of gender, ecology, resistance (agency), counter-resistance, trade (volume, velocity, value, and substance), and many other factors. It is only through many such comparisons that analysts will be able to tease out the roles of the various contexts noted above.

There also needs to be more research attention paid to the impacts of newly incorporated areas on systems doing the incorporating, typically but not exclusively core states. The impacts of crops discovered in the New World and exported to Europe and Asia is a familiar example (e.g., Crosby 1972, 1986). As Hollis (2005) noted, newly incorporated areas are where states often develop new strategies of expansion, sometimes bringing those lessons back home. Yang (2009) shows how the incorporation of Yunnan into China over two millennia played a significant role in shaping modern Chinese culture and ethnic identities. Careful study of the consequences of incorporation may well reveal many very important, but often unnoticed, unintended consequences of system expansion.

Finally, detailed case studies (e.g., Blackhawk 2006; Bush 2005a; Dunaway 1996a; Hall 1989a; Reséndez 2005; Yang 2009) will remain extremely valuable for teasing out new processes, factors, and contexts that shape the processes of incorporation. All the refinements discussed here derived from attempts to understand specific instances of incorporation that required modifications of existing concepts and theoretical understandings. Comparative studies will be useful in assessing the limits of applicability and generality of many incorporation processes. While the expansion of the European "modern world-system" gave rise to many instances of incorporation, there are thousands of instances of incorporation by other systems in other times.

References

Allen, M. (1997) "Contested Peripheries: Philistia in the Neo-Assyrian World-System", unpublished PhD Dissertation, Interdepartmental Archaeology Program, UCLA.

Allen, M. (2005) "Power is in the Details: Administrative Technology and the Growth of Ancient Near Eastern Cores", pp. 75–91 in C. Chase-Dunn and E. N. Anderson (eds) *The Historical Evolution of World-Systems*, New York: Palgrave.

Blackhawk, N. (2006) *Violence over the Land: Indians and Empires in the Early American West*, Cambridge, MA: Harvard University Press.

Boswell, T. and Chase-Dunn, C. (2000) *The Spiral of Capitalism and Socialism: The Decline of State Socialism and the Future of the World-System*. Boulder, CO: Lynne-Rienner.

Bush, C. (2005a) "Land, Conflict, and the 'Net of Incorporation': Capitalism's Uneven Expansion into the Navajo Indian Reservation 1860–2000", unpublished PhD Dissertation, Department of Sociology, Binghamton University.

Bush, C. (2005b) "Reconsidering Incorporation: uneven histories of capitalist expansion and encroachment, Native America", *Studies in Political Economy*, 76: 83–109.

Carlson, J. D. (2001) "Broadening and deepening: systemic expansion, incorporation and the zone of ignorance", *Journal of World-Systems Research*, 7: 225–63.

Carlson, J. D. (2002) "The 'Otter-Man' empires: the Pacific fur trade, incorporation and the zone of ignorance", *Journal of World-Systems Research*, 8: 389–442.

Carlson, J. D. (2011) "Externality and incorporation in the world-system: abyssinia—anomaly or palimpsest?", *Journal of World-Systems Research*, 17: 165–98.

Chase-Dunn, C. and Babones, S. J. (2006) *Global Social Change: Comparative and Historical Perspectives*. Baltimore, MD: Johns Hopkins University Press.

Chase-Dunn, C. and Hall, T. D. (eds) (1991) *Core/Periphery Relations in Precapitalist Worlds*, Boulder, CO: Westview Press.

Chase-Dunn, C. and Hall, T. D. (1997) *Rise and Demise: Comparing World-Systems*, Boulder, CO: Westview Press.

Chase-Dunn, C., Hall, T. D., Niemeyer, R., Alvarez, A., Inoue, H., Lawrence, K. and Carlson, A. (2010) "Middlemen and marcher states in Central Asia and east/west empire synchrony", *Social Evolution and History*, 9: 1–29.

Chase-Dunn, C. and Mann, K. M. (1998) *The Wintu and Their Neighbors: A Very Small World-System in Northern California*, Tucson: University of Arizona Press.

Chew, S. C. (2001) *World Ecological Degradation: Accumulation, Urbanization, and Deforestation 3000 B.C.–A.D. 2000*, Walnut Creek, CA: AltaMira Press.

Chew, S. C. (2007) *The Recurring Dark Ages: Ecological Stress, Climate Changes, and System Transformation*, Lanham, MD: AltaMira Press.

Chew, S. C. (2008) *Ecological Futures: What History Can Teach Us*, Lanham, MD: AltaMira Press.

Crosby, A. W., Jr. (1972) *The Columbian Exchange: Biological and Cultural Consequences of 1492*, Westport, CT: Greenwood.

Crosby, A. W., Jr. (1986) *Ecological Imperialism: The Biological Expansion of Europe, 900–1900*, New York: Cambridge University Press.

Dunaway, W. A. (1994) "The southern fur trade and the incorporation of southern Appalachia into the world-economy, 1690–1763", *Review*, 18: 215–42.

Dunaway, W. A. (1996a) *The First American Frontier: Transition to Capitalism in Southern Appalachia, 1700–1860*, Chapel Hill: University of North Carolina Press.

Dunaway, W. A. (1996b) "Incorporation as an interactive process: Cherokee resistance to expansion of the capitalist world-system, 1560–1763", *Sociological Inquiry*, 66: 455–70.

Dunaway, W. A. (1996c) "The incorporation of mountain ecosystems into the capitalist world-system", *Review*, 19: 355–81.

Dunaway, W. A. (1997) "Rethinking Cherokee acculturation: women's resistance to agrarian capitalism and cultural change, 1800–838", *American Indian Culture and Research Journal*, 21: 231–68.

Dunaway, W. A. (2000) "The international fur trade and disempowerment of Cherokee women, 1680–1775", pp. 195–210 in T. D. Hall (ed.) *World-Systems Reader: New Perspectives on Gender, Urbanism, Cultures, Indigenous Peoples, and Ecology*, Lanham, MD: Rowman & Littlefield.

Fenelon, J. V. and Defender-Wilson, M. L. (2004) "Voyage of domination, 'purchase as Conaues, Skakawea for savagery': distorted icons from misrepresentation of the Lewis and Clark expedition", *Wicazao sa Review*, 19: 85–104.

Ferguson, R. B. and Whitehead, N. L. (eds) (1992a) *War in the Tribal Zone: Expanding States and Indigenous Warfare*, Santa Fe, NM: School of American Research Press.

Ferguson, R. B. and Whitehead, N. L. (1992b) "The violent edge of empire", pp. 1–30 in R. B. Ferguson and N. L. Whitehead (eds) *War in the Tribal Zone: Expanding States and Indigenous Warfare*, Santa Fe, NM: School of American Research Press.

Foran, J. (2005) *Taking Power: On the Origins of Third World Revolutions*, Cambridge: Cambridge University Press.

Frank, A. G. (1966) "The development of underdevelopment", *Monthly Review*, 18: 17–31.

Frank, A. G. and Gills, B. K. (eds) (1993) *The World System: Five Hundred Years or Five Thousand?* London: Routledge.

Guy, D. J. and Sheridan, T. E. (eds) (1998) *Contested Ground: Comparative Frontiers on the Northern and Southern Edges of the Spanish Empire*, Tucson: University of Arizona Press.

Hall, T. D. (1986) "Incorporation in the world-system: toward a critique", *American Sociological Review*, 51: 390–402.

Hall, T. D. (1987) "Native Americans and incorporation: patterns and problems", *American Indian Culture and Research Journal*, 11: 1–30.

Hall, T. D. (1989a) *Social Change in the Southwest, 1350–1880*, Lawrence: University Press of Kansas.

Hall, T. D. (1989b) "Is historical sociology of peripheral regions peripheral?", pp. 349–72 in M. T. Martin and T. R. Kandal (eds) *Studies of Development and Change in the Modern World*, New York: Oxford University Press.

Hall, T. D. (1998) "The Rio de La Plata and the Greater Southwest: A view from world-system theory", pp. 150–66 in D. J. Guy and T. E. Sheridan (eds) *Contested Ground: Comparative Frontiers on the Northern and Southern Edges of the Spanish Empire*, Tucson: University of Arizona Press.

Hall, T. D. (2000) "Frontiers, ethnogenesis, and world-systems: rethinking the theories", pp. 237–70 in T. D. Hall (ed.) *A World-Systems Reader: New Perspectives on Gender, Urbanism, Cultures, Indigenous Peoples, and Ecology*, Lanham, MD: Rowman & Littlefield.

Hall, T. D. (2006) "[Re] Periphalization, [re]incorporation, frontiers, and nonstate societies: continuities and discontinuities in globalization processes", pp. 96–113 in B. K. Gill and W. R. Thompson (eds) *Globalization and Global History*, London: Routledge.

Hall, T. D. (2009) "Puzzles in the comparative study of frontiers: problems, some solutions, and methodological implications", *Journal of World-Systems Research*, 15: 25–47.

Hollis, S. (2004) "Crafting Europe's 'clean slate' advantage: world-system expansion and the indigenous Mississippians of North America", *American Indian Culture and Research Journal*, 28: 77–101.

Hollis, S. (2005) "Contact, incorporation, and the North American Southeast", *Journal of World-Systems Research*, 11: 95–130.

Hopkins, T. K. and Wallerstein, I. (1987) "Capitalism and the incorporation of new zones into the world-economy", *Review*, 10: 763–79.

Hopkins, T. K., Wallerstein, I., Kasaba, R., Martin, W. G. and Phillips, P. D. (1987) "Incorporation into the world-economy: how the world system expands", *Special Issue of Review*, 10: 761–902.

Kardulias, P. N. (2007) "Negotiation and incorporation on the margins of world-systems: examples from Cyprus and North America", *Journal of World-Systems Research*, 13: 55–82.

Lattimore, O. (1962) "La Civilisation, mère de Barbarie?" *Annales E.S.C.*, 17: 95–108 (cited in Wallerstein [1974: 98]).

McMichael, P. (1990) "Incorporating comparison within a world-historical perspective: an alternative comparative method", *American Sociological Review*, 5: 385–97.

McMichael, P. (2000) "World-systems analysis, globalization, and incorporated comparison", *Journal of World-Systems Research*, 6: 668–90.

Ragin, C. (1987) *The Comparative Method: Moving Beyond Qualitative and Quantitative Strategies*, Berkeley: University of California Press.

Ragin, C. (2000) *Fuzzy-Set Social Science*, Chicago, IL: University of Chicago Press.

Reséndez, A. (2005) *Changing National Identities at the Frontier: Texas and New Mexico, 1800–1850*, Cambridge: Cambridge University Press.

Turchin, P. and Hall, T. D. (2003) "Spatial synchrony among and within world-systems: insights from theoretical ecology", *Journal of World-Systems Research*, 9: 37–64.

Wallerstein, I. (1974) *The Modern World-System: Capitalist Agriculture and the Origins of European World-Economy in the Sixteenth Century*, New York: Academic Press.

Wallerstein, I. (1989) *The Modern World-System III: The Second Era of Great Expansion of the Capitalist World-Economy, 1730–1840s*, New York: Academic Press.

Wallerstein, I. (1993) "The time-space of world-system analysis: a philosophical essay", *Historical Geography*, 23: 5–22.

Wells, P. S. (1999a) "Production within and beyond imperial boundaries: goods, exchange, and power in Roman Europe", pp. 85–101 in P. N. Kardulias (ed.) *World-Systems Theory in Practice*, Lanham, MD: Rowman & Littlefield.

Wells, P. S. (1999b) *The Barbarians Speak: How the Conquered Peoples Shaped Roman Europe*, Princeton, NJ: Princeton University Press.

Yang, B. (2009) *Between Winds and Clouds: The Making of Yunnan Second Century BCE to Twentieth Century CE*, New York: Columbia University Press.

2.2

The great transition debate and world-systems analysis

Eric Mielants

Since its creation in the 1970s, world-system analysis has challenged the way social scientists think about the past, the present, and the future (e.g., Hopkins and Wallerstein 1982). In doing so, this approach has raised significant epistemological questions about existing academic structures (such as disciplines and social science journals) in terms of whether the traditional modes of gathering and evaluating knowledge about the modern world really made sense (see Wallerstein 2004a). In particular, the nineteenth century artificial construction of separate disciplines institutionalized in different departments in the modern university setting, leading political scientists to study the state, sociologists civil society, economists the market, historians the past, and anthropologists the non-Western world as if separate logics were at work (Wallerstein 2004a: 124), was challenged by this new perspective. This challenge implied not only embracing a unidisciplinary approach to social science (see Wallerstein 2001), but also a critical deconstruction of traditional perspectives' meta-narratives explaining how, why, and when the modern world had become modern, or, alternatively, how capitalism had emerged in previously pre-capitalist geographical areas. In doing so, world-system analysis essentially dismissed sociological analyses that looked for cultural and religious variables to explain the origins of capitalism (e.g., Stark 2005; Tawney 1926; Weber 1930) while opening up a two-front war on two historical materialist perspectives that had for decades dominated previous debates over the emergence of capitalism and modernity: liberal Smithianism and Orthodox Marxism.

Wallerstein (1974, 2004b), considered by many as the founding father of world-system analysis, essentially defined the capitalist world-economy, or modern world-system, as a socio-economic entity that emerged in sixteenth-century Europe and subsequently expanded to incorporate more areas of the world such as the surrounding mini-systems and world-empires, which operated according to a different logic. Characterized by a global division of labor resulting from intense yet unequal bulk trade linkages and illustrated by commodity chains between different zones that he labeled core, periphery, and semiperiphery, this capitalist world-system gradually expanded from 1492 CE until the end of the nineteenth century when the entire world had become capitalist.

As stated earlier, this theoretical position ran counter to the Orthodox Marxism embraced by the Soviet Union and Communist Parties as well as intelligentsia allied to them. Marx himself was always much more interested in providing a scientific analysis of nineteenth-century capitalism than he was in exploring the specific origins of capitalism. Many of his statements were essentially "contextual observations dependent on his analysis of the capitalist production" (Guérreau 1980: 57) during the industrial era. The focus of Marx and Engels on Britain (and their subsequent characterization of non-European areas of the world as Oriental Despotism) was continued in the twentieth century by renowned Marxist scholars such as Maurice Dobb (1976: 59) and Robert Brenner (1985: 275), who also juxtaposed the general socioeconomic feudal crisis "on most of the continent" with the "critical breakthrough to self-sustaining growth in England." More recently, Georges Comninel (2000) and Ellen Meiskins Wood (1999) have updated this Orthodox Marxist perspective. For the latter, the seventeenth-century world-economy was mainly trade in luxury goods being bought cheaply and sold dearly to make profit—so-called merchant capitalism, not real industrial capitalism using wage labor. England was a unique place with unique agricultural conditions, its "internal market" and its "capitalism in one country." This exceptional island was capable of becoming on its own the first capitalist nation-state (Wood 1999: 72–130).

Despite Orthodox Marxism and Smithian liberalism being intellectual adversaries, from a world-system perspective they are only *frères ennemis*: members of the same family, pushing in different political directions. Like Marxism, liberal Smithianism also created a meta-narrative in which nothing novel or important happened in the so-called long sixteenth century. In addition, in common with the great ideological counter-narrative of Marxist theory, the Industrial Revolution of the late eighteenth and early nineteenth centuries was looked on as the most important historical watershed, opening the gates of modernity and unleashing capitalism. Like Marxists, modernization theorists have also predominantly focused on England in order to understand modernity (e.g., Wrigley 1988). Consequently, the Smithian's main interest in preindustrial economies and societies lies with understanding the constraints that allegedly prevented them from becoming modern. Therefore, modernization theorists problematize specific political, social, or cultural features of certain (mostly non-European) societies as backward or primitive, and similarly, they problematize current "undeveloped" or "emerging" economies without providing a rigorous and critical analysis of the complex interactions between different but mutually constitutive parts of the world-system. Despite their intellectual differences, liberal Smithian theorists as well as Orthodox Marxist scholars often characterize medieval Europe as well as contemporary "medieval-like" pre-modern and pre-capitalist societies as predominantly feudal, hierarchical, deeply conservative, and religious, with economies composed of self-sufficient autarchies populated by passive, self-subsistent peasants (see Bois 2000: 41).

Essentially, both Orthodox Marxist and liberal Smithian approaches have, over time, created a venerable social science tradition in which most Western scholars focus mainly on Europe (and occasionally also on European settler states). This scholarly activity on the origins of capitalism does not frame itself with reference to developments in, or in dialogue with, "the rest of the world" (e.g., Emigh 2008; Lachmann 2000).

To challenge this intellectual legacy of Eurocentrism, leading world-system scholars such as Immanuel Wallerstein, Andre Gunder Frank, and Samir Amin have written extensively about the pitfalls of Eurocentrism and have attempted in their own work to criticize parochial social science that disguises itself as objective, unbiased, and universalist. Studying the interconnectedness of the entire planet before the Industrial Revolution, world-system scholars have, since the 1970s, reframed the "great transition debate" about the origins of modernity. They analyze uneven socioeconomic development over several centuries, while also raising important questions about contemporary public policies and methodology. In terms of Eurocentrism, Wallerstein (1999: 169) has remarked on how social science has, from its inception, suffered from biases in the construction and continuous reproduction of its historiography, the parochialism of its universalism, and its assumptions about (Western) civilization, Orientalism, and an incorrect theory of progress. This stance holds that the main meta-theoretical perspectives used to explain the origins of the modern world, as well as its continuing dynamics, are fundamentally flawed. They have also served particular colonial and imperialist projects—be they of a Soviet or Western variety—while masquerading under the label of scientific objectivity.

According to world-system analysis, it was in the context of European colonialism and imperialism that unfolded in the so-called long sixteenth century (1450 to 1640) that the entire world gradually and irrevocably became interlinked through trade patterns constitutive of unequal exchange. This ushered in a genuinely profound transition: that of the genesis of a modern capitalist world-system, initially centered around Europe and the New World but subsequently incorporating ever more geographical zones. By stressing the importance of economic cycles and the commodity chains of leading sectors, world-system theorists study the upward and downward mobility of hegemonic polities that occurred long before the Industrial Revolution (see Chase-Dunn and Hall 1997; Frank 1998; Modelski and Thompson 1996). The Industrial Revolution was, from this point of view, just another instance in which a rising hegemon was able to attain a comparative advantage in new lead industries and to develop new capabilities for leading the capitalist world-economy toward greater consolidation (e.g., Arrighi 1994: 175, 188).

This downplaying of the significance of the Industrial Revolution as a watershed event in world history runs counter to the narrative predominant in the twentieth century of both Orthodox Marxism and liberal Smithianism (and its outgrowth of modernization theory). This is a major break with these perspectives as well as with mainstream historiography, which classifies time periods according to arbitrary Eurocentric demarcations: the classical, medieval, early modern, and modern period, the latter often only being ushered in once the Industrial Revolution has occurred. While Smithians (e.g., Lal 1999: 211–21; Wrigley 2004) and Marxists (e.g., Mandel 1962: 124) alike consider a "pre-modern," "pre-capitalist," or "feudal" society's overall economic growth to be limited and/or not recurrent, and suggest that the Industrial Revolution changes all of this, the world-system perspective claims that the cumulative windfalls generated by centuries of colonization decreases the need to focus on technological innovations (e.g., Mokyr 1990). Minimizing the significance of the Industrial Revolution, Wallerstein (1989) points out that it should not be interpreted as the major event that brought about recurring growth, but rather as the outcome of modern economic growth in the core. Thus the significance of the Industrial Revolution lies with the fact that it enabled the West to achieve economic growth more rapidly and more intensely (which ultimately widened the gap between the core and the periphery) sometime around 1800 (Guarducci 1983: 695–746; cf. Prak 2001) when a systematic policy of colonization and deindustrialization of the periphery allowed Western powers to accumulate great power and resources (Habib 1985).

Because of its emphasis on the importance of trade linkages based on unequal exchange (exploitation) and their lasting impact on both the core and the non-core over long periods of time, world-system analysts have often been labeled as "Marxist" by their Smithian detractors, who lament the way "leftist political economists and economic historians think in terms of core and periphery" (Landes 1998: 252). Unlike the Orthodox Marxists, who focus on the nation-state, however, world-system theorists regard the existence and movement of free labor and slavery as coexisting within the same mode of production in a large (worldwide) geographical arena interconnected by trade linkages. This is a fundamentally different theoretical position in the "great transition debate." Using the world-economy as their main unit of analysis, world-system analysts insist that modes of labor control as well as production for trade are historically, relationally, and therefore mutually constitutive. It is only because of the low cost remuneration of coerced labor in the periphery, which historically exported raw materials for the world market, that wage remuneration could be relatively high in the core, where high value-added products are produced and exported. The logic of ceaseless accumulation of capital in the system is intrinsically linked with the racial hierarchy constructed during the long sixteenth century

(Quijano and Ennis 2000; Quijano and Wallerstein 1992). If one desires to understand the contemporary dynamics of racial polarization within as well as between countries, it is imperative to comprehend the history of a global polarizing system over several centuries. For world-system scholars, the Smithians are themselves the ideological by-product, if not the apologists, for a highly racial and unequal global order that finds its roots in the long sixteenth century. Denying this reality, together with a failure to acknowledge the colonial features that are still prevalent in many postcolonial states only brings about more "feudalmania," or "temporal distancing" among liberal Western elites who produce false knowledge that denies the coevolution of "developed" and "underdeveloped" parts of the world (Grosfoguel 2003: 12). For world-system analysts, the Smithian pipedream in which all or most of the impoverished areas of the capitalist world-economy will someday catch up with the West in terms of production and income standards so long as they implement the correct policies, must be refuted. What makes the capitalist modern world-system different from mini-systems, world-empires, or other entities is that the quest for ceaseless accumulation is dependent upon and inseparable from a system of unequal exchange. This resembles the dependency school's analysis of the "development of underdevelopment" (see Frank 1967; cf. So 1990). Central to its analysis is the argument that what is so unique about the emergence of a capitalist modern world-system in the sixteenth century, initially centered on Europe, is the transition from a non-capitalist world into a capitalist world during which growth of world trade contributes to the expansion of a polarizing zero-sum game, reproducing and expanding poverty and inequality on a world scale. The crucial concept of semiperiphery was introduced to clarify this idea theoretically: the three zones (core, periphery, and semiperiphery), in which different political entities such as nation-states are located, contain divergent practices (in terms of life expectancy, standard of living, labor control, production of items for sale on the world market, and political regimes) precisely because of their hierarchical location and interconnection within the capitalist world-economy. World-system scholars argue that it is only in analyzing inequality and social mobility on a global scale over long periods of time, and in focusing on their relational interactions, that one can understand the dynamics of a centuries-old system that reproduces a "high inequality equilibrium" (Korzeniewicz and Moran 2009; cf. Bergesen and Bata 2002). It is therefore no coincidence that the origins of capitalism and the construction of "modern" notions of citizenship go hand-in-hand (Mielants 2007).

Despite world-system scholars' disagreements with Smithians and the latter labeling the former as Marxists, Orthodox Marxists claim no affiliation with the perspective. Indeed, the different interpretations surrounding the entire Transition Debate has led Orthodox Marxists such as Robert Brenner to call world-system scholars "neo-Smithian Marxists" who are "circulationists." Rather than focusing on production and the extraction of surplus value derived from wage labor, which according to Orthodox Marxists essentially determines whether a particular nation-state made the transition from feudalism to capitalism, the world-system scholars commit a fundamental error. For example, the fact that scholars such as Wallerstein have considered the Soviet Union not as a communist experiment illustrating the successful transition from capitalism to socialism, but rather as a typical powerful semiperipheral state embracing protectionism (as opposed to the typical state located in the core zone favoring free trade), was seen as anathema. Analyzing workers' revolts in tandem with and co-equal to other anti-systemic movements (decolonization, feminism, environmentalism, etc.) (e.g., Martin 2008) also constitutes unforgivable revisionism.

As Sanderson (1995: 178) has pointed out, however, "[w]hile it must be admitted that much of what goes on among Wallerstein's zones of the capitalist world economy involves exchange, much also involves production. When, for example, core capitalists peripheralize a region in order to extract surplus value from it, they are doing so through the direct establishment of productive activities that they ultimately control. How can such an economic relationship between core and periphery be regarded as based only on exchange?" It is therefore important to note that it is the combination of profits based on the exploitation of wage labor, as well as those derived from unequal exchange, which constitute a capitalist system. After all, profits derived from surplus value are not necessarily "of greater historical significance" (Heilbroner 1985: 66) than those derived from unequal exchange (cf. Norel 2009: 87–104). One should also consider that "the exploitation of labor is not only determined by the wage bundle and the extraction of labor from labor power in the production process, but in substantial measure by the prices at which goods are exchanged between the economies that make up the world-system" (Bowles 1988: 434). World-systems analysis has endeavoured to explain the emergence and expansion of the capitalist system by integrating a Marxist focus on production with an analysis of trade and unequal exchange in trade prices between core and periphery. This unequal exchange has been a continuous process due to differences in rates of labor remuneration that are larger than the differences in productivity. World-system scholars also insist that these differences were not only being produced by colonialism but are currently reproduced by neocolonial institutions such as the World Bank and the International Monetary Fund as well as public policies such as laws that restrict international labor migration (Jones and Mielants 2010). This interdisciplinary approach produced an analysis of "globalization" long before it became a fashionable buzz word in the 1990s (Guillen 2001: 241).

Over the last 25 years the "transition debate" has been revisited within the "world-system school" itself, and published in scholarly journals such as the *Review* (published by the Fernand Braudel Center), the *Journal of World System Research*, the *Journal of World History*, and the *Journal of Historical Sociology*. These more recent debates have reconsidered when and where a transition occurred from a pre-modern/pre-capitalist world to a modern capitalist world-economy. Scholars such as Janet Abu-Lughod (1989) and, in the last years of his life, Andre Gunder Frank (1998), articulated doubts about the temporal and spatial parameters of the Wallersteinian version of world-system analysis, calling for a rethinking of the meta-narrative about the genesis of capitalism in Europe in the sixteenth century. Frank (2000) in particular, in his research on the existence of a five-thousand-year-old world-system and its cycles, considered world-system theory too Eurocentric. The argument about whether there is empirical evidence of a clear break or transition in human history has more recently led to calls for a different kind of historical social science, such as that embraced in the newly created *Journal of Global History* (cf. Beaujard et al 2009).

But the theoretical disagreements between Marxism and Smithianism on the one hand and world-system analysis on the other, in terms of the transition debate, do not only revolve around differing temporal or geographical units of analysis. More fundamentally, the very definition of capitalism is itself substantially different. Whereas Orthodox Marxists consider the extraction of value based on wage labor in an industrial setting key to considering a particular nation-state as "capitalist," Smithians forego the focus on production (Reinert 2007: 68–69) and advocate instead that the invisible hand of free markets constitutes the essence of a global capitalist economy (cf. Noble 2000: 17–100), leading both to believe that there are currently certain areas of the world that are not yet capitalist but which may become so if they experience a particular stage of development. Following from the insights of the French economic historian Fernand Braudel, however, world-system analysis considers capitalism as an "enemy of the market" (Wallerstein 2001: 202–17) in which oligopolies and monopolies are formed. For Braudel, the very "coexistence of the market economy with the older subsistence economy" (Burke 2003: 63) is crucial to understanding not only the reality of the structures of daily life on the ground but also "the wheels of commerce." This, in turn, may ultimately provide an accurate "perspective of the world." For world-system analysis, the capitalistic phenomena identified in sixteenth-century Europe were not only wage labor, the specialization of industries, an advanced division of labor, class struggles, profits from trade (derived from the fact that entrepreneurs who own the means of production are involved in specialized production and competition), and complex financial techniques but also the systematic construction of an exploitable periphery to further the ceaseless accumulation of capital. This further implied that different zones of the world-economy, interlinked through trade, had divergent labor practices (such as slavery or serfdom) because of the inherent capitalist dynamics of the system itself. Therefore, the different life expectancies, labor practices, and political constellations could not be properly explained simply by looking at a particular nation-state at a specific moment in time, but at relational power structures over large parts of the world. These relational core-periphery dynamics that Western Europe enjoyed with the New World and to a certain degree also with Eastern Europe from the sixteenth century onward, were non-existent in other parts of the world and are therefore crucial to explaining the subsequent great divergences between various parts of the world that persist today (e.g., Pomeranz 2000: 268, 289). The way in which we envision, interpret and explain past, present, and future states of affairs is intrinsically interlinked, which means that the great transition debate will continue. It remains to be seen which school of thought will prevail in the twenty-first century.

References

Abu-Lughod, J. (1989) *Before European Hegemony*, New York: Oxford University Press.

Arrighi, G. (1994) *The Long Twentieth Century*, New York: Verso.

Beaujard, P., Berger, L. and Norel, P. (2009) "Prologue. Par-delà l'eurocentrisme: les mondialisations et le capitalisme au prisme de l'Histoire globale", in P. Beaujard, L. Berger and P. Norel (eds) *Histoire globale, mondialisations et capitalisme*, Paris: La Découverte.

Bergesen, A. and Bata, M. (2002) "Global and National Inequality: Are They Interconnected?", *Journal of World System Research*, VIII, 1: 130–44.

Bois, G. (2000) *La grande dépression médiévale, XIVe et XVe siècles*, Paris: Presses Universitaires de France.

Bowles, S. (1988) "Class versus World-Systems Analysis? Epitaph for a False Opposition", *Review* 11, 4: 433–51.

Brenner, R. (1985) "The Agrarian Roots of European Capitalism", in T. H. Aston and C. H. E. Philpin (eds) *The Brenner Debate*, Cambridge: Cambridge University Press.

Burke, P. (2003) "The Annales, Braudel and Historical Sociology", in G. Delanty and E. Isin (eds) *Handbook of Historical Sociology*, London: Sage.

Chase-Dunn, C. and Hall, T. (eds) (1997) *Rise and Demise: Comparing World-Systems*, Boulder, CO: Westview Press.

Comninel, G. (2000) "English Feudalism and the Origins of Capitalism", *Journal of Peasant Studies* 27, 4, July: 1–53.

Dobb, M. (1976) "A Reply", in Rodney H. Hilton (ed.) *The Transition from Feudalism to Capitalism*, London: NLB Publishers.

Emigh, R. (2008) *The Undevelopment of Capitalism*, Philadelphia, PA: Temple University Press.

Frank, A. G. (1967) *Capitalism and Underdevelopment in Latin America: Historical Studies of Chile and Brazil*, New York: Monthly Review Press.

Frank, A. G. (1998) *ReOrient*, Berkeley: University of California Press.

Frank, A. G. (2000) "The Five Thousand Year World System in Theory and Practice", in Robert Denemark et al (eds) *World System History*, London: Routledge.

Grosfoguel, R. (2003) *Colonial Subjects*, Berkeley: University of California Press.

Guarducci, A. (ed.) (1983) "Sviluppo e sottosviluppo in Europa e fuori d'Europa dal secolo XIII alla Revoluzione Industriale." Atti della Decima settimana di studio 7–12 aprile 1978. Instituto Interazionale di Storia Economica "f. Datini," Serie II, 10, Firenze: Le Monnier.

Guérreau, A. (1980) *Le féodalisme: Un horizon théorique*, Paris: Le Sycomore.

Guillen , M. (2001) "Is Globalization Civilizing, Destructive or Feeble?", *Annual Review of Sociologys*, 27: 235–60.

Habib, I. (1985) "Classifying Pre-Colonial India", in T. J. Byres and H. Mukhia (eds) *Feudalism and Non-European Societies*, London: Frank Cass.

Heilbroner, R. (1985) *The Nature and Logic of Capitalism*, New York: Norton.

Hopkins, T. and Wallerstein, I. (1982) *World-Systems Analysis: Theory and Methodology*, New York: Sage.

Jones, T. and Mielants, E. (eds) (2010) *Mass Migration in the World-System: Past, Present and Future*, Boulder, CO: Paradigm Press.

Korzeniewicz, R. P. and Moran, T. P. (2009) *Unveiling Inequality. A World Historical Perspective*, New York: Russell Sage Foundation.

Lachmann, R. (2000) *Capitalists in Spite of Themselves*, Oxford: Oxford University Press.

Lal, D. (1999) "Globalization: What Does It Mean for Developing and Developed Countries?", in S. Horst (ed.) *Globalization and Labor*, Tübingen: Institut für Weltwirschaft an der Universität Kiel & Mohr Siebeck.

Landes, D. (1998) *The Wealth and Poverty of Nations: Why Some Are So Rich and Some Are So Poor*, New York: Norton.

Mandel, E. (1962) *Traité d'Économie Marxiste. Volume 1*, Paris: René Juliard.

Martin, W. (ed.) (2008) *Making Waves: Worldwide Social Movements, 1760–2005*, Boulder, CO: Paradigm Press.

Mielants, E. (2007) *The Origins of Capitalism and the Rise of the West*, Philadelphia, PA: Temple University Press.

Modelski, G. and Thompson, W. (1996) *Leading Sectors and World Powers*, Columbia, SC: University of South Carolina Press.

Mokyr, J. (1990) *The Lever of Riches. Technological Creativity and Economic Progress*, Oxford: Oxford University Press.

Noble, T. (2000) *Social Theory and Social Change*, New York: Palgrave Macmillan.

Norel, P. (2009) *L'Histoire Économique Globale*, Paris: Éditions Du Seuil.

Pomeranz, K. (2000) *The Great Divergence*, Princeton, NJ: Princeton University Press.

Prak, M. (ed.) (2001) *Early Modern Capitalism*, London: Routledge.

Reinert, E. (2007) *How Rich Countries Got Rich and Why Poor Countries Stay Poor*, New York: Public Affairs.

Quijano, A. and Ennis, M. (2000). "Coloniality of Power, Eurocentrism and Latin America", *Nepantla: Views from the South*, 1: 533–80.

Quijano, A. and Wallerstein, I. (1992) "Americanity as a Concept, or the Americas in the Imaginary of the Modern World-System", *International Journal of Social Science*, 134: 549–59.

Sanderson, S. (1995) *Social Transformations: A General Theory of Historical Development*, Oxford: Blackwell.

So, A. (1990) *Social Change and Development. Modernization, Dependency and World-System Theories*, New York: Sage.

Stark, R. (2005) *The Victory of Reason*, New York: Random House.

Tawney, R. (1926) *Religion and the Rise of Capitalism*, New York: Harcourt.

Wallerstein, I. (1974) *The Modern World System*, New York: Academic Press.

Wallerstein, I. (1989) *The Modern World-System. Volume 3*, San Diego, CA: Academic Press.

Wallerstein, I. (1999) *The End of the World as We Know It*, Minneapolis: Minnesota University Press.

Wallerstein, I. (2001) *Unthinking Social Science. The Limits of Nineteenth Century Paradigms*, Philadelphia, PA: Temple University Press.

Wallerstein, I. (2004a) *The Uncertainties of Knowledge*. Philadelphia, PA: Temple University Press.

Wallerstein, I. (2004b) *World-System Analysis*, Durham, NC: Duke University Press.

Weber, M. (1930) *The Protestant Ethic and the Spirit of Capitalism*, New York: Charles Scribner's Sons.

Wood, E. (1999) *The Origin of Capitalism*, New York: Monthly Review Press.

Wrigley, E. A. (1988) *Continuity, Chance and Change*, Cambridge: Cambridge University Press.

Wrigley, E. A. (2004) *Poverty, Progress, and Population*, Cambridge: Cambridge University Press.

2.3

The social foundations of global conflict and cooperation

Globalization and global elite integration, nineteenth to twenty-first Century

Thomas Ehrlich Reifer

Introduction

Rather than ushering in peace and prosperity, the twenty-first century and the much heralded age of globalization began with a bang and not a whimper. Only a little after a decade past the collapse of the Soviet empire in Eastern Europe and the break-up of the USSR itself, the terrorist attacks of 2001 and the declaration of an endless "war on terror" thereafter, replete with the Anglo-American invasion of Iraq plunged the world into a new round of seeming interminable global conflict.

Once again, what Jack Snyder (1991)—drawing on the work of Gustav Schmoller, Eckart Kehr (1973, 1977), and others members of the German historical school—called the "myths of empire" propelled social imperial overexpansion through logrolled political coalitions. With the ratcheting up of the US's aggressive foreign policy, projections of an endless peaceful globalization of the world-economy—like those of the late nineteenth and early twentieth century—faded away. The term imperialism, including thorough analyses of the operations of the corporate press and imperial interest groups in terms of Iraq, much like those offered up by John Hobson during the period of Britain's conquest of South Africa—then home to the world's largest diamond and gold mines—once again gained currency (Marks 2003). The similarities and differences between these two periods, characterized by waves of financialization, militarization, and polarization of the global system, offers crucial insights into the possible trajectories of the global system in the twenty-first century.

The hypothesis put forward here is that recurrent waves of globalization, global elite integration and global conflict and cooperation are related to changes in the compositions of national and transnational elites, reorganizations of class structures and changing modes of citizenship. Waves of globalization have historically led to the formation of transnational ties among elite families, social classes, large firms, and growing movements for democratic citizenship. They have also led to nationalist resurgence, racialized social imperial expansion, the spread of formal colonialism, revolutionary waves, and increasing global conflict. An asymmetrical form of British-led globalization during the nineteenth century, concomitant with a new burst of overseas imperialism by the great powers led to a combination of increased inter-imperial rivalry along with a massive backlash against increasing economic instability, ushering in World Wars, Bolshevism, fascism, and the Great Depression.

After World War II, there was a renewal of transnational ties among elite families, social classes, and large firms, along with movements for democratic citizenship. During this period, the United States remade the global system on new and enlarged social foundations. Eventually, however, contradictions of US hegemony and its systemic cycle of accumulation led to the remaking of the global system once again, albeit this time on increasingly narrow social foundations. These were replete with entwined waves of financialization and militarization as well as the rise of a new region, Chinese-led East Asia, and the Arab Revolutions of 2010 and 2011, both of which portend major future transformations of the global system (Arrighi 2007).

The social foundations of global conflict and cooperation

The British hegemony of the 1800s saw a *tsunami* of international trade, investment, and global integration that peaked in the late nineteenth and early twentieth centuries. Yet growing inequality, intensified uneven development and elite reaction to the growth of democratic movements heightened global instability and violent conflict (Arrighi 2010; Polanyi 2001; Snyder 1991, 2000). During this period, the spread of commercialized trade disrupted economies dependent on domestic production, vastly increasing inequalities between the rich and the poor, and stimulating movements for self-protection against such threats to livelihood and security (Polanyi 2001). The structural contradictions afflicting autocratic states with overgrown military-corporate complexes and narrow domestic economic bases such as Germany, made them fertile grounds for reactionary ideologies such as fascism, and highly vulnerable to democratic disintegration and aggressive war (Arrighi 1990: 38–45). A premium was put on war preparedness and modernization of military-corporate complexes requiring global resources, as the turbulence of world market and interstate competition worked to the temporary advantage of powerful elites that wanted to pursue aggressive foreign policies (Snyder 1991).

By the late nineteenth century, racialized social imperialism, along with elite ideologies and practices of war preparedness and militarized masculinity became the order of the day. The resultant world wars and revolutionary waves were symptomatic of the rising politico-economic and military competition that contributed to the breakdown of the world-economy (Polanyi 2001). Such waves of globalization and global polarization have been closely associated with the narrowing of the domestic and international social bases of declining hegemons.

A large body of literature has demonstrated the close relationship between rising levels of proletarianization, economic development, military conflict, and democratic citizenship rights (Arrighi 1990; Klinkner 1999; Vanhanen 1997; Weber 1961: 240). In terms of military participation, ruling elites have found it necessary to extend citizenship rights to national subjects in exchange for the latter taking on obligations of military service. This was noted long ago by Max Weber (1961: 240) and systematic comparative research has borne him out (Dehio 1962; Downing 1992; Rueschemeyer et al 1992). In the late nineteenth and early twentieth century, the growing socialization of war-making and state-making laid the material basis for the democratization of both citizen-soldiers and shop-floor citizens, whose cooperation became increasingly important for national elites. The expansion and generalization of civil, social, and political citizenship rights in the core countries, and to a lesser extent in the non-core was crucial in the reconstruction of the global system on new and enlarged social foundations under US hegemony. However, the forms of democratic citizenship that emerged were stratified to varying degrees by race/ethnicity, class, gender, and nationality, with profound implications for the form and trajectory of global elite integration (Du Bois 1969; Gordon 1994a, b; Reifer 2007, 2010).

What is important for our purposes is the relationship between these unequal systems of citizenship, forms of integration, and global conflict and cooperation. Elite social groups with interests in overseas expansion or activist foreign policies have traditionally had to form coalitions with other powerful constituencies, including subjects and citizens, in order to pursue their aims. The frequently tenuous nature of these political coalitions often prompted aggressive foreign policies, for as many comparative historical studies have shown, "domestic insecurity of elites is one of the most important causes of war" (Levy 1989; Rosecrance 1963: 306). Specifically, the promotion of strategic myths needed to build logrolled coalitions often increased the power of those groups favoring aggressive foreign policies. Such politico-economic dynamics played central roles in both the British and US cycles of globalization and related forms of militarization (Snyder 1991).

By 1945, many US and European elites had come to understand that volatile currency fluctuations and speculative capital flows played an important contributing role in the breakdown of England's liberal global economic order, recurrent depressions, the rise of fascism, world wars, and global revolutionary waves (Arrighi 1990; Gardner 1969: 75–76; Polanyi 2001). These concerns led the US and European elites to create a liberal international economic order replete with governed markets that provided tangible benefits to workers in the core and a modicum of state-led nationalist development in the Third World. This compact was a fundamental part of the real but uneven expansion of civil, political, and social citizenship rights across the globe (Arrighi 1990, 2010). In this vision, that of a "global New Deal," full citizenship rights were to be generalized to the core. At the same time, development was to allow the world's poorer states to catch up with standards of wealth and ensure that the full extension of citizenship rights were achieved for the largely white working classes and middle strata of this zone. However, US hegemony's promise of full citizenship and high mass consumption in the core and self-determination and development in non-core zones had definite limits (Wallerstein 1995).

The changing social foundations of US hegemony

After World War II, although many in the US labor movement sought universal social provisions, the lack of basic citizenship rights for blacks in the South and people of color more generally posed an insurmountable obstacle to such a vision (Davis 1986; Du Bois 1969; Iton 2000; Lichtenstein 1995). Moreover, the fiscal nationalism of the conservative coalition ruled out the Congressional passage of aid necessary for the continued reconstruction of Europe and Asia after the Marshall Plan (Borden 1984). Yet by playing upon military nationalism, US elites were able to convince Congress to vastly increase military spending during the Korean War, though much of this money was actually used to buy goods from Western Europe and East Asia. Military spending became a form of welfare for corporations and upper class constituents, centrally helped along by the internalization of allied protection costs, as the United States took responsibility for the military security of its allies. Through this crucial innovation of the internalization of allied protection costs via US-led global military alliances, US Cold War military spending provided for the containment of US enemies and allies, the latter as semi-sovereign states and regions, while providing the military forces to protect US allies and to implement a global policy of counterrevolutionary violence to ensure an Open Door for the United States and allied transnational capital in the Third World. Military spending was much preferred by US elites to other forms of social expenditure as it was possible to get it from Congress; moreover, it helped to deflect a vibrant labor movement seeking universalistic social provisions and greater race, class, and gender equality at home, by allowing for its incorporation instead as a junior partner in the overseas expansion of US state-corporate power (Lichtenstein 1995; Lipsitz 1994; Stephan-Norris and Zeitlin 2002).

Instead of universalistic social provisions, organized workers were forced to negotiate private welfare provisions tied to firms, propped up by a permanent war economy (Davis 1986). This ensured that US social provision would be uniquely stratified by race, class, and gender (Reifer 2010). In contrast to universalistic forms of social citizenship achieved in allied advanced capitalist states, military-service related government transfers, notably veterans' benefits, the watershed GI Bill of Rights and low-interest loans, went primarily to white male citizen-soldiers. Entitlements such as Social Security and unemployment went to shop-floor citizens, initially largely white males, who also benefited from military-subsidized private welfare states tied to firms. Second-class, means-tested benefits, stigmatized as "welfare," generally went to the most oppressed—notably women, persons of color, and the white working poor. These different forms of social provision and political attitudes toward overseas intervention among national elites and populations are arguably related. US specialization in providing core military protection subsidized US allies in Western Europe and Japan, leading them to provide more funds for universalistic forms of social welfare provision.

The rise of this US New Deal warfare-welfare state—and its stratification, by race, class, gender, and nationality—has played a key role in shaping elite and popular attitudes toward military versus social spending, based upon the real or expected benefits different groups receive from such programs. It is therefore not surprising that there has long been a race and gender gap on military versus social provisions as well as questions of foreign policy and war and peace more generally, with women and persons of color more supportive of social programs and the peaceful solutions toward international conflict rather than tradition US policies of post-war interventionism. According to two leading political sociologists, although there are gender gaps in policy preferences among men and women cross-nationally, in terms of extent and significance, "No similar 'gender gap' has developed in other democratic polities" (Davis 1986: 284–89; Manza and Brooks 1999: 128).

This appears to be in line with the hypothesis concerning the interrelation between the social base of elites, unequal forms of civil, political, and social citizenship, and their relationship to elite integration and related forms of conflict and cooperation put forward here. US military spending plays an integral role in propping up racialized and gendered structures of the warfare-welfare state and the overseas expansion of US state-corporate power at one and the same time. The resulting cleavages on a variety of social issues and restructuring of the political economy has been integral to the development of hegemonic social blocs—as with the rise of the New Right and the right turn of the Eastern Establishment in the 1970s and 1980s—that have been able to continue high levels of military spending and overseas intervention for domestic and global purposes. Here, elites saw high military spending as a way to curb greater spending on health, education, and human resources, while maintaining US military predominance and global hegemony (Davis 1986).

Thus, just as the old exclusive White Anglo Saxon Protestant (WASP) Establishment reaped the fruits of the inflated capital values of a bygone era in the 1920s, so too did the rise of a broad-based US New Right in the 1970s and 1980s aim to valorize the accumulated gains of corporate capital and the broad propertied strata (Baltzell 1987; Davis 1986: 302, 157–255, 284–89, 2002). This new hegemonic social bloc, solidified by resentments generated in part by the stratification of US social provision by race and gender, included the more privileged segments of the white ethnic working classes and middle strata (notably those in the US South), arrayed against workers of color, labor, the poor, and the Third World (Davis 1986).

Future research on these questions may address Jack Levy's (1998: 143) challenge, heretofore unmet in the literature: "Theories of patriarchy might also help answer the ... question of variations in war and peace, if they identified differences in the patriarchal structures and gender relations in different historical contexts, and if they incorporated differences into empirically testable hypotheses about the outbreak of war." Future research from a world-systems perspective could compare different forms of racialized and gendered social relations, notably in unequal welfare and warfare-welfare states, and relate these to testable hypotheses about forms of elite integration and global conflict and cooperation.

Of course, such studies would have to take into account the changing functions and unintended consequences of elite policies. Take the following example, crucial to the conceptualization of the broadening and narrowing of the social foundations of US hegemony put forward herein. Originally, US military Keynesianism (a form of welfare for the military-corporate complex and upper-class constituents) provided a stimulus to the US and global economy. Military expenditures provided funding for advanced technologies (aerospace, electronics and communications, for example). Such programs thus propped up both US corporate profits and the overseas projection of US military power at one and the same time. Yet over time, the growing costs associated with such policies, in the context of increased global economic competition, led to a fiscal crisis of the New Deal warfare-welfare state and associated world order. During World War II and the early Cold War years, US military spending was based on progressive taxation of corporations and the rich. This was accompanied by limitations on pecuniary accumulation, including federal regulation and anti-trust efforts (Phillips 2002; Reifer 2009–10).

Here was the social basis of the rise of the New Deal world order: reform at home and support for socioeconomic reconstruction and limited forms of nationalist development abroad. Fixed exchange rates, by limiting large fluctuations in currency exchange rates and speculative capital movements, had provided the basis for forms of expansion consonant with the politico-economic and social objectives of the New Deal (Gardner 1969; Helleiner 1994). Yet in the early 1970s, the United States was forced to scrap the Bretton Woods agreements on pegged exchange rates that it had created after World War II. Increasingly, the regressive financing of vastly expanded US military expenditures through borrowing on the global capital markets, rather than expanding the New Deal world order through taxation on corporate profits and the rich, led instead to its ongoing demise (Arrighi 2010; Davis 1986, 2002; Eatwell 1993; Steinherr 1998, 2000). This account suggests a reciprocal causality between changing elite social foundations, including, via unequal systems of citizenship (as in the New Deal warfare-welfare state), and related forms of globalization, global elite integration, and global conflict and cooperation.

Scholars from the "democratic peace school" argue that interstate conflict among core powers is a thing of the past (Brown et al 1996). Recently, however, Jack Snyder (2000) has argued that weakly embedded forms of democratization, in terms of the norms of democratic procedures and governed markets, can promote extreme forms of nationalism and hence intra- and interstate conflict, as with the rise of fascism in Germany and the escalating warfare that has gripped many formerly Communist states of Europe. This raises real questions about Russia, China, and the democratic stability in core powers that jettisoned governed markets, which helped to embed these states within a thick web of institutions.

Conclusion

A growing number of experts contend that the global system recurrently breaks down not so much from rising challengers as declining hegemons resist adjustment and accommodation to new entrants. David Calleo (2003a, b) recently raised concerns that US unilateralism is putting it on a collision course with Europe and Asia in a clash involving disputes about global governance, US military power and competing models of capitalism and the provision of social citizenship.

Hegemonic stability theorists hypothesize that declining hegemons attempt to use their remaining strengths to pursue more unilateral advantages that have fewer collective benefits for the rest of the world. This typically exacerbates rivalry within the core, causing uncertainty, instability, and increasing competition and conflict in the world-economy and the interstate system. This narrowing of the benefits that flow from the policies of the declining hegemon is arguably what we are seeing in the US championing of neoliberal policies and military unilateralism, which some fear could spark another round of intra-core conflict and/or global chaos. Indeed, as Noam Chomsky has pointed out, some analysts argue that the US pursuit of continued military hegemony "carries with it an appreciable risk of ultimate doom" and are thus looking to countries like China to support alternative cooperative visions of global security (Steinbruner and Gallagher 2004: 99). Yet despite these dangers, many theorists today hypothesize that global elite integration and/or related processes of globalization represent a fundamental change in the world-system away from great power conflict. In contrast, other scholars, including world-systems analysts, put more emphasis on continued competition among rival global centers and US power (Gowan 1999).

Thus, future research in this area must take up the study of the analogies and differences in the construction of elite social bases in the recurrent remaking of the global system from the nineteenth to twenty-first centuries, as these changing social foundations of the global system will have important implications for the likely trajectory of global cooperation and conflict in the twenty-first century. A major structural difference is the much greater extent to which the United States has relied on military spending as a means of generating support from domestic constituencies and regional elites for the national and global project of US hegemony. By taking on the role of global policeman through the internalization of allied protection costs, the United States allowed other core states to construct political coalitions on social foundations that were substantially demilitarized, most especially Germany and Japan. On the one hand, this policy difference minimizes the likelihood of intra-core war that dominated the global system in the first half of the twentieth century. On the other hand, America's path-dependent process of militarization and related support for neoliberal forms of globalization is arguably one of the chief causes of global instability in the world today, as the recent uprisings in 2010 and 2011 throughout the Arab World—labeled the Arab 1848 by Tariq Ali—so clearly reveal. Future work will have to trace these changes and try to detect emerging forces that might replace the now discredited Washington Consensus with policy currents that allow for more egalitarian, peaceful, and sustainable futures.

References

Arrighi, Giovanni. 1990. "Marxist Century, American Century: The Making and Remaking of the World Labour Movement," *New Left Review*, no. 179, January/February: 29–64.

Arrighi, Giovanni. 2007. *Adam Smith in Beijing*. New York: Verso.

Arrighi, Giovanni. 2010. *The Long Twentieth Century: Money, Power and the Origins of Our Times*. New York: Verso.

Baltzell, E. Digby. 1987. *The Protestant Establishment: Aristocracy and Caste in America*. New Haven, CT: Yale University Press.

Borden, William S. 1984. *The Pacific Alliance: United States Foreign Economic Policy and Japanese Trade Recovery, 1947–1955*. Madison: University of Wisconsin Press.

Brown, Michael, Sean Lynn-Jones and Steven Miller. 1996. *Debating the Democratic Peace*. Cambridge, MA: MIT.

Calleo, David P. 2003a. *Rethinking Europe's Future*. Princeton, NJ, and Oxford: Princeton University Press.

Calleo, David P. 2003b. "Power, Wealth and Wisdom: The United States and Europe After Iraq", *The National Interest* 72, no. 5, pp. 5–15.

Davis, Mike. 1986. *Prisoners of the American Dream: Politics and Economy in the History of the US Working Class*. New York: Verso.

Davis, Mike. 2002. *Dead Cities*. New York: New Press.

Dehio, Ludwig. 1962. *The Precarious Balance: Four Centuries of the European Power Struggle*. New York: Alfred A. Knopf.

Downing, Brian. 1992. *The Military Revolution and Political Change: Origins of Democracy and Autocracy in Early Modern Europe*. Princeton, NJ: Princeton University Press.

Du Bois, W. E. B. 1969 [1935]. *Black Reconstruction in America: An Essay toward a History of the Part Which Black Folk Played in the Attempt to Reconstruct Democracy in America, 1960*–1880. New York: Atheneum.

Eatwell, John. 1993. "The Global Money Trap: Can Clinton Master the Markets?" *The American Prospect* 12: 118–26.

Gardner, Richard N. 1969. *Sterling-Dollar Diplomacy: The Origins and the Prospects of Our International Economic Order*. New York: McGraw Hill.

Gordon, Linda. 1994a. "Welfare Reform: A History Lesson", *Dissent*, Summer: 323–28.

Gordon, Linda. 1994b. *Pitied but Not Entitled: Single Mothers and the History of Welfare*. Cambridge, MA: Harvard University Press.

Gowan, Peter. 1999. *The Global Gamble: Washington's Faustian Bid for World Dominance*. New York: Verso.

Helleiner, Eric. 1994. *States and the Reemergence of Global Finance: From Bretton Woods to the 1990s*. Ithaca, NY: Cornell University Press.

Iton, Richard. 2000. *Solidarity Blues: Race, Culture, and the American Left*. Chapel Hill: The University of North Carolina Press.

Kehr, Eckart. 1973. *Battleship Building and Party Politics in Germany, 1894–1901: A Cross Section of the Political, Social, and Ideological Preconditions of German Imperialism*. eds R. Pauline and Eugene N. Anderson. Chicago, IL: University of Chicago Press.

Kehr, Eckart. 1977. *Economic Interest, Militarism, and Foreign Policy: Essays on German History*. ed., Gordon A. Craig, Berkeley: University of California Press.

Klinker, Philip. A. with Rogers M. Smith. 1999. *The Unsteady March: The Rise and Decline of Racial Equality in America*. Chicago, IL: University of Chicago Press.

Lichtenstein, Nelson. 1995. *The Most Dangerous Man in Detroit: Walter Reuther and the Fate of American Labor*. New York: Basic Books.

Levy, Jack S. 1989. "The Diversionary Theory of War: A Critique", pp. 259–88 in Manus I. Midlarsky, ed., *Handbook of War Studies*, Detroit: University of Michigan.

Levy, Jack S. 1998. "The Causes of War and the Conditions of Peace." *American Review of Political Science* 1: 139–65.

Lipsitz, George. 1994. *Rainbow at Midnight: Labor and Culture in the 1940s*. Urbana: University of Illinois.

Manza, Jeff and Clem Brooks. 1999. *Social Cleavages and Political Change: Voter Alignments and U.S. Party Coalitions*. Oxford: Oxford University Press.

Marks, Shula, June 2003. *Review Article: Shula Marks on the South Africa War*, H-Africa Review. http://www.h-net.org/~africa/reviews/ReviewEssays/ShulaMarks.htm last accessed, October 31, 2011.

Phillips, Kevin. 2002. *Wealth and Democracy*. New York: Broadway Books.

Polanyi, Karl. 2001 [1944]. *The Great Transformation*. Boston, MA: Beacon Press.

Reifer, Thomas E. 2007. "Blown Away: U.S. Militarism & Hurricane Katrina", pp. 197–223 in Hillary Potter, ed., *Racing the Storm: Racial Implications and Lessons Learned from Hurricane Katrina*, Lanham, MD: Lexington Books.

Reifer, Thomas E. 2009–10. "Lawyers, Guns & Money: Wall Street Lawyers, Investment Bankers and Global Financial Crises, late 19th to 21st Century," *Nexus: Chapman University Journal of Law & Public Policy*, "The 80th Anniversary of the Great Crash of 1929: Law, Markets & the Role of the State" 15: 119–33.

Reifer, Thomas E. 2010. "Beyond Divide & Rule? From the Washington to the Beijing Consensus", pp. 1–52 in Sungho Kang and Ramon Grosfoguel, eds, *Geopolitics & Trajectories of Development*, Berkeley: Institute of East Asian Studies, University of California.

Rosecrance, R. 1963. *Action and Reaction in World Politics*. Boston, MA: Little Brown.

Rueschemeyer, Dietrich, Evelyne Huber Stephens and John D. Stephens. 1992. *Capitalist Development and Democracy*. Chicago, IL: University of Chicago Press.

Snyder, Jack 1991. *Myths of Empire: Domestic Politics and International Ambition*. Ithaca, NY: Cornell University Press.

Snyder, Jack 2000. *From Voting to Violence: Democratization and Nationalist Conflict*. New York: W. W. Norton and Co.

Steinbruner, John D. and Nancy Gallagher. 2004. "Constructive Transformation: An Alternative Vision of Global Security", *Deadalus*, Summer: 83–103.

Steinherr, Alfred. 1998, 2000. *Derivatives: The Wild Beast of Finance*. Chichester: John Wiley & Sons.
Stephan-Norris, Judith and Maurice Zeitlin. 2002. *Left Out*. Cambridge: Cambridge University Press.
Vanhanen, Tatu. 1997. *Prospects of Democracy: A Study of 172 Countries*. London: Routledge.
Wallerstein, Immanuel. 1995. "Response: Declining States, Declining Rights", *International Labor and Working-Class History*, 47, Spring: 24–27.
Weber, Max. 1961. *General Economic History*. New York: Collier Books.

2.4
The East Asian path of development

Alvin Y. So

Introduction

Echoing Andre Gunder Frank's (1998) calling for a non-Eurocentric analysis and drawing upon the work of Giovanni Arrighi (2007) and Immanuel Wallerstein (2005), this chapter will trace the historical development of East Asia in relation to the transformation of the world-system. It will first examine East Asia as the center of the world before 1800. Then it will examine the de-centering and the re-centering of East Asia over the past two centuries.

East Asia before the capitalist world-economy

Asian studies in China, Japan, and the West have revolved around the history of land-based states. However, Hamashita (2003) argues that we need to study East Asia in terms of the interfaces and exchanges that take place within and among maritime zones in order to grasp the "regionalization" in the world-system.

In particular, Hamashita (2003) proposes a "**Tribute/Trade System**" framework to study East Asia. This tribute/trade system was centered upon China and sustained by a hierarchical order defined by the Confucian conception of a "rule of virtue." Like any hegemonic order, it was backed by military force. However, when the system functioned well, principles of reciprocity involving politics and economics permitted long periods of peaceful interactions.

In this China-centered order, tributary states (including Japan, Korea, Ryukyu, Siam, Vietnam, Java, Borneo, the Philippines) sent periodic tribute missions to the Chinese capital. Each time tributary states changed ruler, China dispatched an envoy to officially recognize the new ruler. Tribute relations were not only political but also involved economic and trade relations.

In exchange for the gifts carried to the Chinese court, tribute bearers received silk textiles and other goods from the Chinese emperor. Special licensed traders accompanying the envoy engaged in commercial transactions at designated places in the capital. In addition, more than ten times as many merchants as these special traders exchanged commodities with local merchants at the country's borders and at the designated ports.

In short, lucrative trade was the lubricant for the tributary system defining East Asian regional political, economic, and cultural intercourse. Not only overseas Chinese merchants but also Indian, Muslim, and European merchants participated in this tribute/trade, linking land and maritime zones in East Asia, Southeast Asia, South Asia, Central Asia, and Europe together. This dense and extensive tribute/trading networks forged the coastal areas of maritime East Asia into a diverse and well-integrated economic-commercial realm, making it the most advanced economy in the early modern world-system.

As a result of this tribute/trade system, economic growth in East Asia and intra-Asian trade was much faster and on a much larger scale than European trade before 1800. Indeed, Asian trade was a flourishing and on-going enterprise into which the Europeans only entered as relatively minor players.

The core region of the world-system, especially of industrial production, was in China. While China was in overall balance of trade surplus throughout most of the early modern period due to her export of huge quantities of valuable commodities (silk, tea, ceramics) and importing vast quantities of silver, Europe had a chronic balance of payments deficit due to the bullion drain to Asia.

In addition, China was an indubitable advanced country in the world at least until 1500, and was equal in most respects (productivity in industry; agriculture; transportation; competitiveness; share in world population and production) until 1800. Thus Gunder Frank (1998) argues that East Asia was the center of the early modern world-system.

As such, what explains the great reversal of fortune during the nineteenth century when East Asia and Europe swap their places in the world-system?

The fall of East Asia

The fall of East Asia passed through two phases: (1) Before 1800, when East Asia lost its developmental dynamics, became stagnant and failed to develop further into industrial capitalism; (2) After 1800, when East Asia interacted with European imperialism. Let us examine these two phases one by one in the following paragraphs.

Peter Perdue (1999) argues that the literature tends to provide a "fast-food" explanation that attempts to take a short cut through complex empirical and theoretical issues. The explanatory scheme usually focuses on one factor held to be unique to England and lacking in East Asia. For example, Max Weber uses the factor of the Protestant ethic to explain the divergent development of Europe and East Asia.

From a world-system perspective, Giovanni Arrighi et al (2003) point to the decline of the trade/tributary system as the critical factor in explaining why East Asia lost the dynamics to develop its economic locomotive further. Up to the end of the Song dynasty in 1276, the flourishing maritime trade and robust market economy of the East Asia coastal regions entered a "commercial revolution" characterized by advances in navigation technology, the consolidation of the "sea silk route," and the setting up of new port cities (Quanzhou, Guangzhou) as the centers of tributary trade and private trade for overseas Chinese trading networks.

However, this coastal trading dynamic had largely stopped in the Ming dynasty after the Ming emperor banned Zheng He's voyages in the 1400s. After this, the Ming state turned inwards, strengthened domestic/inland trade, and at times banned foreign trade. To show its determination, the Ming state restricted the number of tributary missions and even banned the building of seagoing ships. After these drastic actions, East Asia maritime trade and commerce lost its dynamic and its economy quickly ran into stagnation. By 1800, East Asia was no longer considered the center of the world-system. In contrast, overseas expansion and colonial conquest had given rise to a new stage of development in Europe, which led to the head-to-head confrontation between Europe and East Asia in the nineteenth century.

Up to 1800, however, China was still a world-empire that dominated the East Asian region. The Chinese empire was economically strong and developed a series of tributary states (Japan, Korea, Vietnam, and Siam) under its control. Thus the Chinese empire should have been able to defend itself from the conquest of the European states. However, two critical events during the nineteenth century served to change the power balance between the Chinese empire and Europe.

The first event was the vast expansion of the opium trade. In the beginning, the rate of Indian opium importation rose drastically in the early nineteenth century, jumping from 200 chests in 1760 to 20,000 chests in the 1820s and 30,000 chests in the 1830s. As the opium trade expanded, it soon came to provide one-seventh of the annual revenue of the British government in India, and it started the outflow of silver from China to Britain, because China had became the major market for "staple" goods produced in British India.

In short, the triangular trade between Britain, India, and China greatly altered the balance of payments between the two regions. Before the opium trade, China was the ultimate sink of the world's silver and maintained a vast trade surplus against the European states. However, after the expansion of the opium trade in the nineteenth century, China suddenly experienced a deficit in its trade balances.

The second event was the Opium War. While the opium trade weakened China economically, the Opium War weakened China politically. Great Britain's political power was greatly enhanced by the resources it obtained in its overseas empire. For example, India's huge demographic resources buttressed Britain's global power both militarily and financially. Paid for entirely by Indian tax-payers, troops were organized in a European-style colonial army and used regularly in the endless series of wars through which British opened up Asia and Africa to Western trade, investment and influence.

In contrast, the Chinese empire was weakened considerably by the closing off of the tributary/trade system and by the massive draining of silver in the opium trade. In 1839, the British sent an expedition to open China up for free trade and diplomacy on the grounds that several thousand chests of British property (opium) had been burned by the Chinese state. This was the beginning of the Opium War and the historic attempts to incorporate the Chinese empire (and its East Asian tributary states) into the UK-centered capitalist world-system.

It turned out that China had no answer to the steam-powered warship that in a single day in February 1841 destroyed nine war junks, five forts, two military stations, and one shore battery. After this disastrous war, China was forced to sign the Treaty of Nanking, which resulted in the opening of treaty ports to foreign trade, the ceding of Hong Kong, and the payment of indemnities to the British. The Opium War signals the fall of the Chinese Empire and the displacement of East Asia from the center of the world-system.

During the Opium War, Britain moved its economic base from Canton (in South China) to Shanghai (in Central China). This sudden shift of foreign trade in the 1840s caused the loss of customs revenue and unemployment in South China, leading to the Taiping Rebellion in the 1850s. Facing foreign domination and internal unrest, the Chinese empire began to decline in the mid-nineteenth century.

In contrast, Japan had a different historical experience during her encounters with Western power. When Japan was confronted with the threat of "market incorporation" from Western core states, the Meiji elites quickly centralized the government, promoted dynamic economic transformations and embarked on overseas imperialist ventures. Why was Japan able to escape the fate of peripheralization in the nineteenth century?

The answer lies in the following factors. First, with regard to the countries' preincorporated legacies, China was a world-empire in the external arena in East Asia, while Japan was not. As a "Middle Kingdom," China established a Sinocentric world order and extended its Confucian civilization to its tributary states. It was only natural that China's bureaucrats and landed upper class were reluctant to borrow techniques and values from the "barbarians." In contrast, Japan had often been at the receiving end of cultural borrowing, beginning with the Taika Reforms in the seventh century.

With regard to agents of incorporation, China was attacked by Britain, while Japan negotiated with the United States. Britain was a hegemonic superpower in the world-economy; it was attracted to the rich resources and huge market in China. Thus Britain was determined to use force to open China. Conversely, Japan was challenged by a Western power (the United States) that had yet to exercise its military domination in faraway East Asia in the nineteenth century. Moreover, when the European powers were preoccupied with China, they did not find Japan attractive in terms of either its resources or its markets. The result was that Japan experienced relatively less foreign domination and draining of resources than China in the second half of the nineteenth century.

Third, with regard to incorporation responses, the Chinese were confronted with the Taiping Rebellion, while the Japanese achieved the Meiji Restoration. In China, the regional shift of foreign trade from Canton to Shanghai resulted in large-scale peasant uprisings. In Japan, after the young samurai saw what the foreigners had done to China, they quickly evoked national sentiments to promote the Meiji Restoration.

Finally, with regard to resultant state capacity, the Chinese state had relatively weaker capacity for the promotion of semiperipheral ascent than the Japanese state. The Chinese state had been weakened by both foreign domination and peasant rebellions and could therefore only put forward at best a mediocre Self-Strengthening Movement. By contrast, the Japanese state, avoiding both strong foreign domination and large-scale peasant rebellions, and spurred by a feeling of the imminent threat of colonialism and the urgency of national survival, was able to carry out the far-ranging Meiji reforms within a short period. Here the historical irony is that in China, the "weak" Qing state was strong enough to suppress the challenges from the peasant rebellion, thus precluding a fundamental restructuring of state organization and power. Conversely, in Japan, the weak Tokugawa government quickly crumbled in face of the nationalistic movement from the outer feudal domains, allowing a strong Meiji state to emerge and institute a variety of reforms.

Obviously, this different mode of incorporation had great impact on the subsequent development of both Japan and China, as will be explained in the next section.

Japan's regionalization effort in the first half of the twentieth century

By the 1890s, this nascent semiperipheral state in Japan was beginning to feel the constraints of limited territory: small domestic market and lack of resources. What new policy, then, should the Japanese state adopt in order to sustain its industrialization toward the twentieth century?

Again, this timing is crucial because it helps to explain how world-system dynamics had influenced Japan's developmental strategy during this period. By the late nineteenth century, the world-economy reached a downward phase, marked by stagnation and contraction of global markets. In addition, Great Britain was no longer a hegemon, as its industrial and military supremacy were increasingly challenged by rival core powers. In order to overcome stagnation, Western core states competed with one another for new colonies and investment rights in the periphery. Therefore, rather than continuing the mid-century mode of "market incorporation," Western core states practiced territorial annexation and colonialism in East Asia at the end of the nineteenth century. Furthermore, instead of advocating the doctrine of free trade, Western core states preached the ideologies of nationalism and racism as they sought to fulfil a "manifest destiny" and assume a "civilizing mission."

Influenced by this global trend toward colonialism, the Japanese state developed a regionalization strategy of empire-building in East Asia. China, in this respect, became a necessary component of the Japanese regionalization strategy because of its cheap labor, abundant resources, and huge market potential. However, as a nascent semiperipheral state and a "late" imperialist, Japanese continental expansion to China was naturally challenged by Western core powers, leading to a new wave of territorial annexation in East Asia at the end of the nineteenth century.

Japan's strategy of regional dominance was halted during the 1920s. The Western core powers at the Washington Conference induced Japan to accept an inferior war fleet and to engage in international cooperation rather than regional rivalry.

The world depression and the dislocations in the economy in the 1930s, however, led to the emergence of ultranationalism and militarism in Japan. The depression testified to the vulnerability of Japan in the world-economy and accentuated Japan's desire for regional integration, and the country stepped up its continental expansion to China and Southeast Asia. The idea of a "Great East Asian Co-Prosperity Sphere" was raised to rationalize Japan's empire-building efforts. When the United States cut off the supply of military-related exports (most notably oil) to Japan, the latter could see no alternative but to fight for a high level of integration with the resources-rich Southeast Asia by eliminating the American presence in the region.

By the time of its defeat by the United States in 1945, Japan had lost three million lives and its economy was in ruin. Nevertheless, the conditions for sustaining post-war economic growth were already sprouting among those ruins. For instance, the wartime munitions industry provided the prototype for the chemical and heavy industrial sector. Links between big factories and small business that developed in the munitions industry also became the basis for the post-war subcontracting system. Administrative guidance by government ministries—a fundamental characteristic of the post-war developmental state—was also a legacy of the wartime controls.

Sino-Japanese wars and the communist revolution in China

Due to geopolitical dynamics in the East Asia region, the historical development of China was closely intertwined with that of Japan from the turn of the twentieth century through the end of the Pacific War.

Japan won the first Sino-Japanese War in 1894 and her economy received a boost from the huge indemnities received from China. Japan recovered tariff autonomy, began to stand on equal footing with Western powers, and started to engage in the regional domination of East Asia. In contrast, China lost the war in 1894. This led to a scramble for concessions, the weakening of the Chinese economy, and finally, the collapse of the Qing dynasty.

However, the second Sino-Japanese War between 1937 and 1945, again altered the paths of development for Japan and China. During the 1930s depression, Japan was prompted by ultranationalism and militarism to invade Manchuria and mainland China in order to build a "Greater East Asia Co-prosperity Sphere." This time Japan lost the war, which resulted in the American Occupation and Japan's forsaking militarism. The second Sino-Japanese War aggravated the Chinese peasant problem, provided more political space for the communist party to operate, and enhanced the "nationalist" calling of the communists. Without the Sino-Japanese War, the Communist Revolution may well still have taken place because of the deep-rooted peasant problem and the active intervention of the communists. However, the Revolution would certainly not have happened so quickly in 1949, and nor would it have been so radical. In this respect, the second Sino-Japanese War set the timing for and acted as a powerful catalyst in speeding up the revolutionary process.

After the second Sino-Japanese War, the development of China and Japan went separate ways. In the 1950s, the nascent Chinese communist regime was forced to withdraw from the capitalist world-economy, while the defeated Japanese government was designated as the junior partner of the American hegemon in Asia. These new paths of Chinese and Japanese development have profoundly transformed the geopolitics of the East Asian region and laid the foundation for East Asia to rise up as the center of the world-system again.

The resurgence of East Asia

The landscape of East Asia in the late twentieth century contrasted dramatically with that in the early twentieth century. Before World War II, mainland China was on the brink of becoming a periphery; Korea, Taiwan, and Hong Kong were colonies; and Japan was merely a semiperiphery. By the 1970s, however, Japan quickly emerged as a core state; mainland China and North Korea became socialist semiperipheries; and Hong Kong, South Korea, and Taiwan moved up to capitalist semiperipheries. By the 1990s, researchers like Gunder Frank began to talk about "ReOrient," the shifting of the center of the world-system from the West back to East Asia.

What then explains the general pattern of ascent in East Asia from 1945 to the mid-1970s? Why did the East Asian states exhibit not one, but three different paths (capitalist core, socialist semiperiphery, and capitalist semiperiphery) of ascent? Can the dynamics of economic growth in East Asia continue?

In order to understand the pattern of ascent in East Asia, it is necessary to bring the changing world-system dynamics in the mid-twentieth century back into the analysis. After World War II, the United States emerged as the new hegemonic power in the capitalist world-economy.

As Cumings (1993: 31) points out, this American hegemony can be called an "open-door empire." On the one hand, it was like an empire "policed by a far-flung naval and military basing system and by penetration of allied defense organizations." On the other hand, the American hegemon advocated "free investment" for transnational corporations. In addition, the United States promoted such new ideologies as democracy and modernization in order to decolonize the third world nations from the European core states. Under American leadership, the post-war world-economy thus became much more liberal, multilateral, and interdependent. This new world order under US hegemony led to an unprecedented expansion of the capitalist world-economy. From 1950 to the mid-1980s, world merchandise exports increased by approximately nine times in volume.

So and Chiu (1995) argue that the main concern of the United States toward East Asia in the late 1940s was how to prevent the further spread of communism from mainland China to other parts of the region. The American strategy of containment called for a polarized East Asian region, divided between the socialist camp of mainland China and North Korea and the capitalist camp of Japan, Taiwan, South Korea, and Hong Kong.

At first, China, after building up a strong Leninist party-state, pursued a mercantilist strategy of heavy industrialization. However, China shifted to a strategy of revolutionary socialism after the Sino-Soviet dispute. At the height of the Cultural Revolution, the ideology of Maoism called for self-reliance, mass mobilization, and absolute egalitarianism.

The communist revolution had a profound impact on the capitalist development of other East Asian states. In order to contain the spread of communism, the United States not only reversed its post-war occupation policies in Japan but also started to build Japan up as its junior partner in East Asia. The US's special concessions to Japan—such as procurement during the Korean War, massive aid, access to US markets, and the relief of the burden of defense—were instrumental in promoting Japan's economic success. Japan's developmental state and capitalist class seized this opportunity, as well as the favorable post-war upward turn of the world-economy, to promote state industrial and financial policies, incorporate labor into enterprise unionism, and engage in transborder expansion.

This transborder expansion of Japan, together with special US concessions and the influx of refugee capital and labor from mainland China laid the foundation for the semiperipheral ascent of Hong Kong, South Korea, and Taiwan (Newly Industrialized Economies [NIEs]).

The economic success of Japan and the NIEs, in turn, lured mainland China back to pursue a reintegration strategy in the late 1970s in order to achieve upward mobility in the world-economy. By the mid-1970s, the golden era of post-war economic expansion came to an end and the United States became a declining hegemon. It was during this period that the East Asian region was turned into a new epicenter of capital accumulation in the world-economy.

This can be seen from the fact that Japan had grown from a regional to a global economic power, challenging the hegemony of the United States. Technologically, Japan's high-tech sectors were closing their gaps with the United States, while Japanese manufacturers have outsold their American competitors. In the 1980s, Japan accumulated a sizable trade surplus and became the largest creditor in the world. Japan also started to play a leadership role in global affairs, and its "flying geese" model of development began to attract followers from the peripheries. In response, the United States imposed protectionist measures against Japanese imports, sought to open up the Japanese market for US imports, and forced the Japanese yen to appreciate, while the US mass media started "Japan-bashing."

The Chinese national unification project was another significant event. Economic integration among the three Chinese semiperipheries proceeded rapidly in the 1980s as mainland China, Taiwan, and Hong Kong faced bottlenecks in their economic growth. The massive relocation of labor-intensive industries from Taiwan and Hong Kong to mainland China led to changing industrial structures and new ethnic/national identities. Should this national unification succeed, the Chinese triangle of mainland China-Taiwan-Hong Kong may further speed up the centrality of East Asia in global capital accumulation.

What, then, is the future of the East Asian region? Will the present trends of economic growth and regionalization continue? Will East Asia become a center of capital accumulation in the world-system in the twenty-first century?

East Asia in the twenty-first century

In the early 1990s, there were indicators that the strong economic growth of East Asia may come to an end. First, the pace of economic growth had slackened in the NIEs. The maturity of these NIEs economies would make it impossible to duplicate or better the high-speed growth of the previous decades. Second, Japan experienced its longest recession after World War II. The burst of the "bubble economy" in the late 1980s caused a major crisis in the Japanese financial system and a slump in domestic demand. Third, the surge of protectionism in the West, the consolidation of the European Community as an economic bloc, and the formation of North American Free Trade Agreement (NAFTA) spelled trouble for the East Asian region because trade with the West still contributed significantly to their economic growth. Finally, the 1997 Asian financial crisis may have signaled the end of the dynamics of the East Asian economy, as they experienced a slowdown in economic growth in the late 1990s.

However, there were also signs of revitalization of the East Asian economy as the region moved toward the twenty-first century. First, the NIEs began to relocate their labor-intensive industries outward, hoping to capitalize on lower production costs abroad. Subsequently, NIEs economic structures were upgraded to high-tech and capital-intensive industries. Domestic demand also played a more significant role in promoting economic growth, especially in the larger economies of Japan, Taiwan, and South Korea. The tertiary service and financial sectors also figured more prominently in the distribution of national products.

Second, there was the emergence of China as the leader of the East Asian economic renaissance. The aggregate size of the Chinese economy in 2002 was more than eight times the size it was in 1978. In 25 years China's per capita income rose by more than 600 percent, from $151 in

1978 to $1,097 in 2003. China's domestic consumer demand burgeoned as its exports continued to grow. China, with its 1 billion strong populations and a fledgling middle class enjoyed the potential for keeping Japan and the NIEs' exports soaring to higher levels. Furthermore, China was admitted to the WTO (World Trade Organization), signifying the accelerated integration of the Chinese economy into the capitalist world-economy. Watching China becoming the locus of economic dynamism in the East Asian region, Arrighi (2007: xi) optimistically predicted that the epicenter of the global political economy will be shifted from North America to East Asia in the twenty-first century. *USA Today* (August 17, 2010: 3B) also reports that China is on track to overtake Japan as the second-largest economy behind the United States in 2010, cementing its status as one of the world's most formidable superpowers.

Third, while the United States and Europe experienced the deepest financial crisis at the end of the 2000s, China and East Asia became the only bright spots that could possibly pull the world-economy out of the slump.

Nevertheless, the prospects for moving East Asia into a new phase of dominance are not non-problematic because of the historical legacy of the Japanese empire. Memories from World War II make it difficult for other East Asian countries to accept Japan's dominance. Mainland China and South Korea, for instance, remains highly skeptical of the revision of Japanese textbooks on the subject of World War II history; the rise of China and the sub-regional integration among mainland China, Taiwan, and Hong Kong competes with Japan for regional dominance. Moreover, the Association of Southeast Asian Nations (ASEAN) states are reluctant to join any Asian regional organization that calls for sole Japanese or Chinese leadership.

In addition, the United States continues to try to suppress the rise of China. Despite China and Japan's strong economic presence in Asia, the region is still under the dual economic domination of the United States, and is still held by a unilateral American security network. The United States formed the APEC (Asia-Pacific Economic Cooperation) as a means of retaining its leadership role in the East Asia region; this explains why the Japanese and Chinese governments have been, so far, reluctant to support APEC wholeheartedly. Furthermore, the United States will continue to resort to trade and financial measures to slow down China's hegemonic advance. For example, the United States might want to appreciate the Chinese dollar still further and to install more "fair trade" measures in China; the United States frequently uses the WTO to harass China on its violation of trade practices; and the United States has declared a "national interest" in preserving freedoms of navigation through the South China sea and strengthening its military alliance with Japan and South Korea to contain the influence of North Korea and China in the Korean Peninsula.

However, due to significant US-Japan cross-investment in high-tech industries and the significant US-China cross-investment in manufacturing industries and geopolitical alliance, US/Japan will continue to be political allies and US/China will continue to be uneasy political partners in the midst of severe economic competition.

As such, East Asia is unlikely to overtake the United States as the next hegemon in the world-economy in the near future. What seems more likely to happen is the continuation of the present trend toward inter-regional economic integration between Japan, mainland China, and the Southeast Asian states. In the twenty-first century, therefore, researchers should be able to speak of a "Greater East Asia" region which links East and Southeast Asian states together economically.

References

Arrighi, G. (2007) *Adam Smith in Beijing: Lineages of the Twenty-First Century*, London: Verso.

Arrighi, G., Hui, P. K., Hung, H. F. and Selden, M. (2003) "Historical capitalism, East and West", in G. Arrighi et al (ed.) *The Resurgence of East Asia*, London and New York: Routledge.

Cumings, B. (1993) "The political economy of the Pacific Rim", in R. Palat (ed.) *Pacific-Asia and the Future of the World-System*, Westport, CT: Greenwood Press.
Frank, A. G. (1998) *ReOrient*, Berkeley: University of California Press.
Hamashita, T. (2003) "Tribute and treaties: maritime Asia and treaty port networks in the era of negotiations, 1800–1900", in G. Arrighi et al (eds) *The Resurgence of East Asia*, London and New York: Routledge.
Perdue, P. (1999) "China in the Early Modern World: Shortcuts, Myths and Realities", *Education about Asia*, 4(1): 21–26.
So, A. Y. and Chiu, W. S. (1995) *East Asia and the World Economy*, Thousand Oaks, CA: Sage Publication.
Wallerstein, I. (2005) *World-Systems Analysis*, Durham, NC: Duke University Press.

2.5
Darfur
The periphery of the periphery

Younes Abouyoub

The most salient feature of the current world-system is the hierarchical power that structures relations between a developed core and the underdeveloped periphery. Contrary to what has been claimed by modernization theories, peripheral countries are structurally constrained by economic and geopolitical pressures to keep reproducing their dominated status to the benefit of the center, which gains from the surplus produced by the periphery. Furthermore, this same unequal core-periphery relationship exists nationally between different provinces of national states at all levels of the nested world-system hierarchy, but the inequalities within national states tend to be much greater in the periphery.

The process of capital accumulation in the global capitalist development underwent three main phases. First, colonies were intensely decapitalized and their developmental capacities were largely weakened. Then, the weak social forces of capitalist development tried in vain to resist the onslaught of the industrial bourgeoisie in the center and its allies in the periphery itself, only to be totally annihilated in the third phase. In the last monopoly capitalist phase, the capacity to develop a peripheral national bourgeois class was severely hampered, consolidating thus the development of underdevelopment in many of the colonies.

Africa, like the rest of world regions, was incorporated into a single socioeconomic system: a capitalist world-economy, whose fundamental dynamic was a major influence on actors in all parts of the global system. The third phase of capitalist development differs from the former two in the sense that the imperial economic system enabled the center to put in place the formal framework for economic activity and, to a large extent, determine the character of development on the global stage. After a nominal independence, peripherized ex-colonies continued to be underdeveloped, often specializing in monoproduction of unprocessed raw materials for export to the core. This situation has contributed to the capital accumulation and development of the core at the expense of the decapitalization and underdevelopment of the periphery.

By 1750, the European metropolis felt the need to expand the outreach of capitalism. As industries were being established in the core, it became necessary to find new areas of primary production that could supply low-priced labor and land. Industrial expansion and the plantations in the periphery resulted in a shortage of labor, which justified the expansion of New World slavery and Africans were used to meet this demand. Eventually there was a growing demand for African land, together with indigenous labor to work on it for primary crop production. European powers eliminated any alternative indigenous sources of employment. Africans were thus deprived of land, their main means of production. This introduced a fundamental change in the mode of production and the social organization of indigenous societies.

In the last part of the nineteenth century and the early twentieth century, British hegemony was challenged by the rise of the United States and Germany and the spread of industrialization elsewhere. The British tried to meet the challenge by strengthening their colonial empire; they also attempted to co-opt the rising German challenge by giving the Germans some colonies in Africa. In 1884 at the Berlin Conference on Africa, the entire continent was divided up among the European colonial powers. The earlier phase of "informal empire," which enabled the peripheral African areas to retain some of their sovereign political institutions, came to be replaced by a formal empire that annihilated all African sovereignty by the end of the nineteenth century. Africa had to produce enough cash crops to sustain the administrative costs of the colonial powers. The British Home Office instituted "indirect rule" based on the principle of Africans ruling Africans for the benefit of the core. This required the preservation and the strengthening of tribal power and traditional chieftaincy. Sudan and Darfur were not exceptions.

By 1916, Darfur was annexed to the Anglo-Egyptian condominium. From the outset, the colonial power focused on the riverine provinces of Sudan and assigned a handful of British officials the task of governing Darfur by Indirect Rule with the *Powers of Nomad Sheikhs Ordinance*. The 1922 *Closed Districts Order* declared Darfur a closed area to anyone without a government permit. This system promoted conservatism and denied education and economic development to the province. Yet, since the Native Administration needed indigenous clerks, some meager educational opportunities were granted to the co-opted tribal chiefs' sons. As for development projects, riverine Sudan received the lion's share, while Darfur and the rest of the South were excluded. The telegraph system did not reach Darfur until after independence in 1956. Roads were nearly non-existent and the modest economic activity focused on a low-quality animal husbandry. Even by Sudanese standards, Darfur's income was very low in the 1930s. Famine and diseases were rampant. The infection and mortality rates were staggering. Politically, the parliamentary system, modeled after Westminster, empowered the Khartoum elite in the Sudanese capital and left little scope for the peripheral regions. Successive Sudanese governments kept reproducing the same political and economic inequalities.

The colonial legacy left a power system that guaranteed domination through tribally organized local authorities. Decentralized despotism, the generic form of the colonial state in Africa, would breed recurring political and economic crisis in post-independence Sudan. In 1956, the first rebellion broke up in the South, while the low-intensity and small scale outbreaks in Darfur would turn into violent battles in the 1980s and an open rebellion against the Sudanese state in 2003.

Part II

Theory and critiques

3. Theoretical frontiers in world-systems analysis

3.1

Externality, contact periphery and incorporation

Jon D. Carlson

I summarize and discuss "incorporation" and the "external arena" in the expanding modern world-system, and their relationship to the mechanisms of systemic expansion. This fleshes-out the middle ground between Wallerstein's (1974, 1980, 1989) analysis (European-focused, state-centric, "inside-out") and Hall's (1986, 1987, 1989, 1998, 1999b, 2000) research on frontiers and incorporation (external, indigenous-oriented, "outside-in"). I add a "*zone of ignorance*" beyond the external arena that prompts the expansion of the European-centered world-system. Building on Chase-Dunn and Hall's work (1991, 1993, 1997) with nested networks in a world-system, the interaction of the "zone of ignorance" with the "information network" primes the pump of expansion. Myths of riches, luxury goods, and fabulous lands fuel systemic expansion; this is what effectively underwrites the high-cost ventures necessary to extend the European world-system. These myths, misinformation and misperceptions are reflected in historical maps, which also reflect effective political control and the mixture of geographical fact and fantasy. One observes considerable social, cultural and political change *prior* to the period during which a zone is fully "incorporated."

A second goal of research on incorporation is to gain insight into "modern" forces of globalization. Case studies provide compelling evidence that the same problems associated with contemporary globalization are inherent in the broadening and deepening of the world-system. "*Protoglobalization*" serves to differentiate early processes from conceptions of modern globalization (Chase-Dunn 1999; Sklair 1999). This is also an area in need of additional research. While the pace of global change has increased, it should be understood in the context of historical cases of systemic expansion, readily evident at the contact periphery.

The contact periphery characterizes the dynamic initial interaction that occurs between two formerly separate world-systems, which involve significant cultural differences. This mirrors Turchin's "metaethnic frontier" (2003) and is where cross-cultural forces of "creative destruction" are unleashed. The nature of incorporation depends on the types of systems coming into contact with one another, and can change over time. These frontiers exist on the edge of any system, like the expanding European system, but also exist at the edge of systems or mini-systems of indigenous peoples (Burch 2005; Chase-Dunn and Mann 1998), imperial systems such as China, or other historical non-European systems.

Why incorporation? The process driving expansion

Why examine "incorporation"? Simply put, to understand the expansion of the international system, one must understand the *process* of that expansion. Examining how actors in the system absorb regions lets us break down the process of expansion and allows for a clearer understanding of categories of interrelated activity (e.g., economic, socio-cultural, political).

The role of multiculturality

Multiculturality implies that economic and socio-political linkages occur between groups that are culturally distinct, with differences in language, religion, normative institutions, mode of governance and other fundamental characteristics of everyday life. How are these differences reconciled? Who decides? What happens when these societies interact?

Wallerstein (1974) identifies three historical modes of production: mini-systems, world-empires, and world-economies. World-empires are two or more culturally distinct groups integrated by the forcible accumulation of surplus ("tribute;" thus the "tributary mode" of accumulation) organized around a single political center. World-economies are integrated via the market rather than a single political center. Left out of this discussion are the mini-systems, which are small-scale economies, integrated through reciprocity-based family relationships (a "lineage" or "kinship mode" of accumulation) (Chase-Dunn and Mann 1998).

"Incorporation" starts much earlier than commonly held, since trade-induced political evolution is a necessary first step toward full incorporation. Therefore, the process begins with the initiation of *contact* between the edges of disparate systems. The primary function behind the expansion of the system may be *economic* (Warner 1998, 1999), while some argue that the "security" of a region is important (Snyder 1991). This presupposes an initial source of cross-cultural contact that needs to be "secured," or that knowledge needs to be exchanged to perceive a "threat" to security. Regardless, this economic function provides the incentive for political action to support possible military adventurism, often to implement a favorable political regime in the region incorporated. This still neglects those areas and polities that are *external* to direct trade linkages. How do they become relevant? The notion of "down-the-line" trade helps expose this relationship. For example, West African gold long impacted Middle Eastern, and then European, markets via the trans-Saharan gold-salt exchange and gave rise to the empires of Western Africa.

The role of bulk vs. prestige goods

In addition to political and cultural differentiation, the type of trade goods matters: are they "necessities" or luxury goods? Chase-Dunn and Hall (1993: 854) note that Wallerstein "originally defined *necessities* in terms of food and raw materials that are necessary for everyday life. Subsequently he has also included bullion and 'protection'." Lane (1979) includes "protection rents," as protection is an important political tool responsible for considerable historic transfers of wealth. Structured trade between polities is not enough to be considered part of the world-system. The *kind* of trade matters, and shapes the incorporation process. For Wallerstein, "non-essential" trade is trade in "preciosities"—prestige goods—which does not produce important *systemic* effects and therefore does not indicate systemic inclusion.

Yet some involvement and process had to occur prior to the emergence of trade in bulk goods, and this stage is underappreciated. Some attention has been paid to luxury goods (Feinman 1999; Peregrine 1992, 2000; Schneider 1977) and Chase-Dunn and Hall (1993: 855) argue that "prestige-goods economies constitute systemic networks because the ability of local leaders to monopolize the supply of these goods is often an important source of stability and change in local power structures." These initial networks have important priming effects. Important social, political, and economic changes are likely to occur well *before* an area is considered "incorporated" into the system by the criterion of bulk goods exchange.

While contact between societies can have immediate socio-cultural impact (e.g., disease, new technology, new precious resources, mere "awareness," identity re-structuring), it is not as clear how luxury trade shapes the political and economic arenas. Yet, a "'prestige system' represents the myriad of ways in which prestige is accrued and maintained in the society. It includes knowledge, rituals, and symbols which convey and display status"(Peregrine 1999: 39). Political power, derived from status, comes from prestige systems.

This introduces two additional points. First, is "knowledge" a luxury good? Dealing with information networks and exchange of knowledge is one area of world-systems theory that needs to be expanded, and should prove fruitful. Second, what roles do symbols and rituals associated with "prestige" play in changes that result after the initiation of trade? Can the exchange of information or ritual create systemic interdependence (Peregrine 2000)? These are worthwhile points to consider when identifying cultural and social change, and they become good measures of cultural and social incorporation.

Spatial boundaries: From "Zone of Ignorance" to core

Another orienting concept of the world-system has to do with how goods and wealth are accumulated. Traditionally, a "tipping-point" for whether or not a region is part of the world-system depends on the division of labor in a region—or "zone"—at a given time. This allows the conceptions of "external arena," "incorporation," and "peripheralization." Indeed, one may view incorporation as a period of "broadening" capitalist development. Since the drive to broaden and deepen the system is inherent in the functioning of capitalism, an area carrying on *any* trade-based relationship with Europe would be effectively "hooked," as incorporation into the system is tacitly inevitable. Incorporation is the continued broadening of a world-system, while peripheralization is the deepening of systemic relations.

The notion of "commodity chains" helps expose linkages, and "describe[s] the production of goods as they move from raw to cooked, slave-cultivated cotton becoming Manchester textiles, peasant-grown Columbian coffee becoming Detroit labor power, and so on" (Gereffi and Korzeniewicz 1994; Goldfrank 2000: 168–9). Capital accumulation—and market responsiveness—are hallmarks of when a commodity chain is "integral to" the world-economy, and a region thereby officially incorporated. Not only is the arena undergoing incorporation seen as "not internal" and yet "not quite external," but as "a given zone is incorporated into the world-economy, this often led to an adjacent *further* zone being *pulled into* the external arena. It is though there were an outward ripple of expansion" (Wallerstein 1989: 167, emphasis added). This delineation is relevant, "As a zone became incorporated into the world-economy, its transfrontier trade became 'internal' to the world-economy and no longer something 'external' to it" (1989: 171).

This indicates that there is some sort of "supra-external" arena that is undefined by the capitalist world-economy and exists as what is essentially a "zone of ignorance" (Carlson 2001, 2002, 2011). What is really an external arena and what is not? One is literally talking about the known universe at any point in history, and that which is not known. Additionally, this indicates that instead of a three-tiered working definition of the world-system (core, semiperiphery, periphery), there are actually six tiers (core, semiperiphery, periphery, incorporating zone, external arena, zone of ignorance).

An external arena is a "zone from which the capitalist world-economy wanted goods but which was resistant (perhaps culturally) to importing manufactured goods in return and strong enough politically to maintain its preferences" (Wallerstein 1989: 167). China and Japan in the eighteenth century both serve as examples. Presumably, the governing agents in any zone bordering the expanding capitalist world-economy have three potential paths: (1) they can be strong enough to maintain their preferences and *not* choose incorporation; (2) they can be strong enough to maintain their preferences yet *choose* incorporation; (3) or they can be too weak to maintain their own preferences. In the latter case, the zone is presumably assimilated and transformed, if it has resources desired by agents in the world-economy. Regardless, any zone adjacent to the world-system will eventually be absorbed, either in the short or middle run.

Various bounding mechanisms are one area Chase-Dunn and Hall (1997; Hall 1999c: 7, 2000: 239) address. They present four types of "bounding mechanisms" within any world-system, which only rarely coincide. The broadest is a boundary of information or cultural flows; an information-exchange network [IN]. This is the "known universe" of a system. The outer edges are "fuzzy," as incomplete information or misinformation become more apparent and the domain of legend or rumor more prevalent. This is the beginning of my "zone of ignorance." The next boundary encompasses luxury or prestige good flows. This is the prestige goods network [PGN]. This is of comparable size, but not coincident with the information network. The third bounding mechanism is political/military interaction [PMN]. The fourth consists of bulk goods, makes up the bulk goods exchange network [BGN], and corresponds to Wallerstein's world-system. Typically, these networks are "nested" within one another.

Frontiers

How are frontiers formed and transformed? What do we mean when activity (e.g., trade, information exchange) is "transfrontier"? Hall (2000: 241) proposes that a frontier is "a region or zone where two or more distinct cultures, societies, ethnic groups, or modes of production come into contact." Like a membrane (cf., Slatta 1998), a frontier's permeability varies with the direction and type of things (information, goods, people) moving through it. While expansion usually posits that the incorporation process is driven from the center outward, a membrane image recognizes that influences can (and often do) flow from the area being absorbed toward the center. How much is inevitable? To what extent can one "negotiate peripherality" (Kardulias 1999, 2007; Morris 1999) or even "negotiate *externality*" and thus resist incorporation?

While any frontier is "narrow and sharp" from a global perspective, "from nearby it is a broad zone with considerable internal spatial and temporal differentiation" (Hall 2000: 240). It is in this zone of historical interaction that conflict often arises from contact, and where "no one has an enduring monopoly on violence" (Baretta and Markoff 1978, in Hall 2000: 241). Monopoly on violence? This is the stuff that defines politics! Frontiers are the zone just beyond the "state," just past the political markings of borders. Frontiers are where the rules of the system are laid bare, and where anything goes.

Modeling incorporation

As has been observed, the very conception of incorporation is problematic: "(a) with respect to incorporating state or system (b) with respect to types of incorporated groups (c) with respect to both timing and degree of incorporation, and (d) with respect to a variety of factors that can affect the process" (Hall 1986: 398; also 1999c, 2000). There are two interrelated aspects involved with the incorporation of an external region. First, we are talking about a series of ongoing *processes*, which may be divided into separate phases. Second, against this background of ongoing process, a zone will experience different "states of being" or conditions within that process.

Processes

Something is involved in the "hooking" of an arena into the external (and ultimately internal) domain. Accordingly, I suggest three sub-processes in the larger incorporation process. First, a "zone of ignorance," or mythic domain, largely unexplored and unknown to the current members of the system is contacted; here the "*grooming*" process of conditioning the area toward capitalist exchange and production is initiated. This may be the process in which most social and behavioral change takes place. Organized production develops, and the notion of "craft specialization" is instrumental in expansion (Kardulias 1990). Consumerism is introduced or develops, along with "induced wants" (Sklair 1991: 131) or "false needs" (Marcuse 1964). Peoples literally learn the values of the world-system: what has value and what does not.

The next phase is one of *incorporation*, wherein an external arena's contact and involvement with the world-system is developed, ultimately producing "nominal incorporation," building toward "effective incorporation." Here, the processes that began during the grooming process become more developed and socially pervasive. Once the incorporation process is advanced, the process of *peripheralization* takes over. An effectively incorporated arena moves into the periphery of the world-system and may proceed into the semiperiphery. Beyond this, additional refinement is also needed to understand the processes behind transition from one "level" to another *within* world-systems.

"States of Being"

Against the background procedural change, a region is contextually fixed in various "states of being" during specific periods of examination. Since the "patterns that we see on a map are actually freeze-frame snapshots of complex processes of incorporation" (Hall 1999c: 11), it makes sense to take this one step further and *use* historical maps to operationalize the phases of incorporation. There is a growing body of literature evaluating "maps as text" (Wood and Fels 1992). Yet "we must remember that the map is a picture, that every picture tells a story, and that every story makes part of the text ... Like any story, maps are propaganda, but maps have the full weight of science behind them" (Mitchell 1999: 40–41). Evaluating maps made during incorporation is a method of reading the text of incorporation, and provides the ability to clarify different "states of being."

For example, some regions exist beyond the bounds of the external arena. While these regions may have trade links, social links, or geographic proximity to regions considered external, they may not be "known" to members of the world-system (e.g., pre-Columbian America, interior Africa). Thus, these regions exist in only a mythic sense beyond the boundary of the information network, or in a *zone of ignorance* outside the information network. This zone would not appear on maps of the members of the world-system, and is characterized by the "*Cave, Hic Dragones*" approach to mapmaking.

The next state of being toward involvement with the system is the *external arena*. A region is known to members of the system, but not productively. The next stage is being in the *IN*. The system members know of a region and can place its name roughly on a map, but this may not reflect geographic accuracy and is commingled with regions that only exist in myth. Priming, luxury trade develops. Trade increases. This includes prestige goods, along with an increase of information exchange and cultural pollination. The region is still considered to be in the external arena (i.e., the *PGN*). Maps begin to reflect a region, but lack significant detail. Outlines of continents should be fairly accurate, but little knowledge of internal features will be evident.

As involvement increases and economic linkages develop with the world-system, a region may be seen as having moved into a state of being *nominally incorporated* (i.e., the political-military network—PMN). Maps have some detail of major cities and geographic features, but will still be incomplete, missing major features beyond the areas central to interaction. Next is a state of being *effectively incorporated*, whereby a given region enters the periphery (i.e., the bulk goods network—BGN). Here, there should be complete detail on maps, as the region is now part of the system. Beyond this, a region may develop out of the peripheralization process into a state of being *semiperipheral.*

To clarify, process overlays states of being:

	Grooming		Incorporation		Peripheralization
I. Process	I-------------------------II		------------------------------------II		----------------------------I
II. State of Being	Zone of Ignorance →	External Arena →	Nominal Incorporation →	Effective Incorporation →	Periphery

Figure 3.1.1 Typography of incorporation
Source: Carlson (2001: 249).

This typography helps visualize the larger processes associated with incorporation. These are not discrete, they segue together and the transition between phases is "fuzzy." Similarly, the transition between states of being is indistinct, even though the core of each may be readily distinguished. In different states of being, particular emphasis on socio-cultural, political, or economic factors may be more relevant than at other times.

Chase-Dunn and Hall (1997: 63) use a similar tool to illustrate the continuum of incorporation. It also helps decipher the verbiage scholars use for incorporation or types of peripheries: external arena, nominal or formal periphery, margin, or hinterland. This is particularly handy for precapitalist settings, but needs refinement:

> First, incorporation is not unidimensional, but multidimensional along the four types of world-system boundaries. Incorporation can be economic (for either bulk goods or luxury goods), political/military, or socio-cultural, which includes all types of information and symbols; Second, incorporation creates multiple frontiers, corresponding to each of the boundaries; Third, *ceteris paribus*, incorporation will begin at the furthest boundaries (cultural, symbolic, informational, or luxury goods) and proceed to narrower, more intense forms along the political-military boundary and finally along the bulk goods dimension; and Fourth, relations among the dimensions of incorporation and the resulting frontiers is complex theoretically and empirically.
>
> *(Hall 1999a: 444)*

Incorporation takes place at different rates and degrees for different regions. Regions may exist only in the information network of one system member, yet be involved in the trade of luxury or bulk goods with another member of the same system. Taken with the recognition of the multiple frontiers involved, incorporation is a complex phenomenon.

Comparative cases and measuring incorporation

A structured, focused comparison seems most promising as a way to study these issues empirically (*cf.*, George 1979; George and Bennett 2004). To examine processes of incorporation, one can trace relevant "markers" over any time period in question. These markers may be degree of political autonomy, type and volume of trade with the system and the core specifically, or alterations in traditional social or cultural behavior. Some type of "pre-contact" baseline for factors should be clarified, so that change may be studied in relation to type of systemic interaction. Certainly processes are interrelated and reciprocal, but by addressing them as separate facets we develop a clearer understanding of the dynamics of incorporation.

Societal change may take the form of alteration of traditional lifestyles, coercion of labor, or other fundamental discontinuity. This may include alteration of migration patterns, gender role shifts, alteration of spirituality, or identity changes. As incorporation increases, changes in classes and mode of production emerge. If early change is present, then an argument for incorporation is strengthened. When looking at the political-military arena, signals include political-structural change (e.g., "state-building," changes in laws, regime change, alteration of power structures), or if a formalized political relationship with the "internal" area has been instituted. Finally, the type and quantity of goods exchanged, including how they are "produced," are measures of the process.

This approach is appealing because it offers a more concrete examination of what is otherwise theoretic posturing. We "see" historical processes in action. Second, as cartographic information was jealously guarded, using maps as a measure provides a method of tracking the transmission of information among system members. Third, maps make it easier to distinguish when states knew certain information about a frontier region. The variable rates of incorporation among relevant actors in a region are distinguishable. Finally, this helps understand not just the important role information plays in the expansion of the system, but the impact of *lack* of information. Geographic ignorance or *dis*-information was actively promoted as a tool of imperial expansion. On the frontier of a system, fanciful myth usually trumps boring fact. Mere rumor of goods, lands or mythic peoples was enough to prompt imperial adventuring, and thereby jump-start the systemic incorporation processes.

Additional theory-building leverage can come from individual case studies, and this is another area for future research. For example, Carlson (2011) looks at the lengthy, cross-systemic incorporation of Abyssinia from a pre-capitalist stage to its peripheralization. The case supports the "pulsation" thesis of world-systems (Beaujard and Fee 2005; Chase-Dunn and Hall 1997; Hall 1987, 1989, 2005), offering insight for future cross-systemic study: are zones along system borders repeatedly incorporated as they fall in and out of neighboring systems, or are they mere geographic socio-cultural bridges between systems? Accordingly, such cases are also applicable to research on "contested peripheries" (Allen 1992, 2005; Cline 2000; Hall 2005). Cases offer insight into the ability of regions to maintain externality or "negotiate peripherality" (Kardulias 1999, 2007). Ethiopia (née Abyssinia) maintained independence from colonial subjugation well into the twentieth century, despite being linked to the African, Eurasian and Indian Ocean systems (Beaujard 2007; Beaujard and Fee 2005). Finally, because of the long historical overview of the case, it lends support to the application of world-system analysis to pre-capitalist settings, as has been argued by Abu-Lughod (1989, 1993), Chase-Dunn and Hall (1991), Frank (1990, 1998), Frank and Gills (1993) and Gills and Frank (1991), and it echoes concerns with globalization in having a considerable historical legacy (Gills and Thompson 2006). The interaction between systems is a useful homologue for contemporary discussions about cultural globalization and civilizational "clashes" (Huntington 1993), and the mini-industry that has arisen around the "clash of civilizations" thesis. Thus, careful attention to incorporation offers a rich research agenda with application to "modern" questions and contemporary global problems. Globalization is not new—it has been going on for millennia. Studying incorporation illuminates this.

References

Abu-Lughod, Janet (1989). *Before European Hegemony: The World-System A.D. 1250–1350.* New York: Oxford University Press.

Abu-Lughod, Janet (1993). "Discontinuities and Persistence: One World-system or a Succession of World-systems?", pp. 278–90 in Andre Gunder Frank and Barry Gills, (eds) *The World-system: Five Hundred Years or Five Thousand?* London: Routledge.

Allen, Mitchell (1992). "The Mechanisms of Underdevelopment: An Ancient Mesopotamian Example," *Reviews*, 15:3, 453–76.

Allen, Mitchell (2005). "Power is in the Details: Administrative Technology and the Growth of Ancient Near Eastern Cores", pp. 75–91 in Christopher Chase-Dunn and E. N. Anderson, (eds) *The Historical Evolution of World-Systems*, New York: Palgrave.

Baretta, Silvio R. D. and John Markoff (1978). "Civilization and Barbarism: Cattle Frontiers in Latin America," *Comparative Studies in Society and History*, 20:4, 587–620.

Beaujard, Philippe (2007). "L'Afrique de l'Est, les Comores et Madagascar dans le Système-Monde Eurasiatique et Africain avant le 16e Siècle", pp. 29–102 in D. Nativel and F. Rajaonah, (eds) *L'Afrique et Madagascar*, Paris: Karthala.

Beaujard, Philippe and S. Fee (2005). "The Indian Ocean in Eurasian and African World-Systems before the Sixteenth Century," *Journal of World History*, 16:4, 411–65. http://www.historycooperative.org/journals/jwh/16.4/beaujard.html.

Burch, Ernest S. Jr. (2005). *Alliance and Conflict: The World-System of the Iñupiaq Eskimos*, Lincoln: University of Nebraska.

Carlson, Jon D. (2001). "Broadening and Deepening: Systemic Expansion, Incorporation and the Zone of Ignorance", *Journal of World-Systems Research*, 7:2, 225–63, http://jwsr.ucr.edu/archive/vol7/number2/pdf/jwsr-v7n2-carlson.pdf.

Carlson, Jon D. (2002). "The 'Otter-Man' Empires: The Pacific Fur Trade, Incorporation & the Zone of Ignorance", *Journal of World-Systems Research*, 8:3, 390–442. http://jwsr.ucr.edu/archive/vol8/number3/pdf/jwsr-v8n3-carlson.pdf.

Carlson, Jon D. (2011). "Externality and Incorporation in the World-System: Abyssinia—Anomaly or Palimpsest?", *Journal of World-Systems Research*, 17:1, 166–99.

Chase-Dunn, Christopher (1999). "Globalization: A World-Systems Perspective," *Journal of World-Systems Research*, 5:2, 176–98.

Chase-Dunn, Christopher and Thomas D. Hall, (eds) (1991). *Core/Periphery Relations in Precapitalist Worlds.* Boulder, CO: Westview.

Chase-Dunn, Christopher and Thomas D. Hall (1993). "Comparing World-Systems: Concepts and Working Hypotheses," *Social Forces*, 71:4, 851–86.

Chase-Dunn, Christopher and Thomas D. Hall (1997). *Rise and Demise: Comparing World-Systems.* Boulder, CO: Westview.

Chase-Dunn, Christopher and Kelly Mann (1998). *The Wintu and Their Neighbors: A Very Small World-System in Northern California.* Tucson: University of Arizona.

Cline, Eric (2000). " 'Contested Peripheries' in World-systems Theory: Megiddo and Jezreel Valley as a Test Case", *Journal of World-Systems Research*, 6:1, 7–16, http://jwsr.ucr.edu/archive/vol6/number1/v6n1_pdf/jwsr-v6n1-cline.pdf.

Feinman, Gary (1999). "The Changing Structure of Macroregional Mesoamerica: The Classic-Postclassic Transition in the Valley of Oaxaca," pp. 53–62 in P. Nick Kardulias, (ed.) *World-Systems Theory in Practice: Leadership, Production and Exchange.* Lanham, MD: Rowman & Littlefield.

Frank, Andre Gunder (1990). "A Theoretical Introduction to Five Thousand Years of World-system History," *Review*, 13:2, 155–250.

Frank, Andre Gunder (1998). *ReOrient: Global Economy in the Asian Age.* Berkeley: University of California.

Frank, Andre Gunder and Barry Gills, (eds) (1993). *The World-System: Five Hundred Years or Five Thousand?* London: Routledge.

George, Alexander (1979). "Case Studies and Theory Development: The Method of Structured, Focused Comparison," pp. 43–68 in Paul Lauren, (ed.) *Diplomatic History: New Approaches.* New York: Free Press.

George, Alexander and Andrew Bennett (2004). *Case Studies and Theory Development in the Social Sciences.* Cambridge, MA: MIT Press.

Gereffi, Gary and Miguel Korzeniewicz, (eds) (1994). *Commodity Chains and Global Capitalism.* Westport, CT: Greenwood Press.

Gills, Barry K. and Andre Gunder Frank (1991). "5000 Years of World-system History: The Cumulation of Accumulation", in C. Chase-Dunn and T. D. Hall, (eds) *Core/Periphery Relations in Precapitalist Worlds*. Boulder, CO: Westview.

Gills, Barry K. and William R. Thompson, (eds) (2006). *Globalization and Global History*. New York: Routledge.

Goldfrank, Walter L. (2000). "Paradigm Regained? The Rules of Wallerstein's World-System Method," *Journal of World-Systems Research*, 6:2, 150–95.

Hall, Thomas D. (1986). "Incorporation in the World-System: Toward a Critique," *American Sociological Review*, 51, 390–402.

Hall, Thomas D. (1987). "Native Americans and Incorporation: Patterns and Problems," *American Indian Culture and Research Journal*, 11:2, 1–30.

Hall, Thomas D. (1989). *Social Change in the Southwest, 1350–1880*. Lawrence: University Press of Kansas.

Hall, Thomas D. (1998). "The Rio de la Plata and the Greater Southwest: A View from World-System Theory," in Donna J. Guy and Thomas E. Sheridan, (eds) *Contested Ground: Comparative Frontiers on the Northern and Southern Edges of the Spanish Empire*, Tucson: University of Arizona.

Hall, Thomas D. (1999a). "Precapitalist World-Systems", pp. 897–900 in Phillip Anthony O'Hara, (ed.) *Encyclopedia of Political Economy*, Vol. 2, New York: Routledge.

Hall, Thomas D. (1999b). "World-System: Incorporation Into," pp. 1265–67 in Phillip Anthony O'Hara, (ed.) *Encyclopedia of Political Economy*, vol. 2, New York: Routledge.

Hall, Thomas D. (1999c). "World-Systems and Evolution: An Appraisal," pp. 1–23 in P. Nick Kardulias, (ed.) *World-Systems Theory in Practice: Leadership, Production, and Exchange*, Lanham, MD: Rowman & Littlefield.

Hall, Thomas D. (2000). "Frontiers, and Ethnogenesis, and World-Systems: Rethinking the Theories," pp. 237–70 in Thomas D. Hall, (ed.) *A World-Systems Reader: New Perspectives on Gender, Urbanism, Cultures, Indigenous Peoples, and Ecology*, Boulder, CO: Rowman & Littlefield.

Hall, Thomas D. (2005). "Mongols in World-Systems History", *Social Evolution & History*, 4:2, 89–118.

Huntington, Samuel (1993). "The Clash of Civilizations?", *Foreign Affairs*, 72:3, 22–49.

Kardulias, P. Nick (1990). "Fur Production as a Specialized Activity in a World-system: Indians in the North American Fur Trade," *American Indian Culture and Research Journal*, 14:1, 25–60.

Kardulias, P. Nick (2007). "Negotiation and Incorporation on the Margins of World-Systems: Examples from Cyprus and North America", *Journal of World-Systems Research*, 13:1, 55–82, http://jwsr.ucr.edu/archive/vol13/Kardulias-vol13n1.pdf.

Kardulias, P. Nick, (ed.) (1999). *World-Systems Theory in Practice: Leadership, Production and Exchange*. Lanham, MD: Rowman & Littlefield.

Lane, Frederic (1979). *Profits from Power: Readings in Protection Rent and Violence-Controlling Enterprises*. Albany, NY: SUNY Press.

Marcuse, Herbert (1964). *One Dimensional Man: Studies in the Ideology of Advanced Industrial Society*. Boston, NY: Beacon Press.

Mitchell, Ken (1999). "Science, Giants, and Gold: Juan de la Cruz Cano y Olmedilla's Mapa Geográphico de la America Meridional," *Terrae Incognitae*, 31:1, 25–41.

Morris, Ian (1999). "Negotiated Peripherality in Iron Age Greece: Accepting and Resisting the East," pp. 63–84 in P. Nick Kardulias, (ed.) *World-Systems Theory in Practice: Leadership, Production and Exchange*. Lanham, MD: Rowman & Littlefield.

Peregrine, Peter (1992). *Mississippian Evolution: A World-System Perspective*. Madison, WI: Prehistory Press.

Peregrine, Peter (1999). "Legitimation Crises in Prehistoric Worlds," pp. 37–52 in P. Nick Kardulias, (ed.) *World-Systems Theory in Practice: Leadership, Production and Exchange*. Lanham, MD: Rowman & Littlefield, pp. 37–52.

Peregrine, Peter (2000). "Archaeology and World-Systems Theory," pp. 59–68 in Thomas D. Hall, (ed.) *A World-Systems Reader: New Perspectives on Gender, Urbanism, Cultures, Indigenous Peoples, and Ecology*. Boulder, CO: Rowman & Littlefield.

Schneider, Jane (1977). "Was There a Pre-Capitalist World-System?", *Peasant Studies*, 6: 20–29; reprinted in C. Chase-Dunn and T. D. Hall, (eds) (1991) *Core/Periphery Relations in Precapitalist Worlds*. Boulder, CO: Westview, pp. 45–66.

Sklair, Leslie (1991). *Sociology of the Global System*. Baltimore, MD: Johns Hopkins University.

Sklair, Leslie (1999). "Competing Conceptions of Globalization," *Journal of World-Systems Research*, 5:2, 141–59.

Slatta, Richard W. (1998). "Spanish Colonial Military Strategy and Ideology", in Donna Guy and Thomas Sheridan, (eds) *Contested Ground: Comparative Frontiers on the Northern and Southern Edges of the Spanish Empire*. Tucson: University of Arizona.

Snyder, Jack (1991). *Myths of Empire: Domestic Politics and International Ambition*. Ithaca, NY: Cornell University.

Turchin, Peter (2003). *Historical Dynamics: Why States Rise and Fall*. Princeton, NJ: Princeton University Press.

Wallerstein, Immanuel (1974). *The Modern World-System I: Capitalist Agriculture and the Origins of the European World-Economy in the Sixteenth Century*. San Diego, CA: Academic Press.

Wallerstein, Immanuel (1980). *The Modern World-System II: Mercantilism and the Consolidation of the European World Economy, 1600–1750*. San Diego, CA: Academic Press.

Wallerstein, Immanuel (1989). *The Modern World-System III: The Second Era of Great Expansion of the Capitalist World-Economy, 1730–1840s*. San Diego, CA: Academic Press.

Warner, Carolyn (1998). "Sovereign States and Their Prey: The New Institutionalist Economics and State Destruction in Nineteenth-Century West Africa," *Review of International Political Economy*, 5:3, 508–33.

Warner, Carolyn (1999). "The Political Economy of 'Quasi-statehood' and the Demise of 19th Century African Politics," *Review of International Studies*, 25, 233–55.

Wood, Denis and John Fels (1992). *The Power of Maps*. New York: Guilford Press.

3.2

Wallerstein's world-system

Roots and contributions

W. L. Goldfrank

Although many of the ideas at the heart of world-systems analysis had been articulated in the 1960s and even prefigured in the World War I era, it was not until the early 1970s that its basic concepts were either first developed or synthesized and then promulgated by Immanuel Wallerstein. He would go on to be its prime mover in an institutional sense and one of its four or five most innovative scholars in an intellectual one. Only by ignoring the contributions of others, amply demonstrated elsewhere in this Handbook, could one overestimate the importance of Wallerstein's impact. He has been an original thinker, a masterful synthesizer, an intellectual coordinator and collaborator, an effective institution-builder, and a tireless activist. He called what he took the lead in developing "world-systems analysis" or the "world-systems perspective," and the names have stuck. (See his website, http://www.iwallerstein.com/ for additional details and a complete bibliography.)

Wallerstein's most significant intellectual contributions have been his reconceptualization of social change as occurring within totalities rather than national societies as units of analysis; his notion of "historical systems," totalities with both systemic (interactive whole-part relationships) and necessarily self-transforming features; his critique of developmentalism, the idea prevalent inside and outside the academy and in both mainstream and Marxist circles that each society can (and should) change in the direction of "modernity;" his discovery of the "semiperiphery," a structurally durable zone of world-systems that helps account for their stability and is one source of long-term change; and his four-volume work *The Modern World-System* (1974, 1980, 1989, 2011), which traces and analyzes the emergence, consolidation, globalization, and transformation of the capitalist world-economy from the sixteenth century into the nineteenth and twentieth centuries (a fifth volume is projected), and which challenges the long-standing social science dogma dating the crucial transition to capitalist modernity in the eighteenth-century industrial and Atlantic revolutions.

Beyond his scholarly publications, Wallerstein's impact on world-systems analysis has occurred in many arenas. He has coordinated and collaborated on multiple research and publication projects, including volumes published as part of the so-called "gang of four," with Samir Amin, Giovanni Arrighi, and Andre Gunder Frank (Amin et al 1982, 1990).

In 1977 he founded and for 28 years oversaw the journal *REVIEW*, one of two major outlets for article publication in the world-systems research community. With Terence Hopkins, he co-authored key works and established both the Fernand Braudel Center at Binghamton University and a doctoral program that became the most important training ground for subsequent cohorts of world-systems scholars and teachers. In 1976 he led the founding of the PEWS section of the American Sociological Association, which has held annual spring conferences since 1977, seeking thereby to include scholars from multiple disciplines kept apart by institutional rigidities in universities. Starting in 1999, he began to publish semi-monthly op-ed articles on the Center's website that receive hundreds of hits and have been widely printed in newspapers around the world, interpreting current political events through the lens of world-systems analysis. During his presidency of the International Sociological Association (1994–98) and through his chairing a Gulbenkian Foundation commission on reforming the social sciences, he vigorously advocated reorienting them toward macrohistorical "unidisciplinarity." Most recently he has been an active participant in the World Social Forum, the global "movement of movements" whose coming together he had anticipated when attempting to understand the long-term institutionalization of twentieth-century revolutionary movements as they were running out of system-transforming steam.

Wallerstein's intellectual formation depended crucially on his New York City upbringing and his formal education there at Columbia University (BA 1951, MA 1954, PhD 1959) in its heyday as a center of cosmopolitanism and critique. He continued as a young professor at Columbia until 1971, thus experiencing the city's rise to global centrality as well as the beginnings of its demise (see Goldfrank 2000; cf Bergesen 2000). Wallerstein's most important Columbia inspiration was probably C. Wright Mills, echoes of whose attacks on what he called "Grand Theory" and "Abstracted Empiricism" can be heard in Wallerstein's repeated efforts to overcome the social science divide between the nomothetic (universalizing theories) and the idiographic (individualizing description). Having led to an MA essay on McCarthyism, Wallerstein's formation then took great leaps forward when he shifted his primary scholarly focus to West Africa, particularly the francophone zones. His route there went via Paris, where Georges Balandier taught him how to study colonial societies as wholes, rather than focusing on tribes as units of analysis; where he came to know at first hand Third World radicalism, especially in the writings and person of Frantz Fanon; and where later he encountered Fernand Braudel and the works of the ANNALES school which were to prove critical in his understanding of capitalism. Wallerstein would eventually and for many years teach one semester at the University of Paris, one of fourteen foreign institutions to award him an honorary doctorate. New York, then Paris, then West Africa: it was in the Ivory Coast and Ghana, where he carried out his doctoral research, and then more broadly around the continent in preparation of his first major book (Wallerstein 1961), that he worked out an explanation for the rise of the indigenous socio-political forces that would lead to the demise of colonial societies. He would later come to understand that the "whole" of each colonial society was but a part of a larger totality with crucial causal consequences, a world context of US hegemony and Soviet challenge and of continuing asymmetrical interdependence between the first world and the third. It is worth emphasizing, however, that alone among the panoply of notable US macrosociologists who in the 1960s and 1970s revived the study of long-term historical change, Wallerstein benefited from an intellectual formation that included a deep immersion in imperialist realities as experienced from below.

Elements of four major intellectual traditions are woven together in Wallerstein's world-systems synthesis, all of them running counter to the hegemonic Anglo-American sociological emphases on universalist liberalism, scientistic positivism and unidirectional modernizationism. The older pair of these traditions, Marxism and German historical/institutional economics, date from the nineteenth century but saw major advances in the twentieth; the younger two, Annales historiography and the dependency school, were twentieth-century creations.

From Marxism comes much of the bedrock that lies beneath the conceptual edifice, at the most general level and in the interpretation of capitalism. Generally, socioeconomic totalities are the most important units of analysis; they are fundamentally conflict-ridden and historically changing; they have systemic yet contradictory properties. Specifically, accumulation and its attendant competitions and class struggles are central to capitalist structure and change, including the attempt to anticipate and encourage class groupings that may be the bearers of movement toward world socialism. Further, Wallerstein borrows from Lenin the importance of labor aristocracies and of inter-imperialist rivalry, and from Mao Ze-Dong the insistence that class struggles continue after soi-disant socialist revolutions. He often drew on the empirical work of European Marxist historians. But Wallerstein also broke from conventional Marxism on one hugely crucial point that it shares with the modernization paradigm when he rejected the idea that each national society goes through a series of stages.

He also borrowed heavily from the Germanic institutional tradition. The critical concepts here include, from Max Weber, modern interstate conflicts as the political framework of capitalism and (ethno-national) status-group struggles as central to all politics; from Joseph Schumpeter, the causal importance of economic cycles and the metaphor of creative destruction, including the long-term impermanence of capitalism itself; and perhaps most important, from Karl Polanyi, the model of three types of socioeconomic totality, reciprocal, redistributive and market, which Wallerstein renamed mini-systems, world-empires and world-economies. This tradition contributed conceptually, but Wallerstein rejected its characteristic and sometimes fatalistic acceptance of inequalities.

A third major source of conceptual inspiration was the ANNALES school, especially the work of its second-generation leader, Fernand Braudel, whose name graces the research center Wallerstein and Hopkins established at Binghamton. It was Braudel who centered his focus on long-term (enduring) deep historical structures rather than on either epiphenomenal and transitory events or on alleged universals such as "economic man." It was Braudel who invented the use of the term "world" in the technical sense of a socioeconomic space with multiple cultures and polities (e.g., "the Mediterranean"). It was the Annalistes who insisted on the importance of rural zones and their peasantries as the underpinning of modern economies, and Braudel especially who homed in on what he called "the long sixteenth century" as the critical transitional period. The Annalistes also provided a large share of the empirical materials which Wallerstein would rely on for documenting his theses in volume one of The Modern World-System (1974). But Wallerstein would emphasize the importance of politics and the state much more strongly than the typical Annaliste.

The final ingredient in the world-systems synthesis came from the newest of these traditions, what came to be known as "dependency theory" in the 1960s. This approach directly confronted both liberal (modernizationist) and Marxist developmentalism by positing different and unequal trajectories of growth and change in the imperialist centers on the one hand and in the dominated peripheries on the other. Where liberals called for technology transfer and foreign capital (and Marxists called for bourgeois revolutions) to jump-start "development" in the Third World, this school called for "delinking" from exploitative center-periphery relationships and for "self-centering" growth. Wallerstein agreed with the diagnosis but found the proposed cure extremely unlikely, perhaps persuaded in part by Frantz Fanon's trenchant critique of the typical neo-colonial governing elite in Third World countries. In any event, the two most extreme attempts at delinking, in Burma (now Myanmar) from the military on the right and in Cambodia from the Khmer Rouge on the left, proved disastrous failures and inspired no imitators. Wallerstein would also depart from his dependentista predecessors in focusing on dominated countries as integral parts of a single systemic whole, rather than as separate and separable units of analysis.

What are Wallerstein's totalities? Historically there have been three types—mini-systems, world-empires and world-economies—each bounded by the basic provision of daily necessities, including protection. Interactions between totalities such as luxury trade and political or cultural borrowing may have causal effects but these are defined as secondary to the slow transformative processes internal to each totality. (Wallerstein also proposes a fourth potential type, in a possible but by no means assured future: a socialist world government.) Empirically, over the course of human history, all mini-systems have been absorbed into world-empires or world-economies and by the twentieth century, all previously existing systems had been absorbed into the global capitalist world-economy, pace the claim of ideologues in the USSR to have created a separate and rival "socialist" world-system. The principal difference between a world-empire and a world-economy is that the former is ruled from a single political center while the latter involves multiple states competing for primacy or dominance but never attaining it for more than the short run. Pre-modern world-economies, such as that in the ancient Mediterranean, typically became world-empires, whereas the secret of both the durability and the dynamism of the capitalist world-economy has been its resistance to conquest by any single power despite occasional 30-to-50-year periods of one or another state's system-wide hegemony.

Although he elaborated a world-historical typology of wholes, Wallerstein has devoted almost all his intellectual efforts to studying the capitalist world-economy, with some attention to its possible future transitions. His tripartite division of the world-economy into core, semiperipheral and peripheral zones is probably his best-known contribution, with his and Hopkins's (1986, see also Gereffi and Korzeniewicz 1994) ancillary concept of commodity chains crossing zonal boundaries inspiring an impressive amount of research. At any given point in capitalist time, core zones feature relatively capital-intensive and monopolized production processes, relatively free and well-paid workers, and relatively strong states both internally and in international politics. Peripheral zones reverse this picture, in part serving as the "countryside" of the world exploited by the core. Monopolization of scientific and engineering institutions enables core zones to develop new products along with new armaments, tending to reinforce existing inequalities once they are established by shifting the location of competitive and hence less profitable processes to semiperipheral and peripheral zones. The history of textiles is exemplary in this regard: a core industry from the sixteenth to the nineteenth century, textile factories began migrating to the semiperiphery in the twentieth century and are now becoming important in the periphery as well. Meanwhile, the incorporation of new, mostly peripheral zones into the capitalist world-economy during its long centuries of expansion from its sixteenth-century origins in Europe and parts of the Americas has made it possible for originally semiperipheral zones like Sweden to rise to the core and originally peripheral ones like Brazil to rise to the semiperiphery.

Identifying the semiperiphery as having major enduring structural importance has proved to be among Wallerstein's most controversial contributions because it challenged all previously existing schools of macrosociological analysis. Modernizationists treated intermediate countries as "transitional" between tradition and modernity. Marxists treated them as semi-feudal or semi-capitalist or sub-imperialist. Dependency theorists treated them as blocked by technological reliance on the core, and the more Manichaean among them as if they were not significantly different from the periphery. In Wallerstein's view, semiperipheral parts of the modern world-system have existed from its origins in the sixteenth century and have continued to be reproduced and newly created ever since. Like the middle class in Aristotle's political sociology, the semiperiphery stabilizes the system. Politically, it often contains peripheral discontent while acting as client or proxy for one or more core states. Economically, it supplies low-skill, low-wage products to the core and/or intermediate-level products to the periphery. In times of profit squeeze in the core, it is a major site of industrial relocation and capital investment; in times of boom in the core, it is a major source of labor migration, both educated ("brain drain") and unskilled. Undeniably, the concept has many fuzzy aspects, for example, the geopolitical weight of large semiperipheral zones like Russia;

however, it has spawned a large volume of interesting scholarship and is likely to be regarded as an enduring idea as well as a non-transitory reality (see, e.g., Martin 1990).

On the political dimension of the modern world-system, Wallerstein took Max Weber's understanding of interstate competition as one of capitalism's necessary foundations and contributed his own conceptions of state strength and of hegemony. In general terms, state strength internationally derives from economic advantage, class alliances, military organization, and geopolitics, while internally it is less a matter of bureaucratic mass than of class compromises that lessen the need for heavy-handed administration. Peripheral states remain weak both because they are dominated from outside and because their ruling groups typically prefer minimal state capacity for interference in their freedom to sell on the world market. Meanwhile hegemony, for Wallerstein (1983), refers to a temporary situation in which one core state dominates and leads the world-system as a whole. In this view, there have been three moments of hegemony: under the Dutch in the mid-seventeenth century, under the British in the mid-nineteenth, and under the United States in the mid-twentieth. Hegemony is achieved through successive but overlapping superiority in productive efficiency, commercial success and financial superiority buttressed by effective military force. After a period of glory, it is lost in the same order, as competitors eat into profit margins, rival centers of accumulation emerge, and the costs of policing and stabilizing the world become overly burdensome. Amsterdam, then London and then New York each had its turn as the world-system's financial capital, symbolizing the transitory hegemonic prominence of their countries.

Wallerstein's structural concepts have proved his most widely recognized contributions to date, but his ideas about dynamics also deserve attention. Three seem especially important: the cyclical rhythms and secular trends of the system as a whole, and the process of incorporation of new zones from external arenas into the capitalist world-economy. Drawing on Schumpeter and on the Russian economist Kondratieff, Wallerstein argues that 50–60 year cycles with an A-phase of expansive growth and a B-phase of stagnation and crisis are constitutive of the accumulation process due to the anarchy of the market. A- and B-phases differentially affect differently located zones, with B-phases especially likely to lead to increased misery in much of the periphery and semiperiphery as well as to a few examples of dramatic ascent—which in its most recent incarnations (Taiwan, South Korea) is hailed as "development" and offered as a model for other countries to follow, as if that were possible. As with Marx and Schumpeter, recurrent crises are necessary parts of the capitalist process rather than failures of monetary or fiscal policy. Caused by overproduction given the scale of the market, crises are overcome when some combination of geographic expansion, technological innovation and downward income redistribution recreates the conditions for profitable investment in production by lowering costs and expanding markets.

In crisis supercession, we see cyclical processes and secular trends intersecting. Geographic expansion is one such trend. The modern world-system has expanded from its Euro-American origins—and with a boost from human pillage in West Africa—to encompass the entire globe in uneven waves of both colonial and trade-based incorporation. A second such trend is the ongoing commodification of land and labor, a key phenomenon highlighted by both Marx and Polanyi. Mechanization of production (increasing the ratio of constant to variable capital), bureaucratization of organizational structures (Weber), and degradation of the natural environment, now entailing ever-higher input costs, are additional trends. Wallerstein's important contributions here are first, that these trends transform the world-system even as they (temporarily) stabilize it, and second, that each of them clearly entails a ceiling, an asymptote that can be approached but never fully reached. What characterizes modern capitalism above all is not so much its tripartite structure as the process of moving toward these asymptotes. Some of these limits have already been reached, or almost, to such an extent that Wallerstein believes the current world-system to be arriving at its final crisis—a crisis "of" the system rather than merely a crisis "in" the system.

A third dynamic contribution worth highlighting is Wallerstein's (1989, Ch. 3) relatively neglected analysis of the process by which new zones become absorbed into—rather than merely trading in luxuries or interacting at a military frontier with—the capitalist world-economy. In a stunning comparative analysis, Wallerstein shows essential similarities in the ways in which in the century after 1750, four hugely different cultural areas—Russia, India, West Africa and the Ottoman Empire—went from being part of the external arena to becoming peripheral zones responsive to and dependent upon world market forces. In all four cases there were three moments: external arena, then incorporation, then peripheralization. In all four cases, new products came to be produced and exported and new imports accepted, with local manufactures declining; the size of enterprises grew; and the coercion of the labor force increased significantly. In all cases, though with more variation due to vastly different initial conditions, states were strengthened internally and weakened externally. Perhaps not incidentally, during the very period when the Atlantic revolutions and British reforms advanced freedom and participatory citizenship in the core, new levels of oppression were being enforced in the periphery.

It is perhaps too soon to appraise two further Wallersteinian contributions to world-systems analysis: his borrowing of Prigogine's chaos theory to argue that the current (or impending) systemic crisis will lead to increasingly wild economic and political fluctuations; and his calls for *Unthinking Social Science* (2001) to reformulate and reshape the institutions and structures of knowledge. Wallerstein's futuristics mimic the environmental predictions of many climate scientists, although they were articulated well before such predictions became current. Some world-systems adherents accept them, even act on them in their political participation, but others find them too apocalyptic or too telescoped, arguing in effect that the asymptotic limits of capitalism have yet to be approached so closely as to presage an imminent age of chaos, turbulence and system transformation. As for the issues raised in Wallerstein's analyses and programmatic recommendations for reorganizing the structures of social scientific knowledge, it is clear that we are witnessing an efflorescence of boundary crossings among academic disciplines and even between the broader areas of natural and social sciences and the humanities. But it seems highly unlikely that these intellectual developments will lead to reduced specialization and compartmentalization, or to movement in the direction of a "unidisciplinary" social science. Possibly such academic and intellectual reorganization will emerge along with other new structures in a post-modern world-system. It is difficult, however, to discern in present power configurations the interests that would be served by such changes.

Polyvocal from the start, the world-systems school would have one most prominent voice for most of the 1970s and 1980s: that of Immanuel Wallerstein. Then, as its concurrent founders began to further develop and differentiate their individual contributions, and as newer generations of scholars added theirs, this polyvocality became more pronounced than before, perhaps even to the point where Wallerstein's later work has been too little pondered or drawn upon. From his roots in New York, capital of the world in the period of his intellectual formation to his contributions at the World Social Forums of the twenty-first century, Wallerstein has been a trailblazing pioneer, a prolific scholar, a mentor, a mensch.

References

Amin, S., G. Arrighi, A. G. Frank, and I. Wallerstein (1982) *Dynamics of Global Crisis*, New York: Monthly Review Press.

Amin, S., G. Arrighi, A. G. Frank, and I. Wallerstein (1990) *Transforming the Revolution*, New York: Monthly Review Press.

Bergesen, A. (2000) "The Columbia Social Essayists", *Journal of World-Systems Research*, VI, 2: 198–213.

Gereffi, G. and M. Korzeniewicz, (eds) (1994) *Commodity Chains and Global Capitalism*, New York: Greenwood Press.

Goldfrank, W. L. (2000) "Paradigm Regained? The Rules of Wallerstein's World-System Method", *Journal of World-Systems Research*, VI, 2: 150–95.

Hopkins, T. K. and I. Wallerstein (1986) "Commodity Chains in the World-Economy Prior to 1800", *Review*, 10, 1: 157–70.

Martin, W. G., (ed.) (1990) *Semiperipheral States in the World Economy*, New York: Greenwood Press.

Wallerstein, I. (1961) *Africa: The Politics of Independence*, New York: Vintage Books.

Wallerstein, I. (1974, 1980, 1989, 2011) *The Modern World-System*, Vols I–IV, New York: Academic Press; Vol. IV, Berkeley: University of California Press.

Wallerstein, I. (1983) "The Three Instances of Hegemony in the Capitalist World-Economy", *International Journal of Comparative Sociology*, 24: 100–108.

Wallerstein, I. (2001) *Unthinking Social Science: The Limits of Nineteenth-century Paradigms*, 2nd edn., Philadelphia, PA: Temple University Press.

3.3

The structures of knowledge

Conceptualizing the sociocultural arena of historical capitalism

Richard E. Lee

Scholars of the modern world-system acknowledge the *longue durée* of historical capitalism as a unique and unitary "world" constituting their unit of analysis. However, as Fernand Braudel noted a half-century ago, "mental frameworks are also prisons of the *longue durée*" (2009: 179), thus recognizing that the human experience is one of phenomenological wholeness which does not end with questions of politics and economics. It is this sociocultural domain of historical capitalism (Braudel's "mental frameworks") that the structures of knowledge approach conceptualizes from a world-systems perspective.

Both analysts and entrepreneurs have generally supposed knowledge to be a vital aspect of the processes through which capital is accumulated. Nonetheless, over most of the past half-century—not incidentally, the period of Kondratieff downturn and the crisis of US hegemony—accounts in both the scholarly literature and the popular press of the "information society" and the "knowledge society" have exploded and the lexicon has expanded to include the likes of "knowledge economy," "knowledge capitalism," and "cognitive capitalism." Now, developments in world-systems analysis allow us to argue that the transformation in this area is deeper and much more fundamental and consequential than most analysts today, or the frameworks from which they work, might imply; indeed, it is associated with the transformation of the system as a whole (see Hopkins and Wallerstein 1996).

Over the past three decades, significant progress has been made in developing modes of integrating the analytically distinct arenas of production and distribution, the economic, and decision making and coercion, the political, in terms of long-term processes and their medium-term fluctuations. Their great overarching structures, the core-periphery axial division of labor and the interstate system, were the first to be widely recognized and examined in depth in reference to a formal conceptualization of their internal relations and the ways in which they related to one another to assure the reproduction of the modern world-system as a whole. However, there was also "a third fundamental aspect to the modern world-system … the broadly 'cultural' aspect … even though little is systematically known about it as an integral aspect of world-historical development … [and] much preliminary conceptual work needs to be done" (Hopkins et al 1982: 43).

The term "systematically" here is crucial. As it has been argued that the modern world-system is constituted inseparably of its regularities and the constant change it exhibits, it became clear that for the structures of knowledge approach, the first step would have to be an elaboration of a conceptual framework analogous to that which had been conceived for the economic and political spheres, particularly the specification of the primary *longue durée* structure or structures of this "third arena"—of cognition and intentionality—and the cyclical rhythms and secular trends of their processes of reproduction that could be recognized over the entire life of the system. The structures of knowledge approach is the outcome of this work (for details and further elaborations see Lee 2003, 2010; Lee and Wallerstein 2000; Wallerstein 1991, 2004).

For centuries, knowledge in the Western world was relegated to either the earthbound or the heavenly. Although each was constituted in different ways, each afforded access to both what was true, or not, and what was good, or not. The late medieval period experienced the beginning of a shift toward measurement in terms of equal quanta and calculations exempt of human values (e.g., double-entry bookkeeping), and away from spiritual connotations of time and space. This was all to the profit of merchants and bankers, artisans and eventually manufacturers, sailors and military men, artists, and musicians, that is anyone who had to actually accomplish something in the real world. As the standing of the commercial classes rose, the authority of the medieval system declined and a radical epistemological divorce of truth (facts) from the good (values) restructuring the acceptable grounding of knowledge claims, unknown at any other time or place in the world, came into being along with the axial division of labor and the interstate system.

The secular trend of the processes reproducing this basic structure, "rationalization," might at times also be called "scientization" or "secularization." The pursuit of objectivity, or the view from nowhere that sidestepped agency, history and subjectivity, progressively privileged formal rationality or disinterested calculation as a generalized means of instrumental action over substantive rationality, the normative pursuit of specifically situated ends. The organization of knowledge into two separate domains, one in which human values are an inseparable component and one excluding human values *a priori*, took the form of the division between the humanities and (what would come to be known as) the sciences. This arrangement was eventually characterized as the "Two Cultures" (Snow 1965) and its internally contradictory nature has been manifested in recurring dilemmas admitting at best medium-term solutions.

The set of intellectual and institutional structures of knowledge defining the geoculture of historical capitalism has been periodically reconstituted through a series of medium-term fluctuations we shall call "logistics." Akin to waves of economic expansion and contraction and cycles of geopolitical hegemony, they express the realignments that have assured the reproduction over the long-term of the basic structure, the intellectual separation of facts from values institutionalized in the sciences and the humanities, that has constantly been under pressure from the material contradictions endogenous to the system as a whole.

The use of force to assure accumulation is expensive. If the interstate system organizes the application of force to assure acculturation, the construction and development of the structures of knowledge put in place will reproduce the (always contested) bases that applied to the much cheaper mechanisms based on consensus that have operated in and through successive medium-term restructurings. That is to say, they will focus on what is possible and legitimate to think (and therefore to do) in a given society and who are the appropriately authoritative agents for which kinds of decisions about which kinds of actions in that society. The image of these repeating patterns of development, the logistics, may be thought of as an elongated S-curve that exhibits slow initial growth, then builds rapidly, and finally flattens out as it approaches an asymptote or limit (rather than a wave-like rise and fall).

The first logistic corresponds to the period of long-term inflation and deflation running generally from the emergence of historical capitalism through the first third of the eighteenth century. The Thirty Years' War, the Westphalian solution, and the establishment of Dutch hegemony mark the middle of the period.

The tenuous movement in the direction of secularization and the incipient separation of facts from values, the true from the good, become increasingly apparent. The synthesis of Baconian induction, Cartesian reductionism and deduction, and a quantified real world manifested in the work of Galileo brought forth a mechanistic model. The model, symbolically and substantively manifested in the work of Isaac Newton, was also interventionist. Bacon noted that "the true ends of knowledge [are] for the use and benefit of life" (1620: 15). He was echoed by Descartes: a "knowledge that is very useful in life and that in place of the speculative philosophy taught in the Schools ... make ourselves, as it were, masters and possessors of nature" (1637: 33). In this first great shift towards secularization, God as final cause or internal governing principle (rather than external goal) is removed from the dominant vision of order in nature and the creative potential of a supreme being is transferred to John Locke's rational individual and sovereign majority.

The second logistic runs from the mid-eighteenth century through the late nineteenth century. It too is marked midway by a thirty-years-long world war and the reestablishment of a state of hegemony in the interstate system, with Great Britain as the leading power. The medium-term resolution of the first logistic was confirmed and consolidated, and sealed when Pierre Simon de Laplace "took the final and crucial step towards an ideology of total determinism" (Hahn 1967: 18), raising "science" to the epitome of authoritative knowledge.

Built on the model of celestial mechanics, classical science held that observable effects were physically determined and the discovery of universal laws would lead to prediction of both future and past—since reversibility was a feature of the mathematics of the laws of motion. Furthermore, when gravity, the motor of motion, was located in matter itself, that quintessential attribute of the Supreme Being—creativity—was bestowed on man. The humanities, at the opposite pole of the structures of knowledge, were concerned with the finitude of the unique and unpredictable rather than the certitude of regularities but they also had to account for change, including emergence. Individual agency was the solution they proffered and imagination served as the connecting link between mind and world.

This period was further characterized by the extension to all of the Enlightenment ideal of rationality. In terms of practice, resistance to coercive labor control made by the oppressed and exploited of the world in the name of freedom became increasingly insistent and effective. One of the legacies of the French Revolution was the idea that the social world was one of change, not of stasis. But if change was the norm and all humanity was capable of rational thought and behavior, the question then became the direction of change, and how to deal with this question in the real world of social upheaval depended on mutually exclusive value orientations. The alternatives available in the early nineteenth century, which made mutually exclusive appeals to values, were either the worldwide extrapolation of the ideals of *liberté*, *égalité*, and *fraternité*, the Rights of Man, Free Trade and Democracy or a return to the *ancien régime* of authenticity, tradition and the organic community, and opposition to laissez-faire liberalism. Neither offered a way forward on which any consensus could be reached to the political confrontations between radicalism and conservatism, or to the threat to capital accumulation of both. From the late nineteenth century, this structural antinomy, the relation between meaning or values and systematic knowledge, became a subject of vigorous debate in both philosophy and economics; eventually, the objective, value-neutral, problem-solving spirit of science was advanced to resolve questions in the domain of human action.

The universal laws that characterize the natural sciences depend on a Newtonian or reversible time associated with absolute determinism and therefore the certainty of predictions. The Human Studies, by contrast, are eminently historical, not time-reversible, and therefore prediction of the future is uncertain. However, the scientific study of society was what many put forward as a framework for a separate, or third, disciplinary domain between the sciences and the humanities.

If one could uncover the laws governing change in this domain (that is, of human action or the social), or complementarily, delineate what could be considered essential features of different human sub-groups (that is, facts of nature) and therefore hardly amenable to change, one could predict within some reasonable and diminishing margin of error and thus control the future without depending on decisions grounded in the relativity of human values. Herein lies the functional link between the social sciences and reformist liberalism. The institutionalization of a set of disciplines to constitute this domain would function to guarantee ordered change—the predictable impact of policy interventions—in the name of (some putatively value-neutral) "progress." "Scientific" control would be exercised by "experts" and based on "hard facts." In political and economic terms, this amounted to liberal incrementalism maximizing accumulation by guaranteeing, that is legitimating, wealth and wage hierarchies based on differentiated status group attributes and minimizing class struggle through the promise of progress from generation to generation.

Despite efforts to preserve the role of values in the production of authoritative knowledge, the eventual consequence of the late nineteenth-century debates determined the intellectual and institutional arrangements for the subsequent construction of knowledge in the social sphere, that is, the grounding by epistemological default of a scientistic third culture, the social sciences, based on the "scientific" criteria of value-neutrality and a neutral or absolute time. The universalist assumptions of the social sciences (either additive or law-given) and their sectorialization (the division of the social world into separate fields treated as private reserves by the separate disciplines) tended to have the political consequence of obscuring underlying organizational arrangements and historical feedback mechanisms, and thereby made organization to produce social change much more difficult.

Thus, the third logistic begins in the late nineteenth century and although exhibiting crisis, does not seem to be exhausted. It is also marked by a thirty-years-long world war that resulted in a new state of hegemony in the system as a whole. The political tensions of the second logistic were resolved in the medium-term with the new liberal consensus of popular sovereignty and public welfare extrapolated worldwide as Wilsonian "self-determination of nations" and Rooseveltian "economic development," the structural equivalents of national-level universal suffrage and the welfare state (Wallerstein 1995: 137). These innovations were supported by the consolidation and institutionalization of the social sciences in disciplines and university departments separated on the basis of what were deemed proprietary theories, methods and subject matters.

The two cultures conception of knowledge, the sciences and the humanities with the social sciences poised precariously in between, was at its apogee during the two decades following the end of the war in 1945. Its dominance was never total, however, and a number of approaches were developed from multiple disciplinary vantage points (e.g., General System Theory, the non-fiction novel) that contested the prevailing model in terms of theoretical frameworks, or methodological practices, or even proprietary subject matters, and sometimes all three. By the end of the 1960s, however, challenges to the liberal order (such as the Vietnam War, the civil rights, feminist and student movements, and the active engagement of third-world scholars and activists) grounded in practice the direct challenge across the disciplines to the dominant two cultures structure that would be embodied in cultural studies, science studies and complexity studies. Moreover, this was not to be simply the beginning of a medium-term, conjunctural, adjustment (a fourth logistic) but a secular crisis and part of the transformation of the modern world-system as a whole.

Cultural studies recognized the inadequacies of received categories of analysis, stressed relationality, decentered and destabilized the naturalized, taken-for-granted separation of the humanities and the social sciences and the divisions among the social sciences, and emphasized values and interpretation in social analysis. What makes this of particular importance is that these attributes collectively undermine the bases of the disciplinary structure defining and organizing what counts as legitimate and authoritative knowledge in the modern world that have come under scrutiny from developments in the social sciences and the sciences as well.

Sciences studies (including "sociology of scientific knowledge," "social studies of science," and "science and technology studies") with bases in the social sciences has taken the development of science itself as an object of analysis. It includes a critique of the dualism of "humans" and "nature" and a close examination of the contingency of scientific knowledge, its (social) constructedness, and its local situatedness. Collateral developments in the social sciences, which attacked the "scientific" underpinnings of the social sciences from within, embrace new disciplinary and departmental groupings that contest the fact-values divide and illustrate how essentialist categories of difference have underwritten the subordination of entire groups; a formal conceptualization of this latter problem is one of the achievements of the structures of knowledge approach. Strands of "feminism and gender studies" have pointed to the constructedness of the categories of "man" and "woman" and have noted how scientific discourse of the female body has functioned to situate women in society. Scholars of "race and ethnicity" in the West have attacked essentialism as well while disputing Western universalism and objectivity, and many studying "non-Western societies" have highlighted alternatives to the Western development model and pointed to the implications of these alternatives for epistemology.

The importance of the above developments is vastly accentuated by what has been happening in the sciences. At the very moment of the worldwide triumph of the Newtonian worldview—a deterministic world of natural laws based on time-reversible dynamics—a new knowledge movement that would challenge its premises directly, and thus the structures of knowledge as a whole, began to emerge. This movement, which eventually came to be loosely known as complexity studies, grew out of concrete research into complex phenomena in nature arising from very simple mechanisms (especially at the humanly perceivable, macro level). It represents a synthetic approach as opposed to a reductionist one in a shift away from a framework emphasizing equilibrium and certainty and defining causality as the consistent association of antecedent conditions and subsequent events; in a word predictability, amenable to experimental replication and hypothesis testing. Work especially in what is known as "chaos," or seemingly random behavior that displays an underlying order—order-in-chaos (strange attractors); order-out-of-chaos (dissipative structures, self-organization); and visual representation of pathological functions and natural forms exhibiting noninteger dimensions (fractal geometry)—either calls for a reappraisal of the assumptions of classical science or actively undertakes a reconceptualization of the objects of study, methods of analysis and goals of inquiry long taken for granted as "scientific" practice.

Implicit in complexity research is a new sense of the meaning of time. In the light of instability and chaos, and the association of the arrow of time with order as well as disorder, Ilya Prigogine has maintained that the laws of nature now express possibilities instead of certainties. Far from being a measure of our ignorance, entropy expresses a fundamental property of the physical world, the existence of a broken time symmetry leading to a distinction between past and future that is both a universal property of the nature we observe and a prerequisite for the existence of life and consciousness. The recognition that probability is more fundamental than trajectories implies what Prigogine calls "the end of certainty," and the end of certainty in scientific prediction connotes an open future of creativity and choice in natural systems and, as a consequence, a vindication of freedom, agency and creativity in the development of social systems, and their study. These reflections take on particular significance in times of crisis when systems have moved "far-from-equilibrium" and are experiencing the instabilities inherent in transitions (see Prigogine

1996). The emphasis in complexity studies—on contingency, context-dependency, multiple, overlapping temporal and spatial frameworks, and deterministic but unpredictable systems whose development displays an arrow of time—suggests, as some scientists are beginning to say, that the natural world as they now see it is beginning to look unstable, complicated and self-organizing, a world whose present is rooted in its past but whose development is unpredictable and cannot be reversed. In short, it is beginning to more closely resemble the social world.

Thus, at the lower end of the hierarchy of the structures of knowledge, cultural studies attests to the disintegration of the boundary between the humanities and the social sciences. At the upper end, the concurrent developments in complexity studies bear striking resemblance to the concerns of those working on questions in the human world, and "objectivity" associated with externalism is called seriously into question by the identification and study of the feedback mechanisms of complex systems, including historical social systems. In this crisis of the two cultures, the ontology itself underpinning the legitimacy and authority of knowledge constructed on the "scientific" model, which is at the foundation of the hierarchy of the structures of knowledge as we have known them, is undergoing a transformation: the vision of the natural world as composed of independent, interacting units—billiard balls or nation-states or individuals suitable for comparison—is giving way to an emerging sense of the world as made up of fundamentally deterministic but unpredictable systems.

Nonetheless, as the epistemological groundings of disciplinary diversity fade, the subject matters we now think of as composing the humanities and the social sciences that "discipline" our thinking and actions will not disappear, nor will those that we think of today as belonging to the natural sciences disappear either. Their future must be that of producing non-contradictory knowledges, for it is the overarching structure itself that is changing and thereby rendering the segregation of subject areas according to their contradictory epistemological premises increasingly problematic. Participating equally in the production and reproduction of the human condition, such fields as music and literature, chemistry and astrophysics will cease to be cultivated in worlds apart. The intellectual sanctions and practical justifications for independent disciplines in the social sciences, where epistemological ambiguities were never put to rest, are collapsing too, with much of the best work impossible to categorize as belonging uniquely to a single discipline.

In the construction of systematic knowledge of human reality, we are thus presented with arguments for the reunification of "is" (the realm of facts, or necessity—the goal of science) and "ought" (the field of values, chance or now more properly possibility—the challenge of the humanities). This will be the hallmark of a social science for our times, a historical social science, necessarily singular and transcending disciplinary boundaries. The structures of knowledge approach suggests that it will have to do two things. First, this historical social science must be premised on the unity of the regularities of social relations, their structure and change, their history. Therefore, process becomes the primary methodological concern. Second, with obvious impact on the debates over Eurocentricism and what have always seemed to be the cross-cutting effects of such status categories as race, gender and class, the unity of structure and change supposes the integration of values as integral to inquiry, not simply as a matter of the personal inclination of the analyst.

The most promising model is that of research that results in, or at least implies, the individuation of possible futures. It combines the human quest to connect with the reality of the world with the realization that we are all moral agents and obliged to work politically to favor the making of a better world as we see it. It is what Immanuel Wallerstein has called "utopistics" (Wallerstein 1998). Social analysts may now make the shift from constructing and verifying theories to imagining and evaluating the multiple possible consequences of diverse interpretative accounts of human reality and the different courses of action they suggest. This then is the barest outline for a unified historical social science as an alternative to the predictive, Newtonian model of social scientific inquiry. Neither nomothetic nor idiographic, historical social science recognizes both those changing features and changeless elements of our world, a world limited nonetheless by a beginning and an end, and the role of values as the key to our differing conceptions of the worthiness of various possible futures and thus of what actions are to be taken—and therefore to the creation of authoritative but not of necessity universal knowledge of human reality for our times.

For instance, one of the most important concrete developments of the many already underway today (and which will have enormous impact in the years to come) is the way the divide between the scholarly and non-scholarly is now facing challenge (indeed, the divide between authority figures and those over which they hold sway). Academics and laymen alike, we have all been impacted by the advances in information technology and communications. From its inception, the modern world-system has been characterized by innovations in communications that have made information more easily disseminated and thus ever more widely and cheaply available. The technological innovations that set this trend in motion did contribute to the conjuncture from which emerged the capitalist world-economy, and today the so widely-hailed developments of the past two decades are both the culmination of a long-term trend and part of the conjuncture that will eventually give rise to a successor system.

One result of this process as it is playing itself out is a state of affairs that is incompatible with the reproduction of the structures of knowledge as we have experienced them over the past five centuries. Indeed, contemporary developments extending the long-term trend are actively dismantling those structures and reorienting the processes of their reproduction. As a result of advances in search engine technology, everything on the internet is equidistant from every other thing in cyber space, as well as from every single individual user. Every classification criteria too, then, is equal to all others; concomitantly, the scholar no longer inhabits a privileged space in the world of knowledge. No literature can remain proprietary; academics and non-academics alike can and now do access literature, and intervene directly in debates without regard for scholarly discipline or institutional status. The disciplines themselves, indeed all institutions and the "status" they enjoy, are thus deprived of much of their gate-keeping function and thereby destabilized. Just as new cultural communities and political constituencies are being created around issues previously segregated in non-communicating areas of knowledge, the consequences of rethinking the opposition between ideas and action are already having an impact on political practice as well as in an expanded conception of market interactions.

In conclusion, for those of us whose intellectual questions concern historical capitalism, giving equal analytic consideration to the construction and reproduction of the two-cultures divide on the same terms as the processes of the axial division of labor and the interstate system integrates the "cultural" realm into analyses of long-term, large-scale social change from the world-systems perspective. This is in line with the injunction that to understand change we must first have a clear idea of that which has remained the same. Over the *longue durée* the hierarchy of knowledge forms, the intellectual and institutional structures of knowledge, separated an epistemology that excluded values from truth claims from one which explicitly included human values in understanding the world. The structures of knowledge stipulated the imaginative horizon of the thinkable and unthinkable, and thus validated and legitimated what might be done and by whom, in terms of intentional action, at any time and place.

Today, however, developments across the super disciplines of the sciences, social sciences and the humanities have undermined the role of the sciences as guarantor of "truth" (even the conception of truth-values governed by the law of the excluded middle), and thus have destabilized the hierarchy of the structures of knowledge, undercut the pretensions to "scientific" authority in the social sciences, and had the effect of devaluing supposedly "scientific" considerations in long standing debates involving such antinomies as structure versus agency, determinism versus freedom, and order versus chaos in the analysis of social reality. The result has been to lay bare the mechanisms of the "naturalization" of status categories, accomplished by giving them "scientific" legitimacy and cultural authority, that ground the underlying consensus associated with the development and deployment of strategies of exploitation and domination, and resistance to them, as a long-term process.

Thus, a historical social science that includes creativity and choice and is no longer at crosscurrents with a holistic experience of social relations is already under construction. The classical considerations of "who, what, when, where, why" and the "view from nowhere" are shifting to the questions of "for whom, for what, for when, for where, from whose point-of-view." Nonetheless, since the final outcome of systemic transformations is unpredictable, the struggle for the future will call for committed, purposeful action in favor of what we consider the most desirable of the possible alternatives. Although from the perspective of the system as a whole, normatively motivated micro-choices and struggles appear quite random, from the point of view of the individual actor they are highly rational and may be focused on specific system outcomes. This is not so much the simple return of agency, but the manifestation of the fundamental relationship between agency and structure. During this period of instability, in which the future is an open future, such small fluctuations—interpretative work meaningful for our times, including that from scholars without institutional affiliations, for instance—will be capable of massive amplification, to the point that eventually some set of irreversible choices will "determine" the emergence of a qualitatively different social system.

References

Bacon, F. (1620) *Novum Organum*; translated and edited by Peter Urbach and John Gibson (1994), Chicago, IL: Open Court.

Braudel, F. (2009) "History and the Social Sciences: *The longue durée*," *Review*, XXXII, 2: 171–203.

Descartes, R. (1637) *Discourse on Method;* trans. Donald A. Cress (1980) *Discourse on Method and Meditations on First Philosophy*, Indianapolis, IN: Hackett.

Hahn, R. (1967) *Laplace as a Newtonian Scientist*, Los Angeles, CA: Clark Memorial Library, UCLA.

Hopkins, T. K. and Wallerstein, I. (1996) *The Age of Transition: Trajectory of the World-System 1945–2025*, London: Zed Books.

Hopkins, T. K., Wallerstein, I. and Associates. (1982) *World-Systems Analysis: Theory and Methodology*, Beverly Hills, CA: Sage.

Lee, R. E. (2003) *Life and Times of Cultural Studies: The Politics and Transformation of the Structures of Knowledge*, Durham, NC: Duke University Press.

Lee, R. E. (2010) *Knowledge Matters: The Structures of Knowledge and the Crisis of the Modern World-System*, St. Lucia: Queensland University Press.

Lee, R. E. and Wallerstein, I. (2000) "Structures of Knowledge," pp. 227–35 in Judith Blau (ed.) *The Blackwell Companion to Sociology*, Cambridge, MA: Blackwell.

Prigogine, I. (1996) *The End of Certainty: Time, Chaos, and the New Laws of Nature*, New York: Free Press.

Snow, C. P. (1965) *The Two Cultures and A Second Look*, New York: Cambridge University Press.

Wallerstein, I. (1991) *Unthinking Social Science: The Limits of Nineteenth-Century Paradigms*, Cambridge: Polity Press.

Wallerstein, I. (1995) *After Liberalism*, New York: New Press.

Wallerstein, I. (1998) *Utopistics: or Historical Choices of the Twenty-First Century*, New York: New Press.

Wallerstein, I. (2004) *The Uncertainties of Knowledge*, Philadelphia, PA: Temple University Press.

3.4

The multiplicity of national development in the world-system

A critical perspective

Nobuyuki Yamada

Introduction

The world-system theory is characterized by a holistic perspective. Such a perspective has many cognitive benefits, but it also has some drawbacks. One of these drawbacks is that the world-system theory tends to overlook the specific dynamism of various units that comprise the system itself. For example, the modern world-system, that is, the capitalist world-economy, is thought to have arisen in the long sixteenth century and to have extended to literally the whole world in the nineteenth century.

Many of the world's regions and societies have formed their social and cultural uniquenesses in their territories before being incorporated into the expanding modern world-system. The analyses of national development by world-systems scholars often ignores these specificities. This chapter tries to remedy this drawback by forging a theoretical model to grasp the dynamism of specificities in national developments in the world-system. In other words, it attempts to theorize the multiplicity of national capitalist development in the world-system. For that purpose, this chapter takes note of specific relationships that existed before each society was consolidated into the capitalist world-economy. Given that most societies were still *pre-capitalist* before being integrated into the world-system, such specificities can be found in pre-capitalist social relationships. Therefore, this chapter employs the concept of the articulation of modes of production in order to revise the world-system theory. However, because the concept of articulation of modes of production is somewhat static, it needs to be reframed as a dynamic concept in order to grasp capitalist development. Therefore, this chapter considers the articulation of pre-capitalist and capitalist modes of production as representative of the transition from pre-capitalist to capitalist society. This chapter also considers the case of Japan, which experienced a specific type of transition, and analyzes the particularities of Japanese capitalist development in the world-system on the basis of this new theory, focusing on the uniqueness of the globalization of capital and labor in Japan.

The holistic perspective in world-systems analysis

There is no doubt that the holistic perspective in world-system analysis has brought many cognitive benefits in sociological research. These are not limited to the analysis of globalization, one of the most prevalent contemporary societal trends, which literally requires holistic analysis. Wallerstein (1974, 1980, 1989) has also attempted to solve some difficult questions in historical capitalism.

For example, taking the holistic view, he tried to explain why the French Revolution and the Industrial Revolution just happened at that particular time in those countries (Wallerstein 1989). His answer was that these "revolutions" could not be understood from the national or local point of view but they should be grasped from the global perspective. He asserted that all the causes and effects of these revolutions, if they really happened, should be regarded as consequences of the operation of the world-system. This holistic perspective was certainly a great contribution in that it suggested a new interpretation, given that the various existing interpretations did not wholly accord with reality.

However, the holistic perspective of world-system analysis has some drawbacks, because it does not fully consider the specificities of various units of the world-system in the analysis of national or local social change. For example, although they are all located in the same zone of semiperiphery, so-called NIEs (Newly Industrialized Economies) in East Asia and Latin America differed greatly from each other in economic growth from the 1970s until at least the 1990s. The explanation for this should be explored through endogenous factors in those countries. While the formation of the developmental state can affect development in these societies, the extent to which the developmental state can be constructed is likely to depend on the history and traditions of the society. For example, Evans (1995) indicated such a point in the comparison of South Korea, Brazil and India.

Furthermore, it is questionable whether or not the uniqueness of national capitalist development in Japan can be fully explained by a holistic world-system analysis. Aside from contingent factors such as geopolitical position, the reason that national capitalist development was possible is not clarified in world-system analysis. In short, Japan's success would be regarded as exceptional from the holistic perspective. It could only be explained as a *conjunction* in the structural persistence or *long dureé* of the world-system.

So how can world-system analysis overcome such drawbacks? For one thing, a more holistic orientation can be pursued temporally and spatially in order to deal with such problems. For example, Chase-Dunn and Hall (1997) explored the modes of operation of *pre-modern* world-systems or world-empires, and their mutual influences. Frank and Gills (1993) also expanded the historical span and territorial area of the *world-system* and insisted that the inception of the *world-system* dates back five thousand years and that it has continued as a single system until the present.

Furthermore, Frank (1998) aimed to relativize the predominance of Europe and criticize the world-system analysis of Euro-centrism. Frank also asserted that among other regions Asia had been dominant for many centuries and, after a brief decline, rose again in the twenty-first century.

These approaches certainly have the potential to solve these problems. However, given that so many data are required for such large processes and so long a period, persuasive explanations are difficult to make at this time.

In addition, a more endogenous approach could be taken to world-system analysis. In other words, with regard to national development, the societal specificities which date back to the period before the incorporation of each society into the world-system, need to be examined. Given that the modern world-system is the capitalist world-economy and, therefore, national development is also capitalist, specific factors that have influenced national development should be taken into account when considering the formation of capitalism in a given society. Of course in the periphery, capitalism would have been *implanted* through the incorporation process rather than forming endogenously.

The recognition of capitalism in world-system analysis

So how should the formation of capitalism be understood? Needless to say, in world-system analysis, the rise of the European world-economy is synchronized with the formation of capitalism, which has operated globally since its inception. Furthermore, Wallerstein (1979) has asserted that the entire world has been a single capitalist system since the nineteenth century. Since the 1970s this view of capitalism in world-system analysis has given rise to fierce controversy.

In those days, *socialist* countries utilized methods of production different from those of capitalism. According to Wallerstein (1979), even in socialist countries, the valorization and accumulation of capital should be sought in production and exchange and the socialist countries are understood to have been parts of the larger capitalist world-system.

Nonetheless, there was a wide range of subsistence economies in the periphery in the 1970s. In such economies, it seems certain that peasants still had claims on the means of production and did not pursue expansive reproduction. In this sense, social relationships, which cannot be fully connected by the movement of capital, exist mostly in developing countries or in the periphery.

In this way, by distinguishing between capitalism and non-capitalism, less distinct world-system analysis lacks sensitivity to capitalism and so to non-capitalism. Capitalism in world-system analysis refers to labor control, households and the like, and so not only to exchange or markets but also to production and reproduction. One of the early criticisms of world-system analysis was that it only handles exchange or market relations rather than class or production relations and is therefore circulationist (Brenner 1977). This criticism is incorrect.

However, such an analysis could not adequately grasp how capitalism is constructed, and operates in relation to other non-capitalist relationships. Above all, this article is concerned with the influence of *pre-capitalist* relationships on the formation of national capitalism.

The transition to capitalism and the articulation of modes of production

How, then, can pre-capitalist social relations, from which the specificities of a society can emerge, affect national capitalist development? The formation of capitalism has conventionally been addressed in terms of the transition to capitalism. For example, on the one hand, Maurice Dobb (1946), Robert Brenner (1985) and other theorists emphasized endogenous factors such as class struggle in the transition from feudalism to capitalism—a discussion of proto-industrialization might also emphasize the endogenous origin of the transition. On the other hand, theorists and historians such as Paul Sweezy (Sweezy et al 1976) emphasized exogenous factors, such as the effects of exchange between distant places.

Furthermore, taking into consideration the situation of peripheral regions in the world-system, this controversy acquired a new dimension—the concept of the articulation of modes of production (Wolpe 1980). At first, the articulation of modes of pre-capitalist and capitalist production was thought to be a stagnant aspect of underdevelopment, primarily in dependency theory (Amin 1973). However, it should be regarded as part of the transition to capitalism in the society and the dissolution process of pre-capitalist modes of production—generally speaking, such pre-capitalist societies are not necessarily feudal and, particularly in the periphery, various types of pre-capitalist societies are plausible.

In addition, if "mode of production" can be defined as the sum of social relationships in various areas such as production, exchange, distribution, and consumption at the most abstract level of analysis, the articulation of modes of pre-capitalist and capitalist production can mean the interchange or replacement of social relationships between two modes of production. In that case, the articulation of modes of production could be a useful conceptual instrument for analyzing the influence of pre-capitalist social relations on national capitalist development.

Conventionally, the concept of articulation was merely a heuristic device, and so static that it tended not to accommodate the dynamic processes of transition. So how should we modify the concept of articulation? Contrary to world-system analysis, the following discussion will proceed dialectically from the abstract and simple level to the concrete and complex level.

The transformation of class relations in the transition to capitalism

In order to mold the concept of articulation as a more dynamic and useful analytical tool, class relations and their transformation must be considered. This is because class relations are not only one of the most definite relations in societies characterized by inequality, but they are also one of the dynamic factors of social change. In the transition to capitalism, how should class relations be transformed?

Before considering this transformation, it is necessary to define the classes present in both societies involved in the transition. In capitalist society, above all at the level of mode of production, capital-labor relations are the most basic; therefore, the capitalist class and the working class are the primary classes. In contrast, in pre-capitalist society, the primary classes are the landowning class, the merchant capitalist class, and the producing class such as peasants and crafts. Given that the primary classes in capitalist society cannot emerge without any origins, they should be derived from the primary classes in pre-capitalist society.

At this time, because the valorization and accumulation of capital predominantly structure social relationships in capitalism, it is critical to identify the pre-capitalist classes from which the capitalist class can emerge. There are two possibilities; the ruling class, such as the landowning class and/or the merchant capitalist class (Type 1), and the dependent class, namely the producing class (Type 2). It is also possible for the capitalist class to derive from other societies. That is, foreign capital rules the structuring of social relations in the society (Type 3).

In short, we have three logical types of class transformation in the transition to capitalism. It is to be noted that these three types are mutually exclusive in theory. It is also to be noted that these three types are theoretically equivalent to each other and that one of them is not a prototype for the others. In the case of Type 1, it is supposed that the ruling class in pre-capitalist society is so strong that it can readily transform itself into the ruling class, namely the capitalist class, in capitalist society and that the dependent class cannot become the ruling class.

In contrast, in the case of Type 2, it is supposed that the dependent class in pre-capitalist society can overthrow the ruling class, which includes landowners and merchants, and replace them in capitalist society. In other words, in this type, social revolution, like civil revolution, is likely to happen. In this way, Type 1 and Type 2 are completely different from each other and cannot logically coexist. Furthermore, in the case of Type 3, it is supposed that indigenous capital cannot emerge or take an initiative in development before foreign capital comes in. In other words, Type 3 is likely to be a late capitalist society and located in the peripheral regions of the world-system. Therefore, Type 3 is incompatible with the other types.

The effects and differences of class transformation

We have already confirmed that the transition to capitalism is characterized by the articulation of pre-capitalist and capitalist modes of production and that the primary classes in capitalist society emerge from those in pre-capitalist society in such processes. So what are the effects of this transition?

In a word, the replacement of capitalist relations with pre-capitalist ones in capital-labor relations would occur. This is because the transition period is not momentary and it takes time to remold class relations. Therefore, pre-capitalist relationships do not vanish in the transition period. In other words, they are present at least during the transition period.

Capitalist relations and pre-capitalist relations

In analyzing the effects of pre-capitalist relationships, it is necessary to clarify the differences between capitalist and pre-capitalist relationships theoretically. These differences can be posited as ideal types. Capitalist class relations are thought to be connected in three spheres: *economic, political,* and *cultural/ideological.* Althusser (1965) *rediscovered* these three spheres in the mode of production, and some theorists, such as Carchedi (1977) and Poulantzas (1975), introduced them into class analysis.

Economic relations are wage relations, which are derived from the fact that workers are deprived of the means of production—they are merely economic, without non-economic coercion (Chase-Dunn 1989: 25). Political relations are related to the control of labor, which is generally bureaucratic and *non-personalistic.* Cultural/ideological relations are related to the knowledge of production process and, at the more concrete level of *social formation*, the consciousness of work, society and class relations themselves, and the like—they are also generally materialist and meritocratic.

What, then, are pre-capitalist relations? Pre-capitalist class relations exist in the same three spheres. Economic relations are characterized by the connection of means of production with producers. Political relations are characterized by the *personalistic (persönlich),* and arbitrary control of work. Cultural/ideological relations are characterized by the full appropriation of knowledge of production by producers and the consciousness of social order, including class relations themselves—such consciousness is often characterized by communitarianism and the paternalistic legitimation of class conflicts.

Class relations in the transition process

In the transition process, two types of class relations (pre-capitalist and capitalist) are mixed and the replacement of capitalist relations with pre-capitalist relations can occur. In the economic sphere, capitalist wage relations are not fully connected because workers (they are generally formed from the producing class, such as peasants) still have some ties with the means of production and thus retain other sources of income.

In the political sphere, the control of work is arbitrary, and workers can be *personalistically* subservient to employers. In the cultural/ideological sphere, the capitalist appropriation of knowledge is not fully achieved; class relations or capital-labor relations can be paternalistically regarded as benign, and communitarianism can conceal class conflicts. It is to be noted that pre-capitalist class relations are not necessarily benign and are often antagonistic.

The end of transition and its aftermath

How can the end of the transition period be defined? Given that the transition to capitalism progresses through the articulation of modes of production, the end of transition can be defined as the dissolution of the pre-capitalist mode of production. It is the consequence of articulation. What kind of situation can be defined as its dissolution? Even though the mode of production consists of the sum of the social relationships in all three spheres, economic relations seem to be the most definite in the formation of mode of production. Therefore, the point of its dissolution can be defined as the period in which class relations in the economic sphere are completely capitalist—all workers, at least those in leading sectors at that time, are fully proletarianized.

This may mean that remnants of precapitalist relations remain. However, as capitalism develops in the society, pre-capitalist relations are sooner or later extinguished because purely capitalist relations are more efficient in the accumulation and valorization of capital.

For example, if all other conditions are equal, full-fledged wage relations are likely to hold more economic compulsion than those which are mediated by such replacement with pre-capitalist relations; the bureaucratic and non-personalistic control of labor is likely to contribute to organizing production more *rationally* than a pre-capitalist arbitrary and personalistic one; the full appropriation of knowledge of production by capitalists is likely to lead to more efficient innovation in production than partial appropriation.

Three types of transition; their differences and effects

However, depending on the type of the transformation of class relations, the length of time required to conclude such replacement is likely to be different. In Type 1, in which the ruling class becomes the capitalist class, the time is likely to be longer. This is because the newborn capitalist class in the society has been benefiting from pre-capitalist relations and retains *more intimacy* with those relations.

In contrast, in Type 2, in which the dependent class becomes the capitalist class, the transformation period is likely to be shorter. This is because the newborn capitalist class in the society has been suffering from pre-capitalist relations and retains *less intimacy* with such relations. Lastly, in Type 3, in which foreign capital takes an initiative in capitalist development, the transition period is indefinite. This is because foreign capital retains *no definite intimacy* or is *neutral* with pre-capitalist relations in the society.

What, then, are the expected effects of these differences on the capitalist development of a society? If other factors are not taken into account, Type 1 should be stagnant; Type 2, in contrast, should be more progressive; and Type 3 should be neutral. However, if this discussion is made more concrete and includes in the analysis other factors such as class conflicts and the state, it could draw the opposite conclusion.

That is to say, Type 1 would be characterized by more progressive capitalist development, given that pre-capitalist relations can persist longer; as a result, paternalistic ideology, which is preferred by the capitalist class in the society, can prevail through the capitalist state and hinder class conflicts. In contrast, Type 2 could be characterized by stagnation in capitalist development, given that pre-capitalist relations can vanish more quickly and as a result, there is nothing to prevent workers from protesting. In short, a harsh class struggle is likely to disrupt the accumulation and valorization of capital. Type 3 could still be characterized as an indefinite transition period, given that foreign capital is supposed to be neutral to pre-capitalist relations in the society.

In this way, if the articulation of modes of production is understood from the dynamic perspective of the transformation of class relations, the endogenous specificity of pre-capitalist relations in the society can be added into the analysis of national development in the world-system. However, the above discussion is merely theoretical and abstract, taking place primarily at the level of modes of production. In order to make it more concrete, we will consider the case of Japan.

The case of Japan: Its particularity

Japan as a case of Type 1

Because there tend to be more anomalies and contingencies at the concrete level of analysis, it might not be so easy to determine which countries belong to which of these three types. Nevertheless, dominant trends can be identified in the transition of each society. Japan and Germany belong to Type 1, while countries such as the United Kingdom and the United States probably belong to Type 2. Many semiperipheral and peripheral countries belong to Type 3. In the following analysis, capitalist development and the pattern of globalization in Japan will be summarized, focusing on the specificity of class relations.

Inception

Japan experienced the dramatic transition to capitalism from the Meiji era, in the late nineteenth century. The capitalist class in Japan was primarily derived from feudal landowners (*Daimyo*), who received vast amounts of money in exchange for their territories, and big merchants of the Edo era. Furthermore, many corporations were formed as family businesses (*Zaibatsu*), many of which were based on merchant families before and during World War I.

Therefore, industrial relations in such corporations were usually paternalistic and characterized by *management familialism* because these relations were not originally based on employment but rather on master-servant relations. The Emperor system (*Tenno-sei*) in Japan, which was the political system before and during World War I, also legitimatized a national paternalist ideology.

The primary institutions of the Japanese industrial relations system—lifetime employment (*Shushin-Koyo*), the seniority wage system (*Nenko-Chingin*), and enterprise-based unions—were forged during the Taisho era (the 1920s through the 1940s) just after World War I (Hyodo 1971). These institutions were often forged by contingent factors, but they have also been endorsed by personalistic relations and paternalist and communitarian ideology—the specificity derived from the type of transition to capitalism in Japan. Thus, they persisted for a long time.

Dissolution and persistence

Of course, pre-capitalist relations were persistently dissolved through capitalist development in Japan. For example, wage relations were fully connected in leading sectors as early as the Taisho era. Furthermore, the labor movement had been active as early as the pre-World War I era—especially in 1917, when the movement arose for seeking decent and humane working (*Jinkaku-Shonin*). After World War II, at least until the 1950s, there were many oppositional movements and strikes.

These labor disputes might indicate that pre-capitalist relations were fully dissolved and unable to soften industrial conflicts. However, until recently, there has been evidence of *company communitarianism* and a preference for personalistic labor control. For example, enterprise unions, though in opposition to the interests of capital, have been a part of the management system and have contributed to the establishment of benign industrial relations.

Furthermore, Japan's evaluation system (*Satei*) for the motivation, attitude and loyalty of employees, which was introduced into many large companies starting in the 1960s and has been utilized in determining promotion, wage increases, and bonuses, can be predicated on the preference for personalistic relations.

Consequently, benign industrial relations prevailed in the late 1960s, at least in large companies. This is because pre-capitalist relations were not completely eradicated and the concrete form of specificity of Japanese capitalism has been persistently maintained.

Japanese uniqueness in globalization

What are the implications of the fact that various institutions endorsed by pre-capitalist relations have preserved benign industrial relations? Some countries classified as Type 2, such as the United States, were faced with challenges regarding their industrial relations. It has been indicated that oppositional labor unions pursued wage increases, and that simple, repetitive, hard work in Fordism stagnated productivity. One of the solutions was the relocation of production to peripheral regions of the world-system.

This relocation also signaled the *requisition* of pre-capitalist relations in peripheral regions, because many workers have rural backgrounds and keep ties with means of production. In addition, many immigrant workers flowing into core regions from peripheral regions often preserve their rural ties and move back and forth between the two regions, forging transnational communities. It could be that employing immigrant workers in the core, who often work in sweatshops, can mean the *mobilization* of pre-capitalist relations.

If the internationalization of capital and labor can transform labor-advantageous industrial relations into capital-advantageous ones, the Type 2 societies are likely to experience such internationalization, above all that of capital, earlier than those classified as Type 1. This is because pre-capitalist relations in Type 2 are supposed to vanish earlier than in Type 1.

Based on this consideration, globalization in Japan is remarkable for the period of its inception. It is true that some corporations in Japan were transnationalized as early as the 1950s, but they were only a minority. People from Korea and China, formerly Japanese colonies, were still living in Japan, and many immigrants already lived in Japan after World War I. Nevertheless, until recently, there were very few immigrants related to the mobility of capital (Sassen 1988).

As late as the late 1980s, the internationalization of capital and labor began swiftly. However, the reason behind it differed from that in such countries as the United States. In fact, there was no reason for many corporations to transnationalize themselves in Japan. Industrial relations were so benign that workers were only able to obtain a smaller share of profit during the 1970s and employers were not pressured to relocate production to peripheral regions.

In fact, such regions were unfamiliar to many Japanese companies and, in some cases, somewhat risky to invest in. In other words, Japanese companies were forced to transnationalize themselves because of trade conflicts and the appreciation of yen—on account of which they were no longer able to export their products as before.

In a word, Japanese industrial relations were replaced by pre-capitalist relations at least in the political and cultural/ideological spheres, until other core countries experienced the internationalization of capital. Therefore, the Japanese economy performed well thanks to benign industrial relations, and produced high-quality products.

This might mean that the transnationalization of Japanese corporations will be more likely to impair harmonious industrial relations than to build them, as other core regions were able to by transforming labor-advantageous or conflictive relations in transnationalization. Through globalization, Japan, a Type-1 society, is likely to transform its industrial relations into fully capitalist ones and, in so doing, undermine its origin of favorable development and less conflictive relations.

Concluding remarks

In this chapter, the world-system theory was criticized for its holistic perspective. It tends to disregard the specific dynamism of various units, particularly countries in the world-system. For the purpose of overcoming this drawback, the theory of articulation of modes of production was revised to account for dynamism and to complement the world-system theory. In so doing, the transformation of class relations in the transition to capitalism can be analyzed and three types of transformation can be established.

Depending on the type of transformation, the length of the persistence of pre-capitalist relations can vary from society to society, and the extent to which specific endogenous factors in each society can affect national development will also be diverse. Therefore, such endogenous factors as pre-capitalist relations can be identified as one of the causes of the multiplicity of national development.

Japan, a Type-1 society, has experienced the long persistence of pre-capitalist relations as a unique form of management style and industrial relations system. The uniqueness of Japanese capitalist development is derived partly from this endogenous factor. Its swift upward mobility to the core of the world-system and its late internationalization of capital and labor were endogenous in origin. However, through the process of globalization, Japan is experiencing the undermining of its advantageous specificity and is becoming a more thoroughly capitalist society.

References

Althusser, L. (1965) *Lire le Capital*, Paris: Francois Maspero; trans. Hitoshi Imamura (1996) *Shihonron O Yomu*, Tokyo: Chikuma Shobo.

Amin, S. (1973) *Le Développement Inégal: Essai sur les Formations Sociales du Capitalisme Périphérique*, Paris: Les Éditions de Minut; trans. Jun Nishikawa (1983) *Fukinto Hatten*, Tokyo: Toyokeizai Shinposha.

Brenner, R. (1977) "The Origins of Capitalist Development: A Critique of Neo-Smithian Marxism", *New Left Review*, 104.

Brenner, R. (1985) "Agrarian Class Structure and Economic Development in Pre-Industrial Europe" in T. H. Aston and C. H. E. Philpin (eds) *The Brenner Debate*, London: Cambridge University Press.

Carchedi, G. (1977) *On the Economic Identification of Social Class*, London: Routledge & Keagan Paul.

Chase-Dunn, C. (1989) *Global Formation: Structures of the World-Economy*, Cambridge: Basil Blackwell.

Chase-Dunn, C. and Hall, T. D. (1997) *Rise and Demise: Comparing World-Systems*, Boulder, CO: Westview Press.

Dobb, M. (1946) *Studies in the Development of Capitalism*, London: Routledge & Kegan Paul.

Evans, P. (1995) *Embedded Autonomy: States and Industrial Transformation*, Princeton, NJ: Princeton University Press.

Frank, A. G. (1998) *ReOrient: Global Economy in the Asian Age*, New York: Cambridge University Press.

Frank, A. G. and Gills, B. K. (1993) *The World System: Five Hundred Years or Five Thousand?* New York: Routledge.

Hyodo, T. (1971) *Nihon Ni Okeru Roshi-Kankei No Tenkai (The Development of Capital-Labor Relations in Japan)*, Tokyo: The University of Tokyo Press.

Poulantzas, N. (1975) *Classes in Contemporary Capitalism*, London: NLB.

Sassen, S. (1988) *The Mobility of Capital and Labor: A Study in International Investment and Labor Flow*, London: Cambridge University Press.

Sweezy, P., Dobb, M. et al (1976) *The Transition from Feudalism to Capitalism*, London: New Left Books.

Wallerstein, I. M. (1974) *The Modern World-System: Capitalist Agriculture and the Origins of the European World-Economy in the Sixteenth Century*, New York: Academic Press.

Wallerstein, I. M. (1979) *The Capitalist World-Economy*, New York: Cambridge University Press.

Wallerstein, I. M. (1980) *The Modern World-System: Mercantilism and the Consolidation of the European World-Economy, 1600–1750*, New York: Academic Press.

Wallerstein, I. M. (1989) *The Modern World-System: The Second Expansion of the Capitalist World-Economy, 1730s-1840s*, New York: Academic Press.

Wolpe, H. (ed.) (1980) *The Articulation of Modes of Production: Essays from "Economy and Society"*, Boston, MA: Routledge & Keagan Paul.

3.5

Crises in the world-system

Theoretical and policy implications

John Barnshaw and Lynn Letukas

So (1990) has noted that the world-system approach has been critiqued for an absence of historical specification in the analysis of global dynamics. This chapter seeks to remedy this critique by demonstrating how short-term crises are best understood through an analysis of the world-system approach and the *longue duree*. Specifically, by understanding how broad historical and economic forces have developed over hundreds of years, we are better able to understand how perceptions influence agents and social structures in policy making during crises. Our conclusion is that if the world-system approach influences the most immediate social occasions, then it probably influences broader social change, as has been claimed by some of the foremost contributors to this volume.

Using the example of the 2004 Indian Ocean tsunami and the 2008 global economic crisis, we develop three propositions that we highlight herein. *First, the initial policy response to crisis is heavily influenced by core entities representing short-term economic interests, with little exploration of long-term implications on semiperipheral or peripheral zones of production.* Following the Indian Ocean tsunami, core interests took the form of corporate, non-governmental entities, and governments offering unprecedented levels of assistance to the impacted regions, with little consideration of those impacted by the catastrophe (Inderfurth et al 2005). During the 2008 economic crisis, core interests sought to stabilize the economy via governments serving as a lender of last resort to troubled financial institutions through commercial paper facilities, direct loans or capital injections, with little discussion of how best to resolve the underlying problems of financial leveraging and liquidity shortfalls.

Second, core intervention tends to result in greater exploitation of semiperipheral or peripheral zones of production. Following the Indian Ocean tsunami, there were numerous positive reports of assistance in the United States media, with the focus on a neo-Marshall Plan that would "help those in need," but what was lost in these reports was the high degree of inappropriate aid (e.g., hamburgers to Hindus) and lack of attention to requested resources by those impacted (Usborne 2005). During the 2008 economic crisis, the United States and Western governments were quick to inject billions to shore up unstable financial institutions, which securitized billions more in Alt-A and subprime loans, but were slow to write down or refinance the affected peripheral areas of global finance that fell into bankruptcy, default, or foreclosure.

Third, as core intervention results in greater exploitation, semiperipheral and peripheral zones become increasingly financially and socially unstable in the long-term, leaving the world-system increasingly vulnerable to future crises. The Indian Ocean tsunami left fishing and tourist communities more vulnerable, as resources needed to stimulate development never materialized. Loans stunted growth and development in ways strategic grants would not have, and the burden of debt they imparted continues to grow. Similarly, following the 2008 economic crises, many households that were not able to roll over their debt went into foreclosure, while neighbors' property value declined, leaving thousands of mortgages under water. In short, when core interests intervene with little understanding of the long-term implications, semiperipheral, and peripheral regions remain compounding areas of surplus extraction. This produces increasing crises in quantity and intensity. However, such policy outcomes are not inevitable, as prior research has shown that policies focused on the efficient coordination of local resources in concert with local interests can effectively remedy crises in a more equitable method for those affected (Easterly 2001; Price 1955).

To date, little research has focused on crises in the world-system (cf. Bergensen and Lizardo 2004; Letukas and Barnshaw 2008). The conventional reason is that historically, disasters, catastrophes, and crises were conceptualized as short-term aberrant occasions with little impact on the long-term development of economic, political, and social institutions. We believe that the world-system approach does influence economic and political decisions following crises, and that further research should explore its influence on short-term crises. For example, if core zones of production are to continue to intervene in future crises, they should coordinate responses with semiperiphery and peripheries in ways that are mutually beneficial. Specifically, this can take the form of offering grants to normalize stability rather than providing loans. Admittedly, this approach may at best offer a softer landing for global exploitation and inequality, but hopefully future research can show how the larger historical forces of the world-system can offer solutions that do not antagonize core-periphery relations when peripheral regions are at their most fragile and unstable in crises.

References

Bergensen, A. and Lizardo, O. (2004) "International terrorism and the world-system", *Sociological Theory*, 22: 38–52.

Easterly, W. (2001) *The Elusive Quest for Growth: Economists' Adventures and Misadventures in the Tropics*, Cambridge, MA: MIT Press.

Inderfurth, K., Fabrycky, D., and Cohen, S. (2005) *The 2004 Indian Ocean Tsunami: Six Month Report*, The Sigur Center Asia Papers. [Online] Available at: http://www.gwu.edu/~elliott/news/tsunamiinderfurth.pdf. [Accessed 27 January 2010].

Letukas, L. and Barnshaw, J. (2008) "A world-system approach to post-catastrophe international relief", *Social Forces*, 87: 1063–1088.

Price, H. (1955) *The Marshall Plan and Its Meaning*, Ithaca, NY: Cornell University Press.

So, A. (1990) *Social Change and Development: Modernization, Dependency and World-System Theories*, Thousand Oaks, CA: Sage.

Usborne, D. (2005) *America Urged to Devise Marshall Plan for Asia*, The Independent. January 3.

3.6

Core, semiperiphery, periphery

A variable geometry presiding over conceptualization

Nicole Bousquet

Among important contributions to the world-system analytical framework is the tripartite division into core, semiperiphery, and periphery. Defining core and periphery raised little debate; not so for the semiperiphery, its most innovative concept coined by Immanuel Wallerstein. Early on, Wallerstein (1979, 1984) had made clear that separate dimensions of analysis were required for positioning countries in the world-economy and the interstate system and that occasionally, countries would not hold identical status in both systems.

The structural foundation of the world-economy was deemed to be the division of labor between its three constituent zones, crisscrossed by systemic commodity chains along which "surplus" was transferred from periphery to center, transiting through the semiperiphery. If the center and the periphery were considered as the locus of respectively highly and poorly rewarded activities along the commodity chains, the semiperiphery was conceived as the locus of a mixed bag of surplus generating activities, some being core-like surplus-wise, while others were of a peripheral type. Measuring the amount of "surplus" benefiting respective world-system zones was a real challenge. It entailed mapping innumerable commodity chains crisscrossing the world-economy at any given time.

In the early 1980s, Arrighi and Drangel (1986) estimated that a shortcut should and could be found to measure the quantum of "surplus" devolved to each world-economy zone. They hypothesized that the differential quantum of surplus reverting to the core, semiperiphery and periphery would translate into different levels of GNP per capita. They then proceeded to verify Wallerstein's postulate that notwithstanding occasional country mobility within the tripartite structure, the relative importance of each region in terms of proportion of world population was a constant.

At about the same time, the historical division of labor between core and periphery came to be conceived as resting on the exchange of "high-wage goods" against "low-wage goods" (Chase-Dunn and Rubinson 1977) rather than manufactured against primary goods, as was previously thought. In addition, a combination of high and low wage goods was coming out of the semiperiphery. This left unanswered the question of the prevailing wage levels inside the semiperiphery: high and low wages versus an average intermediate level of wages.

A further tangent of the definition of the three zones of the world-economy was Wallerstein's proposition that the expansion and deepening of the world-economy would lead eventually to the complete "proletarianization" of the world-labor force. As new regions became integrated into the system, part of their population was to be transformed into "semi-proletarians" thereof partially dependent on the sale of their labor but still having access to some land for part of their subsistence. Eventually, semi-proletarians would become proletarians, that is, totally dependent on the sale of their labor time. Unanswered questions remained: if at any given time the labor force composition in the semiperiphery could be distinguished from that in the periphery in terms of the relative importance of full versus semi-proletarians; and were the growing masses of poor urban dwellers scavenging for a living in self-employment in the informal sector to be considered proletarians or semi-proletarians.

In a nutshell, three defining criteria and empirical indicators of each of the three zones of the world-economy emerged from the world-system analytical framework: the relative wealth of countries (per capita income); the differential level of wages; and the differential labor force composition ("proletarians"/"semi-proletarians").

Preliminary calculations indicate quite a high correlation between the per capita income of countries, the average level of wages (high in high income countries, intermediate in medium income countries and low in low income countries) and the proportion of wage labor in the working age population (high in high income countries, less so in intermediate income countries, low in the poorest countries). This being the case, the level of wages and the type of income source can be considered as defining dimensions of the three regions of the world-economy and an important criteria for classifying states in its tripartite structure.

National economic power is yet another dimension relevant for the classification of countries in the tripartite structure of the world-system. For instance, China and India, given the sheer size of their national economies and their emerging importance in the world-economy have often been considered parts of the semiperiphery. If national economic power is to be a legitimate dimension of the composition of zones in the world-economy, this implies that the relative importance of the semiperiphery and periphery in terms of percentages of world population will be altered to the point that the tripartite structure will end up taking the shape of a diamond instead of its historical pyramidal shape. According to our calculation, the semiperiphery would then be the predominant zone at about 60 percent and the periphery about 25 percent, with the relative size of the core remaining the same.

When the dimension of "economic power" is taken into consideration, a different composition of both the semiperiphery and the periphery pertains. Some countries will inevitably be characterized by status incongruence in terms of belonging to the semiperiphery versus the periphery. Moreover, level of income and economic power are not the only relevant dimensions for classifying countries in the world-system. There are also those dimensions of politics and power in the interstate system (including internal state strength, military power, external political weight, and influence).

What we should perhaps conclude is that from now on it is judicious to conceptualize the tripartite structure of the world-system as characterized by a number of axial dimensions of analysis. We should work with them all with the understanding that the composition and shape of the zones might vary according to the dimension considered. In a nutshell, we should consider the variable geometry.

References

Arrighi, G. and Drangel, J. (1986) "The Stratification of the World-Economy: An Exploration of the Semiperipheral Zone", *Review*, X (1): 9–74.

Chase-Dunn, C. and Rubinson, R. (1977) "Towards a Structural Perspective on the World-System", *Politics & Society*, VII (4): 453–76.

Wallerstein, I. (1979) *The Capitalist World-Economy*. New York: Cambridge University Press.

Wallerstein, I. (1984) *The Politics of the World-Economy*. New York: Cambridge University Press.

3.7

Terminal crisis or a new systemic cycle of accumulation?

Christopher Chase-Dunn

World-systems scholars have presented different explicit or implicit models of the extent to which the modern world-system has evolved since it emerged about 500 years ago. Immanuel Wallerstein (1974) sees the emergence of a capitalist world-economy out of the crisis of European feudalism in the long sixteenth century (1450–1640). Wallerstein's implicit model has the modern system becoming larger but not really changing its basic structure since the beginning. There are cycles and long-term trends and a game of musical chairs with the rise and fall of hegemons. The European system incorporates the rest of the world and the population grows but the basic processes of the system do not change. The expansion of capitalist accumulation and the resistance of workers and farmers to it produce long-term upward trends in the costs of labor, raw materials, and taxes and these are now reaching levels that capitalism cannot accommodate (Wallerstein 2003). The logic of profitable investment cannot continue once the profit rate has declined to near zero. The expansion of the system has now come to the point where it is in a terminal crisis, a period of chaos, in which it will become transformed into a very different kind of system in the next decades—either a kind of tributary world state controlled by a small elite or a form of democratic and socialist global governance in which the Global South will play an important role. Wallerstein stresses the continuities of historical capitalism as a system and locates his understanding of transformation at the beginning and at the end.

Giovanni Arrighi's model starts at almost the same point, but contains important differences (Arrighi 1994, 2006a; Arrighi and Silver 1999). For Arrighi (following Fernand Braudel), the key relationship in capitalism is that between state power and finance capital. This relationship evolves during the history of the modern world-system in a series of overlapping "systemic cycles of accumulation." The first systemic cycle is an alliance between the Portuguese state and the Genoese state in the fifteenth century. The second, which is similar to Wallerstein's portrayal of world history, is the hegemony of the Dutch Republic in the seventeenth century. Arrighi's systemic cycles evolve in the sense that more and more functions of the system are taken under the control of the hegemonic state and finance capital in each successive cycle. Arrighi is careful to specify how the systemic cycles are different from one another and what the implications of the differences are likely to be. So for Arrighi, transformations occur during the process of the evolution of the modern system. These theoretical differences have important implications for the analysis of the current conjuncture.

Arrighi (2006b) sees the rise of China as the harbinger of the arrival of a new systemic cycle of accumulation that he characterizes as "market society." Arrighi and several scholars with whom he worked saw the East Asian world-system as operating according to a somewhat different logic than the system that emerged from Europe. The Peoples Republic of China, because of the Chinese Revolution and the institutional heritage of the East Asian system, may be able lead the world toward a new order that is less warlike than the European version of capitalism has been and is less under the control of finance capitalism than has been the case with the hegemons of the Europe-centered system. Wallerstein's model is more millenarian and may be more useful for mobilizing transnational social movements to strongly challenge (once again) the powers that be in the contemporary world-system. But Arrighi's evolutionary model may be more useful in the long run, or may at least provide a good start, for figuring out what parts of capitalism are really bad and what parts, such as markets, are not so bad.

References

Arrighi, Giovanni 1994 *The Long Twentieth Century*. London: Verso.

Arrighi, Giovanni 2006a "Spatial and other fixes of historical capitalism" pp. 201–13 in C. Chase-Dunn and S. Babones (eds) *Global Social Change*. Baltimore, MD: Johns Hopkins University Press.

Arrighi, Giovanni 2006b *Adam Smith in Beijing*. London: Verso.

Arrighi, Giovanni and Beverly Silver 1999 *Chaos and Governance in the Modern World-System: Comparing Hegemonic Transitions*. Minneapolis: University of Minnesota Press.

Wallerstein, Immanuel 1974 *The Modern World-System*, Volume 1. New York: Academic Press.

Wallerstein, Immanuel 2003 *The Decline of American Power*. New York: New Press.

4. Explicit modeling as a research strategy

4.1

Measuring transition and hierarchy of states within the world-systems paradigm

Raymond J. Dezzani

Introduction

One purpose of the world-systems paradigm is to challenge traditional and conventional theories and perspectives in integrated social science and this goal has been partially accomplished over the intervening years since its inception by Immanuel Wallerstein (Wallerstein 1974, 1991). Many contributors have added to and extended the world-systems paradigm through applications of greater analytical rigor (e.g., Arrighi 1994; Babones 2009; Chase-Dunn 1998). However, much more needs to be done to achieve a fully rigorous and logically consistent approach to the multi-scale problems of world-systems and the processes and mechanisms of political-economy.

Traditional neo-classical economic approaches are flawed by retention of a nineteenth century adherence to continuous deterministic mathematics and by ideas of external valuations of resource bases, continuous logistic growth and simplistic assumptions of information, utility, and mechanistic equilibrium stability. To advance the world-systems perspective, sophisticated ideas of hierarchical, stochastic evolutionary systems applied as probabilistic political-economy might be employed in order to account for the complexity and interactions of economy, politics, and social system necessary to the paradigm. For examples of this approach across a variety of topics see Georgescu-Roegen (1971), Cavalli-Sforza (1981), Nicolis and Prigogine (1989), Dendrinos and Sonis (1990), Arthur (1994), Ayres (1994), and Bodley (2003). A starting point for such an approach is through measurement of the dynamics of polities in the hierarchical world-economy, which forms the intermediate-scale evolutionary dynamics of the world-system.

Most aspects of world-systems research have been viewed through a historical and social lens with the purpose of refining explanations of particular social or economic phenomena. Examples of this research abound and constitute the majority of world-system studies (e.g., Arrighi 1994, 2007; Chase-Dunn 1998; Pomeranz 2000; Sanderson 1995; Wallerstein 1974, 1980, 1989). However, for the world-systems paradigm to become a more functional instrument for policy change and analysis as well as providing more rigorous insights into regional and global-scale aggregate economic, political, and social processes, efforts toward quantification with the goal of prediction must be attempted. The aspects of the world-systems perspective that lend themselves to such statistical and mathematical manipulation are rooted in the economic basis of the world-economy, such as hierarchical zonation classification, trade and interaction modeling, rigorous power/geopolitical analysis of polities both within and among world-economy zones, and temporal and spatial studies of hegemonic and secular transition cycles that induce or reflect evolutionary change in the cycles over the short- and long-term.

Country mobility through the world-economy hierarchy has been a key feature of the world-system paradigm since inception (Chase-Dunn 1998: 199–215; Wallerstein 1974). Three regions, or zones, consisting of core, semiperiphery, and periphery, have constituted the hierarchical structure of the world-economy. Mobility of country-level state economies among these regions was originally conceived as an optimistic reflection of development and progress, and most movement was inferred to be upward from the periphery through the semiperiphery to the core. Two features of this framework are necessary for proper empirical evaluation of the mobility process. The first involves adequate classification of countries or polities into world-economic classes. The second feature requires the measurement of individual transitions of countries or polities across time intervals, to provide estimates of aggregate movement or mobility.

Various scholars have examined the tripartite structure of the world-economy, employing a variety of methods in order to classify and allocate polities and states, and they have produced very similar results. The three-level structure is statistically persistent across time periods and the methods of assessment (Babones 2005). The preponderance of evidence suggests the statistical viability of partitioning the world-economy into three regions. However, the reality suggests a continuous scaling of country-level economies that may be classified with statistical credibility into three or more zones (Chase-Dunn 1998; Dezzani 2001). Within any zone of country-level economies, there always exists a gradation of economic and political behavior (Arrighi and Drangel 1986; Dezzani 2001). The dynamics of the world-economy rely on the veracity of this structure to capture system transitions and evolution. State or country, mobility among regions can then be measured as the basis of systemic transition.

This survey will examine methods for measuring the mobility of states within the hierarchical world-economy as well as the hierarchical structure of the system. These transition cycles occur within other cycles such as long-term change of system cycles and hegemonic cycles. Hegemonic cycles are separate longer-term sequences resulting from core state conflict (Agnew 2005; Arrighi 1994; Arrighi and Silver 1999; Friedman 2005). The state mobility transitions/movements can be perceived as shorter-term, fast-process cycles occurring within both hegemonic cycles and also nested within systemic collapse and change cycles occurring at much slower rates (Arrighi 1994; Chase-Dunn 1998; Chase-Dunn and Anderson 2005; Friedman 2005; Tainter 1988).

The frameworks used to examine shorter-term, fast-process changes in state mobility may also be used to examine longer-term interval, slow-process hegemonic cycles, and system change behavior independently. However, explanatory models should change to suit the scale and causality of the cyclical process. This change may be difficult in the near-term as there is much debate over variable selection providing information on differences among zones or regions (Babones 2005). Several structural and relational economic considerations may help alleviate such difficulties (see Chase-Dunn 1998: 214–27; Dezzani 2001).

Hierarchically structured world-economy

The basis of rational classification of states and polities requires a theoretical framework that is potentially quantifiable and logically consistent with the functional frameworks to be employed. The basic description provides that the hierarchy consists of a high income core that produces goods of high value that has a geographical range covering North America, Europe, and Japan; a middle income semiperiphery producing a mixed range of goods and commodities and composed of non-core countries and polities that are relatively capital rich and sophisticated so as to maintain cross-hierarchy interactions and independence of action; a low income periphery producing primary sector, low value-added commodities consumed by semiperipheral and core states and consisting of the majority of the worlds countries and polities that have little independence of action in the world-economy (Babones 2005, 2009; Chase-Dunn 1998; Wallerstein 1974). This framework requires further explication to justify a statistically valid partition of polities in the world-economy. That is, an ordination or ordering of economies based on income and production types, as well as trade and exchange interaction must be constructed; this will form the basis of statistical partitions for classification of countries into zones or regions.

Dezzani (2001) suggested that a hierarchical framework, such as that developed by Hanink (1991), will provide a useful instrument for quantifying differences in a hierarchical world-economy. Hanink's (1991) hierarchical market model assumes that country utilities are additive functions of commodity consumption and interaction and, as such, are limited by income. The production of goods and commodities are functions of mean income threshold demand, for which there is variation in the world-economy. Countries exhibiting the full range of goods/commodities contain nested markets and thus, a corresponding nested potential production—this defines hierarchical dependence. Hence, the production of the lowest order good/commodity occurs in all countries and with progression up the production potential ranking, fewer countries may participate in common production. If trade, (i.e., exports), is viewed as a function of domestic incomes, (i.e., GDP), or induced demand, then the frequency of potential goods exports can be ordered. Furthermore, if factor homogeneity is assumed across the countries then trade will only occur between countries where incomes are greater than or equal to the order of the commodity/good in the ordered market. With variation in income and demand across the world-economy, country and product threshold demands can be represented as a progression:

$$\sum TC_1 < \sum TC_2 < \ldots < \sum TC_N$$

over N countries for which C_N represents the highest order country (core) and TC is the cumulative frequency of intra-industry trade partners (Dezzani 2001; Hanink 1991). Sub-partitions, permitting overlap, can be used to represent the ordered distribution of a classification of the world-economy and constitute the core, semiperiphery, and periphery, or any other more complex division suiting the needs of analysis. This framework also provides much needed suggestions for independent variable selection to be used in the creation of classifications.

Measuring position in world-economy hierarchy: Classification and assignment

The first step in establishing a rigorous assessment of world-economy zones is to classify countries into categories based on some intrinsic quantifiable criterion of the system. There are a variety of methods to accomplish this goal. Several have been used in recent studies (e.g., Arrighi and Drangel 1986; Babones 2005; Dezzani 2001; Korzeniewicz and Martin 1994; Nemeth and Smith 1985; Smith and White 1991; Snyder and Kick 1979). Economic classifications of countries are of continuing utility for comparative and analytic purposes. However, some methods of arriving at classifications can be *ad hoc*, subjective and imprecise, and do not permit the assignments to be used for closer and more detailed analysis. As previously stated, three main approaches have been used: (1) income trough methodology (Arrighi and Drangel 1986; Babones 2005; Korzeniewicz and Martin 1994), (2) network blockmodeling methods (e.g., Nemeth and Smith 1985; Smith and White 1991; Snyder and Kick 1979), and (3) parametric classification for use with stochastic change assessment (Dezzani 2001, 2002).

The income trough methodology and the network blockmodeling methods are primarily classification by use of case, (i.e., country), assignments. These are extremely useful for within-cohort comparisons but of limited use for explanatory purposes, unless coupled with other compatible methods of analysis. In addition, no independent explanatory variables are incorporated to provide functional assessment and analysis.

The income trough methodology (Arrighi and Drangel 1986; Babones 2005; Korzeniewicz and Martin 1994) is a classification framework that assigns countries or polities to a class based on the distribution characteristics of the income level of the particular country. Basic parametric expressions of mean and variance of the dependent variable provide either the sole or primary assignment criterion. Babones (2005) extends the idea by using Gaussian kernels as the basis of aggregate smoothing for the complete data set. This technique is extremely useful for single criterion assignment of cases to a class under the assumption of continuum (i.e., the distribution is generated by common stationary processes operating in the world-economy). As such, the method provides a useful "feed-forward" classification for further analysis. The method is easy to apply and provides consistent results with other methods.

However, the income trough classification is entirely descriptive of a single variable only. In addition, the method assumes generating process stationarity both within-cohort and also over time if more than one time period is considered. If, in fact, the generating process is changing over time, this method may produce biased results. However, the method has produced consistent representation of the triple zone hierarchy in the world-economy over extended time periods and with variations in application. As such, it has provided sufficient and consistent results and will continue to be used for basic country-level assignments to world-economy zones. Dezzani and Babones (2008) have demonstrated the applicability of this method with temporal mobility analysis on an annual basis for the time period (1975–2005).

The network blockmodeling methodology (Kick 1987; Nemeth and Smith 1985, Smith and White 1991; Snyder and Kick 1979) corrects some of the deficiencies of the income trough methodology through the use of multiple variables based on measurements of interconnection in the world-economy. Partitions are assigned based on an essential characteristic of the world-system: interaction and flow interconnectedness. Network blockmodeling is intended as a within-cohort analysis of structure for determining zonal partition. As it assumes particular structural relations within-cohort, it is not useful for examining the evolution of structure or change in structure over time. Generating process assumptions are relaxed in this method but there is a potential hazard of tautological assignment since structure is assumed to be hierarchical and the method provides confirmation. However, again, the methods have demonstrated a basic three-tiered structure, with some variations and additions, confirming the basic zonal structure of the world-economy. The network blockmodeling approach produced much more detailed structure within zones than had been observed by just assigning income levels. As a result, there appears to be multi-level structures within zones, suggesting zone-dependent processes. This observation has been confirmed by the parametric approach that has produced more detail and greater variation in zones than suggested by income as sole assignment criterion.

The third methodology might be termed linear and non-linear parametric classification due to the use of linear and quadratic discriminant function analysis as the basis of case assignment (Dezzani 2001). However, this method was designed to elaborate generating mechanisms and even to permit non-stationary changes in generating mechanisms over time. The goal was to provide classifications that could be used for temporal structural change and mobility analysis using stochastic change models (Dezzani 2002; Fingleton 1997). The parametric classification does provide potentially explanatory functional relationships between case-class assignment and the independent variables employed for classification. This methodology can incorporate many of the network features of the blockmodeling approach and also add other potentially explanatory endogenous and exogenous variables. An associative relationship exists between the dependent class assignment and the classifying explanatory variables. As such, parametric classification methods, such as linear and non-linear discriminant analysis, provide a relationship using multiple classifiers:

$$EMP_{\text{Class } i} = F_i\ (K_i, V_{i1}, V_{i2}, \ldots, V_{im}); i = 1,\ldots,n$$

where "$EMP_{\text{Class } i}$" represents the empirical world-economy zone/group classification measure for zone i, F is a linear or non-linear function of K, a group block or mean effect measure, and V_{ij} are the m independent explanatory variables used to assess cases of countries into one of the n classes or categories (see Dezzani 2001). This type of classification permits evaluation of the magnitude of incorporating each of the V_j variables for each class. Thus, the classification functions can be compared for the functional relationship between case inclusion in a particular class and the magnitude of the particular independent explanatory variable.

Estimation requires "training" through inputs derived from either *a priori* classifications or possibly, an income trough assignment. The estimation procedure then uses a Bayesian approach for posterior final case assignment based on:

$$\begin{aligned} P(EMP_{class\ i} \mid WS_{class\ i}) &= \frac{P(EMP_{class\ i})}{P(WS_{class\ i})} P(WS_{class\ i} \mid EMP_{class\ i}) \\ &= \frac{P(EMP_{class\ i}) P(WS_{class\ i} \mid EMP_{class\ i})}{\sum_{i=1}^{m} P(EMP_{class\ i}) P(WS_{class\ i} \mid EMP_{class\ i})} \end{aligned}$$

where P(*) defines membership probabilities for *m* world-economy zone classes, EMP is the empirically derived classification for a zone and WS is the corresponding world-economy zonal class defined *a priori* or by income trough methods (Dezzani 2001). This framework can be directly incorporated in a Bayesian discriminant analysis (McLachlan 1992; Ripley 1996).

There are a wide variety of parametric classifiers that may be used for case assignment, from discriminant analysis to neural networks and regression tree techniques that may be used to elucidate key variables reflecting zonal generating mechanisms while providing parametric functions for assigning cases (Ripley 1996). Again, the parametric function methodology provided a basic three-tiered structure of the world-economy, this time with some explanatory variables that change over time, suggesting some non-stationarity in the world-economy even over a 40-year time period (Dezzani 2001).

The result of all of these methods of classification is a state space response consisting of the world-economy region or zone and a count or frequency of countries/polities assigned to each by the classification decision rule. Frequency data is either distributed as a multinomial or Poisson random variable depending on type of analysis. As a result, state space classification data must be handled differently than simple regression models of structure would permit. Parametric function classification methods, such as discriminant analysis, provide the standard basis for formal statistical classification. Discriminant analysis and other confirmatory adaptive discriminators can be used to isolate a time-specific set of economic factors delimiting economic class categories that correspond to core-periphery states.

The preceding classification and world-economy zonal assignment methods can provide some initial information for the rigorous quantitative analysis of structure in the world-economy. However, these methods are all static in that they examine structure for particular time periods.

They provide country membership in particular world-economy zones and thus, if the zone memberships are examined over time, trajectories of specific countries or polities can be followed through the hierarchy over time. This type of analysis provides a basic measure of mobility of countries in the world-economy over time. When aggregated as a structured system, general transition or evolutionary characteristics can be observed.

Measuring stochastic country-level transition

The classification and case assignment results from the previously discussed methods can be used to evaluate change over time if they have been applied over a number of time periods. The temporal interval should usually be fixed to control for systematic temporal variation but the Markov methods suggested do not necessarily require fixed measurement intervals (Bartlett 1978; Bartholomew 1982). The results of the classifications are frequencies of country occurrence in a class at a particular time and the goal for change measurement is to evaluate changes in constituencies across two or more time periods (Bishop et al 1975: 257–79). A Markov model is used for capturing country-level mobility and zonal transition in the world-economy over time (Dezzani 2002; Fingleton 1997). These models can be used to assess convergence or variance reduction in the structure of the world-economy or to assess stationarity of the mobility process and rates of mobility change among zones.

The basis of mobility in a system is to capture the individual changes as part of the entire system mechanism. This means that changes in one zone of the world-economy can induce changes in another because they are connected. Any model of change in the world-system must incorporate this caveat. One way to accomplish this is to model the variable of net transition for each zone of the world-economy. The variable then becomes R_i which is the net mobility or transition change in the state space (i.e., zonal classification for i = 1, 2, …, n). Then the entire system can be described as a series of differential equations in terms of every other zonal component as:

$$\partial R_1/\partial t = f_1(R_1, R_2, \ldots, R_n)$$

$$\partial R_2/\partial t = f_2(R_1, R_2, \ldots, R_n)$$

$$\ldots$$

$$\partial R_n/\partial t = f_n(R_1, R_2, \ldots, R_n)$$

That is, the partial time rate of change of net change is a unique zonal function of change within zone as well as change across all other zones. This is necessary since the world-system and the world-economy represent a single interacting mechanism such that changes that occur in one subsystem induce commensurate changes in other subsystems. As such, any description of change must exhibit this structure. Mobility process stationarity now refers to a condition characterized by the vanishing of the changes $\delta R/\delta t$ such that:

$$f_1 = f_2 = \ldots = f_n = 0$$

then the values of $R_1, R_2, \ldots, R_n$ become constants. This condition is satisfied if the changes across zones in the world-economy are modeled as a Markov transition dependence system.

The outcome of the mobility and transition model is dependent on the classification used to create the zonal assignments. In general, this approach is not a single probability model but a particular perspective on the problem of mobility or movement of specific countries or polities across interacting zones of the world-economy. This approach is stochastic, as the basis of estimation is probability, but also graph-theoretic and interaction influenced; the models generated are primarily descriptive but the outcome probabilities can be coupled with explanatory logistic models to provide explanations of change. This framework can be used to assess bifurcation or change in system regime over time (Dezzani 2002). In keeping with a desire to increase quantitative rigor, this approach also permits predictions of change at future time using the basic Markov recursion conditional on the estimation of a stationary transition probability matrix (Bartholomew 1982: 14–16; Bartlett 1978: 24–28).

As the approach is primarily Markov-based, observations across time are assumed to be path-dependent and thus, consistent with historical outcomes. All observations are considered as part of a systemic partition and consistent with the Hanink (1991) description of a piecewise continuous world-economy. Furthermore, all countries are assumed to be classifiable into one zone with minimum error of estimation; the classification zones are assumed to be independent but countries or polities are assumed to be conditionally independent (Dezzani 2002).

The arrangement allows the direct use of classification outputs as frequencies in a time-based cross-classification such that the zones are represented as the rows and columns of a contingency table, but the rows are defined by a previous time assignment and the columns are the succeeding time assignment. The matrix elements are then the frequency of countries or polities that persist or remain in their zone across both time periods or transition across zones either "up" the hierarchy (lower triangular matrix below diagonal), or "down" the hierarchy (i.e., upper triangular matrix above diagonal), as shown (Dezzani 2002):

$$\begin{pmatrix} F_{cc} & F_{cs} & F_{cp} \\ F_{sc} & F_{ss} & F_{sp} \\ F_{pc} & F_{ps} & F_{pp} \end{pmatrix}$$

F_{ij} are observed frequencies of a country being in zone I at a previous time and zone j at a succeeding time. In this case, the indexes i and j take on values for world-economic zones of core, semiperiphery, and periphery (i.e., "c," "s," and "p," respectively). In this example, the diagonal elements are "persistent," meaning these are countries that are not mobile over the time interval. Maximum likelihood estimates of the Markov transition probabilities can then be computed using the previous time classification as a basis:

$$\begin{pmatrix} F_{cc}/\sum_{\forall i} F_{ci} & F_{cs}/\sum_{\forall i} F_{ci} & F_{cp}/\sum_{\forall i} F_{ci} \\ F_{sc}/\sum_{\forall i} F_{si} & F_{ss}/\sum_{\forall i} F_{si} & F_{sp}/\sum_{\forall i} F_{si} \\ F_{pc}/\sum_{\forall i} F_{pi} & F_{ps}/\sum_{\forall i} F_{pi} & F_{pp}/\sum_{\forall i} F_{pi} \end{pmatrix}$$

In this manner, all transition probabilities may be computed for any two time periods or aggregated across all time periods. As an example, Dezzani and Babones (2008) produced a stationary transition matrix for the time period 1975 to 2005 as follows:

$$\begin{pmatrix} 0.84375 & 0.15625 & 0.0 \\ 0.04348 & 0.82609 & 0.13403 \\ 0.01923 & 0.15385 & 0.82692 \end{pmatrix}$$

This matrix describes "transitions" of countries across zones for the time period 1975 to 2005. The matrix elements are maximum likelihood estimates of the stationary Markov transition probabilities that serve as aggregates of individual country mobility. The diagonal elements represent systemic inertia as individual zone persistence—the likelihood that once countries entered a zone, they were likely to remain in that zone over the time interval. It is quite clear that persistence is quite similar for core, semiperiphery, and periphery (i.e., 0.8438, 0.8261, and 0.827, respectively) (Dezzani and Babones 2008). During the same time period, it was more likely for semiperipheral countries to drop into the periphery (p_{sp} = 0.134) than to advance into the core (p_{sc} = 0.0435). The likelihood of peripheral states entering the semiperiphery (p_{ps} = 0.154) was about the same (statistically equivalent) as the country mobility from the semiperiphery into the periphery. There was also a fairly large chance of core countries losing position in the core to the semiperiphery (p_{cs} = 0.156). Thus, it can be demonstrated that both the structure of the world-economy and the likelihood of mobility can be estimated using these methods.

A statistical test of the matrix using log-linear models and a McNemar test for matrix symmetry suggests the structural arrangement and zonal transitions are marginally stable over the time period of the study (Bishop et al 1975; Boudon 1973). This suggests that no significant change in regime has occurred. This result is not surprising for the time period studied. However, it is highly significant to now possess a method that can provide statistically significant measures of transition and systemic regime shifts.

Validation of stationarity that will permit prediction is accomplished using a log-linear model that examines the components of effects in the matrix (Bishop et al 1975; Dezzani 2002; Fingleton 1997). Such a model may appear as:

$$\log(p_{ij}) = \mu + \mu_{ti} + \mu_{tj} + \mu_{(ti \times tj)}$$

where μ are the log-likelihood of the matrix cell expectation such that the sum of the individual effects is zero and can then serve as a model of row symmetry. If the transition matrix, called T, is stationary then it can be used as a basis for prediction using:

$$Z^t = T\,Z^{t-1}$$

where Z is a vector of classification assignments and T is the stationary transition matrix. The matrix elements of T are the stable p_{ij} transition probabilities. To obtain explanation of change in the transition probabilities p_{ij}, a logistic regression framework can be employed to set the probabilities as dependent variables and use both endogenous zonal, country-based and system change variables to serve as an independent explanatory basis. The logistic model would take the form:

$$P_{ij} = F(V_{ij}) = F\left(\beta_0 + \sum_i \beta_i X_{ij}\right) = (1 + e^{-V_{ij}})^{-1}$$

where P_{ij} are the transition probabilities and F is a function of explanatory variables X_{ij} designed to explain the mobility of countries from zone i to zone j in the specified time interval defined in terms of change from the base classification i. Methods such as these will provide a rigorous basis of transition probability explanation through the use of time-change-based independent variables (Agresti 1990). In addition, these models can provide tests of specific hypotheses regarding specific variable sets and their ability to explain induced change in country-level mobility.

The promise of quantitative approaches in world-systems and future research

It is quite clear that a great deal of progress has been made regarding measurements of country-level mobility in the zones of the hierarchical world-economy as well as the development of methods that delineate the structure of the hierarchical world-economy. These efforts over the last 30 years represent the development of a rigorous, quantitative framework for the world-system perspective. It is now possible to quantitatively fully describe the zonal structure of the world-economy using a variety of valid methods, and estimate movement of country mobility within that structure across any time period. It is also possible to estimate the stationarity of generating processes that produced the structure and induced the country mobility behavior. It is also possible at this time to provide explanations of the general stochastic transition behavior and study the sensitivity of the hierarchical mobility to changes in specific economic, political, and social variables at different scales.

Advances in quantitative methodology, particularly differentiable dynamics, Bayesian statistical models, random field theory, hierarchical stochastic modeling, time series modeling, and spatial and space-time analysis provide ideas for thinking about generating processes that create and modify structures of the world-economy inducing evolutionary change. However, these methods and statistical approaches also provide the means of parameter estimation and quantitative problem solution that were not available 30 years ago. Granted, the learning curve will be steep as these methods require profound knowledge of mathematical and statistical theory and procedures as well as evolutionary theory, information and entropy and general systems frameworks to achieve full fruition and implementation. Perhaps, in the next 30 years, the world-systems community will contribute to a fully developed quantitative framework that is compatible with the rich theoretical and descriptive tradition that the world-systems paradigm has provided thus far.

References

Agnew, John. (2005). *Hegemony: The New Shape of Global Power*. Philadelphia, PA: Temple University Press

Agresti, Alan. (1990). *Categorical Data Analysis*. New York: John Wiley & Sons.

Arrighi, Giovanni. (1994). *The Long Twentieth Century*. London: Verso.

Arrighi, Giovanni. (2007). *Adam Smith in Beijing: Lineages of the Twenty-First Century*. London: Verso.

Arrighi, Giovanni and Beverly J. Silver. (1999). *Chaos and Governance in the Modern World System*. Minneapolis, MN: University of Minnesota Press.

Arrighi, Giovanni and Jessica Drangel. (1986). "The Stratification of the World-Economy: An Exploration of the Semiperipheral Zone." *Review* 10: 9–74.

Arthur, W. Brian. (1994). *Increasing Returns and Path Dependence in the Economy*. Ann Arbor, MI: University of Michigan Press.

Ayres, Robert U. (1994). *Information, Entropy, and Progress: A New Evolutionary Paradigm*. New York: American Institute of Physics Press.

Babones, Salvatore J. (2005). "The Country-Level Income Structure of the World-Economy." *Journal of World Systems Research*, 11: 29–55.

Babones, Salvatore J. (2009). *The International Structure of Income. Its Implications for Economic Growth*. Saarbruken, Germany: VDM.

Bartholomew, D. J. (1982). *Stochastic Models for Social Processes*. 3rd edn., Chichester: John Wiley & Sons.

Bartlett, M. S. (1978). *An Introduction to Stochastic Processes*. 3rd edn., Cambridge: Cambridge University Press.

Bishop, Yvonne M. M., Stephen E. Fienberg, and Paul W. Holland. (1975). *Discrete Multivariate Analysis*. Cambridge, MA: MIT Press.

Bodley, John H. (2003). *The Power of Scale: A Global History Approach*. Armonk, NY: M. E. Sharpe.

Boudon, Raymond. (1973). *Mathematical Structures of Social Mobility*. Amsterdam, The Netherlands: Elsevier.

Cavalli-Sforza, L. L. (1981). *Cultrual Transmission and Evolution: A Quantitative Approach*. Princeton, NJ: Princeton University Press.

Chase-Dunn, Christopher. (1998). *Global Formation*. Lanham, MD: Roman & Littlefield.
Chase-Dunn, Christopher and E. N. Anderson. (2005). *The Historical Evolution of World-Systems*. New York: Palgrave Macmillan.
Dendrinos, D. S. and M. Sonis. (1990). *Chaos and Socio-Spatial Dynamics*. New York: Springer-Verlag.
Dezzani, Raymond J. (2001). "Classification Analysis of World-Systems Regions." *Geographical Analysis*, 33: 330–52.
Dezzani, Raymond J. (2002). "Measuring Transition and Mobility in the Hierarchical World-Economy." *Journal of Regional Science*, 42: 595–625.
Dezzani, Raymond J. and Salvatore Babones. (2008). *Mobility in the World-Economy 1975–2005*. Presentation at the *Andre Gunder Frank's Legacy of Critical Science* conference at the University of Pittsburgh, April 11–13.
Fingleton, Bernard. (1997). "Specification and Testing of Markov Chain Models: An Application to Convergence in the European Union." *Oxford Bulletin of Economics and Statistics*, 59(3): 385–403.
Friedman, Jonathan. (2005). "Plus Ça Change? On Not Learning From History", pp. 89–114 in Jonathan Friedman and Christopher Chase-Dunn (eds). *Hegemonic Declines Present and Past*, Boulder, CO: Paradigm.
Georgesu-Roegen, Nicholas. (1971). *The Entropy Law and the Economic Process*. Cambridge, MA: Harvard University Press.
Hanink, D. (1991). "A Hierarchical Market Model of International Trade." *Geographical Analysis*, 23(2): 147–57.
Kick, E. (1987). "World-System Structure, National Development, and the Prospects for a Socialist World Order", pp. 127–155 in Terry Boswell and Albert Bergesen (eds.) *America's Changing Role in the World System*. New York: Praeger.
Korzeniewicz, Roberto Patricio and William Martin. (1994). "The Global Distribution of Commodity Chains", pp. 67–91 in Gary Gereffi and Miguel Korzeniewicz (eds). *Commodity Chains and Global Capitalism*, Westport, CT: Praeger.
McLachlan, G. J. (1992). *Discriminant Analysis and Statistical Pattern Recognition*. New York: Wiley.
Nemeth, Roger and David A. Smith. (1985). "International Trade and World-System Structure: A Multiple Network Analysis." *Review*, 8: 517–60.
Nicolis, Gregoire and Ilya Prigogine. (1989). *Exploring Complexity*. New York: W. H. Freeman.
Pomeranz, Kenneth. (2000). *The Great Divergence*. Princeton, NJ: Princeton University Press.
Ripley, Brian D. (1996). *Pattern Recognition and Neural Networks*. Cambridge: Cambridge University Press.
Sanderson, Stephen K. (ed.) (1995). *Civilizations and World Systems*. Walnut Creek, CA: AltaMira Press.
Smith, David A and Douglas R. White. (1991). "Structure and Dynamics of the Global Economy: Network Analysis of International Trade 1965–1980." *Social Forces*, 70: 857–93.
Snyder, David and Edward Kick. (1979). "Structural Position in the World-System and Economic Growth, 1955–1970: A. Multiple-Network Analysis of Transnational Interactions." *American Journal of Sociology*, 84: 1096–126.
Tainter, Joseph A. (1988). *The Collapse of Complex Societies*. New York: Cambridge University Press.
Wallerstein, Immanuel. (1974). *The Modern World System I*. San Diego, CA: Academic Press.
Wallerstein, Immanuel. (1980). *The Modern World System II*. San Diego, CA: Academic Press.
Wallerstein, Immanuel. (1989). *The Modern World System III*. San Diego, CA: Academic Press.
Wallerstein, Immanuel. (1991). *Unthinking Social Science: The Limits of Nineteenth Century Paradigms*. Cambridge: Polity Press.

4.2

World-systems as dissipative structures

A new research agenda

Peter E. Grimes

Introduction

When I was an undergraduate, I attended a talk by a philosopher about the social sciences. He opened his presentation with the rhetorical question: "Is a physics of society possible?" He then proceeded to attack the social science pretensions of the early 1970s, pointing out that humans were not inert and predictable particles, but self-conscious actors who shaped their own destiny. Hence the rigid rules of physics could never apply to human behavior.

Ironically, even then research in both physics and sociology was occurring that would later converge. In physics, the study of fluid dynamics would give rise by the late 1970s to "chaos theory"; while at the same time in sociology, a new interpretation of human history, later called "world-systems theory," tried to integrate the social sciences (Arrighi and Goldfrank 2000; Chase-Dunn and Grimes 1995; Gleick 1987; Grimes 2000). Their coincident emergence was independent; yet 40 years on, their findings have linked in unexpected and exciting ways. How these developments overlap is the focus of this contribution.

The first half of this essay will explain the revolutionary departures from classical physics emerging as "chaos" and "complexity" theories, leading to the physics of spontaneous, unpredictable, *self-organization* in "far-from-equilibrium systems of energy flow." The second half will apply these lessons from physics to living societies.

But before I can clarify these developments, I must explain some basic concepts in physics.

Physics and the problem of time

We have all been taught by Einstein, via his famous equation $E = mc^2$, that energy is an expanded form of matter. One could also say that matter is a condensed form of energy. Both observations are true, because it is an equation.

Similarly, the brilliant equations of Newton describing the force of gravity can be applied with exquisite accuracy to predict the orbits of planets around the sun, or the trajectory of a bullet. The predictive triumphs of these masterful achievements dazzled those who understood them, leading some optimists to believe that the history and future of the entire universe could, in principle, be determined given enough information about the component particles, directions, and energies within it (Prigogene 1996: 11).

However, there was a problem: precisely because Newtonian mechanics used equations (as with Einstein's equation), there is no "directionality" to time. The conversion of matter into energy and back is equally probable, and the equations describing planetary orbits would work equally well if the directions of those orbits were reversed. In a completely determined universe, *time has no relevance*. Yet ordinary experience says otherwise. Bullets do not spontaneously jump from the ground back into the gun chamber. Hot objects cool down to room temperature, but objects at room temperature do not spontaneously grow hot. Broken glasses do not re-assemble. Time exists and has a direction, yet this did not appear in the equations of classical physics (Prigogene 1996).

Thermodynamics, entropy, and time

The study of movement as flows of energy emerged in the early nineteenth century. The reason was practical. European industry was trying to improve the efficiency of the steam engine (Atkins 1984). The central problem was how to transfer the heat most effectively from the fuel, via the boiler, to the moving parts. It was observed that with every transfer the density or quality of "energy" (defined as the ability to do "work," to move matter) declined. This study of energy loss created the field of *Thermodynamics*. The conclusion was that steam engines, and indeed, *all* systems of energy transfer must *necessarily lead to energy decay*. After an initial energy input, the more matter is transformed, the more the recipient molecules radiate uncaptured energy, which dissipates/ degrades to less powerful forms, ending as heat. Although the input and output energy in the steam engine is all measured as heat, the temperature of the input heat is higher (containing more energy) than the cooler output (less ability to move matter).

Thermodynamics led to many new observations and laws of physics. The following two laws are the most important:

First Law: Energy Conservation. Energy cannot be created or destroyed, but can only change form. Hence the total energy from the "Big Bang" continues to exist, albeit in degraded forms.

Second Law: Entropy. High-density energy, via inefficiencies in moving matter, loses the power to do work and ultimately degrades into less heat, radiating into the cold void of space.

(Atkins 1984: 8–10)

Through the second law of entropy, physics could at last explain and incorporate time. The reason a broken glass cannot spontaneously re-assemble is that the amount of energy required to create that initial order is far greater than what remains after that order is dissipated in the breakage. (Imagine the human time and machine energy needed to make a glass.) Likewise, hot objects radiate heat into the environment until they are at the same temperature as their surroundings. Once they are, the heat flowing out of the object equals that flowing in. Put simply, entropy asserts that the loss of energy into the environment creates an irreversible direction to time. *Hence time is our human metric applied to entropy* (Prigogene 1996).

Thermodynamics, energy, and self-organization

With these basic principles in mind, we can now return to more modern developments in physics that can apply to the social sciences.

Turbulence is one object of study in physics. Sufficient energy injected into fluids and gases excites them to produce seemingly unpredictable material flows. The study of turbulence gave rise to the "chaos theory" emerging during the 1970s. The eventual successes of this field yielded another area of inquiry called "complexity theory": the study of order "spontaneously" emerging within systems pushed by sufficient energy into "far-from-equilibrium" states (Nicolis 1989).

One simple example is a pot of water quietly sitting at room temperature. It is in a state of "thermodynamic equilibrium": the temperature of the water is equal to the air around the pot. Within the pot, water molecules are bouncing about randomly at a rate reflecting the room temperature. Yet this equilibrium state has some interesting qualities. A hypothetical tiny observer inside the pot could not report either its location or the passage of time. In every direction of space, the molecular bounces are the same, and they stay the same over time. This sameness across space makes the environment appear identical regardless of location. In physics, such homogeneity is called "symmetry." Likewise, as long as the temperature stays the same, the pace of molecular bouncing stays the same over time. Hence time is also "symmetrical."

Now imagine that someone wishes to make tea using that water. They turn on the burner beneath the pot. The temperature rises to a boil: a critical ceiling of 212 degrees (*F*). At that point several changes happen to the water. First, "convection cells" develop. Specifically, water rising from the heat below reaches the surface, releasing heat to the atmosphere. The cooler water falls back to the bottom. These water movements up and down quickly become organized into self-contained circulating "cells" of similar size. Viewed in vertical cross-section, each cell rotates either clockwise or counter clockwise; but whichever the direction of a given cell, its neighbor will rotate oppositely. The emergence of these cells is predictable, but their initial rotation is impossible to predict. Some unknown minor initial condition "encourages" the cells to "choose" a direction. Once the rotation starts, the directions are locked in. This unpredictable "choice" is a central observation of complexity theory: *complex structures are extremely sensitive to initial conditions.*

A second change is that the "symmetry" of both space and time is broken with boiling. The emergence of cells of opposite rotation creates an internal structure to every space within the pot, allowing our observer to report "where" it is in space. Likewise, by moving horizontally across cells at a constant rate, the observer can also measure time by noticing the switches in the direction of water flow. The creation of structure and order in far-from-equilibrium systems *breaks the symmetry of equilibrium.* New shapes emerge in space, and they develop and elaborate over time.

A third change is more subtle. Although water boils at 212 degrees, this is an average. Some water molecules may be ejected from the pot as vapor above that temperature, others below. This is another quality of complexity, called "deterministic chaos." The average boiling temperature is *determined*, but the fluctuations around it are *unpredictable and chaotic* (Nicolis 1989: 316–20).

Note that the emergence of structure within the boiling pot does not violate any physical laws, of either classical physics or thermodynamics. The flow of input energy dissipates exactly as predicted by Newtonian mechanics; and likewise, it is consistent with the laws of thermodynamics: the total amount of energy flowing in and out is *conserved*, even as the energy intensity—the ability to move matter—*degrades into heat* (consistent with *entropy*).

Yet, paradoxically, along the path of energy degradation, this injected energy actually creates and sustains complex structures of water flows inside the pot that are only possible "far-from-equilibrium."

Complexity theory and dissipative structures

The example of turbulence in boiling water as not "chaotic" but actually highly organized was typical of the discoveries of chaos and complexity theories in the 1970s–80s. Given the core insight that sufficiently high inputs of energy can cause complex organizations within liquids and gases, examples pop up everywhere: whirlpools in a stream, hurricanes, tornadoes, candle flames, and even stars. The "life-span" of these examples varies from a few minutes to billions of years. Yet they each share several features in common with the boiling water:

(A) Given sufficient energy input, turbulence becomes far-from-equilibrium, and at a critical point "self-organizes" into material structures.
(B) The channels of energy flow that give shape to the structure are not entirely predictable in advance.
(C) The boundaries of these systems are fuzzy, both "determined" by the volume of energy flow yet "chaotically" jumping around that boundary.
(D) Complex systems have a specific history unique to that particular structure. Because of extreme sensitivity to initial conditions, their emergence and resulting structure cannot be precisely predicted in advance. Nor can the timing and form of their collapse. For example, every star is unique, despite the advanced knowledge we have about how stars work. This historical specificity is another quality separating complex structures from other phenomena studied by physics. Hence complex systems evolve over time in a way that uniquely reflects their respective pasts (path dependence). However, that does not prevent prediction (remember the certainty of convection cells); it only shifts it into the realm of probabilities. Stars may be unique, but all follow the rules of stellar evolution.

These common features are grouped by complexity theorists into a category called *dissipative structures*. They are *dissipative* because the energy they release is equal to that received but less able to do work. Yet while dissipating that energy they spontaneously self-organize into *structures* that are unlikely and can only exist far from equilibrium (Nicolis 1989; Prigogene 1996; Warren et al 1998).

Life as a dissipative structure

Living things are also dissipative structures, although far more complicated than the examples given above. Their energy is chemical and channeled via chemical interactions, yet they share in common the defining features of far-from-equilibrium structures. The medium of life is fluid, containing complex chemicals—themselves far from equilibrium—whose interactions are driven by energy and convey that energy to other molecules in a highly organized way to maintain the overall structure. The main distinction between living and non-living, as used here, is that living structures seek out new energy sources (they eat), and living systems reproduce. Stars and whirlpools do neither. They simply exist until they run out of energy. Yet in all cases, when dissipative structures lose their energy supply, they cease to exist and return to equilibrium (die) (Kauffman 1993; Lehn 2002; Nicolis 1989).

Bifurcation and evolution

Recall that in the example of the boiling water, the initial rotation of the convection cells cannot be predicted. It is an apparently random "choice," based perhaps upon subtle initial conditions. Among complexity theorists, this unpredictable "choice" is called a "bifurcation." As energy is pumped through an evolving complex system, new structures and forms may develop that cannot be predicted in advance: the precise flow of energy could have been dissipated equally effectively "choosing" a different material structural path.

Now consider a population of organisms. It increases quickly until it reaches the ceiling of food supply, which acts like the boiling point of water. On average the population cannot exceed the ceiling, but sometimes it does overshoot before contracting, and at other times it contracts before hitting the ceiling. As with the boiling water, it is in a condition of "deterministic chaos."

However, imagine that these organisms discover a new and different food supply in addition to their usual food. This discovery will potentially raise the population ceiling, but only if it is accompanied by appropriate behavioral changes in the organisms (learning how to capture the new food). This discovery and the capacity to change behavior accordingly are also *bifurcations*. Both the discovery of the new food source *and* the necessary changes in lifestyle are unpredictable. Given the new food source, it is equally possible that the changes in lifestyle will not occur, thus forcing the population back below the original ceiling. But if *both* changes happen, the internal organization and relations among the population will elaborate into new structures of energy flow. In this case, bifurcation evolves new population structures.

In retrospect, *all of evolution can be viewed as a series of bifurcations resulting in new and more complex material elaborations of energy acquisition and flows. All changes to a species, including the evolution of new species, can be thought of as bifurcations in a complex dissipative structure* (Nicolis 1989: 325–28; Prigogene 1996).

Energy flow between organisms

Ecologists have spent decades measuring energy flow between species in a variety of locations. They have discovered that plants use about 2 percent of the solar energy they receive; herbivores and carnivores, around 15 percent of what they eat. These low efficiencies of energy use illustrate the toll taken by entropy. That same toll also imposes a pyramidal structure on the abundance of each category of life: despite the low efficiency of plants, the sheer volume of sunlight is so great that plants are far more abundant than the other two categories. Given that abundance, the collective weight (biomass) of all of the herbivores cannot exceed 15 percent of that of the plants, and the biomass of the carnivores is similarly constrained with respect to the herbivores. Hence the input of solar energy is transferred across organisms via predation and channeled "upward" into a shallow pyramid (Carbone and Gittleman 2002; Colinvaux 1978; Marquet 2002; Raffaelli 2002).

In turn, each layer has evolved/bifurcated over evolutionary time to specialize in a specific method of energy capture. Via this evolved channeling of solar energy across species, ecosystems are also *dissipative structures*. They absorb input energy, conducting it upward into decreasingly probable (far-from-equilibrium) specialized life forms, meanwhile emitting exhaust via heat, sewage, and dead bodies.

Predation is not the only way that energy can flow between organisms. Another common method is *parasitism*. Parasites such as ticks and common diseases rarely kill their hosts; instead, they simply place an additional energy tax on them. However, there are other, more subtle forms of parasitism that exist between organisms of the same species, and that form the core method of energy transfer between humans.

Consider a nesting pair of birds whose chicks have hatched. To feed them, one parent will leave the nest to locate and capture food, flying home with its catch for the babies. The chicks receive not only the energy in the food, but also the energy expended by the parent to *get* the food; it is energy the chicks would otherwise have had to expend *themselves*. Instead, this energy spent by the parent (unused by the chicks getting their own food) can be spent by the chicks on their own growth.

The same parasitic energy flow occurs among humans—not just between adults and children, but also among adults. Suppose you need to move between dwellings. You prevail upon a friend to help you. The labor of that friend is energy that liberates you to do more tasks that day; there is a net transfer of energy from your friend to you.

Human societies as dissipative structures

All sociologists are trained to recognize the major social divisions between groups. In order of sequential historical appearance, these divisions are age/gender, class, ethnicity, and states. For simplicity I will suggest methods of biological parasitism across these social divisions in pre-industrial agrarian societies. That allows us to by-pass the complications introduced by modern fuels.

Age

In pre-literate societies, the elders were the repositories of cultural knowledge. Language and culture accelerate adaptation to environmental change (in contrast to genetic variation), therefore the knowledge of the elders would be highly valuable. Because of their age, their ability to contribute to the food supply would have been small, suggesting that they were net food consumers: biological parasites.

Gender

Humans inherited strategies of sexual selection adopted by other large mammals. For females, the criteria for mate selection are male health and social status. Both improve the chances of survival for their young, genetically and nutritionally (respectively). Unlike females, males cannot be sure that they are truly the biological fathers unless they completely control the reproductive lives of their mates. In complex chiefdoms and agricultural societies, as the inheritance of class status and possessions was fully institutionalized, the reproductive stakes were greatest for males to insure the genetic connection to their young. Hence patriarchy peaked in agrarian societies, women were viewed as property, and control over their daily lives became a form of terror. This patriarchal peak of control also allowed for the exploitation of female labor, creating another parasitic channel of energy flow from women to men.

Class

True social classes emerged in complex chiefdoms. There is a natural tendency for class status to become hereditary. Children learn language, social skills, and practical skills from their parents. If your parents are farmers, potters, or metal-workers, you will absorb those skills with socialization. The pursuit of that profession would be the child's default choice, unless blocked by exterior forces. The caste system is a classic cultural artefact of the inheritance of class/occupation from that era persisting into the present.

Class differences are another opportunity for parasitism—the transfer of energy from one class to another via unequal exchange. The lower the class status, the harder the workload and the lower the social value of one's product. The peasants passed not only food up the social ladder, but also the energy required to grow that food. Generalizing this example across society as a whole, *there is a net flow of energy upward from those with the lowest class status to those with the highest.*

Ethnicity

In agricultural societies, wealth comes from control of arable land. Conversion of land into crops for human use powers the entire social structure. Land *is* power. Thus ambitious agricultural societies are motivated to expand their land area. As a result, successful aggressors eventually incorporated distant societies with alien cultures, languages, and religions. Integration of other cultures gave rise to the first true multi-cultural empires: subsuming under one political authority diverse conquered peoples who were often enslaved and eventually assigned a social status. While slavery was endemic to the agrarian world, it was not necessarily a permanent status. Yet the ethnic divergence was. The conquered peoples and their descendants were consigned to "inferior" occupations requiring more daily energy, which in turn was dutifully passed up the social ladder. Such conquests also reflected the emergence of imperialism and the core-periphery hierarchy of exploitation both within and between states, a hierarchy that persists to this day.

States

States have a governing elite protected by military power. If they were isolated, the elite would be the final repository of all the available energy remaining from the social categories over which they preside. Yet trade between states allows energy to flow, and it is the unequal nature of that trade (and energy) which is axiomatic to world-systems theory (Amin 1974). If such inequality were absent, there would be no cores or peripheries (Hornborg and Crumley 2007).

In summary, my argument here is that just as in ecosystems, every transfer of energy between groups is subject to entropic waste and loss. Hence as energy passes up the social hierarchy, less is available to those at the top. In other words, entropy requires that the numbers of rulers (like those of top carnivores) be far fewer than the numbers of those at the bottom. *The social pyramid is ultimately rooted in the laws of thermodynamics.*

I also argue, like Straussfogel (2000), that collectively, world-systems and all of the social divisions within them are dissipative structures channeling input energy via parasitism to a ruling elite geographically scattered across the core. The energy inputs today are quite large and varied, resulting in the same output as ecosystems but on a much larger scale: heat, waste pollutants, and dead bodies. (It is the misuse of these inputs and outputs that has created the environmental crisis of our time.) As with other dissipative structures, the addition of greater energy enabled and compelled the emergence of bifurcations creating new social structures and institutions which could, in principle, have been scientifically assigned probabilities in advance.

Conclusion

Clearly, measuring and testing these ideas is daunting, and would take much time. So what is the reward? Why view world-systems through the lens of complexity and dissipative structures?

It allows us to use the common energy metric to conduct comparisons across social systems and other dissipative structures, such as ecosystems and organisms. It even enables us to compare ourselves to non-living structures, thereby helping us learn greater efficiency and forewarning us of symptoms of system collapse. Such symptoms may be common across all complex structures. Finally, it can guide policy. The enormous gap between the elite and the bottom costs huge inefficiencies and may contribute to system collapse. For example, history teaches us that great inequalities of energy and diminishing information flows from the bottom to the top of complex structures are precursors of collapse in both human systems (Fischer 1996; Grimes 1999; Tainter 1988) and ecosystems (Neutel et al 2002). We may find that closing that gap, accompanied by decentralization, could make us more sustainable, shorten the feedback between the environment and governance, and improve the quality of our lives. Certainly we have already learned that mapping the world of Newton onto social policy must fail, because complex structures cannot be driven like a bus. That was the failure of nineteenth century social thought. Instead, complexity theory suggests that social structures can be encouraged to direct energy into one particular path that is more acceptable than another, potentially liberating us from the grip of random bifurcations and enhancing human control over our fate.

References

Amin, S. (1974) *Accumulation on a World Scale* (2 vols), New York: Monthly Review Press.

Arrighi, G. and Goldfrank, W. (eds) (2000) "Special Issue: Festschrift for Immanuel Wallerstein", *Journal of World-Systems Research*, 11: (2 and 3) Summer/Fall and Fall/Winter.

Atkins, P. W. (1984, 1994) *The Second Law: Energy, Chaos, and Form*, New York: W. H. Freeman.

Carbone, C. and Gittleman, J. L. (2002) "A common rule for the scaling of carnivore density", *Science*, 295: 2273–76.

Chase-Dunn, C. and Grimes, P. (1995) "World-systems analysis", in J. Hagan and K. Cook (eds) *Annual Review of Sociology*, 21: 387–417.

Colinvaux, P. (1978) *Why Big Fierce Animals Are Rare*, Princeton, NJ: Princeton University Press.

Fischer, D. H. (1996) *The Great Wave: Price Revolutions and the Rhythm of History*, New York: Oxford University Press.

Gleick, J. (1987) *Chaos: Making a New Science*, New York: Viking Penguin.

Grimes, P. (1999) "The horsemen and the killing fields: The final contradiction of capitalism", in A. Goldfrank and A. Szaz (eds) *Ecology and the World-System*, Westport, CT: Greenwood Press.

Grimes, P. (2000) "Recent research on world-systems", in T. Hall (ed.) *A World-Systems Reader*, New York: Rowman & Littlefield.

Hornborg, A. and Crumley, C. (eds) (2007) *The World System and Earth System: Global Socioenvironmental Change and Sustainability since the Neolithic*, Walnut Creek, CA: Left Coast Press.

Kauffman, S. A. (1993) *The Origins of Order: Self-Organization and Selection in Evolution*, New York: Oxford University Press.

Lehn, J. (2002) "Toward self-organization and complex matter", *Science*, 295: 2400–403.

Marquet, P. A. (2002) "Of predators, prey, and power laws", *Science*, 295: 2229–30.

Neutel, A., Heesterbeek, J. A. P. and de Ruiter, P. C. (2002) "Stability in real food webs: Weak links in long loops", *Science*, 296: 1120–23.

Nicolis, G. (1989) "Physics of far-from equilibrium systems and self-organization", in P. Davies (ed.) *The New Physics*, New York: Cambridge University Press.

Prigogene, I. (1996) *The End of Certainty: Time, Chaos, and the New Laws of Nature*, New York: The Free Press.

Raffaelli, D. (2002) "From Elton to mathematics and back again", *Science*, 296: 1035–37.

Straussfogel, D. (2000) "World-systems theory in the context of systems theory: An overview", in T. Hall (ed.) *A World-Systems Reader*, New York: Rowman & Littlefield.

Tainter, J. H. (1988) *The Collapse of Complex Societies*, New York: Cambridge University Press.

Warren, K., Franklin, C. and Streeter, C. L. (1998) "New directions in systems theory: Chaos and complexity", *Social Work*, 43: 357–72.

4.3

Narrating stories about the world-system of the First Global Age, 1400–1800

J. B. Owens

This chapter will deal with how narrative and social network analysis (SNA) can be effectively integrated into a world-systems approach to the First Global Age, 1400–1800. The essay takes as its starting point Andre Gunder Frank's *ReORIENT* (1998). Owens worked closely with Frank during the writing of this book and afterwards. In the book, Frank metaphorically proposed a three-legged stool to represent the factors that must be integrated into a satisfactory world-systems analysis, but because he did not know how to achieve this integration, he concentrated only on the "economic leg," even then neglecting, as he later realized, the environmental aspects of the issue. In response to Frank's challenge to show how the social/institutional and cultural environments (the missing two "legs" of the metaphorical stool in *ReORIENT*) shaped human action, Owens (2005) demonstrated the way that competing institutions and multiple, non-hierarchical social norms in the cultural environment shape human activities, events and processes in a nonlinear fashion. Triggered by discussions with Frank about integrating the "legs," Owens explored geographic information systems (GIS) as an integrative information technology. Prompted by Frank's insistent questions about what kind of system the First Global Age was, Owens created the "Geographically Integrated History" concept. Owens defined this historical period as a complex, dynamic, nonlinear system, and on this theoretical basis, he offers a different interpretation of Frank's "inclination" in the world-system, for which Frank never adequately accounted. Frank argued that the world-system was a complex one, which influenced the histories of all locations all of the time, but he continued to frustrate Owens' attempts to get him to recognize that his failure to consider how local action influenced the system devalued his own activist career. To deal with this aspect of world-system analysis, Owens focused on commercial, political, and other social networks, particularly clandestine ones, to expose how strategic interactions in these networked relationships could produce new forms, which influenced the system of which the networks were a part.

Owens (2007a) has told elsewhere the story of how he was introduced to GIS during a discussion of possible titles for Frank's book. After a period of study, he recognized that GIS offered both the possibility of integrating data from all three legs of Frank's metaphorical stool in a context that also permitted representing the webs of connections among geographic locations during the First Global Age (Owens 2007b). However, current GIS integrates poorly with world-systems analysis. Many speakers and writers have attributed to Albert Einstein the assertion that, "Problems cannot be solved by the same level of thinking that created them." With

Geographically Integrated History, Owens created a level above muddled attempts to harmonize research done, whether using GIS or not, within historical geography and by historians interested in geography (on the muddle, see Baker 2007) and the failed explanations of historians and historical social scientists who have not recognized the importance of geography for an adequate understanding of their problems.

As a research and teaching discipline, Owens founded Geographically Integrated History on the idea that the understanding of historical processes requires an integration of the natural, social, and cultural environments on the basis of place, space, and time and accomplishing this integration poses a challenge, which can be met with modern information management, especially GIS, and visualization techniques. The underlying transformative Geographically Integrated History research paradigm posits that:

1. The history of any place is shaped by the way the place is connected to other places and by the changes in these connections over time (De Blij 2009; Organization of American Historians 2000; Owens 2007b; Subrahmanyam 2004, 2007);
2. Contrary to Frank's assertion of a 5,000-year world-system, historical periods are complex, dynamic, nonlinear systems, which are spatially large, and in more recent centuries, global in extent, and which sometimes become unstable, leading to a phase transition, bifurcation, and the organization of new systems (on such systems, see Puu 2003: especially Ch. 12; Puu 2006: Ch. 9; Rosser 2000);
3. Within such systems, people and places are connected by self-organizing networks, which are the sources of innovation and the emergence of new forms (Burt 2004; Ikegami 2000, 2005).

Because the emergence of new forms in complex systems often moves research from a topic within one discipline to a topic within another, the attempt to understand the past must be interdisciplinary and collaborative.

The First Global Age, 1400–1800, was a complex, dynamic, nonlinear system. The main characteristics of such systems are:

1. First, although many factors are involved in complex systems, system stability depends on a small number of variables (sometimes only one), which are always near instability (Haken 1983). Rosser (2000: 54–61), after a thorough review of theories of complex systems, argues persuasively that Haken provides the most satisfactory understanding of such systems and, therefore, a point of synthesis for them. The term "stability" does not mean that nothing happened during this four-century period, for the system was highly dynamic.
2. Second, such nonlinear dynamic systems admit only limited predictability and research on them should not be directed to the study of long strings of linear causation but to identifying the system's characteristics for the understanding of it, of the emergence of new forms, and of the transition to a different system.
3. Third, within the dynamics of the system, new forms emerged that could compensate for weaknesses in the system caused by the proximity to instability of the stability-maintaining "control" factors. However, these weaknesses remained, and the complex, dynamic, nonlinear system of the First Global Age eventually entered into a period of as yet poorly understood systemic chaos, phase transition, and bifurcation during the period 1750–1850. Out of this transition appeared a new system, characterized by new forms of world interaction, institutional organization, values, and perspectives on the world. The point that after a phase transition to a new human system, the cultural environment would be dominated by

new values and perspectives about the world, impairing the ability of people to understand their ancestors in the earlier system, was emphasized for Owens by Sonis (2007).

The discipline of History depends heavily on narrative explanations, and narrative constitutes a unique and valuable form of knowledge (Bruner 1985; Geertz 1973; Hexter 1971; Hunter 1991; Launer 2002). Within the context of new GIS, narrative can provide an effective means to understand the nonlinear dynamics of complex world-systems. However, Frank challenged Owens to discover a way to connect local action to the world-system of the First Global Age and to explain how such local action influenced system dynamics. In response, Owens focused on social networks, particular commercial and political ones, as the key to the human system, and Owens argues that narratives provide an especially good window for understanding network dynamics and the interactions of social network dynamics with natural systems and geography.

Self-organization largely characterized these social networks because in an open system, their participants most often created and shaped them toward increasing complication, rather than depending on the leadership of some authoritarian entity or individual, independent of the networks themselves. Smuggling networks of the First Global Age, which accounted for a high percentage of world commerce, provide particularly striking examples of self-organization (e.g., Erikson and Bearman 2006; Moutoukias 1988; Mui and Mui 1968; Winius 1983). Because of this self-organization, commercial and related social networks constituted important sources of creativity and innovation, which can be understood as the emergence of new forms (Burt 2004; Ikegami 2000, 2005; Landa 1994; Schulte Beerbühl and Vögele 2004: editors' introduction; Sonis 2009). Although its call to recognize the networks connecting humans to the natural world is useful, Owens largely rejects Latour's (2005) Actor-Network Theory (ANT) because of its insufficient attention to the variable importance or weight of networked connections and to the plasticity of identity, which permits an actor to participate in polymorphous network domains (White 1995, 2002, 2008). For good introductions to SNA, see Easley and Kleinberg (2010), Grabher (2006), and Scott (2000).

Within the world-system of the First Global Age, events periodically disrupted these social networks. Sometimes, disruptions occurred because of processes within the human system itself, such as war, for example, but disruptions also occurred because of the interactions of the human system with complex, dynamic, nonlinear systems in the natural world, such as disease epidemics or earthquakes, to pick two disruptive interactions of recent concern. To maintain system stability, actors, both men and women, had to counter the disruptions by constituting more robust networks, but the resulting denser webs of connections permitted even more widespread future cascades of disturbances. In the later eighteenth century, such a cascade of innovation initiated a transition to a new historical period, the Second Global Age in which we now live. By combining a social network approach with a world-systems analysis, we offer a world historical account that addresses larger issues of network dynamics and provides a solution to the common criticism of world-system analysis that it ignores the relationships between the local and the larger system (Ngalamulume 2009).

Nonlinear dynamics permit only limited predictability. In other words, changes appeared in the First Global Age rather suddenly, without a long period of development. For reasons best explained in narratives, innovations cascaded through social networks with every disruption, threatening community stability and challenging actors, both men and women, to weave greater network robustness in an effort to protect them from disaster and enhance their ability to survive and prosper, however they understood these concepts. Therefore, knowledge about activities, events, and processes during these four centuries will not come from continuous narratives of developments over the entire period. Instead, each narrative would focus on the way that a cascade of disruptions and innovations prompted strategic actions to create more robust commercial and other social networks within an expanding geographic space, a process that permitted the emergence of new forms in the social and cultural environments of human action. This emergence of new forms sustained the system by dealing with weaknesses, which might otherwise have led to system collapse, as opposed to a phase transition. In general terms, this process increased network density and adaptation to geography and made the system more robust or stable. However, the complicated ways that multiple networks interlinked with each other over expanding geographic spaces enhanced possibilities that cascading disruptions would affect ever larger portions of the world-system and entail ever more elaborate efforts, ultimately frustrated, to repair damage and prevent such sizable disruptions in the future (for recent work on such cascades, see Buldyrev et al 2010; Vespignani 2010).

"Ultimately frustrated" because as a consequence of the nature of complex systems, the sources of systemic instability would not be eliminated by such repairs, and the increased network density and interlinking would make it ever more likely that, in an event that could not have been predicted, some disruption involving a "control" variable would cascade so widely that the system would enter a period of chaos and transition to a new system. Sometime in the eighteenth century, one or more of the stability-maintaining variables became unstable, and this situation led to system instability and the development of a period of bifurcation around which a new system organized. The sudden weakness of a "control" variable propagated throughout the system from a cluster of individuals who could spread their influence through commercial and other social networks by their impact on individuals who were near them in the nodes of these networks and wished to enhance their ability to survive and prosper (see Watts 2003: 229–44 and 250–52). Gladwell (2000) suggests that it is adoption by the influential and highly connected that leads to a cascade, but Watts' current work (e.g., Watts and Dodds 2007) shows that such elite influence probably does not exist. Researchers cannot understand the dramatic suddenness, in relative terms, of this process unless they grasp how individuals were involved in multiple social networks and how this involvement required them to shape their social identities depending on the variable contexts of their interactions with others. The global cascade of system instability depended on the irregular linkage of different social networks so that the sources of instability could be propagated before any system-saving repair was possible.

Because SNAs are frequently expressed as graphs, researchers can use evolutionary game theory to explore strategic interaction (e.g., cooperation) among individuals and groups in the creation and maintenance of commercial, political, and other social networks. Gintis (2009) argues that evolutionary game theory can serve as a unifying platform for the social sciences and permit analysis of the interplay of social norms, the integration into models of "gene-culture" co-evolution, and the development of narrative knowledge. The narratives about social networks aid evolutionary game theory in several ways. Lehmann et al (2008) review cooperation research, which is mostly done by evolutionary biologists and economists interested in game theory, and they account for the inadequacy of results by pointing to the failure to deal with the role of individual intentions, which are variables within strategic interactions that historical narratives highlight. By narrating lots of stories of disruptions, world-systems researchers will form an increasingly large catalog of possible interactions as the basis of models and theories. Roca et al (2009) show that so far, the game theory models which have been developed lack generality, for the details of the dynamics and the interactions modify the results in a qualitative manner. For this reason, Sánchez (2009; Sánchez et al 2009) proclaims that we need new models and theories, which are tasks to which world-systems historians can contribute through the generation of narrative knowledge.

In order to produce narratives that are sufficiently complicated to reveal the bases for strategic interactions, we must insert these networks into the dynamic geographic context, which we wish to express in a GIS capable of updating multiple data types. This geographic context accounted for many of the network disturbances, both in terms of the nonlinear dynamics of natural and human systems and of the opportunities provided by the increased space of human activity and connections among locations of human habitation. These disturbances of existing social networks produced activity, by men and women, to elaborate more robust networks, which led to the emergence of new forms of entrepreneurship, gender and political interactions, and cultural expression across a vast geographic landscape. There have been some recent, tentative efforts to integrate SNA into the networks' geographic environment, but available software does not permit results that are satisfactory for geographers or historians (e.g., see Radil et al 2010). Therefore, current research focuses on transforming GIS, which is largely static, deals poorly with temporal factors of change and movement, and does not permit the user to deal simultaneously with the interactions of more than a few variables. Historians require a "Dynamics GIS," to use the phrase of Yuan (2007, 2008, 2009; Yuan and Stewart Hornsby 2008), in order to construct complicated narratives capable of telling the necessary stories of system dynamics.

Although GIS users normally give little consideration to the matter, the database schemes we employ raise significant barriers to a dynamic GIS. We most commonly use relational database schemes, which store information in rows in two-dimensional tables. We store entities as rows, which fixes attributes. If an attribute changes, we must create a new row. This process produces the fragmentation of time within our databases. Almost worse for representing dynamics, we also store relationships in rows, which is a feature that makes it difficult to bundle many relationships to create complicated entities. Recently, Goodchild (2008; Goodchild and Glennon 2008) has advocated employing object-oriented database schemes, but these schemes present a significant barrier to their use by historians. The object-oriented schemes are limited to an arity of two, which means that any entity can only be connected to two other entities, at most. In computer science "arity" refers to the number of arguments that a function or operator can take. Thus, the scheme requires the user to treat all information as though it could be organized ontologically into a strict hierarchy. To destroy these barriers, Owens and his colleagues (Kantabutra et al 2010) now employ a novel database scheme, Intentionally-Linked Entities (ILE). ILE permits the use of entity and relationship sets and effectively removes arity limits. A user may build a complex "object" of an entity and relations. With its great flexibility, ILE permits users to organize dynamic data, expressed in complicated spatial and temporal entities, more effectively, while permitting faster queries and database modifications.

Finally, computers require numerical precision, but much significant historical data are qualitative, imprecise, ambiguous, incomplete, uncertain, contradictory, and otherwise messy and, therefore, difficult to represent in current GIS (Gregory and Ell 2008; Gregory and Healey 2007; Owens 2007a, 2008). The complexity of the human system of the First Global Age and its relation to geography constitutes a veritable fortress against numerical precision. Historians frequently resist any demands that they force their information into fixed categories required by a computational environment because so often, they feel that vital characteristics of some information will be lost when they do not fit precisely into a particular box. As the "domain" experts, after years of reading the surviving documents and examining other period objects, historians develop a sense of the system, of degrees in the characterization of human behavior, and of the meanings conveyed in the natural language of the people they study. Historians find shades of meaning in the evidence that defy conventional numerically based approaches. Animal ecologists and geographers often confront the same ambiguity. For example, where is southeastern Spain? "Essentially, vagueness results in a range of interpretations for a place reference and a range of borderline values for the spatial footprint of each reference" (Hill 2006: 28–29). But we have a way forward.

As a solution to the incompatibility of complexity and precision, Zadeh (1965, 1994, 1997) created fuzzy logic (also, fuzzy set theory) to permit us to think precisely about ambiguous things without losing a record of the ambiguity. Briefly, fuzzy logic possesses several strengths. It provides an effective means of moving efficiently from the available data to developing mathematical models as a basis for analysis and knowledge. Data can be imprecise or vague. Researchers can employ natural language linguistic variables, complete with their "hedges" (predicate modifiers), which are subjective in nature (e.g., "trust") (Zadeh 1972, 1975). Fuzzy logic uses assessment rules, which are created by the researcher and are transparent and thus readily understood and evaluated by other researchers (Zadeh 1973, 1976). Thus, other scholars can debate the appropriateness of both the rules and the method, contributing to a progressively greater understanding of the system and vital aspects of human action. From the application of the method to the rules, the researcher obtains a "crisp" value, which can be utilized within a computational environment such as a GIS or an analysis of a social network.

To conclude briefly, the First Global Age, 1400–1800, was a complex, dynamic, nonlinear world-system. Within this system, the nonlinear dynamics of the human and natural systems periodically disrupted commercial, political, and other social networks, and these disruptions included those caused by the expansion of opportunities and of the system itself. To maintain system stability, actors, both men and women, had to counter the disruptions by constituting more robust networks, but the resulting denser webs of connections permitted even more widespread future cascades of disturbances. In the later eighteenth century, such a cascade of innovation destabilized at least one of the system's stability-maintaining "control" variables, which initiated a transition to a new world-system, the Second Global Age in which we now live. In their narration of revealing geographically integrated stories, historians of the First Global Age can combine, using Dynamics GIS, a dynamics approach to SNA with a complex systems analysis, and in doing so, they will offer a world historical account that addresses larger issues of system and network dynamics and provides a solution to the common criticism of world-system analysis that it ignores the relationships between the local and the larger system.

Author's note

This material is based upon work supported by the US National Science Foundation under Grants No. 0740345 and No. 0941371 and by European Science Foundation project 06-TECT-FP-004. Any opinions, findings and conclusions or recommendations expressed in this material are those of the author and do not necessarily reflect the views of the National Science Foundation or the European Science Foundation.

References

Baker, A. R. H. (2007) "Classifying geographical history," *The Professional Geographer*, 59, no. 3: 344–56.

Bruner, J. S. (1985) *Actual minds, possible worlds*, Cambridge, MA: Harvard University Press.

Buldyrev, S. V., Parshani, R., Paul, G., Stanley, H. E. and Havlin, S. (2010) "Catastrophic cascade of failures in interdependent networks," *Nature*, 464, no. 7291 (15 April): 1025–28; doi: 10.1038/nature08932.

Burt, R. S. (2004) "Structural holes and good ideas," *American Journal of Sociology*, 110: 349–99.

De Blij, H. J. (2009) *The power of place: geography, destiny, and globalization's rough landscape*, Oxford: Oxford University Press.

Easley, D. and Kleinberg, J. (2010) *Networks, crowds, and markets: reasoning about a highly connected world*, Cambridge, UK, and New York: Cambridge University Press.

Erikson, E. and Bearman, P. (2006) "Malfeasance and the foundations for global trade: the structure of English trade in the East Indies, 1601–1833," *American Journal of Sociology*, 112, no. 1 (July): 195–230.

Frank, A. G. (1998) *ReORIENT: global economy in the Asian age*, Berkeley and Los Angeles: University of California Press.

Geertz, C. (1973) *The Interpretation of cultures*, New York: Basic Books.

Gintis, H. (2009) *The Bounds of reason: game theory and the unification of the behavioral sciences*, Princeton, NJ, and Oxford: Princeton University Press.

Gladwell, M. (2000) *The tipping point: how little things can make a big difference*, Boston, MA: Little, Brown.

Goodchild, M. (2008) "Combining space and time: new potential for temporal GIS," pp. 179–97 in A. K. Knowles (ed.) *Placing history: how maps, spatial data, and GIS are changing historical scholarship*, Redlands, CA: ESRI Press.

Goodchild, M. and Glennon, A. (2008) "Representation and computation of geographic dynamics," pp. 13–29 in Stewart Hornsby, K. and Yuan, M. (eds) *Understanding dynamics of geographic domains*, Boca Raton, FL, and New York: CRC.

Grabher, G. (2006) "Trading routes, bypasses, and risky intersections: mapping the travels of 'networks' between economic sociology and economic geography," *Progress in Human Geography*, 30, no. 2: 163–89.

Gregory, I. N. and Ell, P. S. (2008) *Historical GIS: technologies, methodologies and scholarship*, Cambridge: Cambridge University Press.

Gregory, I. N. and Healey, R. G. (2007) "Historical GIS: structuring, mapping and analysing geographies of the past," *Progress in Human Geography*, 31, no. 5: 638–53.

Haken, H. (1983) *Advanced Synergetics: instability hierarchies of self-organizing systems and devices*, Berlin and Heidelberg: Springer-Verlag.

Hexter, J. H. (1971) *The history primer*, New York and London: Basic Books.

Hill, L. L. (2006) *Georeferencing: the geographic associations of information*, Cambridge, MA: MIT Press.

Hunter, K. M. (1991) *Doctors' stories: the narrative structure of medical knowledge*, Princeton, NJ: Princeton University Press.

Ikegami, E. (2000) "A sociological theory of publics: identity and culture as emergent properties in networks," *Social Research*, 67: 989–1029.

Ikegami, E. (2005) *Bonds of civility: aesthetic networks and the political origins of Japanese culture*, Cambridge and New York: Cambridge University Press.

Kantabutra, V., Owens, J. B., Ames, D. P., Burns, C. N. and Stephenson, B. (2010) "Using the newly-created ILE DBMS to better represent temporal and historical GIS data," *Transactions in GIS*, 14, s1: 39–58.

Landa, J. T. (1994) *Trust, ethnicity and identity: beyond the institutional economics of ethnic trading networks, contract law and gift exchange*, Ann Arbor: University of Michigan Press.

Latour, B. (2005) *Reassembling the social: an introduction to Actor-Network-Theory*, Oxford: Oxford University Press.

Launer, J. (2002) *Narrative-based primary care: a practical guide*, Oxford: Radcliffe Medical Press.

Lehmann, L., Foster, K. R., Borenstein, E. and Feldman, M. W. (2008) "Social and individual learning of helping in humans and other species," *Trends in Ecology and Evolution*, 23, no. 12: 664–71.

Moutoukias, Z. (1988) *Contrabando y control colonial en el siglo xvii: Buenos Aires, el Atlántico y el espacio peruano*, Buenos Aires: Centro Editor de América Latina.

Mui, H.-C. and Mui, L. (1968) "Smuggling and the British tea trade before 1784," *American Historical Review*, 74: 44–73.

Ngalamulume, K. (2009) *Is total history possible?* Review of D. R. Wright (2004), in H-Africa, H-Net Reviews (March): http://www.h-net.org/reviews/showrev.php?id=24284, accessed 15 May 2010.

Organization of American Historians (2000) *LaPietra Report: A Report to the Profession*, http://www.oah.org/activities/lapietra/final.html, accessed on 14 May 2010.

Owens, J. B. (2005) *"By my absolute royal authority": Justice and the Castilian commonwealth at the beginning of the First Global Age*, Rochester, NY: University of Rochester Press.

Owens, J. B. (2007a) "What historians want from GIS," *ArcNews*, 29, no. 2: 4–6, and http://www.esri.com/news/arcnews/summer07articles/what-historians-want.html (accessed May 10, 2010); reprinted in *GIS best practices: Essays on geography and GIS* (2008), Redlands, CA: ESRI, 35–46, and in *GIS best practices: Social sciences* (2009), Redlands, CA: ESRI, 35–45.

Owens, J. B. (2007b) "Toward a geographically-integrated, connected world history: employing geographic information systems (GIS)," *History Compass*, 5, no. 6: 2014–40.

Owens, J. B. (2008) "Historical studies, GIS for," pp. 220–21 in K. Kemp (ed.) *Encyclopedia of geographic information science*, Thousand Oaks, CA: Sage.

Puu, T. (2003) *Attractors, bifurcations and chaos: non-linear phenomena in economics*, 2nd edn., Berlin and Heidelberg: Springer-Verlag.

Puu, T. (2006) *Arts, sciences, economics: a historical safari*, Berlin and Heidelberg: Springer-Verlag.

Radil, S. M., Flint, C. and Tita, G. E. (2010) "Spatializing social networks: using social network analysis to investigate geographies of gang rivalry, territoriality, and violence in Los Angeles," *Annals of the Association of American Geographers*, 100, no. 2: 307–26.

Roca, C. P., Cuesta, J. A. and Sánchez, A. (2009) "Evolutionary game theory: temporal and spatial effects beyond replicator dynamics," *Physics of Life Reviews*, 6, no. 4: 208–49.

Rosser, J. B., Jr. (2000) *From catastrophe to chaos: a general theory of economic discontinuities*, Vol. I, 2nd edn., Boston, MA: Kluwer Academic.

Sánchez, A. (2009) "Evolution of cooperation and lack of universality in game theory on graphs," paper presented at the Centro de Física Theórica e Computacional, Universidad de Lisboa, 13 May.

Sánchez, A., Roca, C. P. and Cuesta, J. A. (2009) "Evolutionary game theory on graphs, network reciprocity and emergence of cooperation," paper presented at the conference "Evolution of cooperation: models and theories," International Institute of Applied Systems Analysis (IIASA), Laxenburg, Austria, 17 September.

Schulte Beerbühl, M. and J. Vögele (eds) (2004) *Spinning the commercial web: international trade, merchants, and commercial cities, c.* 1640–1939, Frankfurt am Main, Berlin, Bern, Bruxelles, New York, Oxford, Wien: Peter Lang.

Scott, J. (2000) *Social network analysis: a handbook*, 2nd edn., Thousand Oaks, CA: Sage.

Sonis, M. (2007) "The evolution of complexity," paper presented at the Mathematics Colloquium, Idaho State University, Pocatello, Idaho, USA, 27 February.

Sonis, M. (2009) "Innovation diffusion theory: 100 years of development," pp. 243–71 in M. Sonis and G. J. D. Hewings (eds) *Tool kits in regional science: theory, models, and estimation*, Berlin and Heidelberg: Springer-Verlag.

Subrahmanyam, S. (2004) *Explorations in connected history*, Delhi, India: Oxford University Press.

Subrahmanyam, S. (2007) "Holding the world in balance: the connected histories of the Iberian overseas empires, 1500–1640," *American Historical Review*, 112, no. 5 (December): 1359–85.

Vespignani, A. (2010) "The fragility of interdependency," Nature, 464, no. 7291 (15 April): 984–85.

Watts, D. J. (2003) *Six degrees: the science of a connected age*, New York and London: W. W. Norton.

Watts, D. J. and Dodds, P. S. (2007) "Influentials, networks, and public opinion formation," *Journal of Consumer Research*, 34, no. 4 (December): 441–58.

White, H. C. (1995) "Social networks can resolve actor paradoxes in economics and in psychology," *Journal of Institutional and Theoretical Economics*, 151: 58–74.

White, H. C. (2002) *Markets from networks: socioeconomic models of production*, Princeton, NJ: Princeton University Press.

White, H. C. (2008; 1st ed. 1992) *Identity and control: how social formations emerge*, Princeton, NJ: Princeton University Press.

Winius, G. D. (1983) "The 'shadow empire' of Goa in the Bay of Bengal," *Itinerário*, 7, no. 2: 83–101.

Wright, D. R. (2004) *The World and a very small place in Africa: a history of globalization in Niumi, the Gambia*, 2nd edn., Armonk, NY: M. E. Sharpe.

Yuan, M. (2007) "Adding time," pp. 169–84 in Wilson, J. and Fotheringham, A. S. (eds) *Handbook of geographic information science*, Oxford: Blackwell Publishing.

Yuan, M. (2008) "Dynamics GIS: recognizing the dynamic nature of reality," *ArcNews*, 30, no. 1 (Spring): 1, 4–5, and http://www.esri.com/news/arcnews/spring08articles/dynamics-gis.html, accessed on 15 May 2010; reprinted in GIS best practices: essays on geography and GIS (2008), Redlands, CA: ESRI, 17–24.

Yuan, M. (2009) "Knowledge discovery of geographic dynamics in spatiotemporal data," pp. 347–66 in H. J. Miller and J. Han (eds) *Geographic data mining and knowledge discovery*, 2nd edn., London: Taylor and Francis.

Yuan, M. and Stewart Hornsby, K. (2008) *Computation and visualization for understanding dynamics in geographic domains: a research agenda*, Boca Raton, FL: CRC.

Zadeh, L. A. (1965) "Fuzzy sets," *Information and Control*, 8: 338–53.

Zadeh, L. A. (1972) "A Fuzzy-set-theoretical interpretation of linguistic hedges," *Journal of Cybernetics*, 2: 4–34.

Zadeh, L. A. (1973) "Outline of a new approach to the analysis of complex systems and decision processes," *IEEE Transactions on Systems, Man and Cybernetics*, SMC-3: 28–44.

Zadeh, L. A. (1975) "The Concept of a linguistic variable and its applications to approximate reasoning," *Information Sciences*, 8: 199–249 (Part I); *Information Sciences*, 8: 301–57 (Part II); *Information Sciences*, 9: 43–80 (Part III).

Zadeh, L. A. (1976) "A fuzzy-algorithmic approach to the definition of complex or imprecise concepts," *International Journal of Man-Machine Studies*, 8: 249–91.

Zadeh, L. A. (1994) "Why the success of fuzzy logic is not paradoxical," *IEEE Expert*, 9, no. 4: 43–45.

Zadeh, L. A. (1997) "Toward a theory of fuzzy information granulation and its centrality in human reasoning and fuzzy logic," *Fuzzy Sets and Systems*, 90: 111–27.

4.4

World-systems theory and formal and simulation modeling

Hiroko Inoue

This essay considers the ways in which formal modeling can be used to understand the evolutionary dynamics of world-systems. Developing formal models to explain world-systems dynamics is intertwined with the unique theoretical assumptions of world-systems theorizing. This essay reviews some modeling strategies relevant to world-systems modeling and suggests useful approaches for world-systems analysis.

Unlike most sociological approaches, the world-systems perspective understands institutional change as occurring in whole interpolity systems rather than single polities. It is claimed that social change should be re-conceptualized in terms of the world-system as a unit of analysis. Evolving from regional polities, world-systems have comprised multi-level interaction systems, forming the hierarchical world-systems (Chase-Dunn and Hall 1997).

Research topics and theoretical orientations determine the types of models. Among different types of modeling, the two major contrasting approaches are reviewed: (1) mathematical and simulation modeling, and (2) system dynamics (SD) and agent-based (AB) modeling within simulation modeling. It has been argued that mathematical and simulation models, respectively, have distinct characteristics (Cioffi-Revilla 2009). In mathematical modeling, the target dynamics are explained with a few mathematical formulae. They do not precisely represent detailed processes but are compelling for comprehending processes that are too complex to be described with verbal reasoning (McElreath and Boyd 2007). In contrast, simulation modeling is useful for exploring complicated dynamics by integrating detailed processes in a cumulative way. It may be too detailed for capturing elementary dynamics but it excels at inclusive portrayal and dynamic modeling.

Some approaches use the advantages of mathematical modeling and reveal particular systemic properties of historical evolution (Korotayev 2005; Turchin 2011). Turchin's (2011) modeling, for instance, they do not focus on the entire evolutionary process but on particular historical systems (i.e., agrarian states) with compact modeling. Other modeling approaches apply simulation methods for modeling world-systems, presenting comprehensive pictures with complicated dynamics based on detailed rules and assumptions (Kowalewski and Hoover 1995).

Further, within simulation modeling, there are noticeable differences. One major distinction is whether or not the modeling tries to construct a SD or AB representation. Having feedback structures, the causal reasoning of the SD model is examined with various "what-if" scenarios. SD modeling allows for complicated interactions among elements at different levels to examine how systems behave and evolve. Unlike SD modeling, AB modeling is centered on the emergence of structures from local interactions utilizing simple initial rules. Object-oriented AB programming allows interactions of multiple agents, generating complex dynamics in a bottom-up manner (Epstein and Axtell 1996).

World-systems models encompass feedback processes as their theoretical basis (Chase-Dunn and Hall 1997). SD modeling properly addresses such circular, yet evolving system dynamics (Fletcher et al 2011; Kowalewski and Hoover 1995). Interactions of multi-level structures in world-systems are thus well apprehended by SD modeling.

While it emphasizes the systemic property, world-systems theorizing also appreciates seemingly non-systemic conditions for the emergence of subsystems. Such subsystems dynamics, for instance, includes "semiperipheral development" (Chase-Dunn and Hall 1997). The semi-periphery is considered as the location where innovative technologies are more likely to be implemented, and this often leads to the growth of semiperipheral polities and attainment of greater power and upward mobility. The advantages of the semiperiphery originate primarily from its geographical locations.

The explicitly spatial quality of some multi-agent modeling is usefully applied to the understanding of the geographical impact on mobility. For instance, identifying each unit with an individual agent distributed across a landscape, Kohler and his colleagues (2005) have examined uneven development of the size of settlements among horticultural polities. Such AB modeling supplements SD modeling in capturing the emergence and upward mobility of subsystems.

More importantly, the locational significance for emerging structures is relative to the other locations within a system. Semiperipheral development is relational and can only be understood with a conceptualization of the world-system as a whole. The local dynamics modeled by AB modeling, in fact, interact with system level forces which are typically modeled in SD modeling. The effective modeling of world-systems theorizing thus requires the synthetic approach of SD and AB modeling.

Although not specifically aiming at world-systems modeling, some studies show a synthetic approach. Hanneman (1988) discusses semi-mathematical language which allows the user to handle complex dynamics without losing the explanatory power of formal language. Turchin (2011) models how warfare as a selection mechanism causes the evolution of social hierarchies among multiple agents by applying price equation from evolutionary biology.

World-systems theory seeks a general evolutionary model that explains the fundamental dynamics across time and space (Chase-Dunn and Hall 1997). This is usefully represented with formal mathematical modeling. World-systems analysis also explores the contingency of evolutionary social changes with particular local conditions which interact with world-systemic forces. In developing formal models of world-systems, these unique theoretical assumptions are incorporated into the modeling strategy. This may be most effectively done by integrating the SD and AB approaches.

References

Cioffi-Revilla, C. (2009). "Simplicity and Reality in Computational Modeling of Politics." *Computational and Mathematical Organization Theory*, 15 (1): 26–46.

Chase-Dunn, C. and Hall, T. (1997). *Rise and Demise: Comparing World-Systems.* Boulder, CO: Westview Press.

Epstein, J. M. and Axtell, R. (1996). *Growing Artificial Societies: Social Science from the Bottom Up.* Washington, DC: Brookings Institution Press.

Fletcher, J. B., Apkarian, J., Hanneman, R. A., Inoue, H., Lawrence, K. and Chase-Dunn, C. (2011). "Demographic Regulators in Small-Scale World-Systems." *Structure and Dynamics*, 5(1): 1–31.

Hanneman, R. (1988). *Computer-Assisted Theory Building: Modeling Dynamic Social Systems.* [Online] Available at: http://www.faculty.ucr.edu/~hanneman/ accessed, November 1, 2011.

Kohler, T., Gumerman, G. and Reynolds, R. (2005). "Simulating Ancient Societies: Computer Modeling is Helping to Unravel the Archaeological Mysteries of the American Southwest." *Scientific American*, 293: 76–84.
Korotayev, A. (2005). "A Compact Macro Model of World System Evolution." *Journal of World-Systems Research*, 11(1): 79–93.
Kowalewski, D and Hoover, D. (1995). *Dynamic Models of Conflict and Pacification: Dissenters, Officials, and Peacemakers*. New York: Praeger Publishers.
McElreath, R. and Boyd, R. (2007). *Mathematical Models of Social Evolution: A Guide for the Perplexed*. Chicago, IL: University of Chicago Press.
Turchin, P. (2011). "Warfare and the Evolution of Social Complexity: A Multilevel Selection Approach." *Structure and Dynamics*, 4(3), Article 2: 1–37.

4.5

Mathematical models of world-system development

Andrey Korotayev and Sergey Malkov

Large-scale models of world-system development

The most well-known models of world-system dynamics were developed in the early 1970s on the initiative of the Club of Rome (e.g., Forrester [1971: *World Dynamics*]; Meadows et al [1972: *The Limits to Growth*]; Mesarović and Pestel [1974: *Mankind at the Turning Point*]). The projects by Forrester and Meadows were the first efforts to develop large-scale models of world-system dynamics. Forrester's WORLD-2 model does not assume regional or zonal divisions, but rather the world-system is represented as a single whole. Forrester chose the following five systemic aspects: population, capital funds, that part of investment that is made in agriculture, natural resources, and pollution. There are five model blocks corresponding to those five aspects. They are linked by a system of positive and negative feedback loops. On the basis of this model, Forrester and his colleagues made calculations up to the year 2100. Forrester's model was developed further by the project led by Meadows. The Meadows WORLD-3 model was more complex but it retained many of the features of Forrester's model.

The project by Mesarović and Pestel (1974) was the second (after Meadows') in the series of reports to the Club of Rome. With this model, they tried to achieve a higher level of detail by representing the world-system as ten interacting regions. Each region is described with a system of submodels that is the same for all the regions. The regional differences are produced by values of initial conditions and parameters. The relations among regions are modeled as imports, exports, and migration.

At present there is quite a large number of such mathematical models of world-system development (e.g., Hughes 2008; Reuveny 2008); however, the principles of construction of these models continue the traditions of the above-described approaches.

Compact macromodels of world-system growth

In 1960, von Foerster et al published a striking discovery in the journal *Science*. They showed that between 1 CE and 1958 CE, the world's population (N) dynamics can be described in an extremely accurate way with an astonishingly simple equation:

$$N_t = \frac{C}{t_0 - t} \qquad (4.5.1)$$

where N_t is the world population at time t, and C and t_0 are constants, with t_0 corresponding to an absolute limit ("singularity" point) at which N would become infinite.

Note that the graphic representation of this equation is nothing but a hyperbola; thus, the growth pattern described is denoted as "hyperbolic." It was shown later (Korotayev et al 2006) that the world GDP dynamics up to the early 1970s can be described in a similarly accurate way with a quadratic hyperbolic equation:

$$G_t = \frac{C}{(t_0 - t)^2} \tag{4.5.2}$$

where G_t is the world GDP.

The hyperbolic pattern of the world's population growth (accompanied rather logically by quadratic-hyperbolic growth of the world-system GDP) has been shown to be accounted for by the following nonlinear second order positive feedback mechanism: the more people—the more potential inventors—the faster technological growth—the faster growth of the Earth's carrying capacity—the faster population growth. With more people you also have more potential inventors and hence faster technological growth (Kremer 1993; Korotayev 2005, 2008; Korotayev et al 2006).

For the period prior to the 1970s, the world-system economic and demographic macrodynamics have been shown to be described in a rather accurate way with an extremely simple two-differential-equation mathematical model (Kremer 1993; Korotayev 2005, 2008), for example:

$$\frac{dN}{dt} = aSN \tag{4.5.3}$$

$$\frac{dS}{dt} = bNS \tag{4.5.4}$$

where N is the world population, S is the surplus produced, per person, over the amount minimally necessary to reproduce the population with a zero growth rate in a Malthusian system; a, b are constants.

In order to describe the world-system's demographic dynamics in the last decades (as well as in the near future), it has turned out to be necessary to extend the Equation system (4.5.3)–(4.5.4) by adding to it Equation (4.5.6), and by adding to Equation (4.5.3) the multiplier $(1 - l)$, where l is proportion of literate population, which results in Equation (4.5.5), and produces a mathematical model that describes rather accurately not only the hyperbolic growth of the world-system up to the 1970s, but also its withdrawal from the blow-up regime afterwards (Korotayev et al 2006):

$$\frac{dN}{dt} = aSN\,(1-l) \tag{4.5.5}$$

$$\frac{dl}{dt} = cSl\,(1-l) \tag{4.5.6}$$

The model takes into account earlier empirical findings identifying female literacy as a major factor of fertility decline. The additional equation introduced for literacy growth (4.5.6) has the following sense: the literacy growth rate is proportional to the proportion of literate people in the whole population l (potential teachers), to the proportion of illiterate people in the whole population (1 − l) (potential pupils), and the presence of surplus S, which can be used for educational purposes.

References

Foerster H. von, Mora, P. and Amiot, L. (1960) Doomsday: Friday, 13 November, A.D. 2026. *Science* 132: 1291–95.

Forrester, J. W. (1971) *World Dynamics*. Cambridge, MA: Wright-Allen.

Hughes, B. (2008) "Forecasting Globalization: The Use of International Futures (IFs)", in G. Modelski (ed.) *Globalization as Evolutionary Process*. London: Routledge.

Korotayev, A. (2005) A Compact Macromodel of World System Evolution. *Journal of World-System Research* 11: 79–93.

Korotayev, A. (2008) "Compact Mathematical Models of World-System Development", in G. Modelski (ed.) *Globalization as Evolutionary Process*. London: Routledge.

Korotayev, A., Malkov A. and Khaltourina, D. (2006) *Introduction to Social Macrodynamics: Compact Macromodels of the World System Growth*. Moscow: URSS.

Kremer, M. (1993) Population Growth and Technological Change: One Million B.C. to 1990. *The Quarterly Journal of Economics* 108: 681–716.

Meadows, D. H., Meadows, D. L., Randers, J. and Behrens, W. W. III. (1972) *The Limits to Growth: A Report for the Club of Rome's Project on the Predicament of Mankind*. New York: Universe Books.

Mesarović, M. D. and Pestel, E. (1974) *Mankind at the Turning Point: The Second Report to the Club of Rome*. Laxenburg: IIASA.

Reuveny, R. (2008) "On Forecasting Globalization Using World Models", in G. Modelski (ed.) *Globalization as Evolutionary Process*. London: Routledge.

5. Critical contributions to world-systems analysis

5.1

World-system history

Challenging Eurocentric knowledge

Robert A. Denemark and Barry K. Gills

World-system history, without a hyphen, asks a series of questions about global history that challenge many inherited assumptions.

> [R]ecognizing the common existence of capitalist practices in East and West for much of world history allows us to reassess completely the overarching dynamic of the world system and the capital accumulation process on a world scale. This is the intention of world system theory: to reveal the common rhythms, the competition and rivalry, and the mutual influence among all the zones of the world economy over a period of several millennia, … with an objective structural emphasis on the process of capital accumulation on the world scale and the real patterns of this process, both in terms of the "long cycles" of world economic expansion and contraction, and the "center shifts" between rival economic and political power centers.
>
> *(Gills 2002: 152)*

From this perspective, world-systems analysis (with the hyphen) has led to a set of temporal, spatial, and processual assumptions that occlude our vision and lead to unsupportable conclusions. World-system history is constructed on a critique of those biases.

World-systems analysis emerged in the 1970s positing three important insights. The first addresses the unit of analysis. The state had served as the focal point of social analysis from the seventeenth century onward. States were not just handy social categories, but political units that self-consciously sought power. World historian Frederick Teggart argued that:

> … academic history has not succeeded in liberating itself from the influence of the Romantic period, during which, in every country of Europe, the spirit of nationality demanded the rewriting of history in terms of a new sense of national existence and a new enthusiasm for national achievements in the past … The division of history … into "ancient," "medieval," and "modern" obscures the fact that these terms have reference, not to the world at large, but to a relatively small part of the earth's surface.
>
> *(1925: 40–41)*

A generation later, Fernand Braudel's (1949) history of "the Mediterranean world" rejected the state as the central unit of analysis and put forward the *économie-monde'* that "cut across many political boundaries and was knitted together by its skein of economic interdependencies" (Wallerstein 2008). Wallerstein honored Braudel by naming the research center he founded as a platform for the emerging world-system perspective after him. The error of adopting a state-based analysis was further highlighted by the lack of real progress among most newly independent former colonies in the post war world. The failure of "modernization," which had been a central prediction of state-centric social science, revealed the poverty of this theory. This failure opened the way for new radical social theory. In 1974, Wallerstein published his path-breaking study of the "modern world-system," which rejected the assumptions of modernization and the centrality of the nation-state, and proposed an alternative conception based on the world-system as the unit of analysis. Andre Gunder Frank was preparing a similar work on the emergence of the modern world-system titled *World Accumulation, 1492–1789* (Frank 1978). Frank extended insights established in earlier work on dependency theory and underdevelopment. These included moving the origins of the capitalist world-system back to "the sixteenth century expansion," examining cycles of rivalry by contending core powers, utilizing "long wave" analysis of the world-economy, mapping the "triangular" economy of Europe-America-Africa, and including Asia.

World-systems analysis adopted an important change in temporal perspective. In a system with presumed fixed social units (i.e., states) that developed along pre-determined lines, time could be linear in nature. Once formed, states were believed to "develop" and could be compared in terms of that (inevitable) development. The western capitalist state was assumed to be the universal archetype. Braudel was especially wary of this view. He put forward provocative alternatives in the shape of "structural" time and "cyclical" time. In world-systems perspective, time was cyclical, encompassing cycles of economic expansion and contraction, rise and decline, and worldwide geopolitical cycles (e.g., K-waves, "A" and "B" phases, cycles of hegemony, and rivalry). These cycles punctuate history, reflecting its underlying constitutive structure. In addition, longer-term secular trends created important limits to the changes that could take place within the modern capitalist world-system.

The third facet of world-systems analysis concerned revising our epistemological understanding. Increased emphasis on systemic complexity, and not mechanical Newtonian predictability, revealed the fundamental character of social (as well as physical) reality. Thus Wallerstein (2008) has suggested that "Probabilities are not reflections of our scientific ignorance but describe states of the world, indeed the vast majority of states of the world." World-systems analysis pointed to disjunctures in the "linear progress" assumed by conventional theories. Its scholars argued that we would understand more about historical social systems if we examined not only periods of "normal" order, but the repeated periods of social breakdown (and reformulation), of disequilibrium, disruption, and "crisis," which may contain the seeds of potential transformation.

These insights are important, progressive, and broadly accepted by contemporary scholars of world-system history (without the hyphen). Where world-system history takes issue with world-systems analysis is in the manner that each insight has been applied. Where world-systems analysis focuses on the alleged unique nature of the modern capitalist world-system, world-system history contends that there is much fundamental continuity that allows us to see longer-term dynamics that have been ignored or mis-theorized. The axial division of labor, identified as constitutive of the modern world-system is very real, but its foundation precedes modern capitalism (understood as a system with a requirement for the ceaseless accumulation of capital that emerged in the last few centuries). Conventional liberal, Marxist and world-systems analysis tend to conflate the current manifestations of a long-standing "form," that is, "capital," with the historical process of capital accumulation itself. Furthermore, the changes that world-systems analysis argues characterize the current system are derived from limited (modern European) temporal and geographical horizons. The European social processes commonly identified in nineteenth- and twentieth-century analyses have been confused with universal social processes. It was on the basis of fundamentally endogenous "internal" factors that such changes were said to be driven, while worldwide systemic structural forces were ignored or relegated to marginal significance. World-system history remains fundamentally oriented to a break with the Eurocentric knowledge construction of world history.

Finally, the world-system history perspective is sympathetic to the position taken by world-systems analysis on the inherent limits to predictive capacity, but adds a further caution. So long as traditional concepts of "capitalism" remain at the heart of the analysis, we will remain trapped in essentially Eurocentric social theory. In the course of the development of the new world-system theory, Gunder Frank (1991) argued that such traditional concepts as "feudalism," "capitalism," and "socialism," when understood as "modes of production," were less scientific than "transideological." It is necessary to break with these traditional concepts to make intellectual and theoretical progress.

The continuity thesis in world-system history

The continuity thesis in world-system history expresses the idea of long-term historical continuities, in contrast to the idea of periodic epochal transformations in characterizing social relations (Gills 1996). Central to this view is the argument that "production of commodities, commerce, trade, (price-setting) markets, private 'enterprise,' private capital, capital accumulation, and, yes, the 'international' division of labor have all played much more important roles much further 'back' in world history than has usually been recognized or accepted" (Gills 1996: 228). In European thought, Machiavelli relied on the assumption that humans remained psychologically essentially the same. When faced with similar situations they respond in similar ways. Students of these patterns could learn valuable lessons. Enlightenment *philosophes*' were wedded to a view of human psychology, derived from John Locke and conveyed by Etienne de Condillac, which disallowed any change in human nature. Social change had to come from outside the individual. With fixed internal laws on the one hand, and an obviously fluid set of social and historical situations on the other, the foundation for a linear theory of progress was established.

There are many ways to posit the alternative idea of historical continuity: the continuity of an historical pattern, for example, the "rise and decline" of large political units; the continuity of (deep) structure, grounded in a materialist philosophy of history; the continuity of human psychology which reproduces similar behavior (and social relations) as successive generations replay the same impulses and patterns.

Conceptions of cyclical time and cyclical history have been common in the past (Pollard 1968). These conceptions could take the form of theological or cosmological understandings, as in classical Mayan thought; or a secular form, as in the "dynastic cycle" described by Chinese historians, where "what has been fragmented will be united, and what has been united will fragment." In the medieval world, Ibn Khaldun (1958) retained the idea of patterns centered on a historical dialectic between urban based and rural kinship based social relations.

For world-system history, the point is neither to return to pre-modern conceptions of cyclical history, nor fall prey to modern biases regarding the primacy of change. World-system history simply suggests that too much scholarly emphasis has been placed on the elucidation of discontinuities and epochal transitions in global social relations. There are unrecognized implications to this focus on change. Much of the social order may be stable in nature. Changes that appear significant in the near-term may actually be minor alterations embedded in a broader set of patterns. It is easy to over-emphasize the differences that exist between areas, or to draw a sharp distinction between social forces that exist "before" and "after" some putative historical threshold. Before we make any *a priori* decision to focus our analyses on change, we should consider those significant continuities that exist across time and space. We may come to believe that behaviors that fail to regularly change are part of a fixed social order. So conceived, such continuities may be considered insignificant. There are also important ideological and political issues at stake. A study of "other" societies may be undertaken to facilitate understanding, or it can also be used to disparage "others" and aggrandize the accomplishments of dominant groups. A narrow focus on the here-and-now might be used to falsely credit a given set of policies with having generated positive outcomes, or descend into bigotry to justify injustice and aggression.

From the perspective of the continuity thesis there is no compelling reason to focus on change, nor is there any self-evident reason to declare that some differences are particularly significant, unless they can be placed in the context of a long-term analysis. World-system history seeks to identify those long-term contexts. Only an analysis of the world-system as a whole will provide us with a good understanding of how all the parts of the story fit together. Many of the social forces considered to be unique to the modern era are actually of significant antiquity, and can be better understood when viewed in global and historical perspective.

Material(ist) analysis and (world) systematicity

Both world-systems analysis and world-system history hold to a broadly materialist conceptualization of social relations. World-systems analysis identifies the "modern world-system" as a capitalist system born in Europe in the fifteenth and sixteenth centuries. European capitalism is said to have first endowed social relations with the requirement for ceaseless accumulation. This ceaseless accumulation brought with it an allegedly unique socioeconomic environment where modern classes and nation-states emerged, economic cycles built upon contradictions in the process of capital accumulation played themselves out, and advanced capitalist (core) areas were driven to expansion into the periphery. From this perspective there have been multiple world-systems, but the social processes that existed in "precapitalist" world-systems were fundamentally different, requiring a focus on change and "transition(s)."

This orientation was challenged by Frank and Gills, starting in about 1989. Their joint studies of history, especially outside Europe, caused them to question the idea that significant differences existed between important social processes that took place in allegedly unique social epochs. Wallerstein (1990: 288–89; see also 1992) identified twelve "characteristics presumed to be the description of a capitalist world-economy" including an axial division of labor between core and periphery; semiperipheral areas; continued non-wage labor; cohabitation with a sovereign state system; periods of hegemony and rivalry; cyclical rhythms and secular trends; and a variety of forms of exploitation (i.e., national, racial, ethnic, gender) used to extract value from certain groups. Frank and Gills saw most of these phenomena as having been well entrenched both earlier and elsewhere. If the same kinds of social processes were operating across these temporal contexts, generating the same kinds of outcomes, they could be neither new nor unique to the modern era. From the perspective of the resulting continuity hypothesis, the modern world-system is not fundamentally unique, nor is the rise of capitalism in Europe the appropriate starting point for understanding the present era (Gills 2003, 2010a, 2010b). While we may learn a great deal from the study of comparative world-systems (following the work of Chase-Dunn and Hall 1997), we can expect that the current period will bear a significant resemblance to those of previous periods. If we want to understand the world-system, we should be focusing on both continuities and changes between earlier and later iterations (Denemark 2010).

From the perspective of the continuity thesis, the world-system was created and sustained by an expanding set of intersecting trade and communication routes that emerged some five millennia ago. People and events in areas that are far-removed have influence on one another through "inter-penetrating accumulation" processes (Gills and Frank 1990, 1991, 1992). The antecedents of such systemic routes extend even back into the Neolithic period across significant stretches of Eurasia. The Anatolian settlement of Catal Huyuk is a much studied example of a seventh millennium BCE community that was dependent on long-distance trade.

Perhaps the most famous set of trade and communication routes are the Silk Roads across Central Asia that connected areas as disparate as Britain and Scandinavia with Indonesia and Korea. The crucial gateways from Asia into Europe traversed the "three corridors" of interzonal exchange (Gills 1996: 231; Gills and Frank 1990, 1991). From at least 500 BCE, these and other interconnected routes linked the farthest points of the known world (ArchAtlas 2010; Rowlands et al 1987; Sherratt 1993). Interestingly, the newly renovated Ashmolean Museum in Oxford has been reorganized to illustrate the central importance of these routes throughout world history.

This system of long-distance exchange carried more than just trade-goods. The roles that societies played in the processes of acquiring raw materials, production, transportation, marketing, consumption, and investment served to influence the structure of social orders. By engaging in the mutual intercourse of Eurasia, societies were mutually influenced. The implications of such extensive linkages are traced back 5000 years in the work of Ekholm and Friedman (1982) in the first explicitly formulated thesis on the historical continuity of capital accumulation and "capital-imperialism." There is ample evidence from the ancient world for extensive long-distance trade, the competitive accumulation of capital via market-driven production and exchange, and the development of imperialism and core-periphery relations based on a mixture of economic and geopolitical motives. (See also Ekholm and Friedman 2005; Friedman 2000.)

The structure and significance of world-system linkages, and their continuity during the medieval and early modern period was superbly analyzed by Abu-Lughod (1989). Her review of the thirteenth century Eurasian world-system focuses on eight interlocking and mutually dependent globe-spanning regions. Abu-Lughod illustrates the manner in which fundamental social processes are generated not by local conditions but by the incentives provided by participation in a world-system of production and exchange.

Here is where world-system history parts company with Wallersteinian world-systems analysis. For Wallerstein, the modern world-system is fundamentally different from the non-capitalist organizational forms that came before it. Capitalism, from this perspective, includes private property, wage labor, commodification, and the extension of (market based) commercialization. But the real difference between a capitalist and non-capitalist era, according to Wallerstein, is the imperative that capital accumulation be *ceaseless*, meaning that continuous reinvestment in the means of production is absolutely required. Those who predominate in the process of ceaseless accumulation gain power, which emanates from the creation of wealth. In earlier eras, to the contrary, it was political power that generated wealth (Amin 1999: 300, 2000).

Contrary to the arguments of these scholars, much historical evidence suggests that such processes have long existed, and that capital accumulation has normally been a competitive process requiring continuous reinvestment. Specialist histories suggest that private property and the quest for profit was a fundamental driver of the world-economy far earlier than the classically understood advent of capitalism. Braudel argues that capitalist cities and capitalism itself were already developed by the thirteenth century in Italy. These cities mirrored practices of the Arab world of the second and possibly the third millennia BCE. Rostovtzeff (1926) considered the differences between the modern market economy and the fourth century BCE Hellenistic world as merely quantitative. Students of the Levantine Phoenician cities and of the western Carthaginian empire are familiar with extensive commodity production and long-distance markets used in the pursuit and accumulation of money profits.

Actors and processes that would be uncontroversially labeled as capitalist in the modern era can be traced back to second millennium BCE Assyria, where private owners of joint-capital textile firms employed thousands of wage laborers. The markets in which these firms were situated set prices, and provide evidence for the accumulation of capital, and both circulating and fixed capital assets (Larsen 1976). It is no surprise therefore that students such as Ekholm, Friedman, and Larsen conclude that the relevant processes of capital accumulation can be traced back into the third millennium BCE. Indeed, even Max Weber traced early modern European business contract forms back to ancient Babylonia.

Cycles and trends

Among the most interesting and controversial predictions of world-systems analysis is the hypothesized array of cyclical rhythms and secular trends. The social, political, and economic cycles are among those suggested to be driven by the unique processes of the modern world-system. The trends are likewise either driven by, or the inevitable result of, capitalist interaction (Wallerstein 1990, 1992).

Students of world-system history are particularly concerned with the periodicity of economic rise and decline. If similar cycles existed prior to the onset of the modern system, then they were driven by processes that are not uniquely associated with the modern system. Any attempt to understand them exclusively from within the "modern" period will overlook important world-historical processes. Giovanni Arrighi (1994) identified cycles of economic rise and decline prior to the sixteenth century in his analysis of early Italian city-states. Frank, Gills, and Thompson, among others, look for evidence of large-scale cyclical behavior and find 150 to 200-year economic upswings and downswings in the Bronze and Iron Ages (Frank 1993; Frank and Thompson 2006; Gills and Frank 1992). These economic cycles appear world-system-wide, constituting evidence of "systematicity" in patterned long-term material interactions. The causes that produce these cycles that pre-date the onset of the modern world-system, like the causes of the cycles that exist during the modern period, remain in question.

The cycle of hegemony and rivalry provides another focus. Hegemony is defined as the existence of a significantly hierarchical structure of the accumulation of surplus among polities that is mediated by force, and is hardly unique to the modern era. A hierarchy of polities is established that appropriates a privileged share of surplus, and the power that comes with it, to a given center/state and its ruling/propertied classes. In such a structure, the hegemonic center subordinates secondary centers and their respective systems of production and accumulation (Gills and Frank 1990, 1991, 1992).

From the perspective of world-system history, one must raise questions about conceptualizations of hegemony that begin with modern states like Portugal, the Netherlands, or Great Britain. These actors were not as wealthy or as dominant as coterminous Ming/Qing China or Moghul India, though they outlasted them. The story of the domination of western hegemonic states may be more a function of ideology than historiography (Blaut 1993; Frank 1998). Instead, we find that several core-periphery-hinterland complexes might exhibit significantly unequal flows of value to a small set of core states, yielding a hegemonic actor. Following Abu-Lughod, we should look for the *temporary* rise and decline of some states or group of states relative to others. As Wilkinson (2009) verifies in his latest papers on the polarity structures of "central civilization," we see more periods of rivalry than hegemony, and there is no fundamental change in this regard with the advent of the modern world-system.

Both economic and geopolitical upswings and downswings may be driven by the same process. For Modelski and Thompson (1996), the rise of the market in about 1000 CE gave birth to a set of innovation-driven cycles that define and animate the world-system. Such innovations provide their societies with sources of wealth and accumulation. Innovations then diffuse, returns are reduced and there is a downturn. Leaders may have sufficient resources to continue to innovate, and have incentives to do so in the context of global threats created during economic downturns. What results is a pairing of K-waves driving both economic and political leadership. Innovations give rise to significant resources upon which new powers rise and consolidate their positions (usually after victory in a major war). Their economies eventually slow. A downturn may lead to global violence and another round of innovation. Global leaders may earn another round of leadership if they leverage their initial position into additional innovation, or they may be replaced by those who pioneer new innovations.

Modelski and Thompson trace this pattern back one thousand years. Devezas and Modelski (2003) suggest a broader, longer-term evolutionary learning cycle. (See also Modelski et al 2008). These cycles of economic upswings and downswings extending across temporal and spatial frames, yet constituted by very different, if not incompatible social structures, provides further evidence for the continuity thesis in world-system history.

A conclusion and an invitation

As a result of these considerations, Frank (1991, 1993) came to define as problematic all of the various lines of western thought that identified transitions between "modes of production" as central to the study of global development and global history. According to Frank, there were no such transitions or even discrete modes of production, and Europe was not the exemplar upon which we could construct a globally-relevant social theory. A more complete "humanocentric" perspective can only emerge when we abandon these "transitional ideological modes" and adopt a form of analysis that traces all significant historical processes to their origins, while simultaneously looking to the macro-structure of the world-system as the only appropriate unit of analysis. World systems-analysis is not so much "incorrect" as it is insufficiently world-historical and world-systemic. Social processes that are truly long-term in nature, and not confined to or pre-determined by European experience, must be the focus of additional attention.

Those who are interested in world-systems analysis should also be interested in world-system history. Wallerstein responded to some methodological challenges to world-systems analysis, including those from world-system history, by categorizing the big picture of world-system history as a form of ideographic analysis wherein critical disjunctures are effaced with unhelpful truisms regarding the fact that "... there are always some things that do not seem to have changed" (Wallerstein 1994). If the questions raised by world-system history were about "some things" then this would be particularly apt, but when the "things" that do not seem to have changed are the processes that rest at the center of what is allegedly unique about the modern world-system, then it is worthwhile to spend some time studying the level to which continuity is evident, and what that continuity might teach us. A healthy skepticism regarding criticisms is good, and intramural debates can bring clarity to fuzzy concepts and light to obscure conditions. In the case of world-system history, scholars in world-systems analysis should not lose sight of the fact that these two perspectives have ontological and epistemological foundations and intellectual agendas that are more similar than different. No form of social analysis can long exist under the misperception that it cannot be improved. We have much to learn from one another.

References

Abu-Lughod, J. (1989) *Before European Hegemony*, London: Oxford.

Amin, S. (1999) "History Conceived as an Eternal Cycle", *Review*, 22: 291–326.

Amin, S. (2000) "Economic Globalism and Political Universalism: Conflicting Issues?", *Journal of World-Systems Research*, 6: 582–622.

ArchAtlas (2010) http://www.archatlas.org/.
Arrighi, G. (1994) *The Long Twentieth Century*, London: Verso.
Blaut, J. (1993) *The Colonizer's Model of the World*, New York: Guilford Press.
Braudel, F. (1949; 2nd edn. 1966) *The Mediterranean and the Mediterranean World in the Age of Philip II*, London: Collins.
Chase-Dunn, C. and Hall, T. (1997) *Rise and Demise*, Boulder, CO: Westview Press.
Denemark, R. (2010) "World System History", in R. Denemark (ed.) *The International Studies Encyclopedia*, Oxford: Wiley.
Devezas, T. and Modelski, G. (2003) "Power Law Behavior and World System Evolution: A Millennial Learning Process", *Technological Forecasting and Social Change*, 70: 819–59.
Ekholm, K. and Friedman, J. (1982) "'Capital' Imperialism and Exploitation in Ancient World-Systems", *Review*, 6: 87–110.
Ekholm Friedman, K. (2005) "Structure, Dynamics, and the Final Collapse of Bronze Age Civilization in the Second Millennium BC", in J. Friedman and C. Chase-Dunn (eds) *Hegemonic Declines Past and Present*, Boulder, Co: Paradigm.
Frank, A. G. (1978) *World Accumulation, 1492–1789*, New York: Monthly Review Press.
Frank, A. G. (1991) "Transitional Ideological Modes: Feudalism, Capitalism, Socialism", *Critique of Anthropology*, 11: 171–88.
Frank, A. G. (1993) "Bronze Age World System Cycles", *Current Anthropology*, 34: 383–430.
Frank, A. G. (1998) ReOrient: *Global Economy in the Asian Age*, Los Angeles: University of California Press.
Frank, A. G. and Thompson, W. (2006) "Early Iron Age Economic Expansion and Contraction Revisited", in B. Gills and W. Thompson (eds) *Globalization and Global History*, London: Routledge.
Friedman, J. (2000) "Concretizing the Continuity Argument in Global Systems Analysis", in R. Denemark et al (eds) *World System History*, London: Routledge.
Gills, B. (1993) "Hegemonic Transitions in the World System", in A. G. Frank and B. Gills (eds) *The World System: Five Hundred Years or Five Thousand?*, London: Routledge.
Gills, B. (1996) "The Continuity Thesis in World Development", in S. Chew and R. Denemark (eds) *The Underdevelopment of Development: Essays in Honor of Andre Gunder Frank*, London and New Delhi, India: Sage.
Gills, B. (2002) "World System Analysis, Historical Sociology and International Relations: The Difference a Hyphen Makes", in S. Hobden and J. Hobson (eds) *Historical Sociology of International Relations*, Cambridge: Cambridge University Press.
Gills, B. (2003) "Globalization as Global History: Introducing a Dialectical Analysis", in M. Tetreault et al (eds) *Rethinking Global Political Economy*, London: Routledge.
Gills, B. (2010a) "Going South: Capitalist Crisis, Systemic Crisis, Civilisational Crisis", *Third World Quarterly*, 31: 169–84.
Gills, B. (2010b) "Globalization, Crisis and Transformation: World Systemic Crisis and the Historical Dialectics of Capital", *Globalizations*, 7: 275–88.
Gills, B. and Frank, A. G. (1990) "The Cummulation of Accumulation: Theses and Research Agenda for 5000 Years of World System History", *Dialectical Anthropology*, 15: 19–42.
Gills, B. and Frank, A. G. (1991) "5000 Years of World System History: The Cumulation of Accumulation", in C. Chase-Dunn and T. Hall (eds) *Core/Periphery Relations in Precapitalist Worlds*, Boulder, CO: Westview Press.
Gills, B. and Frank, A. G. (1992) "World System Cycles, Crises, and Hegemonial Shifts, 1700 BC to 1700 AD", *Review*, 15: 621–87.
Gills, B. and Frank, A. G. (1994/2011), "The Modern World System under Asian Hegemony: The Silver Standard World Economy 1450–1750", pp. 50–80 in P. Manning and B. Gills (eds) *Andre Gunder Frank and Global Development: Visions, Remembrances, and Explorations*, London: Routledge.
Ibn Khaldun (1332–1406) (1958) *The Muqaddimah*, trans. F. Rosenthal. Princeton, NJ: Princeton University Press.
Larsen, M. (1976) *The Old Assyrian City-State and its Colonies*, Copenhagen: Akademisk Forlag.
Modelski, G., Devezas, T. and Thompson, W. (eds) (2008) *Globalization as Evolutionary Process*, London: Routledge.
Modelski, G. and Thompson, W. (1996) *Leading Sectors and World Powers*, Columbia: University of South Carolina Press.
Pollard, S. (1968) *The Idea of Progress*, New York: Penguin.
Rostovtzeff, M. (1926) *The Social and Economic History of the Roman Empire*, Oxford: Clarendon Press.
Rowlands, M., Larsen, M. and Kristiansen, K. (eds) (1987) *Centre and Periphery in the Ancient World*, Cambridge: Cambridge University Press.

Sherratt, A. (1993) "What Would a Bronze-Age World System Look Like?", *Journal of European Archaeology*, 1: 1–57.

Teggart, F. (1925/1960) *Theory of History*, New Haven, CT: Yale University Press, reprinted Berkeley: University of California Press.

Wallerstein, I. (1990) "World-System Analysis: The Second Phase", *Review*, 13: 287–93.

Wallerstein, I. (1992) "The West, Capitalism, and the Modern World-System", *Review*, 15: 561–619.

Wallerstein, I. (1994) "Hold the Tiller Firm: On Method and the Unit of Analysis", *Comparative Civilizations Review*, 30: 72–80.

Wallerstein, I. (2008) "World-Systems Analysis", in G. Modelski and R. Denemark (eds) *The Encyclopedia of Life Support Systems* (EOLSS), Oxford: EOLSS Publishers. [http://www.eolss.net, accessed July 6, 2010].

Wilkinson, D. (2009) "The Power Configuration of the Central Civilization/ World System, 100 BC–AD1", paper presented at the International Studies Association Annual Convention, New York, February.

5.2

The failure of the "Modern World-System" and the new paradigm of the "Critical Theory of Patriarchy"

The "civilization of alchemists" as a "system of war"

Claudia von Werlhof

Origins and potential of the new paradigm of the Critical Theory of Patriarchy

The Critical Theory of Patriarchy has been developed in the last 20 years (Werlhof and Behmann 2010). It is based:

- on the earlier feminist analysis of global capitalism/socialism as "capitalist patriarchy," which showed that instead of proletarian wage labor it is the worldwide "housewifeization" of women and their systematically low or completely unpaid labor in private and public that has been the human pre-condition for capitalist accumulation since the end of the witch-hunts (cf. Mies 1986; Mies et al 1988; later Federici 2004);
- on "ecofeminism" as it defines the relationship between society, women, and nature, shows the fundamental importance of nature in general and gives a definition of it as a living organism instead of supposedly "dead matter"; the latter has been suggested by the modern sciences responsible for the "death of nature" (Merchant) since the seventeenth century, following the death of women as "witches" (cf. Daly 1978, Merchant 1980; Mies and Shiva 1993);
- on the historical analysis of "patriarchy" as the "deep structure" of modernity itself; patriarchy is no longer defined as a merely "traditional" remnant of male domination that will sooner or later be abolished by modernity, a position held by most analysts, "gender" studies and the Left (Werlhof 2007a);
- on a different definition of "civilization," comparing the original and in parts of the world still existing matriarchal civilizations to the patriarchal one that developed later. We reject the image of patriarchy as the first or general form of organization of human society and/or its first form of

"high civilization" and the simultaneous wrong definition of matriarchal civilization as women's or mothers rule, or their denial all together—like that of pre-modern indigenous cultures/civilizations. Matriarchy—mater arché—means "in the beginning the mother" and is based on the evidence of the origins of life in the mother (cf. discussion in Göttner-Abendroth 1988; Derungs 2003; Genth 2009; Werlhof 2009a).

Finally, our new paradigm is based:

- on an analysis of techniques (cf. Greek "techné," meaning ruse, trick), as a means to develop patriarchy as a social utopia opposing the natural world, the world of mothers and of matriarchal civilizations. We therefore reject the idea that these techniques developed by patriarchy are "neutral" and the presumably necessary and inevitable "evolutionary progress" of humanity (cf. Genth 2002; Mumford 1977; Wagner 1970; Werlhof 1997a, 2001, 2007b).

Critical Theory of Patriarchy has developed into a new trans-disciplinary paradigm, which has also recently been explored philosophically (Behmann 2009). In contrast to most existing theoretical approaches, Critical Theory of Patriarchy is able to:

- explain the crisis of the West as the logical failure of the development of patriarchal civilization up to modernity; a failure that could not be understood as long as patriarchy was banned to the collective unconsciousness, wrongly defined, a "blind spot" and the apparently unspeakable "secret" of the Modern World-System;
- consider the existence of patriarchy and make us aware of the historical depth of the present global crisis and the strong interconnectedness of its different dimensions;
- discuss the logic for the alternatives to 500 years of the Modern World-System and 5000 years of different forms of pre-modern patriarchy;
- explain the tremendous difficulties of a general conscious and peaceful global change toward post-capitalist-patriarchal social relations, so urgently needed now.

Critical Theory of Patriarchy has identified the struggles of the Left for a new socialism as a project of industrial society and thus of patriarchy. The Left cannot present alternatives to the crises of today. This insight is one of the results of the cooperation with Immanuel Wallerstein, the founder of the Modern World-System theory, which began in the 1980s (cf. last time, Werlhof 2004). However, our concept of patriarchy, capitalist patriarchy, and the critique of modern technology as the "development of the productive forces," and of "progress" as "patriarchal" have not yet been discussed by many Modern World-System theorists. André Gunder Frank, for example, did not want to consider the category of patriarchy, even if it had enriched his far-reaching intent to understand not only 500 but 5000 years of "World-Systems" (cf. Frank and Gills 1999).

We do not see how alternatives to the Modern World-System can emerge from it as the Modern, Capitalist World-System of Patriarchy. We are more related to indigenous alternatives in the global South that stem from pre-colonial, mostly matriarchal civilizations, which are still or again practiced worldwide (by the Zapatistas, cf. Werlhof 1997b, the Bolivian "Rights for Mother Earth" or "Pachamama" Movement, the Indian "Earth Democracy" Movement, cf. Shiva 2005, or Brazilian Liberation Theology, cf. Boff 2010). These alternatives can also be found in the pre-patriarchal history of the global North (cf. Derungs 2003; Eisler 1987; Gimbutas 1991).

Concepts

The Modern World-System as a failed utopia; the definition of patriarchy as an opposition to matriarchy; and the technological project of patriarchy as an opposition to nature

Western modernity can be characterized as a civilization that tries by all means to materialize the utopia of a supposedly "better" and "higher" life. Today, with the "globalization" of neoliberalism as the most advanced form of capitalist/modern world patriarchy, war, economy, and technology, it is evident that this project has failed. Instead of heaven, we are facing hell on Earth, already experienced by masses of people in the global South (Chossudovsky and Marshall 2010).

There exists no explanation for this "development of underdevelopment" (Frank 1967). The confusion can first be met by using a "larger telescope" (Mies 2003). This means to compare the two central forms of civilization in human history—on the one hand the egalitarian and co-operative matriarchal civilization in tune with nature and life, and on the other hand the hierarchical patriarchal one trying to dominate nature and life. The approximately 5000-year-old patriarchal society has reached its climax of material realization with modernity. Nearly all of the relationships of the original matriarchal civilizations of the world, including growing parts of nature itself, have been "replaced" by a "progressive" counter-world and counter-nature in the form of "capital" (the commodity, modern money, machinery, systems of domination). This destructive transformation—that could not have been realized with "tools," but is based on continuous wars, processes of "primitive accumulation" inside and outside of Europe, and the development of the "natural" sciences—has been glorified as "peaceful," "civilized," "productive," "humane," and "creative." It defines "capitalist patriarchy" or the Modern World-System, including "socialism." It is futuristic and utopian as it proclaims the possibility to construct an alternative to life, matter, mothers, nature, and ultimately to the suppressed matriarchies.

The failure of western civilization as the Modern World-System can be observed today on a global level as well as on all levels of human, animal, and plant existence, including the elements themselves, though this is of course not really recognized "officially" and much less "explained," neither by political and social groups and institutions nor by the sciences. On the contrary, in spite of the planetary crises in all aspects of life, which are developing their synergetic effects and may even lead to a general collapse of the fundamental conditions for the maintenance of life on earth as such, nearly all of those who comment on this system (atical) crisis still maintain the idea of infinite progress and development, of industrial society as the best form of civilization in history and of rationality as the highest form of intelligence (critiques by Jaeger 2008; Projektgruppe 2009; O'Leary 2010; Werlhof 2008).

In view of this paradox, we can use our new and newly defined theoretical concepts that are mutually interrelated and can explain the overall "logic" of the failure of modern civilization as a "necessary" one: in the long run, the "war logic" of destruction cannot be a success.

Our approach is based on a new and much broader as well as more complex concept of "patriarchy" that brings a new, "technical," definition and "periodization" of it into the debate. This definition exceeds by far the usual understanding of patriarchy as the domination of the father or the man in the family, society and/or the state and leads to a new understanding of industrial society by linking it to the much longer global historical development of patriarchy.

By doing this, the patriarchal "project"—or delusion—of constructing a new world, which is opposed to the existing one, could be identified. This is the reason why patriarchy is so much occupied with techniques that would help in this respect only.

The utopian "New and Better World" Project that directly and deeply opposes the former matriarchal world and absorbs the whole of nature and Earth itself in the course of its development as a form of "divide, transform and rule!" has been propagated since the beginnings of patriarchy. Its main characteristic is: "Alchemy"!

This was my main "discovery" when I tried to find out which category would define the transformation process from the patriarchal past into capitalism as the patriarchal present (Werlhof 1997a, 2001, 2004, 2009b).

"Alchemy" as the link between the past patriarchy and the present, "capitalist" one: The Modern World-System as a patriarchal "civilization of alchemists" and a "system of war"

In Critical Theory of Patriarchy, "alchemy" is understood as a world view and an "interdisciplinary science," technique, ideology, religion, and psychology in the form of a quest for patriarchal male identity and "individuation." Alchemy—one of its translations being "black mud of the Nile"—has its origins in the matriarchal garden-civilizations of the so-called Paradise (the Persian word for garden). After their conquest by war, the method of the beginnings of patriarchy as a societal order (Dieckvoss 2003), the alchemical tradition is passed over to patriarchs and experiences big changes and perversions. The purpose of alchemy under more or less patriarchal conditions is no longer related to a cooperation with nature and a good life of the community, but defined by trying to realize the "Gnostic" utopia of a "male creation" that is supposed to be higher, better, and more divine than the natural or female, matriarchal one.

The new, alchemical idea of a creation under male control, or even as an independent male "creation," is based on the negation, transformation, destruction, and "replacement" of the female creation, of nature in general, and ultimately of the human being itself. Destruction is becoming the pre-condition of "creation" or "production," and a conscious deterioration the pre-condition for a supposed "improvement"!

Starting in the wars of conquest of matriarchal civilizations and in the laboratories of single alchemists, the idea of a male "creation by destruction" was developed, using the warriors experience and justification, his techniques, the enrichment as a result of war as plunder and experiments on the basis of still pre-patriarchal knowledge. Alchemical experiments were in the beginning very much related to those of shamans, blacksmiths, and metallurgy in service of war (Eliade 1980). These experiments have been meditative and immaterial as well as material and practical. They are systematically separating and dissolving living matter, trying to overcome time by speed and generally aiming at surmounting nature and her cycles, life and death. Alchemy under patriarchy intends to become independent from mothers and nature in bringing death to life/matter—"mortification"—as well as in "creating" life, and in becoming active in transforming the origins and "Gestalt" of existing matter by re-combining separate matters in the alchemist's final "Opus Magnum" (Schütt 2000). The result of the procedure is supposed to be the "philosopher's stone," a symbol of ultimate power over nature.

Alchemy, however, is starting its general and global career only with modernity, embracing in the end the whole world and finally our planet. Most of the literature on alchemy argues that alchemy was a failure already in antiquity, not to speak of the Latin middle ages, as it did never succeed in producing gold and life and has therefore allegedly been abolished until the end of the eighteenth century. My concept of alchemy as a violent patriarchal technique and general method of an artificial male creation, however, proves not only its ongoing existence but also its final "success" insofar as it has become the ideological, religious, psychological, and practical basis of modern technology (machinery), science and "progress," and of capitalism in general, extending to a sort of economic, social, and political alchemy (Werlhof, man.).

In the texts and justifications of modern progress, it is argued that all matter/life has to be appropriated, brought to the laboratory, cut into pieces, put together with other materials, and transformed into something supposedly higher, better, and more civilized. This is nothing else than the alchemical form of destruction—"mortification"—which is followed by the intended "Opus Magnum" of a male creation of matter/life, which is not necessarily born but made—today through machine technology—and which is supposed to be much "higher" than the supposedly "low," bad and uncivilized, if not sinful, creations of mothers and Mother Nature.

This generalization of modern alchemy does not mean, however, that it has not failed at the same time. The project of dominating living nature as a whole, on a micro as well as on a macro-level, its exploitation, dissection, and transformation into "dead capital," and the intents to replace it by a "second nature" (Bruiger 2006) of a "new life" or "industrial society" cannot work. The times of "ostrich policies" are over as the effects have become planetary in their dimensions.

All this has become obvious to the many people who have started a search for an alternative world and relationship to nature, also on a technical level. However, as they do not generally have an analysis of alchemy from the point of view of patriarchy, they even think of pre-modern alchemy as an alternative to modern science and technology! (Wilson et al 2007) Such illusions do not solve our dilemma.

The belief in alchemical principles is a very old and religious one, based on the patriarchal idea of the "Creator" as a male God beyond nature, mothers, the Goddess, and Earth. Following this period (Assmann 2003), apparently only this God was able to "make" matter and life exist, and not only in the beginning, but also afterwards; from the perspective of progress, Man thinks it is his calling to improve creation in a supposedly divine manner as if He could act like God on Earth. This is what alchemy and science or progress are about: the alchemists/scientists/capitalists trying to become divine creators on Earth, godly "Fathers" of a "pure" and better matter—like gold—and of a better life—a "new" human being, or today even a "post-human" one (Rifkin 1983; Schirrmacher 2001; Weizenbaum 1976; Werlhof 2010).

Patriarchy as a utopia can be materially realized only via a successful male "creation by destruction." From this perspective, modernity with industrial society is the huge worldwide intent spreading from Europe by means of colonialism, imperialism, war, and "globalization," in order to develop a "pure" patriarchy without any matriarchal remnants and dependencies, be they related to women, mothers, or Mother Nature.

I therefore call the developed Modern World-System of today the "civilization of alchemists," since nearly everybody has become a part of this "Great Transformation" (Polanyi 1978) of the world into a machine for the globally growing "creation" by destruction: commodity production, modern money, machinery, and command—capital. Capital, as the most important result of modern alchemy, is indeed regarded as the new gold (cf. Binswanger 1985) and even "life." But the alchemical production process, not to mention its most modern forms, is not peaceful at all, though the system tries to hide its systematic violence. Hence I call it a "system of war." This war is occurring every day and does not have to be "declared." "Life," nevertheless, cannot be created by alchemy, neither in earlier times nor today. On the contrary, under the regime of alchemy, life is dying out. This shows that it is possible to kill life, but not to bring it into being artificially. In alchemical terms: mortification is working, but the Opus Magnum is not, and the idea of the philosopher's stone is vanishing into the nowhere!

The West's failure is due to the "alchemical" destruction of the world and the so-called natural "resources"; a destruction that is thought of, theologically proclaimed, and fetishized as "creation." Wherever and whatever alchemy is used for, it is applied by following the same principles.

However, the counter-productivity of the patriarchal project's "necessary" violence can neither be recognized nor given up by its "elites," most of the men and always more women in this world. It would mean the renunciation of not only 500 but 5000 years of patriarchy and patriarchal civilization! The utopian dream of modern civilization to finally become detached from history, nature, and the female part of the species will not become reality.

Critical Theory of Patriarchy helps to leave behind the confusion, ideologies, illusions, belief systems, and propaganda about "the West" and the Modern World-System. It allows for a completely new insight into the real, long-term character of modern civilization at the very moment of its failure, and it is therefore necessary for all those who want to move forward to real, "deep" alternatives. Patriarchy or Life! This has become the question of today.

The war system as a planetary one: A new "military alchemy" and the threat for planet Earth; explaining the existence of technologies using the whole planet as a weapon

Mostly unnoticed by the public and academics, the latest developments in military technology since World War II (Bertell 2000) show that the Modern World-System of Patriarchy has reached its ultimate goal: the intent to subject, change and control not only nature and (wo)man, but also the planet itself. The military tries to convert it from a living "Mother Earth" into a "system" that can be switched on and off like a machine (Werlhof 2011). The Earth is irresponsibly and criminally put at risk by the purposeful use of technologies that are transforming the planet itself into a weapon of mass-destruction: instead of nuclear armament—"nuclear alchemy" (Bertell 1985; Wagner 1970)—the new non-nuclear one is working with "plasma weapons, weather wars and geo-engineering" (Bertell 2010; Hamilton 2010). It seems as if the biblical, lightning-and-thunder-throwing God Jahwe has become incarnated! In "Slowly Wrecking our Planet," Bertell describes the negative effects on the order of life on planet Earth which can be identified in the partial destruction of the atmosphere and atmospheric layers, the ozone hole, a wobbling of the planet's course, and the dangerous interferences in its magnetic field inside as well as outside of it (Bertell 2010). Most of these realities, which we experience as climate change or natural catastrophes, are already undermining the oldest experience of (wo)mankind: our deep confidence in Mother Nature. Is this loss occurring in anyone's interest? For Europeans, this has happened once already, namely, at the beginnings of modern society, when nature—like woman—was presented as a bad and violent, demonic and dangerous stepmother instead of a caring, peaceful, and nurturing mother of all beings (Merchant 1980). In fact, the program to dominate nature by modern science was developed at the same time. The realities we are facing today, on the other side, cannot simply be explained as nature's "revenge" against carbon emissions, though they are presented as such in the climate change debate. They are also, to an unknown percentage, produced by the use of artificially created super-powerful electromagnetic waves through "ionospheric heaters" in installations like HAARP in Alaska, "Woodpecker" in Russia, and in at least ten further places on the globe, known originally as Nicola Tesla's (1856–1943) inventions (Tesla 1997). They were the result of an alternative view of the earth and the solar system as a living—"electromagnetic"—interconnected entity, and were opposed to the mainstream of scientists who since the sixteenth century have had a mechanistic understanding of the planetary system as a dead mechanism. Instead of using the peaceful potential of Tesla's inventions for the urgently needed alternatives (O'Leary 2010), the military in the East and the West took over the patents and used them for the opposite. They seem unable to imagine anything other than seizing power as if "might is right" (Bertell 1985), destroying life and appropriating the "global commons," establishing control over everything by violent transformation, now of the planet itself—and propagating this procedure as a "progress" in the domination and "improvement" of nature and Earth! The entire scale of the negative effects of these experiments is unknown, at least to the public, which is not allowed access to the information. This ignorance is leading the public to the wrong conclusion once again, namely, that even more domination of nature is required! The ultimate step to gain power over humankind is at stake; the totality of the planet and its "alchemical" transformation into a "better" one, or rather, into a destroyed one.

Today, the military and related economic and political interest groups are on the way to establish a literal "system of war"—the totalitarian "New World Order"—as the last outcome of the modern "civilization of alchemists."

At this point of the apparent "success" of the powerful, behind which hide the catastrophes that are already under way and will accumulate further if not stopped in time, the general failure of western civilization as the Modern World-System can be observed on a last dimension, the planetary level. The unimaginable mega-crimes to which the Earth and all living creatures on it are more and more exposed, for instance by a steadily growing number of experiments with the planet and their cumulative effects, could even lead to an overall "Omnicide" (Bertell 1985) or "Terracide" (Hoags 2010). Who can speak of "success" in face of these incredible dangers?

We founded the "Planetary Movement for Mother Earth" in Germany in May 2010, and we are hundreds of women and a minority of men strong. We have to do something about the situation of our planet and we will—together with other already existing social movements—try to inform and mobilize as many people as possible worldwide!

Conclusions

I do not know of the existence of any other theoretical approach that is able to explain the unimaginable horror and megalomania of the new "military alchemy." For Critical Theory of Patriarchy it is "easy," but we do not consider this to be a triumph. We would prefer to be wrong! (cf. Klein 2007).

Our approach has widened the analysis and added something to it, which was necessary in order to fully understand the crisis of today and the existing alternatives that require discussion and practice. The crucial difference between our approach and the Modern World-System analysis is our critique of modernity as "patriarchal," which is why we do not recognize its capacity to lead to the required alternative. On the contrary, we are looking for remnants of our matriarchal past and present as a "second culture" (Genth 1996) underneath patriarchy. This is the source of our endeavors for today and the future.

References

Assmann, J. (2003) *Die Mosaische Unterscheidung oder Der Preis des Monotheismus*, München: Hanser.

Behmann, M. (2009) "Idee und Programm einer Matriarchalen Natur-und Patriarchatskritischen Geschichtsphilosophie. Zur Grundlegung der Kritischen Patriarchatstheorie angesichts der Krise der allgemeinsten Lebensbedingungen", in Projektgruppe "Zivilisationspolitik", 107–77.

Bertell, R. (1985) *No Immediate Danger? Prognosis for a Radioactive Earth*, London: The Women's Press.

Bertell, R. (2000) *Planet Earth. The Latest Weapon of War*, London: The Women's Press.

Bertell, R. (2010) "Slowly Wrecking our Planet", *Epilogue for "Planet Earth"*, German edition in preparation.

Binswanger, H.C. (1985) *Geld und Magie*, Stuttgart: Edition Weitbrecht.

Boff, L. (2010) *Die Erde ist uns anvertraut. Eine ökologische Spiritualität*, Kevelaer: Butzon & Bercker.

Bruiger, D. (2006) Second Nature. *The Man-made World of Idealism, Technology and Power*, Victoria: Leftfield Press.

Chossudovsky, M. and Marshall, A. G. (eds) (2010) *The Global Economic Crisis. The Great Depression of the XXI Century*, Montréal: Global Research Publishers.

Daly, M. (1978) *Gyn/ecology: A Metaethics of Radical Feminism*, Boston, MA: Beacon Press.
Derungs, K. (2003) "Die Natur der Göttin", pp. 12–55 in James E.O. (ed.) *Der Kult der Großen Göttin*, Bern: Amalia.
Dieckvoss, G. (2003) *Wie kam Krieg in die Welt? Ein archäologisch-mythologischer Streifzug*, Hamburg: Konkret Literatur.
Eisler, R. (1987) *The Chalice and the Blade. Our History, Our Future*, San Francisco, CA: Harper.
Eliade, M. (1980) *Schmiede und Alchemisten*, Stuttgart: Klett-Cotta.
Federici, S. (2004) *Caliban and the Witch. Women, the Body and Primitive Accumulation*, New York: Autonomedia.
Frank, A.G. (1967) *Capitalism and Underdevelopment in Latin America*, New York: Monthly Review Press.
Frank, A.G. and Gills, B.K. (1999) *The World System: Five Hundred Years or Five Thousand?* London: Routledge.
Genth, R. (1996) "Matrirchat als zweite Kultur", pp. 17–38 in von Werlhof C., Schweighofer A. and Ernst W. (eds) *Herren-Los: Herrschaft-Erkenntnis-Lebensform*, Frankfurt: Peter Lang.
Genth, R. (2002) *Über Maschinisierung und Mimesis. Erfindungsgeist und mimetische Begabung im Widerstreit und ihre Bedeutung für das Mensch-Maschine-Verhältnis*, Frankfurt: Peter Lang.
Genth, R. (2009) "Zivilisationskrise und Zivilisationspolitik", in Projektgruppe "Zivilisationspolitik", 31–57.
Gimbutas, M. (1991) *The Civilization of the Goddess. The World of Old Europe*, San Francisco, CA: Harper.
Göttner-Abendroth, H. (1988) *Das Matriarchat I. Die Geschichte seiner Erforschung*, Stuttgart: Klett-Cotta.
Hamilton, C. (2010) "The Return of Dr. Strangelove. The Politics of Climate Engineering as a Response to Global Warming", in *Requiem for a Species: Why We Resist the Truth About Climate Change*, London and Washington, DC: Earthscan and Allen & Unwin.
Hoags, Bonnie (2010) *Email 23rd of August, The Bonnefire Coalition*. Available at www.agriculturedefense-coalition.org
Jaeger, M. (2008) *Global Player Faust oder Das Verschwinden der Gegenwart. Zur Aktualität Goethes*, Berlin: WJS.
Klein, N. (2007) *The Shock Doctrine. The Rise of Disaster Capitalism*, New York: Metropolitan Books; Toronto: Knopf Canada.
Merchant, C. (1980) *The Death of Nature. Women, Ecology and the Scientific Revolution*, San Francisco, CA: Harper & Row.
Mies, M. (1986) *Patriarchy and Accumulation on a World Scale. Women in the International Division of Labour*, London: Zed Books.
Mies, M. (2003) "Über die Notwendigkeit, Europa zu entkolonisieren", pp. 19–40 in von Werlhof C., Bennholdt-Thomsen V. and Faraclas N. (eds) *Subsistenz und Widerstand. Alternativen zur Globalisierung*, Wien: Promedia.
Mies, M., Bennholdt-Thomsen, V. and von Werlhof, C. (1988) *Women, the Last Colony*, London: Zed Books; New Delhi: Kali.
Mies, M. and Shiva, V. (1993) *Ecofeminism*, London: Zed Books.
Mumford, L. (1977) *Mythos der Maschine*, Frankfurt: Fischer.
O'Leary, Brian (2010) *The Turquoise Revolution. Innovation and Sustainable Solutions—An Urgent Appeal to Scientists, Environmentalists and Progressives*, http://drbiranoleary@wordpress.com, June 14.
Polanyi, K. (1978) *The Great Transformation*, Frankfurt: Suhrkamp.
Projektgruppe (2009) *"Zivilisationspolitik": Aufbruch aus dem Patriarchat—Wege in eine neue Zivilisation?*, Frankfurt: Peter Lang.
Rifkin, J. (1983) *Algeny*, New York: Viking.
Schirrmacher, Frank (ed.) (2001) *Die Darwin AG. Wie Nanotechnologie, Biotechnologie und Computer den neuen Menschen träumen*, Köln: Kiepenheuer & Witsch.
Schütt, H.W. (2000) *Auf der Suche nach dem Stein der Weisen. Die Geschichte der Alchemie*, München: CH Beck.
Shiva, V. (2005) *Earth Democracy*, Cambridge: South End Press.
Tesla, N. (1997) *Seine Werke. Edition Nikola Tesla*, Peiting: Michaels Verlag.
Wagner, F. (1970) *Weg und Abweg der Naturwissenschaft*, München: CH Beck.
Weizenbaum, J. (1976) *Computer Power and Human Reason. From Judgement to Calculation*, New York: W. H. Freeman & Company.
Werlhof, C.v. (1997a) "Ökonomie, die praktische Seite der Religion. Wirtschaft als Gottesbeweis und die Methode der Alchemie", pp. 95–121 in Ernst U., Gubitzer L. and Schmidt A. (eds) *Ökonomie M(m)acht Angst. Zum Verhältnis von Ökonomie und Religion*, Frankfurt: Peter Lang.
Werlhof, C.v. (1997b) "Upheaval from the Depth. The 'Zapatistas', the Indigenous Civilization, the Question of Matriarchy and the West", *International Journal of Comparative Sociology*, 38: 106–30.

Werlhof, C.v. (2001) "Losing Faith in Progress? Capitalist Patriarchy as an 'Alchemical System'", pp. 15–40 in Bennholdt-Thomsen, V. Faraclas, N. and von Werlhof, C. (eds) *There is an Alternative: Subsistence and Worldwide Resistance to Corporate Globalization*, London: Zed Books.

Werlhof, C.v. (2004) "Using, Producing, and Replacing Life? Alchemy as Theory and Practice in Capitalism", pp. 65–78 in Wallerstein I. (ed.) *The Modern World-System in the Longue Durée*, Boulder, CO, and London: Paradigm.

Werlhof, C.v. (2007a) "Capitalist Patriarchy and the Negation of Matriarchy. The Struggle for a 'Deep' Alternative", pp. 139–53 in Vaughan G. (ed.) *Women and the Gift-Economy. A Radically Different World View is Possible*, Toronto: Inanna.

Werlhof, C.v. (2007b) "No Critique of Capitalism without Critique of Patriarchy! Why the Left is No Alternative", *CNS—Capitalism Nature Socialism*, 18(1): 13–27.

Werlhof, C.v. (2008) "The Globalization of Neoliberalism, its Consequences, and Some of its Basic Alternatives", CNS—*Capitalism-Nature-Socialism*, 19(3): 94–117.

Werlhof, C.v. (2009a) "The Utopia of a Motherless World. Patriarchy as War-System", pp. 29–44 in Göttner-Abendroth H. (ed.) *Societies of Peace, Matriarchies Past Present and Future*, Toronto: Inanna.

Werlhof, C.v. (2009b) "Das Patriarchat: 'Befreiung' von Mutter (und) Natur?", in Projektgruppe "Zivilisationspolitik", 59–103.

Werlhof, C.v. (2010) *Vom Diesseits der Utopie zum Jenseits der Gewalt. Feministisch-patriarchatskritiscehe Analysen —Blicke in die Zukunft?*, Freiburg: Centaurus.

Werlhof, C.v. (2011) *The Failure of Modern Civilization and the Struggle for a "Deep" Alternative*, Frankfurt: Peter Lang.

Werlhof, C.v. and Behmann, M. (2010) *Teoría Crítica del Patriarcado. Hacia una Ciencia y un Mundo ya no Capitalistas ni Patriarcales*, Frankfurt: Peter Lang.

Wilson, P.L., Bamford, C. and Townley, K. (2007) *Green Hermeticism. Alchemy and Ecology*, Great Barrington: Lindisfarne Books.

5.3

Authenticating seventeenth century "hegemonies"

Dutch, Spanish, French, or none?

David Wilkinson

In scrutinizing the history of the modern world-system of great powers, writers have variously attributed "hegemony" (or some cognate) to several different powers at different times. While Spanish, French, British, and American "hegemonies" have been asserted (and contested), perhaps the most contested such imputation is the "Dutch" (or Netherlands) hegemony of the seventeenth century. An examination of claims, counterclaims, and historical evidence in the Dutch case, and in the Spanish and French cases to the extent that they overlap the Dutch, offers us the opportunity to clarify criteria and develop procedures for evaluating "hegemonies."

The "Dutch hegemony"

The claim of a Netherlandish hegemony comes from the world-system writers Immanuel Wallerstein, Terence Hopkins, and others. Wallerstein has provided a definition of hegemony that is generally in accord with standard usage, though perhaps a bit lenient:

> Hegemony in the interstate system refers to that situation in which the ongoing rivalry between the so-called "great powers" is so unbalanced that one power is truly *primus inter pares*; that is, one power can largely impose its rules and its wishes (at the very least by effective veto power) in the economic, political, military, diplomatic and even cultural arenas.
>
> *(Wallerstein 1984: 38)*

> I mean this to be a relatively restrictive definition. It is not enough for one power's enterprises simply to have a larger share of the world market or simply to have the most powerful military forces or the largest political role. I mean hegemony only to refer to situations in which the edge is so significant that allied major powers are *de facto* client states and opposed major powers feel relatively frustrated and highly defensive vis-à-vis the hegemonic power.
>
> *(Wallerstein 1984: 39)*

When did this Dutch hegemony exist? Wallerstein gives tentatively suggested maximal bounding points for the Dutch hegemony, and for British and American hegemonies as well: 1620–72, 1815–73, 1945–67 (Wallerstein 1984: 40). There are some confounding propositions: the period 1651–78 is seen as "the height of Dutch hegemony" (in Wallerstein 1980: 65); while (in Wallerstein 1984: 58), 1650–72 is the period in which the United Provinces fell from "unquestioned supremacy" to "the lesser status of a very powerful state … but nonetheless one state amid several." Regardless of this contradiction, the period within which a Dutch hegemony must be found, if it indeed existed, is 1620–78.

It may be noted that the exponents of a Dutch hegemony have frequently added to their burden of proof by expounding the argument that, *in addition to* its politico-military domination, a hegemon must also meet certain economic criteria: when

> one core power exceeds all others in the efficiency of its productive, commercial and financial activities, and in military strength. We can call this latter "hegemony."
>
> *(Hopkins and Wallerstein 1982: 116)*

Other analysts also accept the Dutch hegemony, for example, Chase-Dunn et al (2005) and Peter Hugill (2009). But there exist alternative readings of seventeenth-century power structures and power transitions.

Alternatives to a Dutch hegemony

The *Political Maxims* (1746) of Pieter de la Court and Johan de Witt, Grand Pensionary of Holland (1653–72) discuss (228–51) "Alliances which Holland might enter into with mightier Potentates than themselves," namely France, Spain, and England; of whom the author says "neither of them [sic] fear us, but we must fear them" (228) and "so long as Holland can stand on its own legs, it is utterly unadvisable to make any alliance with those who are more potent … ," (249), that is, France, Spain, and England. So de la Court (or de Witt), a contemporary of the alleged hegemonies, did not notice any of them. Holland in particular he viewed not as a hegemon but as a second-class power, threatened on all sides by stronger states, yet capable of preserving its independence. (N.B.: Holland was the richest and strongest of the United Provinces of the Netherlands, the Dutch Republic, and the name of the part was and is frequently used as a convenient, if unofficial, synonym for the whole.)

Balance-of-power historians have asserted that in the seventeenth century, and indeed from the sixteenth to the twentieth centuries, there were hegemonists, but never a hegemony. For Ludwig Dehio (1962), the system of states between 1494 and 1945 saw repeated bids for hegemony: by the Habsburgs under Charles V (34–42); by Spain under Philip II (43–62), by France under Louis XIV (72–90), and again under Napoleon (137–80); and by Germany under Wilhelm II (224–46) and again under Hitler (247–68). But each bid was successfully resisted and collapsed. The outcome of the struggle between Anglo-American and Russian power remained uncertain as of 1962 (269–88).

Paul Kennedy (1987) similarly contended that a Habsburg bid for mastery, 1519–1659, just barely failed. French, British, and German empires were, during their bids for mastery, the most powerful states economically and militarily, but were met by countercoalitions. American and Soviet empires produced a bipolar world; the bid of America, the more powerful, also produced a countercoalition. Neither Dehio nor Kennedy recognized a Netherlands hegemonic bid, let alone a full-blown hegemony.

The leading English-School theorist, Martin Wight (1978: Ch. 2), recognized a sixteenth-century hegemony, but none in the seventeenth or following centuries; instead he saw a series of bipolar or tripolar power structures, regional rather than systemwide hegemonies, usually geopolitically divided as between a landpower and a seapower. "Dominance," "hegemony," "ascendancy," or "mastery" of the states-system was, in Wight's view, enjoyed on land and sea by sixteenth-century Habsburg Spain under Charles V. Then it was divided three ways, among Sweden in the Baltic under Charles XII, France in continental West Europe under Louis XIV, and Holland upon the sea. Next it was divided between Napoleonic France on land and Britain on sea. Next it was divided by Kaiserine Germany on land and Britain on sea. Then it was divided between Hitler's Germany on land and the United States on sea. Most recently it was divided between Soviet Russia on land and the United States on sea. In Wight's reading, there was no Dutch systemwide hegemony in the seventeenth century, but rather a regional (seapower) hegemony within a tripolar systemwide structure.

The *Brockhaus* German historical school (Neuhaus 1998; Nicklas 1998) does see seventeenth-century hegemony, and indeed a hegemonic succession, but the succession is from Spain (sixteenth to seventeenth centuries) to France (seventeenth to eighteenth centuries). The Netherlands is not cited as a hegemon. Similarly, *The New Cambridge Modern History* moves directly from "the decline of Spain" 1609–48/59 (Cooper 1970) to "the ascendancy of France" 1648–88 (Carsten 1961).

The nineteenth-century historian Leopold Ranke (1833) saw two rounds of French hegemony in Europe. France achieved "supremacy" or "domination" in Europe under Louis XIV, but was deprived of it by the eighteenth-century rise of the great powers—England, Austria, Russia, and Prussia. France regained supremacy under Napoleon, and then had its predominance broken once more by the revival of the great powers. Ranke did not observe any "supremacy" of the Netherlands.

A similar picture for the seventeenth century, in the context of a longer-term power evaluation 1495–1975, was seen in the late twentieth century by the political scientist Jack Levy: bipolarity 1495–1556 (France vs. Hapsburgs), unipolarity 1556–88 (Spain), multipolarity 1588–1659, unipolarity 1659–1713 (France), multipolarity 1713–97, unipolarity 1797–1815 (France), multipolarity 1815–1945, bipolarity 1945–75 (US vs. USSR). For Levy, "unipolarity" equals "hegemony": "If a single state, unmatched by any other, attains a position of hegemony or dominance over the system, then the system is said to be characterized by unipolarity (e.g., Napoleonic Europe)" (Levy, 1985: 47). In Levy's reading, Spain was then hegemonic 1556–88; there was no hegemon 1588–1659; and France was hegemonic 1659–1713. The Netherlands never appears in a unipolar/hegemonic role, or even as a joint hegemon in a bipolar or tripolar structure.

In the main and in brief, we may declare that four mutually incompatible claims have been made concerning hegemony in the great power system of the seventeenth century: Dutch hegemony in the mid-century; Spanish hegemony in the early century; French hegemony late in the century; and no hegemony at all, early, mid, or late. At least three of these claims must be mistaken.

How can we arbitrate such conflicting claims?

Developing a procedure for weighing hegemonic imputations

Weighing the Dutch candidacy for hegemonic status against those of its rivals can help us develop our understanding of the types and degrees of subaltern subordinations that exist, and of the criteria to be applied in weighing other hegemonic claims.

The existence of a systemwide hegemon necessarily implies that the other powers in the system stand in a relation of inferiority, of subordination, to the hegemon. The term "subaltern" originally denoted a junior officer in the British army, and therefore a person who had authority over others yet was also subject to the authority of one of higher rank. The most inclusive term for describing states which have power over their own people and territories, and perhaps also over lesser states, yet are also subordinated to a systemwide hegemon, would seem to be "subaltern states." Let us use this label to encompass both Wallerstein's "*de facto* client states" and his "relatively frustrated and highly defensive" opposed major powers.

When it is asserted that at some time for some interstate system state X was "hegemon" to that system, the procedure for verifying or falsifying that assertion must involve a survey of the other states of the system, and an inspection of their relations with the candidate hegemon, to determine whether they did indeed stand in relation to it as "subalterns." If all other states are indeed subalterns to the candidate, the candidate is undoubtedly a hegemon. If none are, it certainly is no hegemon. If some are and some are not, there is cause for careful reflection.

As regards the place of the "hegemon" in the system of states, we may reasonably expect that when a state attains hegemony, such of its historic objectives as have in the past been balked by the resistance of other powers will be overcome, and the hegemon will clear its backlog of desire.

Further evidence of hegemonic attainment would be the scaling up of previous ambitions, and the attainment of objectives previously out of the question: objectives once local or regional will be expanded to systemwide scale, as local frustrations are replaced by systemwide achievement.

Appraising Dutch, Spanish, and French "hegemonies"

Let us apply these procedures to the states in the system farthest from the candidates, all three of whom are at what may be called the "West End" of the interstate system of the era, namely the often-neglected "East End" and "North End" states—the Ottoman Empire, Iran, Russia, Sweden, Denmark. (As of the seventeenth century, the "modern" world-system would also include the European overseas colonies as peripheries, but would exclude the coexisting world-systems of India, China, and Japan, as well as the interiors of Africa and the Americas.) Next we can look at the great powers nearer to the candidates, that is, Habsburg Austria and England. What were they doing 1620–78, and how, if at all, do they display *de facto* client status, or high frustration and defensiveness, vis-à-vis the Netherlands (or Spain then France)? Finally, we can examine the mutual relations of the three candidates.

In 1620–78, the *Ottoman Empire* fought Poland, Iran, Russia (via a Crimean vassal), Venice, Transylvania, Austria, Poland again, and Rușsia again, with shifting fortunes. The Ottomans put down rebellious governors or vassals in Syria and Transylvania, and acquired new vassals in Ukraine. The Netherlands, Spain, and France were not involved. There is no evidence that the Ottomans were subaltern to any hegemon, nor frustrated and defensive vis-à-vis any of the proclaimed hegemons, though such terms would at times describe their relation to their Central and West European and Southwest Asian neighbors.

In 1620–78, *Iran* established trade ties with the Dutch, (the English had already arrived) fought the Ottomans and conquered what is now Iraq, then suffered losses and became quiescent. Spain and France were not involved; ties with the Dutch were economic, with the Dutch acquiring no political power. There is no evidence that Iran was subaltern to any hegemon, or that it was frustrated and defensive vis-à-vis any of the proclaimed hegemons, though it was indeed frustrated by and defensive toward the Ottoman Empire late in the subject period.

In 1620–78, *Russia*, having recovered from its "Time of Troubles," generally followed a cautious policy of the status quo, punctuated by wars with Poland to acquire territory in what is now Belarus, until it became involved in the First Northern War (1655–61) to defend Poland against Swedish expansion. Denmark, Transylvania, Brandenburg, Ukraine, Austria, Holland, Crimea, and (diplomatically) France, successively become involved, some to gain territory, some to prevent the gains of others. After this war, Russia resumed its war with Poland and made territorial gains. The Dutch role in the First Northern War helped Denmark, hurt Sweden, and did not affect Russia; French diplomacy restrained Sweden, but again did not affect Russia. Spain was not involved in Russian affairs. There is no evidence that Russia was subaltern to any hegemon whatsoever, or frustrated and defensive vis-à-vis any of the proclaimed hegemons.

In 1620–78, *Poland* fought Russia, sustained a rebellion by Cossacks and Ukrainians aided by Tatar Crimea, fell into chaos under Russian pressure, was attacked in the First Northern War by predators (Sweden, Transylvania, Brandenburg, and Ukraine), then rescued by Russia, Denmark, Austria, and French diplomacy; fought Russia again, fell into civil war and fought Cossacks, Tatars, and Ottomans. Holland and Spain were not involved; France's role was minimal. There is no evidence that Poland was subaltern to, nor frustrated and defensive vis-à-vis, any of the proclaimed hegemons, though there were periods when it was very defensive vis-à-vis Sweden and Russia, and in a client-state position with regard to several rescuers—but not Holland, Spain, or France.

In 1620–78, *Sweden* expanded against Poland, intervened on behalf of the German Protestants in the Thirty Years' War but was frustrated by Habsburg, Imperial-Catholic, and Spanish armies, remained in the war sustained by French subsidies, made major gains in the Peace of Westphalia, attacked Poland (provoking the First Northern War) and after multiple interventions and counter-interventions, gained Polish and Danish territory; joined England and Holland to oppose Louis XIV, but was bought off by France, then fought Brandenburg, Holland, and Denmark with French support, was badly defeated, but rescued by France in the Peace of Nimwegen (1678). It would be fair to say that Sweden was subaltern to France in the last phase of the Thirty Years War 1635–48, and in the last portion of the Franco-Dutch War of 1672–78; Swedish involvement with Holland and Spain was active, but did not show subordination. The major source of Swedish frustration was Austria and Catholic German states.

In 1620–78, *Denmark* intervened 1625–29 in favor of German Protestants, but was frustrated by Imperial and Catholic League forces and made a separate peace. Later, Denmark repeatedly sought to resist the expansion of Sweden, generally being defeated, but was rescued by Holland in 1658–60, aided Holland against England in the Second Anglo-Dutch War 1665–67, and fought Sweden at Holland's behest and to no profit 1675–79. It would be reasonable to treat Denmark as a subaltern client to the Netherlands 1658–79, but not earlier; the sources of its frustrations were Austria, German Catholic states, and Sweden.

In 1620–78, Habsburg *Austria* became fully engaged in the Thirty Years' War (1618–48), in which it struggled to maintain its local hegemony in the Holy Roman Empire against German Protestant rebels, these supported successively by Denmark, Sweden, and France, while Spain, and a German Catholic League supported Austria. The resulting Peace of Westphalia (1648) greatly decentralized the Empire, weakening Austria. In the First Northern War, Austria intervened to stabilize southern Poland. In 1663–64, Austrian designs on Transylvania were frustrated by Turkey; despite support from France, Austria made a conciliatory peace. Austria joined Spain and Lorraine to support Holland against France in the Franco-Dutch war of 1672–78; France made some gains, at Spain's expense. There is no evidence that Austria was subaltern to any hegemon, nor frustrated and defensive vis-à-vis any of the proclaimed hegemons during the epochs of their respective purported hegemonies; it was frustrated and defensive against German Protestant rebels and their successive supporters Denmark, Sweden, and France during the Thirty Years' War (a period of alleged Dutch, and alleged Spanish hegemony), and later followed an independent and cautiously interventionist policy, usually on behalf of a status quo, not as a client of Holland, Spain, or France.

In 1620–78, *England's* internecine political strife reinforced a generally vacillating foreign policy. Spain was fought without gain 1625–28; English support for the French Huguenots

1625–29 did not save them from France; three aggressive Anglo-Dutch wars 1652–54, 1665–67, 1672–74 brought little gain; useful assistance to France against Spain 1657–59 gained Dunkirk, soon sold to France; an alliance with Holland in 1668 persuaded France to limit its gains in the War of Devolution against Spain. There is no evidence that England was subaltern to any hegemon, nor frustrated and defensive vis-à-vis any of the proclaimed hegemons; England alternated between indulging in internal war unmolested by its neighbors, and tentatively asserting a rather "forward" policy that lacked commitment and had minor results.

We come at length to the purported hegemons, *Holland*, *Spain*, and *France*. Their involvement in the world-system's wars and diplomacy was certainly extensive. In the Thirty Years' War, Holland fought Spain 1621–48; France fought Spain 1635–59. Holland defended itself more or less successfully against England in three Anglo-Dutch wars, restricted French advances against Spain in the War of Devolution, and successfully defended itself against France 1672–78.

All three alleged hegemons certainly displayed classic great-power behavior. Toward the end of the period, the growing strength and aggression of France were reflected by hegemonistic policies directed at Spain and Holland, but the targets and coalition partners contained French expansion with minor damage: accordingly, we cannot say that any of these three powers was hegemonic over, or subaltern to, any of the others. As regards the rest of the system, as we have already seen, although Holland did have Denmark as subaltern for a few years, and France had Sweden as subaltern in several engagements, none of the three had hegemony over the Ottoman Empire, Iran, Russia, Poland, Austria, or England.

Did any of the three candidate hegemons attain historic objectives and then scale up local ambitions to systemwide dimensions? *Holland* sought to preserve its independence from Spain, then from France, and was successful in this; it sought to prevent domination of the Baltic by Sweden or Denmark, and was again successful. It failed to reacquire and reunite the Spanish Netherlands, lost before 1590. Having defeated Spain's navy in the Battle of the Downs (1639), Holland acquired naval supremacy in the English Channel, but lost it in the First Anglo-Dutch War (1652–54). *Spain* failed to reacquire the northern Netherland (1621–48), was defeated at sea by Holland at the Battle of the Downs and on Land by France at Rocroi (1643). *France* made significant territorial gains from the Empire (in Lorraine and Alsace 1648) and from Spain (Rousillon 1659, several cities in the Spanish Netherlands 1668, Franche-Comté and additional parts of Spanish Netherlands 1678) but was repeatedly forced to stop short of its war objectives, last of all in its attempt to conquer the entire Spanish Netherlands and the Dutch republic in the Franco-Dutch War (1672–78). None of the candidates was fully successful in its expansionist aims, and none scaled up its ambitions and attainments to systemwide dimension.

Accordingly it cannot be maintained that Holland, Spain, or France was a systemwide hegemon at any time between 1620 and 1678. The hegemonic hypotheses are disconfirmed; the alternative hypothesis that is confirmed is that the system was multipolar throughout the period.

A note on the economics of hegemony

The exponents of a Dutch hegemony have frequently added to their burden of proof by expounding the argument that, *in addition to* its politico-military domination, a hegemon must also meet certain economic criteria: when

> one core power exceeds all others in the efficiency of its productive, commercial, and financial activities, and in military strength. We can call this latter "hegemony."
>
> *(Hopkins and Wallerstein 1982: 116)*

If we had found a Dutch, Spanish, or French hegemony to have existed in the seventeenth century, in the classic politico-military sense, we might reasonably have gone on to inquire concerning its economic basis or aspects. Had we found a Dutch hegemony, we might well have associated it with the economics of Dutch "Golden Age" sea trade, global reach, capital investment, merchant concentration, banking, insurance, market competition, advances in productive technology (textiles, shipbuilding, wind power, land reclamation), and labor skills (de Vries and van der Woude 1997: 668–74). Had we found a Spanish hegemony, we might well have associated it with the economics of Spanish "Golden Age" financial dominance through the inflow of gold and silver bullion, used to acquire credit and translated into enormous Spanish armies and fleets (Anderson 1988: 17, 25–26, 36–37; Spooner 1968: 26–28). Had we found a French hegemony, we might have associated it with the ability of the finance minister Colbert to extract the maximum revenue from France's imposing population and wealth in a grand national mobilization for war, and pass it on to the expanded army of the great military administrators Le Tellier and Louvois (Anderson 1988: 17, 60–61, 77–79, 83, 96). But as there was no politico-military hegemon, the advantages of the Netherlands in capitalist development, of Spain in specie and borrowing power, and of France in economic mobilization, in essence canceled one another out in the struggle for hegemony, and merely provide a different economic basis for the great power status of each.

Conclusion

In appraising claims that impute "hegemony" to one state or another at one time or another, it is essential to take account of competing claims and to weigh evidence. The methodological flaw that is shared by the otherwise contradictory interpretations of the seventeenth-century power structure of the interstate system is a failure to come to grips with, or even to notice, the existence of the controversy.

In weighing evidence regarding hegemonic claims, it is equally essential to avoid a regional bias, a "View from 9th Avenue." Saul Steinberg's famed *The New Yorker* cover of March 29, 1976, "View of the World from 9th Avenue," well illustrates a restricted world-image centered on Manhattan in which China, Japan, and Russia are dwarfed in size, detail, and significance by—10th Avenue! In examinations of hegemonic claims, it is important to avoid a focus on a single state or region that overlooks important areas beyond its reach; this can be managed by scrutinizing the world-system state by state and region by region, in search of great powers which behaved independently (i.e., were not under the hegemony of the "hegemon"). In the case of the alleged Dutch (and Spanish and French) hegemonies of the seventeenth century, scholars from the "West End" of the world-system seem to have underestimated the size and power of "East End" states such as the Ottoman Empire, Russia, and Iran, and even of "Central" Habsburg Austria, and to have neglected the importance of non-West-End zones of significant struggles for power such as Iraq, Poland, and Transylvania; had these been attended to from the start, surely all three hegemonic claims would soon have been abandoned.

The key method for evaluating assertions of hegemony must then be the examination of the external relations of each other major power in the system of states with the purported hegemon, without favoring any particular region with research attention. Where there are no relations at all, or economic relations without politico-military connections, there can be no hegemony. Where there is continued fighting without clear victory, there is no hegemony. Where states ally without subordination as a result of coinciding but independent calculations of interest, there is no hegemony. But where one great power commands or buys the obedience of another, or extorts tribute from it, or coercively compels it to abandon its policy objectives (without being itself reciprocally coerced, balanced, and deterred), we find a bilateral hegemonic relationship; where that great power exerts such control over all other major powers, we find a systemwide hegemony.

References

Anderson, M. S. (1988) *War and Society in Europe of the Old Regime, 1618–1789.* London: Fontana Press.

Carsten, F. L., (ed.) (1961) *The New Cambridge Modern History. V. The Ascendancy of France 1648–88.* Cambridge: Cambridge University Press.

Cooper, J. P., (ed.) (1970) *The New Cambridge Modern History. IV. The Decline of Spain and the Thirty Years War 1609–48/59.* Cambridge: Cambridge University Press.

Chase-Dunn, C., Jorgensen, A. K., Reifer, T. E., and Lio, S. (2005) "The Trajectory of the United States in the World-System: A Quantitative Reflection," *Sociological Perspectives*, 48(2: Summer): 233–54.

De Vries, Jan and Ad van der Woude (1997) *The First Modern Economy: Success, Failure, and Perseverance of the Dutch Economy, 1500–1815*, Cambridge: Cambridge University Press.

Dehio, L. (1962) *The Precarious Balance: Four Centuries of the European Power Struggle.* New York: Knopf.

Hopkins, T. K. and Wallerstein, I. et al (1982) "Patterns of Development in the Modern World-System," in T. K. Hopkins, I. Wallerstein et al (eds) *World-Systems Analysis: Theory and Methodology*, Beverly Hills, CA: Sage.

Hugill, P. (2009) "The American Challenge to British Hegemony, 1861–1946," *Geographical Review*, 99(2): 403–25.

Kennedy, P. (1987) *Rise and Fall of the Great Powers: Economic Change and Military Conflict from 1500 to 2000*, New York: Random House.

Levy, J. S. (1985) "The Polarity of the System and International Stability: An Empirical Analysis," in A. N. Sabrosky (ed.) *Polarity and War: The Changing Structure of International Conflict*, Boulder, CO: Westview Press.

Neuhaus, H. (1998) "Spaniens Hegemonie im 16. und 17. Jahrhundert: Katholische Vormacht," in Vera Buller (ed.) *Brockhaus—Die Bibliothek/Weltgeschichte. 3, Um Glaube und Herrschaft (600–1650)*, Leipzig: Brockhaus.

Nicklas, T. (1998) "Frankreichs Hegemonie im 17. und 18. Jahrhundert: Europa im Zeichen der Lilie," in Sabine-Walburga Anders, (ed.) *Brockhaus—Die Bibliothek/ Weltgeschichte. 4, Wege in die Moderne (1650–1850)*, Leipzig: Brockhaus.

Ranke, L. (1833) "Die Grossen Mächte"; trans. Hildegarde Hunt Von Laue. (1950) "The Great Powers," in T. H. Von Laue (ed.) *Leopold Ranke: The Formative Years*, Princeton, NJ: Princeton University Press.

Spooner, F. C. (1968) "The Economy of Europe 1559–1609," in R. B. Wernham (ed.) *The New Cambridge Modern History. III. The Counter-Reformation and Price Revolution 1559–1610*, Cambridge: Cambridge University Press.

Wallerstein, I. (1980) *The Modern World-System II: Mercantilism and the Consolidation of the European World-Economy, 1600–1750*, New York: Academic Press.

Wallerstein, I. (1984) *The Politics of the World-Economy*, Cambridge: Cambridge University Press.

Wight, M. (1978) *Power Politics*, New York: Holmes & Meier.

5.4

Challenges of globalization theory to world-systems analysis

Leslie Sklair

Introduction

Globalization has become a very unwieldy idea. In some senses it has replaced twentieth-century staples of social thought such as social change and development. A sampling of the variety of approaches to, and theories of, globalization reveals an overlap that is probably less than you would find in equivalent volumes covering most of the recognized sub-fields in the social sciences. This probably reflects the relatively chaotic (unparadigmatic) nature of globalization studies. My own version of globalization theory—a theoretical framework for analyzing social change and development in their widest applications in the global (post 1950s) age—is one among many attempts to make sense of globalization. It presents some challenges to world-systems analysis (hereafter w-sa). My theory is based on the following distinctions intended to deconstruct and then reconstruct the concept of globalization in a dialectic of thesis (generic globalization), antithesis (capitalist globalization), and synthesis (alternative globalizations). The argument is that the emancipatory potential of generic globalization has been subverted by the repressive class-based exploitations of capitalist globalization (paralleling the ways in which the progressive aspects of the capitalist mode of production have been undermined by the social relations of actually existing capitalisms), and that the way forward is to be found in a number of non-capitalist alternative globalizations, of which socialist globalization appears (to me, at least) to be the most promising in the long term.

Challenge Number 1: Dating globalization—A world-historical phenomenon versus a late twentieth-century phenomenon

Generally, w-sa treats globalization as a long-standing feature of historic world-systems. "Globalization, in the sense of the expansion and intensification of larger interaction networks, has been increasing for millennia, albeit unevenly and in waves. … Rather than a wholly unique and new phenomenon, globalization is primarily international economic integration, and as such it is a feature of the world-system that has been oscillating as well as increasing for centuries. Recent research comparing the nineteenth and twentieth centuries has shown that trade globalization is both a cycle and a trend" (Chase-Dunn 2007: 1061). However, many if not most globalization theorists and researchers conceptualize it as a much more recent phenomenon. I have argued that we need to deconstruct the almost uselessly general term *globalization* into generic, capitalist, and alternative globalizations. The productive force of generic globalization dates from the inception of the electronic revolution, around the 1960s, which transformed systems of production, distribution, and exchange—capitalist globalization, in effect, creates a new mode of production. Also fundamental to generic globalization are postcolonialisms, the creation of transnational social spaces, and qualitatively new forms of cosmopolitanism. Capitalist globalization undermines the emancipatory potential of the four elements of generic globalization, resulting in a new type of class polarization crisis and the crisis of ecological unsustainability (Sklair 2009). All these factors, I argue, challenge the rather dated conceptual framework of w-sa.

Challenge Number 2: Capitalist globalization—transnational practices versus state centrism

My approach to capitalist globalization is based on the concept of transnational practices, practices that cross state boundaries but do not necessarily originate with state agencies or actors. The concept of transnational practices is an attempt to make more concrete the central issues raised in the debate over globalization: notably, the role of the national economy, the state, and local and global cultures. Analytically, transnational practices operate in three spheres: the economic, the political, and the cultural-ideological. The whole is what I mean by the global system. The global system at the beginning of the twentieth-first century is not synonymous with global capitalism, but the dominant forces of global capitalism are the dominant forces in the global system. To put it simply, individuals, groups, institutions and even whole communities, local, national, or transnational, can exist as they have always done outside the orbit of the capitalist global system but this is becoming increasingly more difficult as capitalist globalization penetrates ever more widely and deeply. The building blocks of the theory of the capitalist global system are the transnational corporation (TNC), the characteristic institutional form of economic transnational practices, a still-evolving transnational capitalist class in the political sphere and in the culture-ideology sphere, the culture-ideology of consumerism (Sklair 2002). W-sa, for all its global potential, seems unable to distinguish clearly between transnational and state-centrist levels of analyses, notably in the still central core/semiperiphery/periphery country framework which, I would argue, is rendered redundant by the transnational practices approach to globalization.

Challenge Number 3: Culture (again)

The long-standing and often damaging criticism that w-sa either ignores cultural factors or when it gives them attention, does so in an "economistic" fashion, is also valid in terms of the w-sa approach to globalization. In particular, all the practices, institutions, beliefs, and aspirations that swirl around what we can term the culture-ideology of consumerism—a central issue for non-economistic research on globalization—have little resonance in the w-sa corpus of work.

Conclusion: Can the world-systems approach be globalized?

In my sense of the term, it is difficult to see how w-sa can be adequately globalized. Having said that, it is only fair to acknowledge that some globalization scholars appear to agree that there is nothing essentially new about globalization. This is ironic due to the fact that without the stimulus of w-sa from the 1960s onwards, the more recent flowering of globalization theory and research (in sociology at least) might have been far less fruitful. As things stand, however, there is every chance that w-sa may eventually take its place as one of the many approaches to globalization as well as a dynamic, if flawed, general approach to historical change in its own right.

References

Chase-Dunn, C. (2007) "Dependency and World-Systems Theories," in Ritzer, G. (ed.) *Blackwell Encyclopedia of Sociology*. Oxford: Blackwell.

Sklair, L. (2002) *Globalization: Capitalism and Its Alternatives*. Oxford: Oxford University Press.

Sklair, L. (2009) "The emancipatory potential of generic globalization", *Globalizations* 16/4: 525–39.

Part III

The contemporary world-economy

6. Markets and exchange

6.1

Surplus drain and dark value in the modern world-system

Donald A. Clelland

Stephen Hawking once said "the histories of the universe depend on the precise question asked." Following that logic, I am convinced that the most fundamental questions in world-systems analysis are: What is surplus drain? Why, how, and where does it happen? What are its consequences? The central idea of this perspective is the notion that world-systems are systems of surplus drain. Thus, the history of civilization is the history of surplus drain from one area to another. The first two sections that follow provide a theoretical overview of the concepts of surplus and surplus drain and explain how capitalist surplus extraction differs from previous systems. The third section focuses on those visible mechanisms of surplus drain that have been most often analyzed by world-systems thinkers. In the fourth section, I revise and extend earlier conceptualizations by focusing on those elements of surplus drain that capitalists render invisible.

What is surplus?

Baran and Sweezy (1966: 29) define economic surplus as "the difference between what a society produces and the costs of producing it." Their surplus is not the same thing as Marx's (1967) *surplus value*, that is, the value added to commodities by labor power that is greater than the costs of reproducing that labor through provision of basic survival necessities. I am convinced that these ground-breakers did not go far enough in their conceptual challenge and that we must push their ideas in five additional directions. First, the economic surplus should be viewed as a foundational component of the capitalist world-system as a whole. By its very nature, such a system is driven toward continuous expansion of the surplus. Second, we need to pinpoint the origins of surpluses. On the one hand, surplus can be extracted from any of the factors of production (land, resources, labor, energy, knowledge, technology, capital). On the other hand, we must recognize that the bulk of surplus extraction is realized through *degrees of monopoly* (Kalecki 1954), not through the *competitive* markets emphasized by Braudel (1982). By degree of monopoly, I mean the control of any mechanism that reduces the costs of production or increases sales prices, in variance from a fully competitive market. The capitalist world-economy is a degree of monopoly system because capitalists seek to avoid competition through the construction of quasi-monopolistic commodity chains. Third, we must rethink how we calculate surplus to reflect the points of origins of the *hidden* surpluses that capitalists expropriate to sustain their degrees of monopoly. In actuality, the total world surplus is far greater that the cumulative GDPs. To arrive at a realistic estimate of surplus, we must take into account the value of all reproductive costs. We should think of the modern world-system as an ice-berg economy (Mies et al 1988) in which uncosted dark value comprises the thicker submerged ice layers that are blocked from view beneath a thin top stratum that is counted as the visible official economy.

Fourth, we need to determine who collects the surplus, how it is expropriated, and how it is used. The capitalist world-system is structured to ensure the concentration of surplus into the hands of capitalists and state elites (in the form of profit, rents, interest, and taxes). On the one hand, these groups are motivated to expand the surplus in order to broaden their consumption of luxury goods. On the other hand, the key to the continuation of the world-system is reinvestment of surplus into expanded reproduction of the system, that is, accumulation. In addition, capitalists share part of the surplus with those who perform the functions of capital, the managerial-professional class, and even with sections of the working class. Finally, capitalists may surrender "surplus opportunity" by passing on hidden surplus in the form of lower prices to commodity buyers.

The centrality of surplus drain

I employ *surplus drain* and *surplus transfer* as synonyms. Similar terms that appear in the literature include economic drain, surplus extraction, and capital drain or transfer. *Unequal exchange* is often used as a synonym, but the originator of that term (Emmanuel 1972) and most subsequent analysts (Amin 1974; Kohler and Tausch 2002; Raffer 1987) intend more narrow usages. Historically, all world-systems have been driven by relations of surplus transfer. Precapitalist systems structured mechanisms of surplus drain between regions, primarily through plunder and tribute. What, then, distinguishes the capitalist world-system as a structure of surplus drain? In contrast to earlier systems, the capitalist world-system is characterized by the highly-rationalized extraction of surplus by means of commodities. World polarization into rich and poor zones is a consequence of routine forms of commodity production and distribution across regions. Capitalists and core states structure the world-economy to insure the continuous flow of economic surplus from peripheral zones, including internal peripheries. Semiperipheral zones are drained by the core and, in turn, drain peripheries. Foundational thinkers (Amin 1974; Frank 1969, 1979; Wallerstein 1974) emphasize surplus drain as the driving force that causes impoverishment and dependency of the periphery in order to accumulate wealth and dominance at the core.

A second distinguishing feature of capitalist surplus drain lies in the need to widen the reach of the system in order to insure its growth and survival. Since the system is based on minimizing costs of production in order to maximize profits, capitalists are compelled to identify and integrate locations where they can take advantage of the lowest costs of production to generate commodities for marketing in areas where prices are highest. Thus, most capitalists want to "buy or produce cheap in the periphery and sell dear in the core." Because this system of commodified production requires territorial expansion, the modern world-system was borne of colonialism and incorporation of noncapitalist arenas, a system of cheap factors and near-monopoly capital. For that reason, the zones of the trimodal structure of the modern world-system are defined by exploitative relationships. Without surplus drains, there is no core, no periphery, no world-system as we know it.

In contrast to other scholars, world-systems foundational thinkers envisioned core and periphery as relationships of surplus transfer, not as geographical or national categories. Frank (1969) conceived the modern world-system to be a "hierarchy of core-periphery *complexes*, in which surplus is being transferred." Wallerstein (1983: 31–32) defines core and periphery in terms of the "transfer of part of the total profit (or surplus)" from one zone to another. "Such a relationship is that of coreness-peripherality. By extension, we call the losing zone a 'periphery' and the gaining zone a 'core.'" He insists that we should define *core* as "those who are living off the surplus value produced by others" and *periphery* as "those who are not retaining all of the surplus value they are producing." Wallerstein (2000: 4) is convinced that core growth is derived from these surplus drains. "The shift of surplus towards the core concentrated capital there and made available disproportionate funds for further mechanization, both allowing producers in core zones to gain additional competitive advantages in existing products and permitting them to create ever new rare products with which to renew the process."

Bright value mechanisms of surplus drain

The first category of extracted surpluses, which I term *bright value*, is distinguished by two features. First, the flow is from peripheral labor/capitalists to core capitalists (or between internal peripheries and their cores). Second, the production costs are rendered visible through formal accounting for economic inputs and outputs. In the sections that follow, I will briefly explore eight sources of bright value surplus drains.

The first major source of bright value is Foreign Direct Investment (FDI) that stimulates transnational repatriation of profits, either immediately or after years of expanded reinvestment. The vast majority of FDI has been between core on core, but it is rarely noted that much of this investment is concentrated in the internal peripheries of those areas. Before 1980, FDI was primarily oriented toward the acquisition of resources and energy for core industries and toward peripheral production of agricultural exports. Subsequently, there was a shift toward FDI in non-core manufacturing. For example, there has been such increased FDI in China that it has become the consumer goods workshop of the world (Petras and Veltmeyer 2007).

The second type of bright value drain occurs in trade monopolies. Significant drains flow from periphery to core through transnational trade that is primarily effected, not through market competition, but through degrees of monopoly. In practice, such trade occurs in complex commodity chains (see entry, this *Handbook*). At every node of a commodity chain, capitalists seek to acquire degrees of monopoly over markets and terms of trade. Thus, commodity chains are "degree of monopoly chains" designed to transfer surplus from competitive capitalists to quasi-monopolist capitalists, generally higher up the chain and in the core. The difference between one peripheral capitalist's costs of production and a competitor's costs is one source of monopsony profit to buyers higher in the trade chain. By combining lower costs of peripheral production with global control over transport and distant markets, core corporations structure quasi-monopolies in consumer goods. For example, companies like Wal-Mart have established monopsony control over peripheral producers in order to maximize bright value drains. Through their monopolistic strategies, corporations simultaneously effect a high degree of monopoly over core retail sales and brand-name dominance and higher prices in peripheral production areas (Lichtenstein 2005).

The third bright value mechanism occurs when core capitalists exert monopolistic control over production inputs through collection of fees for intellectual property rights and/or external services. A fourth bright value drain occurs in commodity flows when foreign personnel repatriate savings or expend their salaries on imports from their home countries. These drain mechanisms shift control over the investment surplus outside the country, causing significant bright value drain from periphery to core.

The fifth form of bright value transfer results from core lending to the periphery. In theory, loans are a mechanism whereby a fraction of the surplus generated within the core is used for expanded reproduction in the periphery. These investments are intended to expand surplus (profits) in the periphery and in the core (through interest). In practice, the results have been spotty for the periphery but satisfactory for the core. Most of the periphery faces overwhelming external debt, amounting to about $400 per citizen. Less than 3 percent of this accumulated debt derives from the original principal, since the vast majority has accrued from "runaway compound interest" (Shah 2005; Smith 1994). In the 1970s, core lending to the Global South expanded after the historical conjuncture of increased dollars held in European banks, rising investment of OPEC profits in core banks, and deregulation of lending practices. Between 1980 and 2002, the external debt of the periphery increased fourfold. Because of the inability of many countries to pay down the principal, higher interest rates and compound interest resulted in repayments of the original loans many times over (Toussaint 2005). For example, Nigeria borrowed \$5 billion, repaid \$16 billion, but still owes \$28 billion accumulated from foreign interest (Smith 1994). As a result, many countries now expend for debt service 50–60 percent of every dollar earned from exports (Shah 2005). A related measure of bright value is *net transfer of financial resources*, that is, net capital inflows after interest and other foreign investment payments are deducted. Between 2002 and 2008, this flow from poor to rich countries escalated from 208 to 891 billion dollars (United Nations 2010).

World-systems scholars (Dunaway 2003: 91–108; Suter 1992) point to the historical regularity of debt crises that seriously limit and interrupt accumulation of surplus in the periphery, but are of lesser consequence for the creditors who are protected by the system. In the Latin American debt crisis of the 1980s and the Asian debt crisis of the 1990s, for instance, debtor countries were bailed out by the transfer of private loans to the IMF, indirectly bailing out core private banks. Stated more bluntly, the surplus drain system was bailed out (Wade and Veneroso 1998). As a defense against future crises, several semiperipheries, most notably China, have used internal surpluses to build reserves in dollars. Such reserves are withheld from investments in the expansion of national economies. Moreover, sizeable portions of these surplus reserves are lent to the United States at such low real interest rates that the surplus drain, measured in terms of nominal interest against inflation, is still to the core. The "investment" in US treasury notes will be repaid at a loss in future cheaper dollars, so the drain continues, but in a new form (Clelland 2010).

Capital flight, exchange rate manipulation, and portfolio investment comprise three additional forms of bright value financial drains. Capital flight is the transfer to the core of peripheral surplus by peripheral capitalists. The damage of such drains to the periphery vastly exceeds any advantages to the core. Between 1979 and 1983, for example, capital flight from Mexico amounted to \$90 billion. In this same period, real wages declined by 60 percent for Mexican workers (Smith 1994). Exchange rate speculation can rapidly devalue a national currency and diminish export revenues, leading to massive surplus drains through currency flight and a wider trade imbalance. Though less significant than other forms of bright value drain, portfolio investment in peripheral enterprises redirects dividends and realized capital assets to the core.

Dark value mechanisms of surplus drain

The second category of extracted surpluses is much more clandestine than bright value transfers. Inspired by the recognition by physicists that imperceptible "dark matter" and "dark energy" account for 90 percent of the universe, I term this invisible surplus flow to consumers *dark value*, a hidden force that fuels the structure and expansion of the capitalist world-system. Unlike bright value that is accumulated as profit by the capitalist, the second type of surplus drain is collected by buyers in the form of cheap commodity prices. Because these transfers are externalized from economic accounting, a full calculation of dark value would require analysis of millions of surplus drains that are derived from unpaid household labor, under-paid labor, under-valued resources, and savings through uncosted human and environmental externalities. Dark value is deeply embedded in every economic transaction or commodity, making it the silenced partner that renders every bright value drain more profitable. Dark value is distinctive in three ways. First, these surplus drains are "free" inputs to capitalists, that is, they are externalized from calculation of the costs of production. While world-systems foundational thinkers did not employ the dark value metaphor, Wallerstein called attention to this "dirty secret" of capitalism, arguing that "an essential element in the accumulation of capital is [that] capitalists [do] not to pay their bills" (Goldfrank et al 1999: 9). In this way, commodity chains are more profitable because of the extraction of hidden surpluses. These chains are "efficient" because they "constitute a method that minimizes costs in terms of output" (Wallerstein 1997).

A second distinguishing feature of dark value is that these drains occur as uncosted contributions to production. Moreover, they represent "opportunity surpluses" for the controllers of production that is beyond the surplus of bright value drains. Assuming a degree of monopoly, the hidden value may be retained by the capitalist as "extra surplus value" (Marx 1967). Alternatively, the dark value may be passed on to the consumer in the form of cheaper commodity prices. It is advantageous to the capitalist to expand the degree of monopoly by underpricing competitors, thereby increasing sales volume and long-term profits. Such a degree of monopoly is most often found near the beginning or end of a commodity chain. Dark value can be differentiated from costed bright drains in a third way. Since dark value is dependent on the existence of mass consumption, its relative economic significance increases over time. Expanded consumption in the core provides the opportunity for expanded drain of value. Logically, the transfer of dark value to the core increases over time as the volume of trade expands, and it will continue to expand in spite of any marginal tendency toward wage convergence.

How, then, is dark value produced? Since dark value may be found in all factors of production (capital, labor, land, natural resources, energy, knowledge), a hidden transfer occurs any time a capitalist obtains a component of production at less than the average world-market price. It is the embodiment of at least four types of hidden subsidies to commodity chains: (1) under-compensated formal labor, (2) under-remunerated or unpaid inputs from households or the informal sector, (3) cheap natural resources, and (4) ecological and human externalities that are "economically free" to capitalists. Since employers have a high degree of monopoly over peripheral labor markets, their payments to labor often fall near or below household subsistence. It is these unpaid hours that are embedded in cheap core consumer prices and concealed in profit accumulation. Workers who do the same tasks with similar skills and equipment earn hourly wages that differ by as much as a ratio of fifty to one between zones, making the core working class an *aristocracy of labor* (Communist Working Group 1986). The core worker becomes an unwitting beneficiary of this exploitative system when he/she uses one waged hour to purchase a product that embodies many more lower-waged hours of peripheral labor (Emmanuel 1972). The typical consumer is aware that the commodity would be more expensive if produced in the core. In that sense, dark value is not truly hidden at all, even though its actual economic value is officially denied.

While Marx (1967) linked commodity value to production labor time, he ignored other elements of value that are deeply hidden. The actual contribution of labor to a commodity consists of the total work hours—costed and unpaid—that are embedded in production and reproduction of labor processes. Peripheral workers survive in households that comprise the underlying substratum of the capitalist world-economy. Household labor and resources subsidize the below-subsistence wages that are paid to peripheral workers (Dunaway 2012). Capitalists are able to drain hidden surpluses from households because a majority of the world's workers earn only a portion of their livelihoods from waged labor. Indeed, these *semiproletarianized households* pool the greater proportion of their resources from nonwaged activities, inadvertently encouraging capitalists to pay below subsistence wages (Wallerstein 1983).

Women inequitably bear the brunt of dark value transfers. In millions of commodity chains, the dark value of unpaid household labor, mostly women's work, sustains the polarized world-system (Mies et al 1988). Peripheral households provide *direct* subsidies to capitalist commodity chains through allocation of unpaid labors and resources to sustain under-paid waged workers. In addition, peripheral households provide *indirect* subsidies to capitalist commodity chains when they absorb the costs of reproducing, maintaining, educating, and socializing the labor force from which capitalists benefit (Dunaway 2001). These types of dark value are repeatedly embedded in a commodity at every node of the production and distribution chains, until it reaches the core consumer with a price reduced by pennies per embodied hour.

In addition to dark value accrued from under-paid waged workers and from semiproletarianized households, capitalists drain hidden surpluses from several types of nonwaged peripheral workers. Beyond each waged worker is a long dark value chain of food producers and informal sector activities that contribute to the reproductive capacity of this worker. By supplying low-cost survival needs to the waged worker, these poorly-remunerated laborers subsidize low capitalist wages. The daily life of the under-compensated peripheral waged worker entails the unequal exchange of one work hour for greater labor time from nonwaged producers. For example, she may drain dark value from a lower-paid child caregiver who makes it possible for her to work for wages outside her household. This flow of dark value cheapens the reproduction costs of peripheral labor and, thus, the wage level that capitalists pay. In addition to these kinds of indirect subsidies to capitalist production, there are several other forms of labor from which dark value is expropriated. Through contract farming and labor subcontracting, capitalists externalize costs of production to peripheral households whose bare subsistence level lowers prices for exports. Dark value is also extracted from informal sector workers who receive low remuneration for goods and services that are absorbed directly into commodity chains. Capitalists also drain dark value from export commodity producers when they generate debt bondage through credit or financing. Transnational labor flows represent labor power and brain drains to the core because the periphery underwrites reproduction and education costs (Dunaway 2012).

The movement of resources is the most visible form of surplus drain. For example, one can readily observe the tons of coal being exported from Appalachia, the trains and trucks returning empty to haul away successive loads of mineral wealth. Such bright value profits fit within the concept of "ecological unequal exchange" (Jorgenson, this *Handbook*). In addition, this emerging framework addresses hidden, uncosted (and plundered) resource drains (Martinez-Alier 2002). Just as seizure and underpayment for resources was crucial to colonizers, core consumers continue to benefit from absentee control over commodity chains based on peripheral ecosystems. Should ownership have been retained in peripheries, the continuing cost of resources would be much higher than it is today, in order to provide "resource rents" to the owners, for example, OPEC. Instead, a large share of the world's peripheral resources are indirectly controlled or contracted at low prices. These dark value flows are not only an unpleasant result of past imperialism, but they also represent "continuing dispossession" in the contemporary world-system (Harvey 2003).

Dark value routinely accrues from ecological costs that are externalized from production accounting. When considering externalities, surplus can no longer be defined simply as that which is available after the reproduction costs of the system have been met, for surplus is maximized through the destruction of the ecosystem itself. Natural capital is withdrawn from the ecological world bank, with no intent of replacement. Consequently, many commodities have even larger ecological footprints (the total ecological base needed for its production and distribution) than is suggested by the popular environment footprint concept (Wackernagel et al 2002). To the extent that the footprint is not fully costed, the consumer collects dark value. Such commodity production often threatens the survival of local communities, as peripheral areas absorb the side effects of the capitalist's unpaid ecological damage through health risks to residents, loss of access to natural inputs into local food security, and taxes for cleanup. Moreover, damage to world ecosystems (especially through global climate change) is disproportionately borne by peripheral areas, reflecting another deeply hidden form of dark value drain (Jorgenson, this *Handbook*).

Conclusion

Dark value drains occur because of the super-exploitation of peripheral labor, households, and ecological resources. Once unequal exchange and surplus drains are initiated, their consequences are historically cumulative in peripheries (Amin 1974). On the one hand, bright value drains remove surpluses that might be invested in expanded reproduction and economic growth in the periphery. On the other hand, extraction of dark value threatens ecological sustainability and lowers the quality of life for worker households and women.

In a very rough sense, whole states that contain major surplus collection points may be designated as core states (Wallerstein 1997). Despite rivalry, these states have established rules that minimize surplus drains among themselves. Because most world commodified economic activity occurs within and between these states, external surplus drain is important, but not the major source of surplus accumulation in the core. Indeed, surplus drains represent a much greater proportion of a peripheral economy than its contribution to the total wealth pool of a core area. Even though these drains are only a small percentage of total core accumulation, these transfers represent massive economic and ecological losses to the periphery. In reality, surplus drains remove nearly one-fifth of the total economic wealth of peripheries (Amin 1974), amounting to more than the capital that is annually invested in expanded reproduction (Kohler and Tausch 2002).

In the US Watergate scandal, "Deep Throat" advised investigators to follow the money. My advice to world-systems analysts is to follow the surplus! The "ecological footprint" model (Wackernagel et al 2002) provides insights we might consider. The ecological footprint provides a proxy for the amount of renewable natural resources needed to maintain a lifestyle. Metaphorically, it is a measure of dark value, since the environmental impact of consumption is far from apparent. The average core resident leaves a huge footprint that is not sustainable. The task for world-systems scholars is to investigate the question of where core footprints fall and with what consequences. A second task is to pinpoint the mechanisms and quantify the amount of underpayment to the (semi)periphery that results in bright value (profit) for core capitalists and in dark value (cheaper prices) for core consumers. These projects may be addressed within the framework of ecological unequal exchange (Jorgenson, this *Handbook*).

A parallel task would be the construction of an "invisible handprint," the amount of labor hours needed to maintain the lifestyle of an average core person. This metaphor toys with Adam Smith's "invisible hand" and recalls earlier historical references to workers as "hands." The world-systems task would be to track where the "invisible handprints" fall geographically and with what consequences for paid and unpaid laborers, followed by determination of the mechanisms and amount of bright and dark value surplus drains. Starting points for a research agenda about dark value are feminist economists (e.g., Antonopoulos and Hirway 2010; Beneria 1999) who emphasize the significance of unpaid labor to world capitalism, as well as world-systems scholars who have conceptualized the semiproletariat, hidden household subsidies to capitalism, the importance of nonwaged labor, and unequal exchange (Dunaway 2012; Kohler and Tausch 2002; Tabak and Crichlow 2000).

Such new research directions would provide clear, concrete answers to contested questions about the benefits of surplus drain to the core and about costs to the periphery. The proposed world-systems application of the ecological footprint model implies that the victim is not just Mother Earth, but real people in real places. It also implies that the beneficiaries of surplus drain include ordinary people, the working class of the core. The distaste for such a suggestion for core radicals and mainstream social scientists probably explains the rather shallow pursuit of the questions of unequal exchange and surplus drain after the passage of a generation of world-systems scholarship. If we rediscover the centrality of surplus drain, world-systems analysis can become a standpoint historical social science that provides a critique of modern capitalism from the vantage point of the periphery.

References

Amin, S. (1974) *Accumulation on a World Scale: A Critique of the Theory of Underdevelopment*, New York: Monthly Review Press.

Antonopoulos, R. and I. Hirway (2010) *Unpaid Work and the Economy: Gender, Time and Poverty in Developing Countries*, New York: Palgrave and Macmillan.

Baran, Paul and Paul Sweezy (1966) *Monopoly Capital: An Essay on the American Economic and Social Order*, New York: Monthly Review Press.

Beneria, Lourdes (1999) "The enduring debate over unpaid labor," *International Labour Review*, 138: 287–309.

Braudel, Fernand (1982) *The Wheels of Commerce: Civilization and Capitalism, 15th-18th Century*, New York: Harper & Row.

Clelland, D. (2010) "The Current Economic Crisis: What Insights Does the World-Systems Perspective Offer?" Paper presented to the Annual Meeting of the American Sociological Association, available at http://filebox.vt.edu/users/wdunaway/SurplusDrain/Clelland.doc.

Communist Working Group (1986) *Unequal Exchange and the Prospects of Socialism*, Copenhagen: Manifest Press.

Dunaway, W. A. (2001) "The double register of history: situating the forgotten woman and her household in capitalist commodity chains," *Journal of World-System Research*, 7: 2–31.

Dunaway, W. A. (ed.) (2003) *New Theoretical Directions for the 21st Century World-System*, Westport, CT: Praeger.

Dunaway, W. A. (2012) "The semiproletarian household over the *longue duree* of the world-system," in R. Lee (ed.) *The Longue Duree of the Modern World-System: In Memoriam to Fernand Braudel*, Albany, NY: SUNY Press.

Emmanuel, A. (1972) *Unequal Exchange: A Study of the Imperialism of Trade*, London: New Left Books.

Frank, A. G. (1969) *Latin America: Underdevelopment or Revolution?* New York: Monthly Review Press.

Frank, A. G. (1979) *Dependent Accumulation and Underdevelopment*, London: Macmillan Press.

Goldfrank, W., D. Goodman, and A. Szasz (eds) (1999) *Ecology and the World-System*, Westport, CT: Greenwood Press.

Harvey, David (2003) *The New Imperialism*, New York: Oxford University Press.

Kalecki, M. (1954) *Theory of Economic Dynamics: An Essay on Cyclical and Long-run Changes in the Capitalist Economy*, London: Allen and Unwin.

Kohler, G. and A. Tausch (2002) *Global Keynesianism: Unequal Exchange and Global Exploitation*, New York: Nova Science.

Lichtenstein, N. (ed.) (2005) *Wal-Mart: The Face of Twenty-First Century Capitalism*, New York: New Press.

Martinez-Alier, J. (2002) *The Environmentalism of the Poor*, Northampton: Elgar.

Marx, Karl (1967) *Capital*, New York: International Publishers.

Mies, M., V. Bennholdt-Thomsen and C. von Verholf (1988) *Women: The Last Colony*, London: Zed Books.

Petras, J. and H. Veltmeyer (2007) *Multinationals on Trial: Foreign Investment Matters*, Hampshire: Ashgate.

Raffer, K. (1987) *Unequal Exchange and the Evolution of the World System*, New York: St. Martin's.

Shah, Anup (2005) "The scale of the debt crisis," *Global Issues*, www.globalissues.org/article/30/the-scale-of-the-debt-crisis

Smith, J.W. (1994) *The World's Wasted Wealth*, London: Institute for Economic Democracy.

Suter, C. (1992) *Debt Cycles in the World-Economy*, Boulder, CO: Westview.

Tabak, F. and M. Crichlow (eds) (2000) *Informalization: Process and Structure*, Baltimore, MD: Johns Hopkins University Press.

Toussiant, E. (2005) *Your Money or Your Life: The Tyranny of Global Finance*, Chicago, IL: Haymarket Books.

United Nations (2010) *World Economic Situation and Prospects*, New York: United Nations.

Wackernagel, M. et al (2002) "Tracking the ecological overshoot of the human economy," *Proceedings of the National Science Academy*, 99 (14): 233–57.

Wade, R. and F. Veneroso (1998) "The Asian financial crisis," *New Left Review*, 28: 13–42.

Wallerstein, I. (1974) *The Modern World-System I: Capitalist Agriculture and the Origins of the European World-Economy in the Sixteenth Century*, New York: Academic Press.

Wallerstein, I. (1983) *Historical Capitalism*, London: Verso.

Wallerstein, I. (1997) "States? Sovereignty? The dilemmas of capitalists in an age of transition," Keynote address at conference on States and Sovereignty in the World-Economy, University of California, Irvine.

Wallerstein, I. (2000) "Cultures in conflict? Who are we? Who are the others?" Y.K. Pao Distinguished Chair Lecture, Center for Cultural Studies, Hong Kong University of Science and Technology.

6.2

The silence of finance and its critics

Portfolio investors in the world-system

Aaron Z. Pitluck

In a classic text, Albert O. Hirschman (1970) observed that stakeholders in organizations typically face a conflict between *exit* and *voice*. We, who can easily exit an organization, have few incentives to invest in changing it; in contrast, when our exit is blocked, we are incentivized to exercise our voice, even when to do so is difficult or politically or socially dangerous. In finance, this social calculus between exit and voice is conceptualized as an opposition between liquidity and control (Coffee 1991). Historically, sociological research has focused on the causes and consequences of financiers' control of corporations (Stearns and Mizruchi 2005). This article instead focuses on a frontier of existing theory: professional money managers who almost exclusively invest in liquid securities, and therefore putatively have few incentives to exercise control.

Although portfolio investors may not seek control, their behavior can be socially consequential. As of early 2011, many housing and financial markets have not yet recovered from a financial crisis that began over three years earlier, caused in large part by portfolio investors seeking liquid investments, without a desire for either control or long-term ownership.

The article begins by demystifying global capital flows by outlining the parties involved. It proceeds by surveying finance studies to outline social forces shaping portfolio investors' behavior. Next, it examines the social consequences of financialization and financial innovation, particularly their role in generating financial crises. The conclusion outlines two social movements advocating reform of financial markets.

Demystifying global capital flows

Scholars of cross-border capital flows implicitly draw on Hirschman's (1970) distinction to define two types of investment. *Foreign direct investment* seeks to build corporate subsidiaries overseas or purchase control of foreign corporations. Such investors profit by building or buying into businesses and claiming a portion of their revenue streams. In contrast, *foreign portfolio investment* is the purchase of relatively liquid securities such as government debt, foreign corporations' debt or equity, or financial instruments derived from such instruments (i.e., derivatives). Such investors seek to profit by buying low and selling high; with an exit strategy planned before every purchase, they have few incentives for exercising voice.

How much of a corporation must a portfolio investor accumulate before she has lost liquidity and gained control of a corporation? In *Global Development Finance*, the World Bank (2008) sets the minimum threshold for control at 10 percent of equity ownership. Based on earlier research

(Pitluck 2011), I find that ownership below this 10 percent threshold is often an illiquid ownership stake. This suggests that social research using World Bank data on global capital flows overestimates the size of portfolio flows and correspondingly underestimates the size of foreign direct investment.

With this conventional definition in mind, we can make several generalizations regarding cross-border portfolio capital flows in the world-system. Firstly, money managers in core countries primarily invest in corporations in their own country—a phenomenon known in the finance literature as "home bias" (Karen 1999). Secondly, when core countries' money managers invest overseas, they primarily invest in one another, or in offshore financial centers that also primarily invest in core countries.

Thirdly, when high-income countries invest outside of Europe, the United States, and Japan, they invest in a narrow selection of countries. In the global South over the past decade, nearly every dollar flowed to only twenty so-called "investable emerging markets," identified as such by Standard & Poor's and other finance industry institutions. These countries are: Argentina, Brazil, Chile, China, Columbia, Egypt, Hungary, India, Indonesia, Lithuania, Malaysia, Mexico, Morocco, the Philippines, Poland, the Russian Federation, South Africa, Turkey, and Venezuela (World Bank 2008). Fourthly, throughout the global North, a growing share of wealth is held outside of the banking system and is managed by portfolio investors employed by investment funds (esp. Austria and Ireland), insurance companies (esp. Belgium and Sweden), and pension funds (esp. Australia, Canada, Iceland, the Netherlands, and Switzerland) (Gonnard et al 2008: 7 and Figure 1).

A valuable step in demystifying high finance in order to understand the social forces shaping global capital flows is to reformulate it as a prosaic work practice. Portfolio investors labor under multiple aliases: portfolio managers, fund managers, asset managers, investment advisers, money managers, and institutional investors. Portfolio investors may be employed by nonfinancial firms in corporate treasury operations. More typically, they are employed by finance firms such as investment banks, commercial banks, mutual funds, pension funds, insurance companies, and hedge funds. The type of finance firm determines the regulations constraining the portfolio investor and the organizational constraints on his or her trading choices. Portfolio investors working in finance firms may manage clients' money or engage in proprietary trading (i.e., trading on behalf of their own firm) (Davis and Steil 2001).

Portfolio investors trade legal contracts that represent ownership of real assets and the cash flows produced by these assets. These contracts (called financial instruments) may be standardized and easily traded on public exchanges, or they may be custom-designed bilateral contracts running to many hundreds of pages.

Social forces shaping portfolio investor behavior

Modal studies in finance emphasize that portfolio investors are profit-maximizing, boundedly-rational agents responding to exogenous information signals. In this framework, to understand why or where portfolio investors make their investment and disinvestment decisions, analysts focus on the destination country's investment climate (e.g., World Bank 2010). The behavioral finance literature retains this methodological individualism to model psychological constraints on investors' rationality. In contrast, sociological approaches emphasize how portfolio investors are agents in social cognitive networks acting in markets constituted by cultural fields and state power.

Social cognitive networks is a term that emphasizes the interpersonal and interorganizational nature of cognition in finance; the decision-making process takes place both inside and outside of professional investors' heads (Pitluck 2011). Some portfolio managers work alone, or in small groups. For example, some hedge funds are small organizations composed of a fund manager and a small collaborative staff of a half dozen people. More typically, when an investment manager works alone or with a small staff, their principal duty is to hire and monitor external portfolio managers. Portfolio managers typically work in large, complex organizations. Within these organizations, traders typically gain information, deliberate, and decide in teams at a "desk," narrowly focused on a single financial instrument or a narrow trading strategy (Beunza and Stark 2005; Clark and Thrift 2005). In social interaction with brokers and competitors, portfolio investors interpret and co-create the market (Knorr Cetina and Bruegger 2002b; Pitluck 2011). Even intra-day arbitrage traders working in crowds on trading floors gain and process information in interpersonal networks (Baker 1984; Zaloom 2006).

Portfolio investors seek to gain a unique interpretation of investment opportunities from public information, but such information is always constructed in interaction with others. For example, "raw" information and numbers regarding corporations is provided and interpreted by corporations' investor relations offices and public filings (Useem 1996; Westphal and Bednar 2008), by market analysts inside and outside of the firm (Lounsbury and Rao 2004; Zuckerman 1999), by private-sector credit rating agencies (Sinclair 2005), and by the mass media (Warner and Molotch 1993). Financial information and how it is interpreted is constructed from prosaic work practices and technologies, such as the use of financial models and proprietary trading technologies (Beunza and Stark 2005; Knorr Cetina and Bruegger 2002a; MacKenzie 2006; MacKenzie and Millo 2003; Preda 2009). Cognition, data and price construction are social activities (Beunza et al 2006; Knorr Cetina and Preda 2005; Pixley 2004).

These investment decisions are made in the context of shifting government regulation. A fundamental insight of world-systems analysis is that all markets are political. All products are bundles of rights and obligations, the value of which is shaped by the legality and enforcement of these property rights (Wallerstein 2004). This theory provides four insights to aid the understanding of financial markets. Firstly, financial instruments are legal contracts outlining rights and obligations to an income stream. The value of such contracts is not only shaped by public interpretations and expectations of the underlying income stream, but also by current and future expectations of the contract's judicial interpretation and regulatory enforcement. As elaborated below, this is particularly germane in the area of financial innovation, where novel contracts are often designed to take advantage of ambiguous legal or regulatory infrastructure, such as tax shelters.

Secondly, even in the trading of standardized financial products, such as shares on stock exchanges, there is a great deal of legal and normative ambiguity regarding the distinction between licit aggressive opportunism that capitalizes on information asymmetries and illicit insider information, fraud, or market manipulation. As Mitchel Y. Abolafia (1996) documented in his ethnographic research, where this legal distinction is drawn shifts over time and is shaped by state regulatory practices in interaction with financial workers' industry culture.

Thirdly, all corporations are artifacts of state power, but this is particularly germane for financial firms. For example, since 1933, what financial instruments US financial firms are legally permitted to invest in and the degree of risk that they are permitted to purchase has been dictated by securities regulation; these regulations have shifted dramatically over the course of the twentieth century up until the present, thereby dramatically altering the size of financial firms, their scope of operations, and their employees' investment behavior (Burk 1992; de Goede 2005). Lastly, as elaborated below, financial markets are marked by periodic crises (Arrighi 2010; Harvey 1982; Reinhart and Rogoff 2009), which provide an opportunity for social movements to advocate for regulatory reform to constrain the power of financial firms (Abolafia 1996; Burk 1992).

Social consequences of professional investors' behavior

Over the past 40 years, alongside observed trends in wealthy countries of deindustrialization and the growing importance of the service sector, numerous scholars and activists have argued that the financial sector has grown more significant in social life. This trend has been broadly termed *financialization*. Financialization is frequently perceived as a conflict between Wall Street and Main Street. For example, some researchers have argued that corporate behavior is increasingly designed to raise short-term stock market prices, and decreasingly oriented to benefit employees, geographic communities, or even the long-term survival of the firm. Others point to the expanding economic, political, and cultural power of financial sector elites. Yet other scholars have argued that diverse banks' influence on corporations—particularly in Japan and the European Union—is diminishing, but is being replaced by the opinions and power of portfolio managers, market analysts, and credit rating agencies.

Pulling together these empirical observations with the theoretical work of Giovanni Arrighi (2010), Greta Krippner (2005: 174) defines financialization as "a pattern of accumulation in which profits accrue primarily through financial channels rather than through trade and commodity production." She describes a post-War trend of US *nonfinancial* corporations increasingly resembling financial corporations, in that financial activities are a growing source of their profits and cash flow. In particular, manufacturing corporations are increasingly dependent for their survival on providing or trading financial products. In other words, the distinction between Main Street and Wall Street is eroding as both sectors pursue liquid investments and eschew illiquid commitments to workers and communities.

Earlier scholars have emphasized the long-term systemic relationship between financialization and financial crises. Drawing on Marx's theory of the tendency for the profit rate to fall and thereby produce crises of overaccumulation, David Harvey (1982:esp. Ch. 6 and 7, 2003: Ch. 3) argues that capitalists seek a "spatio-temporal fix" of investing surplus capital into long-term projects (e.g., infrastructure, education, research) and/or investing abroad. For Harvey, financialization is one path through which capital flows to overcome capitalism's chronic crises.

Drawing inspiration from the work of Fernand Braudel, Giovanni Arrighi (2010: 372) argues that in the history of the world-system, one finds intense periods of financialization (referred to as "financial expansions") marking the transition between a waning cycle of accumulation in one region and the rising cycle of accumulation in the next. He argues that in all such cycles to date, the rapid expansion of world trade and production has resulted in increased competitive pressures and diminishing returns. This encourages the capitalist class to "keep in liquid form a growing proportion of their incoming cash flow" (372) rather than invest it in production. (See Silver [2003] for the related idea of a financial fix, and Block [1977] on capital strikes.) This creates the supply conditions for overaccumulation of capital. Arrighi traces the demand conditions:

> to the tendency of territorial organizations to respond to the tighter budget constraints that ensue from the slowdown in the expansion of trade and production by competing intensely with one another for the capital that accumulates in financial markets. This tendency brings about massive, system-wide redistributions of income and wealth from all kinds of communities to the agencies that control mobile capital, thereby inflating and sustaining the profitability of financial deals largely divorced from trade and production. All the *belles époques* of finance capitalism—from Renaissance Florence to the Reagan and Clinton eras—have been characterized by redistributions of this kind.
>
> *(Arrighi 2010: 372)*

For Arrighi, like Harvey, financialization is a bandage that allows capitalists a profitable respite before the capitalist system slides into a periodic crisis of world-historical significance.

Financial innovation has been argued to either alleviate or cause financial crises. Financial innovation is the "act of creating and then popularizing new financial instruments as well as new financial technologies, institutions and markets" (Tufano 2003: 310). The finance literature and much economic history depict financial innovation in functionalist terms. Financial innovation promotes economic growth and acts as a prophylactic against financial crises. It does so by solving three problems for firms: raising funds (e.g., junk bonds in the 1980s), redistributing risk (e.g., Mortgage Backed Securities with a "waterfall" structure, Langley [2008]; Coval et al [2009]), and circumventing or minimizing taxation or regulatory constraints (Miller 1986). Economy-wide, financial innovation is argued to create new markets, thereby leading to improved information and more efficient allocation of capital throughout the economy, and consequently producing financial stability and economic growth (e.g., Allen and Gale 1994: Ch. 2).

Political economists with a functionalist bent can expand this list, for financial instruments have been created to solve many additional problems—often with unintended harmful consequences. For example, new financial instruments have been designed to redistribute wealth or control within a corporation, such as the "poison pill securities" created in the 1980s (Allen and Gale 1994: 22–24). Lawyers have worked with financial engineers to create new instruments to circumvent new court decisions (Tufano 2003: 320). Islamic scholars have worked with financial engineers to create financial instruments that circumvent Islam's prohibition on interest (El-Gamal 2006). Moreover, in recent years financial engineers have worked with analysts in credit rating agencies to create mortgage backed securities and other collateralized debt obligations that circumvent risk models in order to create financial instruments rated as having exceptionally low risk (Coval et al 2009).

Elsewhere in finance studies, we find constructionist and political economic arguments that financial innovation contributes to parasitism and financial instability. Karen Ho (2009) documents the ways in which investment banks provide self-serving consultation services to corporations that identify "problems" and "needs" and then market risky new financial instruments and strategies as "solutions." Other scholars emphasize how financial firms seek to reduce competition by creating and marketing putatively differentiated financial instruments, or are pressured to create destructively risky "me too" financial instruments in order to compete in under-regulated market niches, such as Collateralized Debt Obligations (CDOs) and Credit Default Swaps (Ashton 2009; Best 2010; Erturk and Solari 2007; Engelen et al 2010; Gotham 2006; Langley 2008; Lounsbury and Hirsch 2010).

In the recent financial crisis, these financial derivatives created a large and rapidly growing demand for conventional and subprime mortgages. Financial firms increasingly originated loans simply in order to re-sell them to financial engineers, who repackaged and then re-sold them as CDOs. When the demand for derivatives began outstripping the supply of potential borrowers, since these mortgage markets were under-regulated, lenders found themselves in a vicious cycle of competitively declining lending standards, in which they created mortgages with increasingly risky terms to borrowers increasingly unlikely to be able to afford repayment. Such mortgage products were disproportionally peddled to racial minorities, residents in historically redlined neighborhoods, and to residents in communities with the most severe real estate price inflation (Hernandez 2009; Langley 2008).

Social movements advocating reform

What alternative institutional arrangements exist to minimize the destructive aspects of financial innovation, financialization, and portfolio investors' behavior? In the spirit of Erik Olin Wright's (2010) real utopias project, here I reflect on two such projects. The social investment social movements seek to reform corporate behavior by using portfolio investors as our progressive political representatives. Islamic finance social movements share some of these goals, and additionally seek to reform financial markets based on religious and moral reasoning. Both social movements advocate for portfolio investors to behave less like absentee owners fetishizing liquidity, and more like owners exercising a voice for social justice. Both social movements, in other words, contest the morality of liquidity.

Social investment is a normative vision of finance seeking "the allocation of capital to advance the social and economic well-being of people" (Bruyn 1987:13). In many countries, as share-ownership had become more widespread through pension funds, mutual funds, and insurance schemes, social investment advocates have sought for the money managers of these funds to invest using not only financial criteria, but nonfinancial political criteria as well. For example, the pension funds of labor unions could promote positive industrial relations using their money manager's voice and threat of exit (Fung et al 2001). Public pension funds could direct their money managers to meet the needs of community residents, including potentially investing in local infrastructure and disadvantaged communities (Clark 2000).

Social investment faces numerous hurdles. Firstly, corporate governance rules are currently crafted to give shareholders (including money managers) limited control over corporations (Black 1991; Coffee 1991). Secondly, savers typically have limited mechanisms for democratic control over their money managers (Pitluck 2008). Ensuring that money managers attend to our financial interests rather than engage in self-interested excessive trading or charge excessive fees is already problematic. Through what mechanisms can we communicate (and enforce) our nonfinancial investment interests? Pension funds in the United States, United Kingdom, and Canada illustrate these challenges (Clark 2000; Fung et al 2001).

One potential solution is socially responsible investment (SRI) financial products that can be purchased by individuals or indirectly via their portfolio managers. Akin to "fair trade" claims in nonfinancial products, SRI accounts have their investments screened to meet minimum nonfinancial criteria. In the United States, these are typically left-leaning political criteria, and/or criteria inspired by Christianity, such as foregoing "sin stocks" such as liquor or tobacco corporations. In early 2010, over $2.51 trillion in US assets were held in such accounts. If we add to this number funds invested with institutional investors who do not practice social screening but who practice shareholder advocacy, and/or invest in community projects that are typically underserved by traditional financial institutions, we find that the size of SRI assets rises to $3.07 trillion. This amount is non-trivial; it represents nearly one out of every eight dollars in the $25.2 trillion under professional management (Social Investment Forum 2011).

Another extant institutional alternative is Islamic finance, also known as Islamic banking. Like SRI, individuals and portfolio investors can invest in Islamic accounts and financial instruments. "Islamic financial institutions are those that are based, in their objectives and operations, on Koranic principles" (Warde 2002: 5). These principles are contested and transformed over time by debate within and between diverse groups, including Islamic (*Shari'a*) scholars, secular lawyers, financial engineers, and regulators (El-Gamal 2006). As a consequence, as one scholar concludes, "No definition of Islamic finance is entirely satisfactory. To every general criterion…one can find some significant exception" (Warde 2002: 5).

Nevertheless, one can posit several core tenants of contemporary Islamic finance (Iqbal and Molyneux 2005: 4–17; Usmani 2002; Warde 2002: esp. 55–72). Islamic finance overlaps with SRI in attending to the behavior of corporations. In Islamic finance, corporations must engage in socially beneficial activities and may not engage in activities forbidden by the religion. The precise list of such activities varies, reflecting the religion's global diversity, flexibility and pragmatism (see Warde 2002: 15; 41–42; Ch. 16). For example, financial firms may only invest in real assets and inventories and may not make money from money (Usmani 2002: xiv–xvii). Nonfinancial corporations may not engage in some activities, such as selling liquor or gambling. Like SRI advocates, Islamic finance seeks to set a pragmatic ceiling of acceptable involvement in unacceptable industries.

In contrast to SRI, Islamic finance additionally critiques and seeks reform of the role of finance in society. Firstly, *riba* is prohibited, a concept frequently interpreted as prohibiting interest-earning transactions (but see El-Gamal 2006: Chapter 3). Secondly, *gharar* is prohibited, a concept that Iqbal and Molyneux (2005:12–15) equate with asymmetrical information. All economic and financial contracts should be transparent, minimize risk, and not seek to profit from uncertainty. Thirdly, all kinds of gambling and games of chance are prohibited (*maysir*). While profiting from entrepreneurial risk or utilizing insurance against natural disasters is permitted, one may not profit from "uncertainties that are not part of everyday life," including speculation (Iqbal and Molyneux 2005: 15).

For scholars and activists, Islamic finance provides a sophisticated and significantly broader critique than SRI of diverse practices in the finance industry, including financial innovation, the morality of making money from money, and alternative conceptions of risk and uncertainty (e.g., Pitluck 2008). Moreover, many practitioners seek to create new "Islamic" financial instruments that shift economic rewards and risks to empower the disempowered. On the other hand, Mahmoud A. El-Gamal (2006) is correct that much scholastic and industry work in Islamic finance is often in practice focused not on seeking a more socially just financial system, but in regulatory and religious arbitrage—replicating the status quo using complex, structured financial products that Islamic jurists can approve and that can therefore be marketed as "Islamic." This is particularly problematic in light of the role that structured financial products played in causing the 2007 financial crisis (Coval et al 2009; Langley 2008).

Conclusion

As Severyn T. Bruyn (1987: 14) has observed, the word *investment* is defined not only as employing money to make a profit, but also to endow or furnish with power, authority or privilege. Until recently, world-systems analysts have focused on the social consequences of financiers' control but have neglected to examine the consequences of financiers' desire for liquidity. Scholars therefore have much to learn from activists in social investment and Islamic finance who have for decades contested the morality of liquidity. By re-politicizing finance (de Goede 2005) and conceiving of financial markets as cultural fields shaped by state power (Abolafia 1996), we position ourselves to better understand and change where portfolio capital flows, how financial and nonfinancial corporations behave, and how to intervene in periodically destructive social processes such as financialization and financial innovation.

References

Abolafia, M.Y. (1996) *Making markets: Opportunism and restraint on Wall Street*, Cambridge, MA: Harvard University Press.

Allen, F. and Gale, D. (1994) *Financial innovation and risk sharing*, Cambridge, MA: MIT Press.

Arrighi, G. (2010) *The long twentieth century: Money, power, and the origins of our times*, London: Verso.

Ashton, P. (2009) "An appetite for yield: The anatomy of the subprime mortgage crisis", *Environment and Planning A*, 41(6): 1420–41.

Baker, W.E. (1984) "The social structure of a national securities market", *American Journal of Sociology*, 89(4): 775–811.

Best, J. (2010) "The limits of financial risk management: Or what we didn't learn from the Asian crisis", *New Political Economy*, 15(1): 29–49.

Beunza, D., Hardie, I. and MacKenzie, D. (2006) "A price is a social thing: Towards a material sociology of arbitrage", *Organization Studies*, 27(5): 721–45.

Beunza, D. and Stark, D. (2005) "How to recognize opportunities: Heterarchical search in a trading room", in K. Knorr Cetina and A. Preda (eds) *Sociology of financial markets*, Oxford: Oxford University Press.
Black, B.S. (1991) "Shareholder passivity reexamined", *Michigan Law Review*, 89: 520–608.
Block, F. (1977) "The ruling class does not rule: Notes on the Marxist theory of the state", *Socialist Review*, 7(3): 6–28.
Bruyn, S.T. (1987) *The field of social investment*, Cambridge and New York: Cambridge University Press.
Burk, J. (1992) *Values in the marketplace: The American stock market under federal securities law*, New York: Aldine de Gruyter.
Clark, G.L. (2000) *Pension fund capitalism*, Oxford and New York: Oxford University Press.
Clark, G.L. and Thrift, N. (2005) "The return of bureaucracy: Managing dispersed knowledge in global finance", in K. Knorr Cetina and A. Preda (eds) *Sociology of financial markets*, Oxford: Oxford University Press.
Coffee, J.C., Jr. (1991) "Liquidity versus control: The institutional investor as corporate monitor", *Columbia Law Review*, 91(6): 1277–1368.
Coval, J., Jurek, J. and Stafford, E. (2009) "The economics of structured finance", *Journal of Economic Perspectives*, 23(1): 3–25.
Davis, E.P. and Steil, B. (2001) *Institutional investors*, Cambridge, MA: MIT Press.
de Goede, M. (2005) *Virtue, fortune and faith: A genealogy of finance*, Minneapolis: University of Minnesota Press.
El-Gamal, M.A. (2006) *Islamic finance: Law, economics, and practice*, Cambridge and New York: Cambridge University Press.
Engelen, E., Erturk, I., Froud, J., Leaver, A. and Williams, K. (2010) "Reconceptualizing financial innovation: Frame, conjuncture and bricolage", *Economy and Society*, 39(1): 33–63.
Erturk, I. and Solari, S. (2007) "Banks as continuous reinvention", *New Political Economy*, 12(3): 369–88.
Fung, A., Hebb, T. and Rogers, J. (2001) *Working capital: The power of labor's pensions*, Ithaca, NY: ILR Press, Cornell University Press.
Gonnard, E., Kim, E.J. and Ynesta, I. (2008) "Recent trends in institutional investors statistics", *Financial Market Trends*, 95(2): 1–22.
Gotham, K.F. (2006) "The secondary circuit of capital reconsidered: Globalization and the U.S. real estate sector", *American Journal of Sociology*, 112(1): 231–75.
Harvey, D. (1982) *The limits to capital*, Chicago, IL, and Oxford: University of Chicago Press and Blackwell.
Harvey, D. (2003) *The new imperialism*, Oxford and New York: Oxford University Press.
Hernandez, J. (2009) "Redlining revisited: Mortgage lending patterns in Sacramento 1930–2004", *International Journal of Urban and Regional Research*, 33(2): 291–313.
Hirschman, A.O. (1970) *Exit, voice, and loyalty: Responses to decline in firms, organizations, and states*, Cambridge, MA: Harvard University Press.
Ho, K. (2009) *Liquidated: An ethnography of Wall Street*, Durham, NC, and London: Duke University Press.
Iqbal, M. and Molyneux, P. (2005) *Thirty years of Islamic banking: History, performance and prospects*, Houndmills and New York: Palgrave Macmillan.
Karen, K.L. (1999) "Trying to explain home bias in equities and consumption", *Journal of Economic Literature*, 37(2): 571–608.
Knorr Cetina, K. and Bruegger, U. (2002a) "Global microstructures: The virtual societies of financial markets", *American Journal of Sociology*, 107(4): 905–50.
Knorr Cetina, K. and Bruegger, U. (2002b) "Traders' engagement with markets: A postsocial relationship", *Theory, Culture & Society*, 19(5–6): 161–85.
Knorr Cetina, K. and Preda, A. (2005) *The sociology of financial markets*, Oxford: Oxford University Press.
Krippner, G.R. (2005) "The financialization of the American economy", *Socioeconomic Review*, 3(2): 173–208.
Langley, P. (2008) "Sub-prime mortgage lending: A cultural economy", *Economy and Society*, 37(4): 469–94.
Lounsbury, M. and Hirsch, Paul M. (eds) (2010) *Markets on Trial: The Economic Sociology of the U.S. Financial Crisis, Part A*. Bingley: Emerald Group Publishing LTD.
Lounsbury, M. and Rao, H. (2004) "Sources of durability and change in market classifications: A study of the reconstitution of product categories in the American mutual fund industry, 1944–85", *Social Forces*, 82(3): 969–99.
MacKenzie, D. (2006) *An engine, not a camera : How financial models shape markets*, Cambridge, MA: MIT Press.
MacKenzie, D. and Millo, Y. (2003) "Constructing a market, performing theory: The historical sociology of a financial derivatives exchange", *American Journal of Sociology*, 109(1): 107–45.
Miller, M.H. (1986) "Financial innovation: The last twenty years and the next", *Journal of Financial & Quantitative Analysis*, 21(4): 459–71.

Pitluck, A.Z. (2008) "Moral behavior in stock markets: Islamic finance and socially responsible investment", in K. E. Browne and B. L. Milgram (eds) *Economics and morality: Anthropological approaches*, Society for Economic Anthropology Monographs, Lanham, MD: AltaMira Press, Rowman & Littlefield Publishers.
Pitluck, A.Z. (2011) "Distributed execution in illiquid times: An alternative explanation of trading in financial markets", *Economy and Society*, 40(1): 26–55.
Pixley, J. (2004) *Emotions in finance: Distrust and uncertainty in global markets*, Cambridge and New York: Cambridge University Press.
Preda, A. (2009) *Framing finance: The boundaries of markets and modern capitalism*, Chicago, IL, and London: University of Chicago Press.
Reinhart, C.M. and Rogoff, K.S. (2009) *This time is different: Eight centuries of financial folly*, Princeton, NJ, and Oxford: Princeton University Press.
Silver, B.J. (2003) *Forces of labor: Workers' movements and globalization since 1870*, Cambridge and New York: Cambridge University Press.
Sinclair, T.J. (2005) *The new masters of capital : American bond rating agencies and the politics of creditworthiness*, Ithaca, NY: Cornell University Press.
Social Investment Forum (2011) *2010 report on socially responsible investing trends in the United States*, Washington, DC: Social Investment Forum.
Stearns, L.B. and Mizruchi, M.S. (2005) "Banking and financial markets", in N.J. Smelser and R. Swedberg (eds) *The handbook of economic sociology*, 2nd edn., Princeton, NJ, and Oxford: Princeton University Press; New York: Russell Sage Foundation.
Tufano, P. (2003) "Financial innovation", in G. M. Constantinides, M. Harris and R. Stulz (eds) *The handbook of the economics of finance*, Amsterdam: Elsevier North Holland.
Useem, M. (1996) *Investor capitalism: How money managers are changing the face of corporate America*, New York: Basic Books.
Usmani, M.T. (2002) *An introduction to Islamic finance*, The Hague: Kluwer Law International.
Wallerstein, I.M. (2004) *World-systems analysis: An introduction*, Durham, NC: Duke University Press.
Warde, I. (2002) *Islamic finance in the global economy*, Edinburgh: Edinburgh University Press.
Warner, K. and Molotch, H. (1993) "Information in the marketplace: Media explanations of the '87 crash' ", *Social Problems*, 40(2): 167–88.
Westphal, J.D. and Bednar, M.K. (2008) "The pacification of institutional investors", *Administrative Science Quarterly*, 53(1): 29–72.
World Bank (2008) *Global development finance*, Washington, DC: World Bank.
World Bank (2010) *Doing business 2011: Making a difference for entrepreneurs*, Washington, DC: World Bank.
Wright, E.O. (2010) *Envisioning real utopias*, London: Verso.
Zaloom, C. (2006) *Out of the pits: Traders and technology from Chicago to London*, Chicago, IL, and London: University of Chicago Press.
Zuckerman, E.W. (1999) "The categorical imperative: Securities analysts and the illegitimacy discount", *American Journal of Sociology*, 104(5): 1398–1438.

6.3

Debt crises in the modern world-system

Christian Suter

Introduction

This chapter deals with the world-system literature on global debt crises, that is, on situations where sovereign borrowers, mostly peripheral and semiperipheral countries, are confronted with debt-servicing difficulties and are forced to restructure the outstanding external debt owed to core lenders. The relevance of sovereign debt crises for world-system research arises from three kinds of interaction between external debt, and world-system dynamics and structures. Firstly, the evolution of global debt and financial crises is related to the general evolutionary dynamics of the modern world-system, that is, trends and cycles. Thus, global debt problems and financial crises have repeatedly occurred in the history of the modern world-system; several scholars have developed models of causal relationships between global debt cycles and financial crises, on the one hand, and long waves of economic growth and innovation, world leadership cycles and processes of state formation, on the other. Secondly, the evolution of the main structures of the modern world-system, the core/semiperiphery/periphery hierarchy, is linked to the processes of periphery indebtedness and debt crises. Thus international lending booms with flows of financial capital from the core to the periphery and the subsequent (periphery and semiperiphery) debt crises have contributed both to the integration of peripheral and semiperipheral regions into the world-economy as to the reproduction and deepening of the core-periphery hierarchy. Thirdly, debt crises impact on the evolution of international financial regimes and global financial governance. Thus, as a consequence of successive sovereign debt crises and debt restructurings, a global regime dealing with sovereign debt has gradually evolved over the past 200 years.

Since the global debt crisis of the early 1980s, there is a growing amount of literature, within and outside the world-system perspective, dealing with the dynamics, causes, and consequences of periphery indebtedness and global debt crises. This chapter focuses on four major issues that have been addressed by world-system research: Firstly, the historical evolution and dynamics of global sovereign debt booms and crises and their causal mechanisms; secondly, the economic, social, ecological, and political consequences and implications of debt dependence and debt crises; thirdly, the relationships between debt crises and world leadership cycles, and, finally, the evolution of a global regime dealing with sovereign debt problems. In what follows, the contribution of the literature to each of these four topics is described and discussed in more detail.

Global debt booms and global debt crises

There is a large body of literature dealing with the historical evolution and dynamics of international lending booms and global financial crises. The fact that sovereign debt crises have repeatedly occurred over the past centuries has already been mentioned in the pioneering works of Winkler (1933), Borchard and Wynne (1951) and Kindleberger (1978). Large-scale sovereign borrowing and great amounts of sovereign default and restructuring did not begin, however, until the early nineteenth century. This was linked, on the one hand, to the expansion of the international state system, that is, the increasing number of sovereign states, particularly the newly established independent states in Latin America, which considerably increased the number of potential sovereign borrowers. On the other hand, financial innovations and the development of international capital markets, contributed to the rise of sovereign debt from the early nineteenth century onwards. Accordingly, most of the literature discussed below refers to the era of industrial and financial capitalism, or to specific historical periods and debt crises within this era.

Worth mentioning in this regard are several studies in economic history published in the 1950s and 1960s dealing with particular historical debt crises or debt-service incapacities of specific debtor countries. In the context of this economic history perspective, the Atlantic economy approach was developed; its basic argument is that within the specific spatio-temporal framework of the Atlantic economy before World War I, cyclical fluctuations of external borrowing by less-developed and underpopulated debtor countries were (inversely) related to Kuznets cycles of domestic investment in the core countries.

In the wake of the debt crisis of the early 1980s, the literature dealing with the problem of periphery indebtedness and debt crisis soared (cf. Suter 1992 for an overview). Most of this literature is rather descriptive, limited to the developments of the late 1970s and early 1980s, confined to single aspects or specific cases, and often considering sovereign debt difficulties as isolated policy problems. However, more comprehensive and systematic approaches to the long-term evolution of periphery indebtedness and sovereign debt have been developed as well (see notably Pfister and Suter 1986, 1987; Eichengreen and Lindert 1989; Kowalewski 1989; Marichal 1989; Eichengreen 1991; Suter 1992, 2009; Reinhart and Rogoff 2009). The subsequent discussion focuses on these more comprehensive studies.

There is widespread agreement in the above-mentioned literature concerning the repeating phases of debt booms and debt crises and on the fact that all lending booms eventually ended in waves of debt-servicing incapacity on the part of sovereign borrowers (cf. Reinhart and Rogoff 2009; Suter 1992). The literature distinguishes nine phases of sovereign debt booms and global debt crisis that are summarized in Table 6.3.1. The extent of the respective debt booms and subsequent debt crises varies, and there is a shift in the geographical focus of sovereign debt crises: the European periphery and Latin America in the 1820s; the US states in the 1840s; the European periphery, the United States, and Latin America in the 1870s; Latin America and Southern Europe in the 1880s; Central and Eastern Europe and Latin America before World War I and during the 1930s; Latin America, the European periphery, and Africa in the 1980s; Eastern Europe, East Asia, and Latin America in the 1990s; and Eastern, Southern, and parts of Central Europe in the current financial crisis.

Table 6.3.1 Periodization of global lending booms and subsequent sovereign debt crises, 1820–2010

Booms	1822–25	1834–38	1834–38	1864–72	1885–90	1905–13	1924–28	1970–81	1993–97	2003–07
Crises	1826–32	1826–32	1838–45	1872–85	1890–95	1914–22	1931–46	1981–95	1999–02	2008–10

Sources: Suter (1992, 2009).

While there is a consensus concerning the existence of these sovereign debt-boom-bust cycles, there is no agreement as to the causal mechanisms. Some authors (e.g., Aggarwal 1996) emphasize single factors like excess liquidity (for the explanation of debt booms), inefficient and consumptive use of foreign resources by debtor countries, external economic shocks, and credit rationing (for the explanation of debt crises). Others argue that lending booms and debt crises are causally linked (Eichengreen 1991; Kindleberger 1978; Marichal 1989; Reinhart and Rogoff 2009). Scholars in the tradition of the world-system perspective, however, favor structural and systemic arguments (Kowalewski 1989; Pfister and Suter 1986, 1987; Suter 1992; Thompson and Reuveny 2009). Thus, in my book *Debt Cycles in the World-Economy* (Suter 1992) I developed the concept of "global debt cycles:" these are characterized by consecutive phases of (1) global lending booms or expansion of foreign credits, (2) the outbreak of payment crises, and (3) the resolving of debt crises, that is, settlement between debtor countries and their creditors in the form of debt restructuring agreements. Besides the fundamental thesis that such global debt cycles exist, I argued that these sovereign debt-default-restructuring cycles are associated with global economic growth and innovation cycles within the framework of Kondratieff and Kuznets cycles, world leadership cycles, and secular trend movements, and lead to increasing institutionalization of international financial relations, as well as—on the level of peripheral and semiperipheral states—with the dynamics and the constellations of (domestic) political regimes.

Empirical analyses of the causal mechanisms of booms in periphery and semiperiphery indebtedness suggest that cycles in economic growth (both Kuznets and Kondratieff cycles) and liquidity in core capital markets are important factors (Kowalewski 1989; Marichal 1989; Suter 1992; Thompson and Reuveny 2009). In addition to economic factors, political factors also played an important role—for example, the decolonization of Latin America and the end of the Napoleonic wars in the early nineteenth century, the end of World War I with the German reparation payments that triggered the lending boom of the 1920s, the decolonization of African and Asian countries that was partly linked to the boom of the 1970s, and the collapse of communism in Eastern Europe that was closely associated with the boom of the 1990s. Additional explanatory factors (for the 1970s boom period) are foreign trade (export fluctuations), foreign direct investment dependency, investment rates and the presence of a development-oriented political regime.

Cross-sectional evidence for determinants of debt-servicing incapacity suggests that debt crises are triggered by a combination of several factors (Reinhart and Rogoff 2009; Sturzenegger and Zettelmeyer 2006; Suter 1992; Thompson and Reuveny 2009 for an overview): first, declining export revenues (due to the collapse of prices for major export commodities, terms-of-trade deterioration, and economic recession in core countries); second, deteriorating prospects of economic growth; third, rising interest payments and/or increasing costs of external borrowing; and fourth, a stop or reduction in new lending by creditors and declining liquidity in international capital markets. This last factor (credit rationing), however, seems to play only a minor role. Furthermore, debt-servicing performance in the past remains an important determinant (after controlling for the other explanatory factors). In addition to these economic factors, political factors played an important role as well. Thus, World War I and revolutions triggered the debt crisis of 1915–22 (defaults by Austria-Hungary, the Ottoman Empire and Bulgaria and by Mexico and Russia, respectively).

Consequences of debt dependence and debt crises

Since the mid-1980s, the economic, social, and political consequences and implications of debt burden and financial crises were increasingly debated. Two interrelated factors contributed to the rising interest of world-system scholars in these topics.

Firstly, several scholars claimed that the boom of core lending and of external borrowing by peripheries and semiperipheries had fundamentally changed the structure of economic dependence. Thus, they maintained that traditional forms of economic dependency characterizing core-periphery relationships during the 1960s and 1970s (trade, foreign direct investment) were increasingly replaced by foreign debt dependence—such as, for instance, debt burden, stocks of external debt, or debt restructurings and the corresponding creditor-mandated structural adjustment programs. These new dependency structures were expected to adversely affect the economic, social, and ecological development of periphery debtor countries.

Secondly, the debt crisis of 1982–83 was not overcome rapidly but turned out to be rather protracted, particularly in Latin America, Africa, and Eastern Europe. The governments of the heavily indebted countries were forced to adopt structural adjustment programs and austerity measures implemented in the context of the multilateral debt restructuring regime of the Washington Consensus (IMF conditionality). These politics of adjustment also influenced the research agenda of development studies, international political economy and world-system studies that increasingly began to explore the structural adjustment programs' economic, social, political, and ecological effects.

In the context of the world-system perspective, four topics were of particular interest: (1) the effects of debt and debt crises on economic growth, and the relationship between debt dependence and traditional forms of economic dependency; (2) the consequences of debt and structural adjustment for quality of life, social development, and sustainability; (3) the impact of debt restructuring and structural adjustment on political protest and the mobilization of anti-systemic movements; and (4) the relationship between debt crises and the evolution of domestic political regimes.

The first group of studies that empirically explored the relationships between trade, foreign direct investment and foreign debt dependence, as well as their impact on economic growth, was inspired by the debate on the (negative) growth effects of foreign direct investment in peripheral countries (e.g., Bornschier and Chase-Dunn 1985). Pfister (1984) and Rothgeb (1984, 1986) found a positive relationship between the penetration of transnational corporations and debt dependence; Rothgeb argues that the repatriation of profits by private foreign capital in the manufacturing sector, in particular, contributed to the accumulation of external public debt. Several studies empirically investigated the impact of external debt on growth: most of them found negative growth and development effects, although these often varied according to the type of debt and the geographical region of the debtor country and/or country group. Stocks of non-concessional external debt vis-à-vis private creditors and IMF-imposed austerity measures, in particular, were negatively related to economic growth (Bradshaw and Huang 1991; Bradshaw and Wahl 1991; Glasberg and Ward 1993; Pfister 1984; Pfister and Suter 1986, Przeworski and Vreeland 2000).

The second group of cross-national studies extended the analysis of the developmental consequences of external debt by focusing on the social and ecological dimensions of development and underdevelopment (basic needs' satisfaction, poverty, physical quality of life, life expectancy, infant mortality, school enrollment, deforestation). Their results, however, are mixed depending on the dimensions of social development. Thus, while Bradshaw and Huang (1991) and Bradshaw and Wahl (1991) report non-significant relationships between foreign debt and IMF conditionality on the one hand and physical quality of life on the other, Sell and Kunitz (1986), Bradshaw et al (1993), Buchmann (1996), Schafer (1999), and Burns (2004) found generally negative effects on educational enrollment and the survival of children, particularly concerning IMF conditionality. Similarly, studies exploring the effects of debt crises on sustainable development showed mixed results. Thus, the findings of Marquart-Pyatt (2004) suggest no relationships between debt and deforestation, whereas according to Shandra et al (2008) debt dependence and IMF conditionality both negatively affect natural forest areas. The more recent analysis of Shandra et al (2004) suggests interaction effects of political democracy, that is, the deleterious effects of debt crises are particularly harmful in countries with low levels of democracy. In addition to these quantitative studies using a cross-national research design, a large number of case studies demonstrated the social and environmental costs of structural adjustment programs (e.g., George 1992).

The third topic concerns the effects of the global debt crisis on political processes in peripheral and semiperipheral countries. The important study by Walton and Ragin (1990) focuses on impacts on political protest. The authors demonstrate that IMF conditionality (but not debt dependence) is an important predictor of the occurrence and severity of social unrest, political protest, and political violence. This confirms the results of other studies showing similar mobilization effects of economic dependence (e.g., for foreign investment, London and Robinson [1989]; Rothgeb [1996]). Another important topic addressed by many studies during the 1980s and 1990s was the impact of the debt crisis on domestic political regimes and political transition (and vice versa the impact of the type of political regime on the implementation of debt restructuring and adjustment policies). This particularly concerns the discussion in Latin America, Africa, and Eastern Europe, where processes of democratization and regime change evolved parallel to the debt crisis. Several studies highlight the variety of domestic political regimes, the importance of starting points and trajectories (i.e., "path dependency") and the autonomy of political processes (for Latin America, cf. Stallings and Kaufman [1989]; Suter [1999]).

Debt crises and world leadership cycles

World leadership cycles or hegemonic sequences refer to the cyclical rise and fall of hegemonic core powers. Several scholars argued that periphery and semiperiphery indebtedness and debt crises are related to the functioning of world leadership cycles (see notably Rasler and Thompson 1983; Thompson and Reuveny 2009). In my book *Debt Cycles in the World-Economy* (Suter 1992), I hypothesized that during hegemonic sequences debt problems are moderated (since the hegemonic power acts as an international lender of last resort) and that debt settlements are more efficient but less favorable for debtor countries. A similar argument has been developed by Reuveny and Thompson (2004). The empirical analyses presented by Suter (1992, 2009), Reuveny and Thompson (2004, 2007) and Thompson and Reuveny (2009) clearly confirm these relationships. Thus, based on annual data from 1870 to 1989, Reuveny and Thompson (2004) and Thompson and Reuveny (2009) demonstrated that debt problems increased with declining world leadership, rising conflict between core countries, and declining world economic growth. Furthermore, several studies on sovereign debt restructuring showed that the crisis settlement was quickest during periods of hegemonic leadership, particularly during the US-dominated post-war period in the context of the restructuring regime of the Washington Consensus, but also (although less pronounced) during the British hegemony between 1870 and World War I (Kamlani 2008; Suter 1992, 2009).

The evolution of a global debt restructuring regime

One fundamental issue in the analysis of sovereign debt problems is the question of exactly how debt crises have been settled. When a debtor country is forced to suspend interest and/or amortization payments, the originally established loan contract between core creditors and periphery debtor country is unilaterally violated. The settlement of debt-servicing incapacity therefore implies a restructuring of the outstanding debt, that is, a renegotiation of the loan contract.

As demonstrated by Suter (1992) and Tomz (2007), the great majority of these restructurings were negotiated within the context of cooperation and formal agreements between creditors and debtor countries. These agreements dealt with the settlement of arrears of interest payments, the renegotiation of interest rates and amortization payments, the reassessment of the outstanding debt and/or of the face value of foreign bonds, and finally, the provision of collateral (for future interest and amortization) payments. The legal and political background of these debt settlements reveals the lack of an international bankruptcy law for sovereign debt and the corresponding contractual enforcement mechanisms, analogous to bankruptcy procedures that exist at the national level (i.e., the principle of sovereign immunity).

The world-system and international political economy perspectives have a particular interest in analyzing the evolution of sovereign debt restructurings, since this process concerns patterns of creditor-debtor and core-periphery cooperation and the issue of (global) financial governance and/or international regime formation. The analysis of sovereign debt and debt crises from the perspective of international cooperation and regime building received growing attention from the mid-1980s onwards (Biersteker 1993; Helleiner 1994, 2005, 2006; Kamlani 2008; Lipson 1985; Pfister and Suter 1987; Porter 1993; Roodman 2001; Soederberg 2005; Tomz 2007; Wood 1986). Only a few of these studies, however, have attempted to develop a long-term perspective dealing with the evolutionary and structural dynamics of a global sovereign debt restructuring regime (see notably Pfister and Suter 1987; Tomz 2007; Kamlani 2008; Suter 2009).

Based on the data and analysis of Suter (1992, 2009), Soederberg (2005), Sturzenegger and Zettelmeyer (2006), Tomz (2007), and Kamlani (2008), the evolution and institutionalization of global sovereign debt restructuring regimes can be described in five consecutive phases, characterized by different degrees of how creditor coordination and cooperation are institutionalized: (1) the weakly institutionalized debt restructuring regime lacking creditor coordination in the early nineteenth century; (2) the debt restructuring regime based on the British Corporation of Foreign Bondholders increasingly institutionalizing creditor cooperation, combined with a high level of sanctions by creditors (1868–1925); (3) the debt restructuring regime based on the US Foreign Bondholders Protective Council decreasing the degree of institutionalization of creditor cooperation and with a low level of sanctions by creditors (ca. 1925/45–ca. 1960); (4) the multilateral debt restructuring regime of the Washington Consensus characterized by high institutionalization and strong creditor coordination within and between creditor clubs, as well as with international financial institutions (ca. 1960–98); and, (5) the post-Washington Consensus regime based on (neo-liberal) market-based debt restructurings (Collective Action Clauses) and decreasing institutionalization of creditor coordination (1999–2008).

A preliminary empirical analysis of the outcome of debt restructurings presented by Suter (2009) demonstrates that the efficacy of the crisis management and the degree of debt relief granted by creditors is related to the strength and the degree of institutionalization of creditor cooperation. Highly institutionalized creditor cooperation (1868–1925 and 1960/70–1998) generally results in more efficient debt restructuring and minor concessions in favor of debtor countries. The new market-based debt restructuring regime of foreign bonds that evolved in the early 2000s, however, seems to be more efficient than expected. This surprisingly efficient pattern of sovereign bond restructuring may be explained by the fact that the seemingly uncoordinated take-it-or-leave-it regime is still embedded in the larger cooperative framework of the Washington Consensus.

An open question is how the regulatory reforms proposed and discussed in the wake of the current international financial crisis will impact on the global debt restructuring regime. The establishment of an international debt restructuring court, as recommended by the 2009 Stiglitz

Commission, seems unlikely. The recent experience of the Greek and Irish debt restructurings rather suggests more decentralized, regional solutions.

Concluding remarks

Sovereign debt crises only became an important issue in world-system analysis from the mid-1980s onwards. Over the past 20–30 years, there has been a certain thematic shift from analyzing crisis mechanisms and the long-term historical dynamics of debt crises (mainly during the 1980s and early 1990s), to dealing with the consequences of debt dependence and debt crisis (mainly during the 1990s), and to dealing with the management of debt crises and global financial governance (1990s and 2000s). This debate has greatly improved our understanding of the modern world-system in three important respects: Firstly, it has provided detailed knowledge of the cyclical pattern of debt crises and its causal relationship with the dynamics of world economic growth and hegemonic leadership and challenged, from the very beginning, the implicit assumption of most (non-world-system) analysts that debt crises are singular (i.e., non-systemic) phenomena. In addition to new theoretical insights into world-system dynamics, this research has provided remarkable new historical data and time series on sovereign debt and debt crises that has inspired research far beyond the world-system perspective (cf. the recent work of Reinhart and Rogoff 2009). Secondly, it has demonstrated the relevance of debt-related dependency for the reproduction of the core/periphery hierarchy and thus considerably added to our understanding of these processes. Thirdly, it has provided insights into the evolution of a global debt restructuring regime and its relationships with world leader leadership cycles.

Despite these advancements, the potential of world-system research on debt crises has not yet been fully exploited. Three topics would particularly merit further research.

A *first* direction for future research concerns the systematic examination of the long-term consequences of the different responses to debt crises for the evolution and consolidation of different development models. As mentioned in Section 3, debtor countries differently responded to debt crises and differently implemented restructuring and adjustment policies—despite the seemingly uniform recipes ordered by core lenders and global governance institutions (like the "orthodox" IMF austerity measures and in the context of the multilateral debt restructuring regime). Debt crises seem to be crucial moments where the future development model and the prospects of a given debtor country are redefined. Analyses on the relationship between debt and development trajectories may significantly contribute to the recent discussion on the evolution of variants of capitalism at the periphery and semiperiphery.

A *second* topic for future research refers to the relevance and the impact of sovereign debt crises at the core of the modern world-system. Historically, sovereign debt crises have principally broken out in peripheral and semiperipheral states which have been hit most often and most hard by debt problems. The current financial crisis and the subsequent sovereign debt crisis, however, not only originated in the core, but also most severely affected core countries (together with Eastern and Southern European, and Latin American semiperipheries). These peculiarities might have an important impact on the working mechanisms of current and future debt cycles.

Closely related with this specificity of the current debt crisis is a *third* underresearched topic, namely the impact of the current financial and sovereign debt crisis on the structure of global financial governance. As hypothesized above in Section 5, important transformations are taking place in the global debt restructuring regime as a result of the current financial crisis. Thus policy-makers of the crisis-affected core countries responded to the financial and sovereign debt crisis in a different, that is, in a much more interventionist way than they recommended to the periphery and the semiperiphery during the crises of the 1980s and 1990s. This contributed to the collapse of the neo-liberal consensus and the partial disintegration of the multilateral debt restructuring regime. Thus, although the IMF could, thanks to the crisis, overcome its severe weakness of the early 2000s, it has not (yet) become the leader defining the agenda of new global regulation. Rather, an informal and more decentralized system of global financial governance, characterized by different regional practices and limited regulative bodies, seems to be emerging in the wake of the 2008–09 crisis.

References

Aggarwal, V.K. (1996) *Debt Games*, Cambridge: Cambridge University Press.

Biersteker, T.J. (1993) *Dealing with Debt*, Boulder, CO: Westview.

Borchard, E.M. and Wynne, W.H. (1951) *State Insolvency and Foreign Bondholders*, 2 vols, New Haven, NC: Yale University Press.

Bornschier, V. and Chase-Dunn, C. (1985) *Transnational Corporations and Underdevelopment*, New York: Praeger.

Bradshaw, Y.W. and Huang, J. (1991) "Intensifying global dependency: foreign debt, structural adjustment, and Third World underdevelopment," *The Sociological Quarterly*, 32: 321–42.

Bradshaw, Y.W., Noonan, R., Gash, L. and Buchmann Sershen, C. (1993) "Borrowing against the future: children and Third World indebtedness," *Social Forces*, 71: 629–56.

Bradshaw, Y.W. and Wahl, A. (1991) "Foreign debt expansion, the International Monetary Fund, and regional variation in Third World poverty," *International Studies Quarterly*, 35: 251–72.

Buchmann, C. (1996) "The debt crisis, structural adjustment and women's education: implications for status and social development," *International Journal of Comparative Sociology*, 37: 5–30.

Burns, J.R. (2004) "Debt dependency, debt relief, and macroeconomic policies," unpublished thesis, Texas A&M University.

Eichengreen, B. (1991) "Historical research on international lending and debt", *Journal of Economic Perspectives*, 5: 149–69.

Eichengreen, B. and Lindert, P. (eds) (1989) *The International Debt Crisis in Historical Perspectives*, Cambridge, MA: MIT Press.

George, S. (1992) *The Debt Boomerang*, Boulder, CO: Pluto Press.

Glasberg, D.S. and Ward, K.B. (1993) "Foreign debt and economic growth in the World system", *Social Science Quarterly*, 74: 703–20.

Helleiner, E. (1994) *States and the Reemergence of Global Finance*, Cornell, NY: Cornell University Press.

Helleiner, E. (2005) "The strange story of Bush and the Argentine debt crisis," *Third World Quarterly*, 26: 951–69.

Helleiner, E. (2006) "The long and winding road towards a sovereign debt restructuring mechanism", unpublished paper, University of Waterloo.

Kamlani, Dirdre Shay (2008) "The four faces of power in sovereign debt restructuring", unpublished thesis, London School of Economics.

Kindleberger, C.P. (1978) *Manias, Panics, and Crashes: A History of Financial Crises*, London/Basingstoke: Macmillan.

Kowalewski, D. (1989) "Global debt crisis in structural-cyclical perspective: 1794–1984", in W.P. Avery and D.P. Rapkin (eds) *Markets, Politics, and Change in the Global Political Economy*. International Political Economy Yearbook, Vol. 4, Boulder, CO: Lynne Rienner.

Lipson, C. (1985) "Bankers' dilemmas: private cooperation in rescheduling sovereign debts", *World Politics*, 38: 200–225.

London, B. and Robinson, T.D. (1989) "The effect of international dependence on income in-equality and political violence", *American Sociological Review*, 54: 305–07.

Marichal, C. (1989) *A Century of Debt Crises in Latin America*, Princeton, NJ: Princeton University Press.

Marquart-Pyatt, S. (2004) "A cross-national analysis of deforestation, debt, state fiscal capacity, and the environmental Kuznets curve", *International Journal of Sociology*, 34: 33–51.

Pfister, U. (1984) "Debt and development: a cross-sectional study of the relationships between external debt and economic growth in developing countries", unpublished research paper, University of Zurich.

Pfister, U. and Suter, Ch. (1986) "Verschuldung im Weltsystem", unpublished report, University of Zurich.

Pfister, U. and Suter, Ch. (1987) "International financial relations as part of the world-system", *International Studies Quarterly*, 31: 239–72.

Porter, T. (1993) *States, Markets, and Regimes in Global Finance*, Basingstoke: Macmillan.

Przeworski, A. and Vreeland, J.R. (2000) "The effect of IMF programs on economic growth", *Journal of Development Economics*, 62: 385–421.

Rasler, K. and Thompson, W.R. (1983) "Global wars, public debts, and the long cycle", *World Politics*, 35 (4): 489–516.

Reinhart, C.M. and Rogoff, K.S. (2009) *This Time Is Different*. Princeton, NJ: Princeton University Press.

Reuveny, R. and Thompson, W.R. (2004) "World economic growth, systemic leadership and southern debt crises", *Journal of Peace Research*, 41: 5–24.

Reuveny, R. and Thompson, W.R. (2007) "How global is economic globalization", in M. Herkenrath (ed.) *The Regional and Local Shaping of World Society, World Society Studies*, Vol. II, Wien: LIT.

Roodman, D.M. (2001) "Still waiting for the jubilee", *World Watch Paper 155*, Washington, DC: World Watch Institute.

Rothgeb, J.M. (1984) "Investment penetration in manufacturing and extraction and external public debt in Third World states", *World Development*, 12: 1063–75.

Rothgeb, J.M. (1986) "Testing alternative conceptions of the relationship between direct foreign investment and external debt in poor countries", *Comparative Political Studies*, 19: 130–76.

Rothgeb, J.M. (1996) *Foreign Investment and Political Conflict in Developing Countries*, New York: Praeger.

Schafer, M.J. (1999) "International nongovernmental organizations and Third World education in 1990: a cross-national study", *Sociology of Education*, 72: 69–88.

Sell, R.R. and Kunitz, S. J. (1986) "The debt crisis and the end of an era in mortality decline", *Studies in Comparative International Development*, 21: 3–30.

Shandra, J.M, Nobles, J., London, B. and Williamson, J.B. (2004) "Dependency, democracy, and infant mortality", *Social Science & Medicine*, 59: 321–33.

Shandra, J.M., Shor, E., Maynard, G. and London, B. (2008) "Debt, structural adjustment, and deforestation", *Journal of World-Systems Research*, 14: 1–21.

Soederberg, S. (2005) "The transnational debt architecture and emerging markets", *Third World Quarterly*, 26: 927–49.

Stallings, B. and Kaufman, R. (eds) (1989) *Debt and Democracy in Latin America*, Boulder, CO: Westview.

Sturzenegger, F. and Zettelmeyer, J. (2006) *Debt Defaults and Lessons from a Decade of Crises*, Cambridge, MA: MIT Press.

Suter, Ch. (1992) *Debt Cycles in the World-Economy*, Boulder, CO: Westview.

Suter, Ch. (1999) *Gute und schlechte Regimes*, Frankfurt: Vervuert.

Suter, Ch. (2009) "Financial crises and the institutional evolution of the global debt restructuring regime, 1820–2008", paper presented at the PEWS Conference, University of San Francisco, August 7, 2009.

Thompson, W.R. and Reuveny, R. (2009) *Limits to Globalization*, Oxford: Routledge.

Tomz, M. (2007) *Reputation and International Cooperation*, Princeton, NJ: Princeton University Press.

Walton, J. and Ragin, Ch. (1990) "Global and national sources of political protest: Third World responses to the debt crisis", *American Sociological Review*, 55: 875–90.

Winkler, M. (1933) *Foreign Bonds, An Autopsy*, Philadelphia, PA: Swain.

Wood, R.E. (1986) *From Marshall Plan to Debt Crisis*, Berkeley: University of California Press.

6.4

Economic-political interaction in the core/periphery hierarchy

Mikhail Balaev

In World-Systems Theory (WST), economic exchange over the centuries is viewed as the central mechanism of the capitalist world-system (Wallerstein 1974). Throughout history, more powerful countries (the core) created various political arrangements to facilitate the economic exploitation of the weaker states (the periphery) through the extraction of resources, division of labor, and trade. From the position of a neo-Marxist argument elevated onto a global scale, WST considers the accumulation of capital in the core as the primary goal of the interstate exploitative mechanism. This unequal nature of trade not only reproduces, but also reinforces the hierarchical structure of the interstate system: the core states remain at the top of the hierarchy and benefit by exploiting the peripheral states. By means of international trade, the rich countries become richer and the poor become poorer.

In the past, hegemonic powers (e.g., Great Britain) could enforce the unequal nature of international trade on their colonies. Today, however, the main economic powers lack the ability to impose administrative orders on the periphery. According to Chase-Dunn et al (2000) the post-colonial core powers no longer have direct administrative control of the periphery and have limited capacity to enforce agreements militarily. Therefore, in order to sustain the exploitative nature of economic exchange, the core countries must expend their political capital and impose structural constrains upon the political-economic arrangements of peripheral and semiperipheral countries (Chase-Dunn and Grimes 1995). Such arrangements may come in the form of bilateral economic agreements or as multilateral organizations (e.g., the International Monetary Fund). Hence, economic and political processes are never separate, and political capital is used as a resource to reinforce neocolonial global economic exploitation.

However, the emphasis on how political power enables unequal exchange risks ignoring the ways in which trade and investment dependence are used to extract noneconomic resources from the noncore. International trade can be used as a tool to increase the noneconomic power of a core state. Considering a broader relationship between power and trade, international trade may not necessarily lead to a direct economic benefit for the core state. Instead, it can be perceived as a mechanism of safeguarding the economic dependency of the periphery on the core, thus increasing the national power and political capital of the core state. Using this power, core states achieve their larger geopolitical goals, which may or may not include extraction of economic benefit from the periphery. For example, by exploiting the economic dependency of peripheral countries, core states influence the domestic political processes of the periphery and pursue their security-oriented goals in the region. Even though these economic agreements are beneficial to the periphery, the direct investment into the economy of a peripheral state and various economic subsidies are patterns of trade policies that core states ultimately undertake in order to pursue their own specific goals and interests.

I argue that these interaction effects between international trade and political power do exist, that these effects are identifiable in the modern core-periphery hierarchy, and that they are common in the core-periphery political-economic relationship. Of course, there is a long tradition of studying how various kinds of international economic dependence reproduce global inequalities by slowing the growth of dependent countries (Bornschier and Chase-Dunn 1985; Galtung 1971; Hirschman 1980). How, then, does the nature of international economic dependence change the current era of global capitalism and US hegemonic decline?

A brief analysis of recent political events supports the idea that economic dependence is important in building up political capital. Through offering various kinds of economic subsidies to the elites of Kyrgyzstan and Uzbekistan, the United States established a military presence in these countries that was essential for the operation "Enduring Freedom" in Afghanistan in 2001 (Beehner 2005). Similarly, in 2002–08, the US government provided significant economic support—totaling around US $15 billion—to the government of the Pakistani President Musharraf (Baker and Robinson 2007). Such extensive US aid to Pakistan caused Pakistan's highly volatile economy to become profoundly dependent upon these economic subsidies (Human Rights Watch 2007: 307). In return, the United States received an important political ally in the Middle East, even though this relationship compromised the democratization of the domestic political process of Pakistan.

In another example, Balaev (2009) showed that since the breakup of the USSR, Russia has been maintaining a policy of economic investments and subsidies to former Soviet states, thus carefully nurturing the development of specific political elites. These new political elites depend on Russia for sustaining their political power and for continuing economic subsidies, in particular cheap oil and gas, which are central to their countries' economies. In exchange, such elites concede to Russian-engineered international trade policies and increase the political capital of Russia.

These types of international trade relationships are examples of the core-periphery economic-political trade-off, wherein economic capital is expended to gain political capital in strategically important geographic areas. The interactions between economic and political power continue to be important at several levels of the core/periphery hierarchy.

References

Baker, A. and Robinson, S. (2007) "Who Lost Pakistan? Why a Key U.S. Ally in the War on Terrorism is Letting Al-Qaeda Get Stronger", *Time*, July 30, p. 31.

Balaev, M. (2009) "The Effects of International Trade on Democracy: A Panel Study of the post-Soviet World-System", *Sociological Perspectives*, 52(3): 337–62.

Beehner, L. (2005) "Asia: U.S. Military Bases in Central Asia", Council on Foreign Relations. Available at: http://www.cfr.org/publication/8440/#10 accessed August 31, 2010.

Bornschier, V. and Chase-Dunn, C. (1985) *Transnational Corporations and Underdevelopment*. New York: Praeger.

Chase-Dunn, C. and Grimes, P. (1995) "World-Systems Analysis", *Annual Review of Sociology*, 21: 387–417.

Chase-Dunn, C., Kawano, Y. and Brewer, B.D. (2000) "Trade Globalization since 1795: Waves of Integration in the World-System", *American Sociological Review*, 65: 77–95.

Galtung, Johan (1971) "A Structural Theory of Imperialism", *Journal of Peace Research*, 8: 81–117.

Hirschman, A.O. (1980 [1945]) *National Power and the Structure of Foreign Trade*. Berkeley: University of California Press.

Human Rights Watch (2007) *World Report*.

Wallerstein, I. (1974) *The Modern World-System*, Vol. 1. New York: Academic Press.

6.5

The other side of the global formation

Structures of the world lumpeneconomy

Zbigniew Galor

The attitude of world-system analysis assumes the impossibility of understanding the modern world without taking it into consideration as a whole, over a longer period (Chase-Dunn and Hall 1997).

The way we understand this self-contained whole has fundamental meaning for understanding world social structures. The basis for thinking about these concepts is rooted in the philosophical paradigms of Plato and Hegel. The precursors of world-system analysis, who categorize the system of world class diversity depending on property like Karl Marx, or emphasize the level of work skill in the process of production like Max Weber, treat the whole in the Hegel way, which means sharp distinctions between the common and the different. This makes it impossible to differentiate "the other side" of phenomena such as the existence of the lumpenproletariat beside "the usual classes" (i.e., engineers can posses the ability to become, for example, thieves in addition to their working skills).

Plato believed that the general things are also common, which tends to pass the differences over. From this perspective, informal economy, as the other side of the formal one, is not the goal of scientific reasoning.

I developed the concept of the lumpeneconomy from foundations in the socioeconomic theory of ownership and the idea of "the other side" of society, developed by Stanislaw Kozyr-Kowalski (2004) and the research of Jacek Tittenbrun, in which lumpenownership is understood as the free use of goods, particularly via tax avoidance (Tittenbrun 2006: 9). As proved by Kozyr-Kowalski (2004), drawing profits from tax avoidance, from the point of view of the socioeconomic theory of ownership, means that the national community is "cheated" because the community is the legal state owner. In modern societies, the community, via the tax system, is the co-owner of all types of private means of production and labor force.

Such relations of lumpenownership can be jointly existing, together with the informal economy, the social margin and social pathology (Galor 2006: 11). The cumulative results of researchers like Andre Gunder Frank, Recep Varcin, and Hernando de Soto, along with the manifesto of Eldridge Cleaver (1972), prove the global dimension of the relation.

Lumpenwork is an element of world lumpeneconomy. Lumpenwork can be described as each kind of activity that gains access to life resources related to the lumpenownership. This kind of work can neither produce new values, like theft or begging (Tittenbrun 1986: 2), nor produce new functional values typical of informal economies, such as illegal moonlight carpentry (Galor 2006). Lumpenwork also includes syndicates, international mafia, and other activities that are not only related to production, but are specific to informal economies.

Such a broad range of lumpenwork constitutes the lumpeneconomy that it reflects the negatives of the mainstream economy. There are lumpencopies of every aspect of production, exchange, transportation, and service. The lumpeneconomy varies depending on its relation to law and morality and can even comprise explicitly criminal activities such as robbery, theft, or fraud.

The basis of the "free rider problem" is the way in which goods and public services are used, which is in the sphere of lumpenownership. Easy riders do not pay for them. Examples of the "problem of easy rider" concern their world dimension.

The syndrome of "free rider" is used as a sociological explanation for the lack of activity of certain members of social movements as "a brake on participation" in mass activities and social movements which consider public values, for example, freedom, independence, and democracy. Sztompka (2002) believed that the above syndrome can be explained via the following reasoning: if the movement wins I will benefit from it anyway and if it does not win, I will not suffer the consequences; at most, nothing will change.

Stiglitz (2000) emphasized that in the economic dimension, the "free rider problem" encompasses the lack of will in individuals to volunteer to finance public goods production. A characteristic of such goods is that it is impossible to exclude people from using them (e.g., public health services); on the other hand, it is impossible to verify whether someone has paid for them or not.

The "free rider problem" in lumpenownership relations can be broad. As described by the media, it can be knowing how to make the most of every opportunity in various world organizations, including the economic and the political. An example of such "free riders" were some EU members that refused to help Greece when the country was in difficult straits during 2010.

Under these circumstances, one more phenomenon should be mentioned, which has been described by Zygmunt Bauman (2004) as blurring the line between "legal" and "illegal." The phenomenon is not "that global mafia organizations profit from lacunas in the law in the meetings of law systems supervised by national states. I mean rather that all operations in the global space, liberated from limitations forced by the state and depending on the current lineup only seem to resemble (consciously or not) the patterns of mafia or syndicate way of work and relation to the law." This phenomenon is becoming increasingly widespread in the world and is characteristic of a developing global lumpeneconomy.

References

Bauman, Z. (2004) *Życie na przemiał*, Kraków, Wydawnictwo Literackie.

Chase-Dunn, C. and Hall, T.D. (1997) "Ecological degradation and the evolution of world-systems", *Journal of World-Systems Research*, 3: 403–31.

Cleaver, E. (1972) *On Lumpen Ideology*, Oakland: The Black Scholar (rep.).

Galor, Z. (2006) *Lumpenwłasność: szara strefa i margines społeczny*, Poznań: Akademia Rolnicza.

Kozyr-Kowalski, S. (2004) *Socjologia, społeczeństwo obywatelskie i państwo*, Poznań: Wydawnictwo UAM.

Stiglitz, J. (2000) *Economics of the Public Sector*, New York: W. W. Norton & Company.

Sztompka, P. (2002) *Socjologia. Analiza społeczeństwa*, Kraków: Wydawnictwo Znak.

Tittenbrun, J. (1986) *Wprowadzenie do materialistyczno-historycznej teorii społeczeństw*, Warszawa: Universitas.

Tittenbrun, J. (2006) "Przedmowa", in J. Tittenbrun (ed.) *Własność i interesy w dobie transformacji ustrojowej*, Poznań: Wydawnictwo Rys.

7. Networks and chains

7.1

Global cities, global commodity chains and the geography of core-ness in the capitalist world-system

Christof Parnreiter

Introduction

World-systems analysts have always been guided by a concern with the ways in which uneven development is organized. It is therefore no surprise that two of the main globalization literatures, which aim at scrutinizing the structures and the *modus operandi* of the world-economy, have been engendered by world-systems analysis: global commodity chain and global city research. In the former, the world-systems analysis legacy is direct. Departing from the insight that the unit of analysis is a key strategic research decision, Hopkins and Wallerstein conceptualized the world-economy as cross-border, firm-based transactions rather than as inter-state transactions. They "invented the term 'commodity chains' to underline a basic process of capitalism: that it involved linked production processes that had always crossed multiple frontiers and that had always contained within them multiple nodes of controlling labor. Further, we believed that a close study of such chains would indicate how and why surplus-value was distributed among its appropriators, and hence explain how the system of 'unequal exchange' worked in practice" (Wallerstein 2000: 221).

In global city research, the legacy of world-systems analysis is less direct (probably because the city has been ignored by many world-systems analysts). While some early contributions to the global city debate did embed their research into world-systems analysis, it was largely absent from Sassen's work on that topic (Sassen 1991). She did, however, reject the nation-state as a "naturally given" unit of analysis. This idea is also central to Taylor's (2004) examination of the world city network, which is defined as an interlocking network in which the relations between world cities are created by the intra-firm flows in the producer service sector (comprised by financial, legal, consulting, insurance etc., services). These services are crucial to global city research, because they are seen as key instruments for the management and control of the cross-border flows that characterize the world-economy (Sassen 1991). Building on world-systems reasoning about core-periphery relations, explorations of the world city network have defined producer services as core activities and world cities as places where these are located (Brown et al 2010; Taylor 2004).

Yet, despite their common theoretical background, the two literatures have developed with little cross-referencing. This mutual unawareness is unfortunate, because it is only through integration that the limitations of each research strand can be overcome. I maintain that integrating global commodity chains with global city research can improve world-systems analysis by illuminating the geographies of core-ness.

Limitations and ways to overcome them

The literatures on global commodity chains and on global cities are amongst the most effective tools for a critical analysis of contemporary globalization. While it is beyond the scope of this paper to enlist all their insights, I shall briefly outline some of their weaknesses in order to expound how and why integrating the two literatures will advance our understanding of uneven development (for more detailed accounts and references see Brown et al 2010; Parnreiter 2010).

As regards the global commodity chains approach, a first limitation is the poor assessment of the chains' geographies. Although the critique that "despite the spatial connotations of the concept, there has been little work on systems of provision from a geographical perspective" (Leslie and Reimer 1999: 403) was raised more than a decade ago, and despite some recent efforts to fill the void (e.g., Hughes and Reimer 2004), spatialized analyses of commodity chains are still rare. Due to this deficit, two of the three dimensions identified by Gereffi (1994: 96–97) as being constitutive for this research field remain underexplored: we know little about the territoriality of commodity chains, and our understanding of input-output structures of value-adding activities is constrained by the lack of information on the geographies of these input-output structures. Even in the case of the best explored dimension—the governance of global commodity chains—the question of where governance is exercised is commonly reduced either to an identification of so-called headquarter-cities or to a discussion of the different types of the lead firms' power and their impacts on upgrading. A second main shortcoming is the limited scope of analysis. Most studies have explored primary goods or industrial commodities, while the service sector has been treated only marginally. Producer services in particular have barely received attention. This is a significant omission with conceptual repercussions, because "without the integrating and coordinating function fulfilled by services, global commodity chains would not be viable" (Rabach and Kim 1994: 123).

As regards global city research, one main shortcoming is that many studies are devoted to major cities in high income countries, while the nodal functions of cities in middle or low income countries have received as little attention as have the connections between global and non-global cities. It is a significant void that the analysis of connections between cities has not transcended the scale of the world city network, which in 2008 comprised 129 Alpha, Beta, and Gamma World Cities. While Alpha world cities are, due to their sizeable and globalized cluster of producer service firms, "very important" to link major economic regions and states into the world-economy, Beta World Cities are "important" in that respect and Gamma World Cities either articulate smaller regions or states (GaWC 2010). Certainly global city studies need to address these leading cities and their interrelations, because it is from there that the world-economy is managed and controlled (Sassen 1991). Yet, it is equally important to grasp the many links between the global and the countless non-global, but yet globalized cities. I understand these to be the countless cities around the world that are shaped by their role to produce for global markets, but which have no management or control functions. Ciudad Juárez in Northern Mexico, for example, has been thoroughly shaped by the automotive, computer, and electronic industries located there, though the "city" has no hand in the governance of these chains. Thus, if the basic idea of a global city is that it articulates local, regional, and national economies into the world-economy (Friedmann 1986; Sassen 1991) then global cities are not only mutually constituted, as stressed by Taylor (2004). Rather, the world city network is built upon ramifications that link global cities to the many "ordinary cities" (Robinson 2006; for an assessment of Robinson's critique of global city research see Parnreiter 2011), where much of the low paid work is supplied that is as vital to the reproduction of global capitalism as is the provision of high-paid producer services.

To support the contention that the shortcomings of both literatures can be overcome by integrating global commodity chains and global city research, I will briefly turn to a case study. In the last three decades, productive capacities in Mexico (e.g., in the automotive and electronic industries) have been transformed into export platforms that serve the United States and Canadian markets. This deep integration of plants and cities in Mexico into global production networks has been analyzed by a number of authors who applied the global commodity chains (or a related) paradigm. While these studies offer many insights, the aforementioned shortcomings of global commodity chains research are also evident.

In the case of automotive chains, for example, producer services are widely ignored, except a brief reference Lee and Gason (1994: 235–36) make to advertising expenses. The authors, however, do not treat advertising as an essential component of the automotive chain, nor do they examine wherefrom it is supplied. The second main objection—the limited geographical assessment—also applies in this case. Most studies think of space as a container (either "Mexico" or specific cities) which is "filled up" with the automotive industry. Additionally, geography comes into play as physical distance or, more precisely, as the cost of overcoming it. In sum, the geographies of input-output relations within the automotive industry are at best cursorily addressed. We lack hence a spatialized analysis of the integration of different labor processes into the chains, and that is why the studies fail to comprehend the complexities of the geographies of core- and periphery-forming processes along the automotive chains in Mexico.

Before addressing how global city research can mitigate this deficit, I will briefly delineate global city formation in Mexico City. Since the 1980s, when the economy in Mexico was once again being reoriented toward the world market, Mexico City underwent a deep transformation, which is characterized by a concurrent decline of manufacturing and the rise of producer services. Concomitant with the growth and concentration of producer services in the country's capital, the headquarters of the main companies operating in Mexico have become ever more centralized in Mexico City (Parnreiter 2002, 2010). In addition, Mexico City's producer service sector is firmly inserted into the world city network, which is what makes Mexico City an "Alpha-World City" (GaWC 2010). Mexico City has thus been changing from a predominantly national production center, catering to and integrating the domestic market, to a hinge between economic activities carried out in Mexico and the world market.

Now, why should integration of global city and global commodity chains research mitigate the deficits of each literature? To begin with, I contend that the voids left by the research on the Mexican segments of global automotive chains can—at least to a certain extent—be filled by resorting to the core ideas of global city research. Firstly, this literature draws attention to producer services. For example, the Mexico City offices of the "Big Four" global accountancy firms (Deloitte, Ernst & Young, KPMG, PricewaterhouseCoopers) provide various services for automotive companies (Deloitte works for GM, Ernst & Young for Nissan, KPMG for Chrysler, PricewaterhouseCoopers for Volkswagen, Ford and Grupo Carso, a Mexican agglomerate that amongst other things produces auto parts). In legal services, the Mexico City office of Holland & Knight, one of the biggest law firms worldwide, has worked, among others, for Toyota, Volvo, Grupo Carso, and for the global electronic company Jabil, which supplies the automotive industry in Mexico. Thus, a "side glance" at the global city research would help commodity chains analysts to comprehend that the automotive industry in Mexico makes use not only of peripheral labor processes, but also of core activities such as producer services, which are supplied not from one of the new industrial cities, but from the country's capital.

It therefore becomes clear why integrating global city research into global commodity chains analysis also helps to overcome the second identified weakness of global commodity chains research, namely the poor treatment of the chains' geographies. The aforementioned examples draw attention to specific inputs into the automotive chains *and* to the particular places wherefrom these inputs are provided. In other words: the forward linkages of global producer service firms to companies in automotive chains can be spatialized as service flows from clearly delineated "global city spaces" within Mexico City to cities like Hermosillo, Saltillo, Puebla, Aguascalientes, or Toluca, where the plants for export production have been built. Though this spatialization of one particular sequence of the value-adding chain is, of course, not the "whole geography," the example refines our geographical knowledge, demonstrating that the automotive chains running through cities in central and Northern Mexico also run through Mexico City. In summary: because producer services are provided in Mexico City to the automotive industry, the city is a *critical* node in all these chains. Due to this input of core activities in chains that in general rely on "cheap labor power and imported inputs," (Dussel Peters 2008: 24), Robinson's (2006) critique of global city research can be countered thus: yes, Mexico City is *on* the map of global cities, and it is there because firms in Mexico City contribute to the fulfilment of global cities functions.

Firstly, the service flows from firms in Mexico City to companies in other Mexican cities, which operate in various global commodity chains that emanate, run through or end in Mexico, contribute to the management of these production networks. They also contribute to the second main global city function, namely the governance of these chains. Mexico City is the center of governance for production networks within Mexico and for the few global commodity chains emanating from Mexico. As regards influences on value creation and distribution within the "Mexican segments" of other global commodity chains, the scope of influence of "Mexico City" (i.e., of local affiliates of TNCs and local offices of global producer service firms) is, however, restricted, namely to issues such as taxes, labor laws, and unions, where the need for in-depth knowledge of local conditions is required (Parnreiter 2010).

As to the aforementioned critique that global city literature "privileges" a few cities in high-income countries, the broadening of research to a place like Mexico City certainly represents an advancement. This is, however, not sufficient, because the thesis of global city formation remains an assumption as long as there is no evidence for connections reaching from the global city sectors in Mexico City to all the Mexican cities where "material" production for the world market is carried out. Put differently: a high concentration of producer services does not *per se* make a global city. As Sassen has frequently indicated, it is only a city's capacity to *export* headquarter, financial, legal, and other producer services which points toward capabilities for servicing and controlling global operations of firms. Yet, if it is critical to empirically identify such *city-to-city service flows*, then it is obvious why integration with global commodity chain research is beneficial: the consumers of producer services are companies that need financial, legal, auditing, and other services in order to sustain their operations in the different global commodity chains they are integrated into. Bringing the perception of global commodity chains analysts closer to global city researchers allows the latter to grasp that global cities articulate regional, national, and international economies because they are service nodes in numberless commodity chains. Therefore, global city formation takes place precisely because—and *only* because—these cities are linked through the export of producer services to all the other places, where non-core activities (e.g., the assembling of auto parts) are carried out.

The conclusion derived from the empirical analysis of global commodity chains and global city formation in Mexico bear important analytical implications that go well beyond an accurate geographical description of production networks. In particular, the finding that core-forming labor processes are supplied from Mexico City points to the need to resume the debate on the geography of core-ness in the world-system and to deepen it. The most appropriate way to do that is bringing cities back into world-systems analysis (Smith 2003), and, above all, bringing global city and global commodity chains research together.

Specifying the geography of core-ness

In world-systems analysis, "core" and "periphery" result from divisions of labor, which split up production processes into distinct segments that are assigned different valences (expressed in a differentiated share in the whole value-adding process). For Wallerstein (2000: 139–140), the decisive factor in this unequal distribution of the accumulated surplus is that some producers manage to "achieve various kinds of temporary monopolies." The important point here is that this social division of labor goes along with geographical division of labor. Core and peripheral processes are neither evenly nor randomly distributed. Because producers in some places are for several reasons more successful in establishing and defending monopolies than producers elsewhere, the geography of core-periphery relations is highly uneven, split into places where core activities are concentrated and places where peripheral activities predominate (Brown et al 2010).

Due to their relative monopolization, core activities are rare and highly demanded by producers in many different commodity chains. Accordingly, many commodity chains coalesce in the places where core activities cluster. This idea allows us to conceptualize what in geographical terms makes a center: "[S]ome areas appear as cores or centers, in virtue of the many relational sequences leading from or to them, while others appear as the hinterlands of these centers in virtue of the small number of relations leading from or to them and locating them as arenas of world-system activities." (Hopkins and Wallerstein 1977: 114) After having introduced the concept of commodity chains, Wallerstein (1983: 30) specified that "commodity chains have not been random in their geographical directions. Were they all plotted on maps, we would notice that they have been centripetal in form. Their points of origin have been manifold, but their points of destination have tended to converge in a few areas. That is to say, they have tended to move from the peripheries of the capitalist world-economy to the centers or cores."

Yet, Wallerstein did not elaborate what was precisely meant by "zones" or "areas." This sketchiness in his geographical diction is astonishing, particularly in the light of his plea to carefully consider the unit of analysis. The fact that Wallerstein did not expand on the role of cities in the organization of uneven development is particularly striking considering that by the time world-systems analysis was introduced, Andre Gunder Frank had already put forward a *spatialized* concept of the "development of underdevelopment": "Just as the colonial and national capital (…) become the satellite of the Iberian (and later of other) metropoles of the world economic system, this satellite immediately becomes a colonial and then a national metropolis with respect to the productive sectors and population of the interior. (…) Thus, a whole chain of constellations of metropoles and satellites relates all parts of the whole system from its metropolitan center in Europe or the United States to the farthest outpost in the Latin American countryside. (…) we find that each of the satellites (…) serves as an instrument to suck capital or economic surplus out of its own satellites and to channel part of this surplus to the world metropolis of which all are satellites" (Frank 1969: 6).

Unfortunately, this very lucid account of the role of cities in the making of uneven development found little resonance in world-systems analysis (see Smith 2003: 114 for this "short history of a promising perspective"). Nevertheless, it deserves little fantasy to grasp how well Frank's hint at cities matches up with Wallerstein's notion that commodity chains come together in the "centers." What Frank suggests (unfortunately without empirical backing) is, firstly, that the transfer of resources along commodity chains is organized in cities. Secondly, he evokes that this organization of uneven development is not accomplished from a limited number of core cities. Rather, Frank insinuated that the transfer of resources is based on innumerable connections between cities of different size, importance, and reach. This is exactly what has recently been proposed more elaborately by Parnreiter (2010) and by Brown et al (2010: 29), who argue that the "creation and (unequal) distribution of value along commodity chains is organized in and governed from world cities."

This brings me back to the aforementioned question as to why core activities cluster in a limited number of places, and to the related question of why these places are cities. If core-ness is the result of the successful formation of (temporary) monopolies, then the question of what kind of entry barriers can be erected in order to keep potential rivals out becomes critical. Wallerstein (1999: 63) refers primarily to the state: "What are the services that capitalists need of the state? The first and greatest service is protection against the free market." In a free market, high profits at a specific node in a commodity chain would attract further entrepreneurs, whereby competition would be increased and profits decreased. A free market is, thus, an adverse environment for the maintenance of monopolies. Among the things a state does to protect its "own" producers are legal constraints (e.g., patents, import/export prohibitions), direct or indirect (e.g., for infrastructure) subsidies, the offer of not paying certain costs (e.g., pollution) and, last but not least, the deployment of direct violence. Though the recent state interventions to save big business from the effects of the crisis 2008/9 have proven Wallerstein correct that the "state still matters—to the entrepreneurs above all" (ibid.: 75), it is important to emphasize that the state is not the only unit and scale that matters with regard to entry barriers. The contention here is that *cities* are key places for keeping potential rivals out, because important entry barriers are intimately related to cities.

A first city-related entry barrier exists where producers need access to capital intensive infrastructure. Harvey has frequently highlighted that capital accumulation is based not only on the mobility of capital (i.e., on overcoming distance), but also on the production of space. It is therefore that "capitalism creates a physical landscape of roads, houses, factories, schools, shops, and so forth in its own image" (Harvey 1985: xv). Yet, because this physical landscape is expensive, its elements are sensitive to economies of scale and, consequently, clustered at specific places, namely cities. According to Harvey (ibid.: 190), it "is impossible to imagine such a material process [the circulation of capital, C.P.] without the production of some kind of urbanization as a 'rational landscape' within which the accumulation of capital can proceed. Capital accumulation and the production of urbanization go hand in hand." Cities are thus seen as resource complexes of socially created assets, which support production, consumption and hence further accumulation: "The stock of fixed capital and of consumption fund assets does (...) provide a solid form of wealth that can be used to produce and consume more wealth" (ibid.: 144).

If access to capital intensive infrastructure is a decisive factor governing whether or not a firm can incorporate itself into commodity chains, then Harvey's line of reasoning suggests that the entry barrier "control over capital intensive infrastructure" has a specific geography. Because producers in cities can rely on infrastructure provided by others (the national and/or local state), they enjoy advantages vis-à-vis non-urban rivals. Yet, the more capital intensive the infrastructure is, the fewer cities will provide this advantage, and the more competitors will be excluded from commodity chains. Monopolization, which results from controlling expensive infrastructure, is thus bound to specific cities.

A similar point can be made regarding innovative capacity. Rather than dealing with the question of how important innovation has historically been to the (geographical) shifts of the centers of the world-economy (for different accounts see Modelski and Thompson 1988; Arrighi 1994), I am concerned here with the issue of why some producers in some places are more efficient than others in launching what Schumpeter (1987 [1934]) called "new combinations."

Similarly to Harvey (though departing from a different theoretical origin), Jacobs has argued that cities are the drivers of economic growth because of the greater incentives to address ever new problems resulting from size and density: "[T]hese grave and real deficiencies [of cities; C.P.] are *necessary* to economic development and thus are exactly what makes cities uniquely to economic life" (Jacobs 1970: 86; emphasis added). Cities are vibrant because they face continuous challenges, and the resulting innovative capacity is what distinguishes them from towns and rural areas: "[A]t the core of all processes of city growth is this root process … [of] adding new kinds of work to other kinds of older work" (ibid.: 50–51).

The important point here is the relationship between innovations ("adding new work") and monopolies. In his attempt to "wed" Jacobs' theorizing of city-ness with world-systems analysis, Taylor (2006: 1987) highlights that the search for "new work" is driven by the promise of extra-profits that accrue to innovations: "New work is not available to competitors; it is in innovation monopolies where the largest profits are made." From this reading of Jacobs' work, a similar conclusion to the one drawn above may follow. There is a specific geography of innovations, because entrepreneurs in Jacobs' "dynamic cities" will breed more of Schumpeter's "new combinations" than their competitors in less vibrant cities, towns, or rural areas. Conversely, the literature on product life-cycles suggests that a product that moves toward standardization creates less technological rent and is therefore shifted to peripheries. Otherwise, cities/regions that stick to this standardized product will descend to peripheries. In sum, monopolization, which results from innovation, is as bound to specific cities as is monopolization developed from access to capital intensive infrastructure.

In brief: monopolization of economic activities—and hence core-ness—is in many cases a city-specific process. Accordingly, many commodity chains coalesce in cities in search of these rare core labor processes, and that is why cities constitute the centers of the capitalist world-system. This does not mean that "city" is the only adequate geographical scale for defining centers (the deployment of direct violence to protect entrepreneurs against competition is, for example, strongly related to the "national territory," even if entrepreneurs are clustered in cities). Nor does it mean that "city" refers exclusively to global cities. Global commodity chains are sustained through the inputs of various core labor processes, of which some—namely producer services—are supplied from global cities, whereas others (e.g., Research & Development) might stem from other cities.

Concluding remarks

In this article, I have argued that integration of global commodity chains and global city research is needed in order to overcome the weaknesses of each of the research strands. Summarizing the arguments put forward here, my first assertion is that all commodity chains run through global cities, because they necessarily include those core-formation processes (producer services) which are produced and traded in global cities. From this follows, secondly, that each global city constitutes a node for numerous commodity chains. This multi-nodal function is the reason for the overall centrality of global cities in the world-economy. Thirdly, I contend that all cities are connected via global commodity chains to the world city network (even if their own input into the chains is constituted by peripheral labor processes), because cities are unavoidably integrated into commodity chains (e.g., city-hinterland connections and/or inter-urban flows), and because all commodity chains run through global cities (see above). It results, fourthly, that the World City Network is built upon ramifications of operations across different scales through macro-regional and national to local.

My second main contention is that integration of global commodity chains and global city research is apt to deepen our understanding of the geographies of core-ness in the capitalist world-system. Though I totally agree with Smith (2003: 113) that "it is so important to bring cities and urbanization back into world-system studies," Frank's claim suggests the need to specify this call. Frank (1969: 6) did not point to cities in general, but to those cities that serve as instruments to suck economic surplus from one place and to channel it to another. If global cities are "highly concentrated command points in the organization of the world economy" (Sassen 1991: 3) which have the "capacity to create the means of control of value creation and distribution and the power to deploy them" (Brown et al 2010: 29), then it becomes obvious that they constitute the nodes wherefrom the "development of underdevelopment" is governed.

References

Arrighi, G. (1994) *The Long Twentieth Century. Money, Power, and the Origins of Our Times*, London: Verso.

Brown, E., Derruder, B., Pelupessy, W., Taylor, P. and Witlox, F. (2010) "World city networks and global commodity chains: towards a world-systems' integration", *Global Networks*, 10 (1): 12–34.

Dussel Peters, E. (2008) "Global commodity chains and development: a conceptual and empirical review", *Competition and Change*, 12 (1): 11–27.

Frank, A. G. (1969) *Latin America: Underdevelopment or Revolution: Essays on the Development of Underdevelopment and the Immediate Enemy*, New York: Monthly Review Press.

Friedmann, J. (1986) "The world city hypothesis", *Development and Change*, 17: 69–83.

Gereffi, G. (1994) "The organization of buyer-driven global commodity chains: how retailers shape overseas production networks", in G. Gereffi and M. Korzeniewicz (eds) *Commodity Chains and Global Capitalism*, Westport, CT: Praeger.

Globalization and World Cities (GaWC) (2010) *The World According to GaWC 2008*, http://www.lboro.ac.uk/gawc/world2008t.html accessed November 01, 2011.

Harvey, D. (1985) *The Urbanization of Capital*, Oxford: Blackwell.

Hopkins, T. K. and Wallerstein, I. (1977) "Patterns of development of the modern world-system. Research proposal", *Review*, 1 (2): 111–45.

Hughes, A. and Reimer, S. (2004) *Geographies of Commodity Chains*, London: Routledge.

Jacobs, J. (1970) *The Economy of Cities*, New York: Random House.

Lee, N. and Gason, J. (1994) "Automobile commodity chains in the NICs: A comparison of South Korea, Mexico, and Brazil", in G. Gereffi and M. Korzeniewicz (eds) *Commodity Chains and Global Capitalism*, Westport, CT: Praeger.

Leslie, D. and Reimer, S. (1999) "Spatializing commodity chains", *Progress in Human Geography*, 23 (3): 401–20.

Modelski, G. and Thompson, W. R. (1988) *Seapower in Global Politics, 1494–1993*, Houndmills: MacMillan.

Parnreiter, C. (2002) "Mexico: The making of a global city?", in S. Sassen (ed.) *Global Networks, Linked Cities*, London: Routledge.

Parnreiter, C. (2010) "Global cities in global commodity chains: exploring the role of Mexico City in the geography of global economic governance", *Global Networks*, 10 (1): 35–53.

Parnreiter, C. (2011) "More than an ordinary city: The role of Mexico City in global commodity chains", in B. Derudder, M. Hoyler, P. J. Taylor and F. Witlox (eds) *International Handbook of Globalization and World Cities*, London: Edward Elgar.

Rabach, E. and Kim, E. M. (1994) "Where is the chain in commodity chains? The service sector nexus", in G. Gereffi and M. Korzeniewicz (eds) *Commodity Chains and Global Capitalism*, Westport, CT: Praeger.

Robinson, J. (2006) *Ordinary Cities: Between Modernity and Development*, London: Routledge.

Sassen, S. (1991) *The Global City: New York, London, Tokyo, Princeton*, Princeton, NJ: Princeton University Press.

Schumpeter, J. (1987/1934) *Theorie der wirtschaftlichen Entwicklung*, Berlin: Duncker & Humblot.

Smith, D. A. (2003) "Rediscovering cities and urbanization in the 21st century world-system", in W.A. Dunaway (ed.) *Emerging Issues in the 21st Century World-System*, Vol. II, Westport, CT: Praeger.

Taylor, P. J. (2004) *World City Network: A Global Urban Analysis*, London: Routledge.

Taylor, P. J. (2006) "Jane Jacobs (1916–2006): an appreciation", *Environment and Planning A*, 38: 1981–92.

Wallerstein, I. (1983) *Historical Capitalism*, London: Verso.

Wallerstein, I. (1999) *The End of the World as We Know It. Social Science for the Twenty-First Century*, Minneapolis: University of Minnesota Press.

Wallerstein, I. (2000) *The Essential Wallerstein*, New York: The New Press.

7.2

Trade, unequal exchange, global commodity chains

World-system structure and economic development

David A. Smith

Introduction: Unequal exchange

The question of what determines the wealth of nations is a very old and basic issue in social science (Smith 1776). Almost as ancient is the discussion of how nations might gain some "comparative advantage" in trade (Ricardo 1817). Skepticism about the idea that countries will automatically benefit and experience economic growth and development if they specialized in trading the products in which they have a putative advantage led to the formulation of the idea of "unequal exchange" by the progenitors of the world-system approach.

In contemporary sociology and social science, these questions are more familiarly framed as issues of economic "development" and a bundle of other social changes that are often presupposed to accompany this process. Why did certain countries and regions become wealthy and "industrialized" over the past two centuries, while others fell behind? In the twenty-first century we still see yawning chasms of material inequality between different parts of the world (as well as within them). What explains these world inequalities? And, perhaps more pertinently, what can be done to overcome them? Is "industrialization" still a viable economic growth strategy for poor countries in today's era of globalization and worldwide "restructuring"? Or does the contemporary conjuncture demand new strategies that go beyond a transition to manufacturing and increasing "value added" production? Do the experiences of some large and very "successful" nations (like China and India) tend to obscure the continuing economic stagnation in many other underdeveloped countries and areas? This brief essay examines these debates about the dilemma of development and the perpetuation of global inequality in today's world-system, with a particular focus on trade, global commodity chains, and unequal exchange.

Emmanuel (1972) and Amin (1974) argued that the essence of core exploitation of the poor, underdeveloped regions of the periphery is transmitted through the unequal trade of low-value and less-processed goods from the periphery in exchange for expensive finished products from the core. Mandel (1975) accepted the idea that "the average productivity of labor" between countries is crucial, but categorically rejected the claim that particular commodities are intrinsically important, since it is not the material product itself that is crucial, but the labor process that produces it. Later, Stephen Bunker (1984) directly challenged this labor-based approach to unequal exchange, arguing that the environmental costs in peripheral economies dependent on resource extraction should also be considered:

> I believe that the unbalanced flows of energy and matter from the extractive peripheries to the productive core provide better measures of unequal exchange in a world economic system than do flows of commodities measured in labor and prices.
>
> *(Bunker 1984: 1018)*

This approach re-emphasized particular commodities, claiming that extractive product flows are important in core exploitation of the periphery for the extrinsic reason that they led to environmental drain and depletion.

Industrialization and the NIDL

Another late twentieth-century turn in the debate on the nature of unequal exchange came from arguments about an emergent "new industrial division of labor" (NIDL) (Frobel et al 1980). Consistent with the main contours of this argument were descriptions of a new "global assemblyline" (Feuntes and Ehrenreich 1984) arguing that there was a de-industrialization of the previously advanced core nations (Bluestone and Harrison 1982; Caporaso 1981) through "capital flight" to the low-wage peripheral or semiperipheral areas beginning as early as 1960 (Ross and Trachte 1990; see also Dicken 2007). Today the existence of a NIDL in the late twentieth century is a widely accepted fact. Poor countries are the now the undisputed centers for manufacturing, particularly for low-cost consumer goods (with China as the veritable "workshop of the world"); there is little doubt that there has been a meteoric rise in industrial production, employment, and exports from these regions in recent decades.

But what does this *mean* for economic growth in these places? Post-World War II era "modernization theory" uncritically assumed that industrialization was the motor of economic development—modernization and industrialization were inextricably intertwined. But today the idea that more investment and/or employment in manufacturing signals the beginning of some sort of "take off" into development (cf. Rostow 1960) seems increasingly implausible as the search for low-wage workers for export production to cost-conscious US consumer markets seems more like a "race to the bottom" (Chan 2003; Korten 1995). Instead, some see empirical evidence that the rise of manufacturing in non-core regions of the world since 1960, instead of leading to "manufacturing miracles" of economic growth and material convergence between the First and Third World, resulted in a persistant North-South divide in per capita income (Arrighi et al 2003). Others suggest that data show poor countries are closing the gap between themselves and rich nations (Firebaugh 2003; but see Korzeniewicz and Moran 2009 for a counter argument) and claim that industrialization is, indeed, still a key generator of economic growth (Firebaugh 2003). The problem with measuring global shifts in manufacturing is that everything from garment sewing factories to semiconductor production, shipbuilding and aerospace, or chemical processing is conflated into the "manufacturing" category. But, of course, we know that the global geography of these disparate industries is very distinct (see Dicken 2007); presumably their developmental impacts where they are located are also very different. Perhaps because of data limitations, there is a tendency to oversimplify this debate (is industrialization good or not?) when the real issue concerns the *qualitative* nature of industrialization as it occurs in historically poor countries.

Of course, the rise of non-core manufacturing for export, albeit in specific industrial niches, did create a very *different* pattern of worldwide commodity trade in the past half century, with low-wage manufacturing goods, in particular, tending to flow from the poor peripheral regions of the world-economy toward the core markets. It was during the emergence of this NIDL that we saw the rise of a number of newly industrializing countries (NICs). Manufacturing for export was hailed as the secret to rapid economic growth of countries like South Korea and Taiwan in the so-called East Asian economic miracle (Gereffi and Wyman 1990; World Bank 1993). However, we need to closely examine the *nature* of the manufacturing activities occurring in particular economies and the specific *linkages* between manufacturing enterprises, global markets, and local, state and transnational capital. Some of the NICs (particularly in East Asia) successfully engaged in "industrial upgrading" in which there is a shift from commodities like textiles, apparel, and footwear to "higher value-added items that employ sophisticated technology and require a more extensively developed, tightly integrated local industrial base" (Gereffi 1992: 92). This might include production of computers, semiconductors, numerically controlled machine tools, VCRs, televisions, etc. However, many peripheral countries remain primarily export platforms for simple low-technology, labor-intensive goods made by low-wage unskilled workers.

Global commodity chains

This focus on an increasingly integrated global economy where countries fill distinct export niches and where industrial upgrading seems to be the only viable option, led Gereffi to argue that global commodity chains (GCCs) should be the key analytical construct (Gereffi 1990; Gereffi and Korzeniewicz 1990). There are parallels between the idea of GCCs and the value chains of economist Michael Porter (1990) or the production chains of geographer Peter Dicken (2007). But Gereffi grounds his initial conceptualization in world-system analysis, drawing on Hopkins and Wallerstein's (1986: 159) definition of a commodity chain as "a network of labor and production processes whose end result is a finished commodity." Elaborating further, "A GCC consists of sets of interorganizational networks clustered around one commodity or product, linking households, enterprises and states to one another within the world-economy … (it is) the sequential stages of input acquisition, manufacturing, distribution, marketing and consumption" (Gereffi et al 1994: 2).

A thorough GCC analysis of particular commodity chains requires some detailed knowledge of the *specific* qualities of the commodity itself, which is why much research on commodity chains are case studies (see Gereffi and Korzeniewicz 1994 op. cit.; similar for case studies of extraction in "the Bunker vein" see Cicantell et al 2005, op. cit.) In network terms, these commodity chains are described as consisting of a number of "nodes" that comprise the pivot points in transformation sequences: extraction and supply of raw materials, the stages of industrial processing, export of goods, and final marketing (Gereffi and Korzeniewicz 1990). Each node is connected to other nodes of related activity and the local, regional, and world economies are seen as ever more intricate web-like structures of these GCCs, creating spatially bounded structures of varying scales.

Some commodity chains are simple: the production of a soft drink can involves extraction of bauxite from a mine, a smelting sequence that produces first alumina, then aluminum, and the fabrication of the container itself. Others, however, are complex. In an early illustration of a commodity chain, Hopkins and Wallerstein (1986) detailed the raw materials and processes of industrial transformation necessary to build an eighteenth-century wooden sailing ship. The process involves the procurement of several key raw materials that make up the hull, mast, sail, ropes, etc.—and then production sequences of varying complexity and length to make the components, leading to final assembly in a shipyard, and distribution to various end users. Of course, the commodity chain (now assuredly "global") for something like a modern automobile is even more intricate.

Most recent research on production networks and commodity chains tends to focus on consumer goods sold in retail stores (Appelbaum and Smith 2001; Gereffi 1994). Shining the spotlight on these final stages of particular GCCs has been extremely fruitful, as scholars have learned a great deal about the promise and perils of industrial upgrading strategies. But we also need to go back to the beginning of the commodity chain process and focus on the initial procurement of raw materials. Bunker reminded us that extraction (via either mining, harvesting of forest products, or agriculture) is extremely critical since this not only provides essential inputs to all finished goods but is also the source of basic energy production and infrastructural construction in contemporary society (Bunker 1985; Bunker and Ciccantell 2005). The emerging "modes of extraction" in peripheral areas of the global system also promotes a form of unequal exchange and leads to what Bunker calls "progressive underdevelopment" (Bunker 1985). John Talbot (2004) recently provided a particularly comprehensive and elegant argument about how the political economy of global coffee production could be grounded in precisely this sort of "from the beginning" global commodity chain analysis.

Ultimately, global inequality and Third World development/underdevelopment are defined by the positions societies (or their firms, localities, extractive reserves, or "industrial districts") occupy in these multiplex networks of worldwide economic production and exchange. All commodities undergo a sequence of transformation from raw materials to finished products. Heavy industrial products (like metals) may end up as girders in buildings, the bow of a ship, the foundation of a highway, or the pilings of a deep-water port. Some strategic raw materials may be processed into military hardware or weapons systems. For consumer products (the overwhelming focus of the existent GCCs literature), commodities end up as packaged and marketed goods. In all of these cases, the processes of transformation and transportation lead to geographic linkages and connections that in a real material sense create a spatially bounded structure for the world-economy. In the global economy of the late twentieth century, these "export networks and export niches are becoming key units of analysis in the contemporary global manufacturing system" (Gereffi 1992: 90).

Differential profit and surplus are generated at various nodes along these commodity chains. But these patterns are not uniform. While there may be a tendency for the highest profits and the most surplus extraction to occur at the later points along the commodity chain, this is *not* always the case. Instead, this pattern can vary across particular commodities (or even across nodes producing the same commodities in different ways). There may also be a tendency for the production nodes with the highest process technologies, turning out the most sophisticated and innovative products, to be the places where the most profits/surplus are captured.

However, while this may have been the assumption of some early conceptual approaches to "the commodity question" and unequal exchange (Emmanuel 1972) and some efforts to measure "levels of processing" in international trade (Firebaugh and Bullock 1986; Jaffee 1985), Gereffi and other recent proponents of global commodity chain approaches, claim that things are not quite this simple. Global commodity chains are more than just sequences of *production*. There is a need to understand how both *marketing* of products (particularly to highly profitable core consumer markets) and *coordination and control* of the integrated global production/marketing networks are crucial. Indeed, it is now clear that for leading edge consumer goods, the marketing and "command and control" functions are where the big profits are made, while actual manufacturing is much less lucrative. On the other hand, it is also important to understand that there are a number of "logistical" processes that must be coordinated and controlled for those heavy industrial goods that, in a very concrete sense, make up the foundation of the contemporary global economy. Here "marketing" may not be critical in the same way—but "command and control" (tempered by things like the physical properties of the materials, the cost of transporting them, business strategies to "corner the market," and ensure supply of strategic commodities, etc.) may still be paramount.

Levels of processing and structural inequality in the world-system

World-systems and dependency theorists argue that the structure of the world-economy creates international inequality. Generalizing from a long-term macro historical comparison, Immanuel Wallerstein claimed that dominant (core) countries rose to ascendancy as a direct result of their exploitation of non-western (peripheral) regions. Through colonization, core states enriched themselves and simultaneously retarded the growth of peripheral states by forcing them onto political and social paths that made technological advancement and competition with the core difficult, if not impossible (Wallerstein 1974). The contemporary mechanisms of growth (or the lack thereof) are bound to the notion that the world-system is composed of hierarchically organized zones (core, semiperiphery, and periphery) that perform different functions in the international division of labor.

> The defining characteristic of world-system zones is the extent to which they are involved in core or periphery production processes. According to Chase-Dunn, the underlying analytic basis of this territorial hierarchy is the distinction between core production and peripheral production. Core production is relatively capital intensive and employs skilled, high wage labor; peripheral production is labor-intensive and employs cheap, often politically coerced labor.
>
> *(Chase-Dunn 1998: 77)*

Thus, historically, core countries' production regimes are the ones that pioneered the move into industrial production and are primarily capital intensive, while peripheral countries regimes are primarily labor-intensive or based on the export of raw materials. Conceptually, the core/periphery distinction is one of a continuum, such that those countries that have a relatively equal mix of core- and peripheral-production processes are labeled semiperipheral and reside between core and periphery countries in the hierarchical structure.

Prior to 1960, this international division of labor was relatively simple: core countries specialized in manufactured goods that were exported abroad, and peripheral countries specialized in raw materials and food stuffs for core markets (Dicken 2007; Mittelman 2000). The effects of this international division of labor were also relatively simple. Countries that specialized in manufacturing remained on the cutting edge of technological innovation and industrial capacity, which gave them an advantage in generating wealth through product innovation. Countries that specialized in raw material and agricultural extraction were disadvantaged in terms of their ability to generate wealth and develop modern industry, and were vulnerable to highly fluctuating prices for raw materials on the world market (Galtung 1971).

But with the rise of the NIDL this long-term historical pattern was fundamentally altered. Today the spread of manufacturing to peripheral areas is very uneven, and the consequences of industrialization in the places where it occurs are less likely to lead to sustainable economic growth and development than in the past. While "manufacturing" is a variegated phenomenon, the spread of factories to peripheral areas of the world-system is often associated with low wages in these regions, perpetuating a "race to the bottom" rather than industrial upgrading.

My colleagues and I have used formal multiple network analysis to analyze the structure of the world-system, in general, and the patterns of international commodity trade, in particular, beginning with data from the 1960s. Some of the earliest network analyses of international trade yielded empirical confirmation of the core-periphery hierarchy (Nemeth and Smith 1985; Smith and White 1992; Snyder and Kick 1979). These and other studies also found a positive relationship between structural position in the world-system and economic growth (Kick and Davis 2001; Kick et al 2000; Nemeth and Smith 1985; Snyder and Kick 1979). They also demonstrated empirically that a country's structural position in trade networks is related to its specific economic role in the international division of labor (Nemeth and Smith 1985; Smith and White 1992).

More recent findings suggest that the structure of the world-economy still conforms to this overall core-semiperiphery-periphery layering (Mahutga 2006; Mahutga and Smith 2011). In these analyses, the structure of international trade conforms to a clear core/periphery structure from 1965 to 2000. Contrary to claims that globalization is primarily characterized by the homogenous spread of "industry" to historically poor countries, the main dimension of core/periphery hierarchy continues to be differentiated by relatively advanced industry at the upper end of the hierarchy versus less advanced industry at the lower end. Confirming world-system expectations, upward mobility is exceedingly rare, even during recent decades of massive change and global restructuring. Indeed, there are also changing mobility patterns over time. During the last two decades of the twentieth century there appears to be *less* upward mobility than occurred in the 1960s and 1970s, suggesting that "globalization," rather than spreading wealth and facilitating development, may be creating barriers for nations that are at the lower rungs of the global economy (Mahutga 2006; Mahutga and Smith 2011).

In fact, we find two types of "peripheralization" in recent years: first, there is one based on a specialization in low-wage, labor-intensive manufacturing (as predicted by the NIDL thesis and the idea of a global assemblyline). But we also find that some peripheral areas are becoming increasingly specialized in raw material exports. Like the light manufacturing regimes highlighted in the NIDL literature (Frobel et al 1980), these peripheral extractive economies can be very exploitative, underpinning a particular pernicious form of unequal exchange (Bunker 1985; Bunker and Ciccantell 2005).

Conclusion

Our results also have implications for broader debates about industrialization's effectiveness as a development strategy across the peripheral zone of the world-system. One of the current "hot" debates within comparative sociology is over the trends and directions in global inequality (Firebaugh 2003; Korzeniewicz and Moran 2009). Consistent with some research focusing specifically on China and India (Hung and Kucinskas 2011), our results suggest that the rapid growth of these two countries may distort the overall picture of attenuated world-system polarization and inequality—and may *not* be very practical "models" for other so-called "less developed countries." Indeed, the trade profiles of China, India (and a few other East Asian NICs) demonstrate that they could become obstacles to the growth of other poor countries by effectively "out-bidding" them for inclusion into the global production chains of transnational corporations (Choi and Smith 2010; Gereffi and Memedovic 2003; Schrank 2004). While it may be possible to statistically manipulate available data to show that there is declining "total world inequality" or at least, some overall reduction in between-country GDP per capita (Firebaugh 2003), it is almost certain that the effect disappears when China and India are removed from consideration (Hung and Kucinskas f2011). And by promulgating a rosy image of "declining world inequality," we are potentially obscuring the fact that recent globalization has been characterized by very slow economic growth in most poor countries (Milanovic 2005). The dynamism of the few should not be confused with the stagnation of the many, otherwise sociologists become apologists for the global neoliberal project, which is essentially about corporate profits, not about genuine sustainable development.

References

Amin, Samir. 1974. *Accumulation on a World Scale*. New York: Monthly Review.

Arrighi, Giovanni, Beverly Silver, and Benjamin D. Brewer. 2003. "Industrial Convergence, Globalization, and the Persistence of the North-South Divide." *Studies in Comparative International Development* 38(1): 3–31.

Appelbaum, Richard and David A. Smith. 2001. "Governance and Flexibility: The East Asian Garment Industry," in Frederick C. Deyo, Richard F. Doner, and Eric Hirshberg (eds), *Economic Governance and the Challenge of Flexibility in East Asia*. Lanham, MD: Rowman & Littlefield,

Bluestone, Barry and Bennett Harrison. 1982. *The Deindustrialization of America*. New York: Basic Books.

Bunker, Stephen. 1984. "Modes of Extraction, Unequal Exchange, and the Progressive Underdevelopment of an Extreme Periphery: The Brazilian Amazon, 1600–1980." *American Journal of Sociology*, 89(5): 1017–64.

Bunker, Stephen. 1985. *Underdeveloping the Amazon: Extraction, Unequal Exchange, and the Failure of the Modern State*. Urbana: University of Illinois Press.

Bunker, Stephen and Paul Ciccantell. 2005. *Globalization and the Race for Resources*. Baltimore, MD: Johns Hopkins University Press.

Caporaso, James. 1981. "Industrialization at the Periphery: The Evolving Global Division of Labor." *International Studies Quarterly*, 25: 347–84.

Chan, Anita. 2003. "A 'Race to the Bottom': Globalisation and China's Labor Standards." *China Perspectives*, 46 (online) Available at: http://chinaperspectives.revues.org/259.

Chase-Dunn, Christopher. 1998. *Global Formation: Structures of the World Economy*. New York: Rowman & Littlefield.

Choi, Wai Kit and David A. Smith. 2010. "China Rising? The Global Apparel Commodity Chain, Labor and Inequality." *Peace Review: A Journal of Social Justice*, 22(4): 416–22.

Cicanntell, Paul, Gay Seidman, and David A. Smith (eds). 2005. *Nature, Raw Materials, and Political Economy*. Research in Rural Sociology and Development: Volume 10. New York: Elsevier.

Dicken, Peter. 2007. *Global Shift*. 5th edn, New York: Guilford Press.

Emmanuel, Arghiri. 1972. *Unequal Exchange: A Study of the Imperialism of Trade*. New York: Monthly Review.

Feuntes, Annette and Barbara, Ehrenreich. 1984. *Women in the Global Factory*. Boston, MA: South End Press.

Firebaugh, Glenn. 2003. *The New Geography of Global Income Inequality*. Cambridge, MA: Harvard University Press.

Firebaugh, G. and B. Bullock. 1986. "Levels of Processing of Exports: New Estimates for 73 Less-Developed Countries in 1970 and 1980". *International Studies Quarterly* 30: 33–50.

Frobel, Folker, Jurgen Heinrichs, and Otto Kreye. 1980. *The New International Division of Labor*. Cambridge: Cambridge University Press.

Galtung, Johan. 1971. "A Structural Theory of Imperialism." *Journal of Peace Research*, 2: 81–117.

Gereffi, Gary. 1990. "Paths of Industrialization: An Overview," Gary Gereffi and Donald L. Wyman (ed.) *Manufacturing Miracles: Paths of Industrialization in Latin America and East Asia*. Princeton, NJ: Princeton University Press.

Gereffi, Gary. 1992. "New Realities of Industrial Development in East Asia and Latin America: Global, Regional, and National Trends." pp. 85–112 in Richard Appelbaum and Jeffrey Henderson (eds) *States and Development in the Asian Pacific Rim*. Beverly Hills, CA: Sage.

Gereffi, Gary. 1994. "The Organization of Buyer-Driven Global Commodity Chains: How U.S. Retailers Shape Overseas Production." pp. 95–122 in Gary Gereffi and Miguel Korzeniewicz (eds) *Commodity Chains and Global Capitalism*. Westport, CT: Greenwood Press.

Gereffi, Gary and Donald L. Wyman. 1990. *Manufacturing Miracles: Paths of Industrialization in Latin America and East Asia*. Princeton, NJ: Princeton University Press.

Gereffi, Gary and Miguel Korzeniewicz. 1990. "Commodity Chains and Footwear Exports in the Semiperiphery." pp. 45–68 in William Martin (ed.) *Semiperipheral States in the World-Economy*. New York: Greenwood Press.

Gereffi, Gary and Miguel Korzeniewicz (eds). 1994. *Commodity Chains and Global Capitalism*. Westport, CT: Greenwood Press.

Gereffi, Gary and Olga Memedovic. 2003. *The Global Apparel Value Chain: What Prospects for Upgrading by Developing Countries?* Vienna: United Nations Industrial Development Organization.

Gereffi, Gary, Miguel Korzeniewicz, and Roberto Korzeniewicz. 1994. "Introduction: Global Commodity Chains." pp. 1–14 in Gary Gereffi and Miguel Korzeniewicz (eds) *Commodity Chains and Global Capitalism*. Westport, CT: Greenwood Press.

Hopkins, Terence and Immanuel Wallerstein. 1986. "Commodity Chains in the World-Economy Prior to 1800." *Review*, 10(1): 157–70.

Hung, Ho-fung and James Kucinskas. 2011. "Globalization and Global Inequality: Assessing the Impact of China's and India's Rise, 1980–2005." *American Journal of Sociology*, 116(5): 1478–513.

Jaffee, David. 1985. "Export Dependence and Economic Growth: A Reformulation and Respecification." *Social Forces*, 64: 102–18.

Kick, Edward L. and Byron L. Davis. 2001. "World-System Structure and Change: An Analysis of Global Networks and Economic Growth across Two Time Periods." *American Behavioral Scientist*, 44(10): 1567–78.

Kick, Edward, Byron Davis, Marlene Lehtinen, and Liya Wang. 2000. "Family and Economic Growth: A World-System Approach and a Cross-National Analysis." *International Journal of Comparative Sociology*, 41(2): 225–44.

Korten, David. 1995. *When Corporations Rule the World*. San Francisco, CA: Kumarian Press.

Korzeniewicz, Roberto and Timothy Moran. 2009. *Unveiling Inequality: A World Historical Perspective*. New York: Russell Sage.

Mahutga, Matthew C. 2006. "The Persistence of Structural Inequality? A Network Analysis of International Trade, 1965–2000." *Social Forces*, 84(4): 1863–89.

Mahutga, Matthew C. and David A. Smith. 2011. "Globalization, the Structure of the World Economy and Economic Development." *Social Science Research*, 40(1): 257–72.

Mandel, Ernest. 1975. *Late Capitalism*. London: New Left Review.

Mittelman, James H. 2000. *The Globalization Syndrome*. Princeton, NJ: Princeton University Press.

Milanovic, Branco. 2005. *Worlds Apart: Measuring International and Global Inequality*. Princeton, NJ: Princeton University Press.

Nemeth, Roger J. and David A. Smith. 1985. "International Trade and World-System Structure: A Multiple Network Analysis." *Review*, 8(4): 517–60.

Porter, Michael. 1990. *The Competitive Advantages of Nations*. New York: Free Press.

Ricardo, David. 1817. *On the Principles of Political Economy and Taxation*. London: John Murray.

Ross, Robert and Kent Trachte. 1990. *Global Capitalism: The New Leviathan*. Albany: State University of New York Press.

Rostow, Walt W. 1960. *The Stages of Economic Growth: A Non-Communist Manifesto*. Cambridge: Cambridge University Press.

Schrank, Andrew. 2004. "Ready to Wear Development? Foreign Investment, Technology Transfer, and Learning by Watching in the Apparel Trade." *Social Forces*, 83(1): 123–56.

Smith, Adam. 1776. *An Inquiry into the Nature and Causes of the Wealth of Nations*. London: Strahan and Cadell.

Smith, David A. and Douglas R. White. 1992. "Structure and Dynamics of the Global Economy: Network Analysis of International Trade 1965–80." *Social Forces*, 70(4): 857–93.

Snyder, David and Edward L. Kick. 1979. "Structural Position in the World System and Economic Growth, 1955–70: A Multiple-Network Analysis of Transnational Interactions." *American Journal of Sociology*, 84(5): 1096–1126.

Talbot, John. 2004. *Grounds for Agreement: The Political Economy of the Coffee Commodity Chain*. Lanham, MD: Rowman & Littlefield.

Wallerstein, Immanuel. 1974. *The Modern World System I: Capitalist Agriculture and the Origins of the European World-Economy in the Sixteenth Century*. Burlington, VT: Academic Press.

World Bank. 1993. *The East Asian Miracle: Economic Growth and Public Policy*. New York: Oxford University Press.

7.3

Global cities and world city networks

Michael Timberlake and David A. Smith

"World cities," according to Peter Hall, who introduced the concept in his book of that title (1966), are the metropolitan regions in which is conducted a disproportionately large share of the world's business. They are typically the centers of government, they are big in terms of population, they are wealthy, and they are centers of high culture. Because world cities are centers of trade and government, business and finance, and attract disproportionate numbers of professionals and the wealthy, they also generate transportation hubs, such as large airports, which provide global linkages. Hall illuminated the world city (WC) concept with seven qualifying city-regions: London, Paris, Randstad Holland, Rhine-Ruhr, Moscow, New York, and Tokyo. Most of the early WC research was case study oriented (cf. Abu-Lughod 1999; Hall 1966; King 1990; Sassen 1991) but these world cities (Hall 1966; Friedmann 1986) and "global cities" (Sassen 1991) are typically categorized, at least implicitly, on the basis of their roles in various relational networks that are planetary in scope. The more recent quantitative studies (cf. Alderson and Beckfield 2004; Smith and Timberlake 1995b; Taylor 2004) are explicit about the networked character of the world urban system and the WC hierarchies that emerge based on those links. At a general theoretical level, the key network connections usually fall into a few functional categories: economic, cultural, social, and, less frequently, political, and they may take the form of material, human, and informational exchanges.

The research identifying empirical WC networks is based on various, imperfect representations of these general constructs. It shows how the world's great cities are increasingly interconnected through organizational ties (e.g., as sites for headquarters and regional offices for global firms) and by the flows of capital, information, and people. Thus, researchers conceptualize these cities as members of a single system of cities, and like most systems, the world's city system is hierarchical, with some cities filling more globally central roles in these organizational and flow networks than others. Systematic WC researchers began with basic mapping of global urban hierarchies but moved to explaining (a) why and how the hierarchy changes and (b) why and how cities are different in relation to their place in the hierarchy. After more than 15 years of systematic, comparative, and (recently) longitudinal research, we know what the world city network (WCN) looks like at multiple points in time, using multiple measures of connectivity, understanding that specific patterns of centrality and hierarchy differ with time period and the particular network relation which the research considers. However, the shifts are ordered, constrained, matters of degree, and tied to familiar world-system processes and structures.

An emerging perspective on world cities in a global hierarchy

Central to the rise in WC research was the growing influence of the world-system perspective on urbanization and development through the late 1970s and 1980s. Foreshadowed by Gunder Frank's imagery of "chains of constellations of metropoles and satellites" in Latin America (1966: 6) and Castells' framing of "dependent urbanization" (1977: chapter 3), Walton (1977) and Roberts (1978) developed arguments about how city growth in the poorer world regions is shaped by the historical context of a region's initial incorporation into the capitalist world-system, as well as subsequent changes in its role in the world-economy. By the early 1980s, Walton (1982) and Chase-Dunn (1984) published essays that systematically laid out the "urbanization in the world-system" approach and established a research agenda. Timberlake's edited volume (Timberlake 1985) provided the most comprehensive effort to develop this perspective with contributors exploring how either national urban patterns (e.g., Nemeth and Smith, 1985) or particular cities' growth trajectories (e.g., Chase-Dunn 1985) were linked to unequally structured economic and power relations operating across countries and globally; Sassen's (1985) essay on capital mobility and labor migration prefigured her later work on global cities. Almost simultaneously, this world-system approach to framing urbanization and development issues began appearing in journals in the form of historical case studies (Feagin 1985; Smith 1987) and a plethora of quantitative cross-national analyses (Bradshaw 1987; Kentor 1981; London 1987; London and Smith 1988; Meyer 1986; Timberlake and Kentor 1983).

In the meantime, Friedmann (1986, 1995) and Sassen (1991) developed more nuanced analyses of WCs. A 1993 conference organized by Knox and Taylor, brought together sociologists, geographers, and urban planning scholars working on world cities (Knox and Taylor 1995). The resulting edited volume established an interdisciplinary project focused on the study of WCs and WCNs. Building on Friedmann's landmark 1986 paper, "The World City Hypothesis," the editors argued that world cities are, to varying degrees, integrated into a broader world-economy which is capitalist, with cities being "key" to the extent that they are important "basing points" for capital. The nature of integration varies according to the different characteristics of each city (e.g., immigration, economic polarization, etc.). Other crucial insights included: the global system of cities is hierarchical in terms of the relative network centrality of the constituent cities (e.g., Friedmann 1995; Lyons and Salmon 1995; Smith and Timberlake 1995a, 1995b); WCs compete for centrality in this hierarchy (e.g., Friedmann 1995; Knox 1995); and the WC hierarchy imperfectly and inconstantly mirrors the world-system position of national societies in the core/periphery hierarchy (Sassen 1995; Taylor 1995). Some authors (e.g., Sassen 1998: 197–202) suggest a secular decoupling over time while others (e.g., Taylor 1995) claim that there have been cyclical realignments of the two hierarchies. Cities are sites where organizations, such as firms, may choose to locate production and command/control functions. When more firms site more activities in particular cities, those cities become more central to the global system of cities. With centrality in globally important organizational networks, key cities become increasingly important as global centers for other commonplace human activities: commerce, immigration, tourism, business travel, consumption, and so on. These activities often "hang together" spatially, rendering an overall global hierarchy of cities.

Measuring world city-ness: From attributes to relations

Research on city systems has a long history (e.g., Duncan et al 1960; Ross 1987), and though it has always relied on theories featuring relational concepts (e.g., dominance), its execution was mainly based on cities' attributes, usually population size, but sometimes some other characteristic, such as their number of Fortune 500 headquarters (e.g., Cohen 1981). A *relational* operationalization of far-flung urban systems was more difficult because it is harder to find quantitative measures of interaction than to estimate attributes such as city population sizes. Meyer (1986) made a valuable early effort, examining international banking headquarters and subsidiaries in Latin American, North American, and European cities. Smith and Timberlake (1995b, 2001) located data on international inter-city air travel to derive a global network with London, Paris, Frankfurt, and New York at the top. Soon, Taylor and other researchers at the Globalization and World Cities research center (GaWC; http://www.lboro.ac.uk/gawc/) developed a relational world city data set based on the locations of headquarters and back offices of top producer services firms (cf., Taylor et al 2002) while Alderson and Beckfield (2004) compiled data on multinational enterprise (MNE) headquarter-subsidiary linkages across more than 3,000 cities. Carroll (2007) operationalized the global urban hierarchy in terms of corporate board of director interlocks among Fortune 500 firms in order to study the relationship between cities and global class formation. Derudder and Witlox improved upon our air passenger flow data with a more complete representation of the city-to-city volume of travelers across a larger number of cities for 2000 (see Derudder and Witlox 2005). Presently, researchers are working on assembling network data representing a myriad of city-to-city linkages for multiple time periods.

Critiques from within and without WC Scholarship

World Cities versus Global Cities? Many researchers working within this tradition freely exchange the terms "world city" and "global city." Derudder (2006) suggests that the empirical research on world cities is often based, unwittingly, on divergent theoretical guideposts. While we have used the term "world cities" to describe places that are integrated into a worldwide network of urban areas and are relatively and to varying degrees important to the operation of the global economy, Sassen uses the term "global city" and consciously distinguishes it from world cities, especially as developed by Friedmann. Derudder argues that whereas Friedmann is concerned about identifying cities that are relatively dominant in terms of global economic power, Sassen is focused on cities that control production because they are the sites for advanced producer services—financial, creative, and highly skilled professional services that not only operate globally, but in fact make "globalization" as we know it possible (cf., Sassen 1998, Ch. 7; Taylor 2011). Friedmann envisages global urbanization in terms of economic power in general (including through manufacturing/ production) while Sassen emphasizes global cities as key sites for global control (particularly as providers of advanced producer services). Derudder goes on to show that the empirical studies of the hierarchical international networks of cities often draw haphazardly on these two divergent theoretical paths to justify their analyses and frame their findings. While it *is* important to specify carefully how the particular relational data that scholars use is connected to key concepts in this literature, there are good conceptual rationales for examining various city-to-city global flows that transcend the Friedmann/Sassen conceptualizations and can capture dimensions of WC-ness not limited to the original formulations. WC scholarship is still at a stage where it is more important to be catholic and inclusive of different efforts than to impose a theoretical orthodoxy.

World Cities off the Map. The research on the WCN creates "maps" locating cities of rich countries atop the world-system's city system. This raises the suspicion that the WCN project is biased in leaving huge cities of the global South "off the map" (Robinson 2002). While non-core world cities were *not* the central focus of most early WC research, the theoretical thinking behind this research offers opportunity to understand *all* cities. Conceptualizing the whole WCN gives us an excellent way to explore how the "megacities" of poor countries "fit" into the global hierarchy—and gives us some new traction in exploring the meaning (or "effects") of how these places are located both in terms of the WC system and the core/periphery hierarchy of national states. Urban areas that hold places on lower rungs of the WCN and are also located in the peripheral zone of the world-system may be doubly disadvantaged. WC scholars have begun conceptualizing city dynamics in terms of the articulation of their structural positions in these other global structures.

The WC Polarization Thesis. Friedmann's "world city hypothesis" (1986) argued that polarization in terms of social inequality was a necessary concomitant of WCN position. The claim that the most dominant places in the global urban hierarchy are also centers of some of the global economy's most severe and intractable contradictions—home, simultaneously, to enormous wealth and power and to marginalization and poverty—is also echoed in other scholar's work (e.g., King 1990; Sassen 1991). Indeed, growing socioeconomic polarization in New York, London, or Tokyo should give pause to those "boosters" who argue that WC status is "good" for most urban residents—data seem to belie this rosy view. However, the precise mechanism linking high WC status to inequality is rather vague: Sassen claims that the presence of so many upper-income professionals in global cities creates pressure for "gentrification" (which displaces poorer people from affordable housing) while also generating demand for low-wage, even "informalized" jobs to service the needs of those wealthy elites. But there is also recent evidence that there is a long-term secular rise in inequality (associated with "global neoliberal" policies) all over the world today (Korzeniewicz and Moran 2009). Is the polarization really more pronounced in world cities, and even if it is, what is the evidence specifically linking this condition to WCN position?

Abu-Lughod's book *America's Global Cities: New York, Chicago, and Los Angeles* (1999) documents each city's deep global connections through the very sorts of mechanisms described by Sassen as defining her quintessential global cities. Abu-Lughod builds the case that all three make the grade as global cities by Sassen's criteria and all are highly polarized, but she locates the sources of these inequalities in the culture and history of each city rather than its role in the WCN. Hamnett (1998) also challenges the WC polarization thesis, arguing that WCs are experiencing increased demand for high wage professional occupations, but there is, in fact, weakening demand for low-wage work. "Professionalization" rather than "polarization" characterizes more globally central cities in the current international division of labor. Hill and Kim (2000), studying East Asian cities, also object (and question the notion of convergent structural features for global cities), and Hill and Fujita (2003) concur, arguing that differences in the nature and policies of the state provide better explanations of relative levels of WC polarization than WCN position.

Recent empirical research

Growing Inclusiveness. A significant amount of recent research focuses on world cities that are beyond the core. For instance, Pirie (2010) shows how the South African cities of Johannesburg and Capetown are integrated into the global urban hierarchy via airline connections and shows that these cities are increasingly linked with cities all over the world, but particularly with cities within Africa, Asia and the Middle East. Córdoba Ordóñez and Gago (2010) perform a similar analysis of Latin American aerial connectivity to show the growth and the degree of cohesion of the region's "internal" urban networks, as well as its extensive ties to the world urban network in the period 1970–2008. Still, African cities have been "neglected" relative to the great number of studies on world cities that are being produced, as are Middle Eastern cities (but see Bassens et al 2009 for a recent exception). Except for some key places (e.g., Hong Kong) Asian cities have been neglected as well, but this is rapidly changing. David Meyer's (2000) masterful case study of Hong Kong as a global metropolis was followed by Shin and Timberlake's work (2000) showing that from the 1990s onward, there was rapid mobility for East Asian cities. Recent scholarship also documents global city formation processes in Taipei (Kwok 2005), Shanghai (Wassertrom 2008), and re-examines Hong Kong's long history as a Chinese global city. Ma and Timberlake (2008) note the rapid rise of Beijing and Shanghai from 2000 to 2005.

New relational measures over more time periods. Following the relatively early work that systematically operationalized the WC networks mentioned above, many other efforts are underway to assemble network-like data representing additional city-to-city linkages. Matthiessen et al (2010) estimate the significance of cities as "knowledge centers" in terms of the locational affiliations of authors of important scientific journal articles. Their research gives us a sense of the relative importance of different city-regions as knowledge centers and looks at this in the mid-1990s and again several years later to get some sense of change. The empirical urban hierarchy they find is distinctive in some ways—but London and Tokyo-Yokohama are at the top in both time periods and just below in the more recent period are the San Francisco area, New York, Boston, and Los Angeles in the United States, Osaka in Japan, and Paris and the Amsterdam city-region (with Beijing rising rapidly in the most recent period). Several years ago, Sassen (2002) and Garcia (2002) called for increased scrutiny of the roles of global cities in the emerging architecture of global digital networking. Choi et al (2006) took up this challenge in their analysis of WC air transport and internet backbone links, finding that the two networks are highly correlated, with London in the most dominant position in both. Vinciguerra et al (2010) recently developed a spatial model of network evolution to examine the European internet infrastructure. It is very likely that digital networks will be a fertile topic for future WC network research.

WC network studies are increasingly longitudinal as well as involving more and different types of linkages. One of the first over-time studies of WC networks was our paper documenting shifts in cities' ranks based on changing patterns of airline passenger flows among pairs of the top 30 world cities for six one-year time periods beginning in 1977 and ending in 1997 (Smith and Timberlake 2001). London, Paris, and New York were consistently in the top four cities, joined by Frankfurt (a major air "hub," to be sure, but probably also a significant global city, too). Tokyo ranked high in 1991 but declined thereafter, while other East Asian cities ascended the hierarchy in the 1990s, consistent with that region's economic dynamism at the end of the twentieth century. Data limitations restricted such studies until recently. In 2010 a special issue of the journal, *Urban Studies* (Derudder et al 2010), published several longitudinal WC network studies using relational data. Alderson et al (2010) examine the MNE headquarter-subsidiary network over a 26-year time span and find that the hierarchical roster of world cities as measured by various types of network centrality "reshuffle" considerably, (though the very top cities remain consistently dominant) with some "control functions" becoming somewhat more centralized in the final two decades of the twentieth century. They also examine changes in the correspondence between cities' location in the world-system of nations (e.g., core, semiperiphery, and periphery) and their hierarchical positions in the WCN and find a close correspondence in each time period, with almost no significant change across the time periods studied (1981–2007) except that semiperpheral cities are more likely to be upwardly mobile. Using data on air passenger flows among about 200 cities from 1977–2007, Mahutga et al (2010) also examine the extent to which the global city hierarchy is increasingly centralized and changes in the correspondence between global cityness and world-system position of the countries in which they are located. They found that the network of airline passenger flows is becoming less centralized. More passengers are flying to and from more cities in the world, producing a less hierarchical network. Consistent with Alderson et al they find that core cities are far more likely to rank higher in the system of cities than cities located in countries of the periphery and semiperiphery. In addition, cities in countries of the periphery are less likely to be upwardly mobile than cities elsewhere, but cities in the semiperiphery are somewhat more likely to have become more globally central over time than cities in the core (and far more likely than cities in the periphery). Moreover, East Asian cities are more likely to have increased their world cityness than other cities, and this effect is independent of world-system position. Derudder et al (2010) map shifts in the degree of global network connectivity of some three hundred cities, using the GaWC data on advanced producer services. From 2000 to 2008 the chief findings, in summary fashion, are that cities in the United States declined, and key cities in Asia, the Middle East, and Eastern Europe became significantly more connected over this period. Orozco-Pereira and Derudder (2010) use the same GaWC data to explain the changing positions of cities in the WCN from 2000 to 2004 in terms of several hypotheses that are represented in their analysis by different independent variables. They show that cities located in countries with higher levels of international trade are more likely to advance in the WCN, as are cities with greater air passenger and telecommunications connectivity and cities larger in population size. Interestingly, "being there first" does not seem to help explain increases in hierarchical position: the lagged dependent variable, measured four years earlier, has no effect on the measure of change.

Conclusions

WC research has exploded in recent years, with scholars producing numerous case studies of specific cities, comparative studies of a few cities, and quantitative comparative studies of a great number of the world's prominent cities. Most of the theoretical underpinnings of this research rely on notions of how cities are involved in global processes and how this involvement links them to global systems of cities. The conceptual schemes for exploring world cities are relational, and explicitly or implicitly evoke social networks. The objectives of the research include describing how particular cities embody the characteristics of world cities, how cities compete for WC position, how the cities of the world map onto the world-system's city system, the extent to which the ordering of cities changes on the global hierarchy, why it changes, and whether relative position on the hierarchy matters for social structure and social relations within cities. This research increasingly employs diverse indicators of how cities are networked and the more recent analyses are more likely to examine changes over time.

Across these dynamic studies we see ascendance of some cities, including the rise of particular Asian and Middle Eastern cities, for example. Other studies stress that the "reshuffling" in the WCN is taking place without as much true change in terms of cities from the bottom tiers of the world-system rising to the top tiers of the WCN. For example, a recent study of the regional and global connectivity of Latin American cities is consistent with the "reshuffling-without-change" conclusion (Córdoba Ordóñez and Gago 2010). While cities may be more intensively interconnected, the patterns characterizing these connections even in the most recent periods are consistent with old notions of "dependent urbanization." Likewise African cities are, in general, still "off the map," although specific cities are becoming increasingly networked to other neglected but dynamic regions. On the other hand, some Asian cities are rising in terms of WC hierarchies, with a few likely to contend for global city status in the coming years. Yet these rising cities are hardly evidence for the sort of decoupling of the global city system from the world-system of nation states posited by some. The most mobile cities of Asia and the semiperiphery are located in strong, developmental states that are deliberately pushing their leading cities toward the top of the global hierarchy.

References

Abu-Lughod, J. (1999) *America's Global Cities*. Minneapolis: University of Minnesota Press.

Alderson, A. S. and J. Beckfield (2004) "Power and Position in the World City System." *American Journal of Sociology*, 109: 811–51.

Alderson, A., J. Beckfield and J. Sprague-Jones (2010) "Inter-City Relations and Globalization: The Evolution of the Global Urban Hierarchy, 1981–2007." *Urban Studies*, August 47 (9).

Bassens. D., B. Derudder and F. Witlox (2009) "Searching for the Mecca of Finance: Islamic Financial Services and the World City Network." *Area*, 42(1): 35–46.

Bradshaw, Y. (1987) "Urbanization and Underdevelopment: A Global Study of Modernization, Urban Bias, and Economic Dependency." *American Sociological Review*, 52(2): 224–39.

Carroll, W. (2007) "Global Cities in the Global Corporate Network." *Environment and Planning, A* 39: 2297–2323.

Castells, M. (1977) *The Urban Question: A Marxist Approach*. Cambridge, MA: MIT Press.

Chase-Dunn, C. (1984) "Urbanization in the World-System: New Directions for Research." pp. 111–20 in M. Smith (ed.), *Cities in Transformation*. Beverly Hills, CA: Sage.

Chase-Dunn, C. (1985) "The System of World Cities, A.D. 800–1975." Chapter 12 in M. Timberlake (ed.), *Urbanization in the World-Economy*. Orlando, FL: Academic Press.

Choi, J. H., G. A. Barnett and B.-S. Chon (2006) "Comparing World City Networks: A Network Analysis of Internet Backbone and Air Transport Intercity Linkages." *Global Networks*, 6: 81–99.

Cohen, R. B. (1981) "The New International Division of Labor, Multinational Corporations, and Urban Hierarchy." in M. Dear and A. Scott (eds), *Urbanization and Urban Planning in Capitalist Society*. New York: Metheun.

Córdoba Ordóñez, J. A. and C. Gago García (2010) "Latin American Cities and Globalization: Change and Permanency in the Context of Development Expectations." *Urban Studies*, 47(9): 2003–2021.

Derudder, B. (2006) "On Conceptual Confusion in Empirical Analyses of a Transnational Urban Network." *Urban Studies*, 43: 2027–46.

Derudder, B. and F. Witlox (2005) "An Appraisal of the Use of Airline Data in Assessing the World City Network." *Urban Studies* 42: 2371–88.

Derudder, B., P. Taylor, P. Ni, A. De Vos, M. Hoyler, H. Hanssens, D. Bassens, J. Huang, F. Witlox and X. Yang (2010) "Pathways of Growth and Decline: Connectivity Changes in the World City Network, 2000–2008." *Urban Studies*, 47: 1861–77.

Duncan, O., W. R. Scott, S. Lierberson, B. Duncan and H. Winsborough (1960) *Metropolis and Region*. Baltimore, MD: Johns Hopkins University Press.

Feagin, J. (1985) "The Global Context of Metropolitan Growth: Houston and the Oil Industry." *American Journal of Sociology*, 90(6): 1204–30.

Frank, A. G. (1969) *Latin America: Underdevelopment or Revolution?* New York: Monthly Review.

Friedmann, J. (1986) "The World City Hypothesis." *Development and Change* 17: 69–84.

Friedmann, J. (1995) "Where We Stand: A Decade of World City Research." pp. 21–47 in P. L. Knox and P. J. Taylor (eds), *World Cities in a World-System*. Cambridge: Cambridge University Press.

Garcia, D. (2002) "The Architecture of Global Networking Technologies." pp. 39–69 in S. Sassen (ed.), *Global Networks, Linked Cities*. New York: Routledge.

Hall, P. (1966) *World Cities*. New York: McGraw-Hill.

Hamnett, C. (1996) "Social Polarisation, Economic Restructuring and Welfare State Regimes." *Urban Studies*, 33(8), 1407–30.

Hill, Richard C. and J. W. Kim (2000) "Global Cities and Developmental States: New York, Tokyo and Seoul." *Urban Studies*, 37: 2167–95.

Hill, R. C. and K. Fujita (2003) "The Nested City: Introduction." *Urban Studies*, 40: 207.

Kentor, J. (1981) "Structural Determinants of Peripheral Urbanziation: The Effects of International Dependence." *American Sociological Review* 46(2): 201–11.

King, A. (1990) *Global Cities: Post-Imperialism and the Internationalization of London*. London: Routledge.

Knox, P. (1995) "World Cities in a World-System." Chapter 1 in P. Knox and P. Taylor (eds), *World Cities in a World-System*. Cambridge: Cambridge University Press.

Knox, P. and P. Taylor (eds) (1995) *World Cities in a World-System*. Cambridge: Cambridge University Press.

Korzeniewizc, R. and T. Moran (2009) *Unveiling Inequality: A World Historical Perspective*. New York: Russell Sage.

Kwok, R. (2005) *Globalizing Taipei*. London: Routledge.

London, B. (1987) "The Structural Determinants of Third World Urban Change: Ecological and Political Economic Analysis." *American Sociological Review*, 52(1): 28–43.

London, B. and D. Smith (1988) "Urban Bias, Dependence, and Economic Stagnation in Non-Core Nations." *American Sociological Review*, 53(3): 454–63.

Lyons, D. and S. Salmon (1995) "World Cities, Multinational Corporations, and Urban Hierarchy: The Case of the United States." Chapter 6 in Knox and Taylor (eds), *World Cities in a World-System*. Cambridge: Cambridge University Press.

Ma, X. and M. Timberlake (2008) "Identifying China's Leading World City: A Network Approach." *GeoJournal*, 71: 19–35.

Mahutga, M. C., X. Ma, D. Smith and M. Timberlake (2010) "Economic Globalization and the Structure of the World-City System: The Case of Airline Passenger Data." *Urban Studies*, 47(9): 1925–47.

Matthiessen, C., A. W. Schwarz and S. Find (2010) "World Cities of Scientific Knowledge: Systems, Networks and Potential Dynamics. An Analysis Based on Bibliometric Indicators." *Urban Studies*, 47 (9): 1879–1897.

Meyer, D. R. (1986) "World-System of Cities: Relations between International Financial Metropolises and South American Cities." *Social Forces*, 64(3): 553–81.

Meyer, D. R. (2000) *Hong Kong as a Global Metropolis*. Cambridge: Cambridge University Press.

Nemeth, R. and D. Smith (1985) "The Political Economy of Contrasting Urban Hierarchiesin South Korea and the Philippines." Chapter 9 in M. Timberlake (ed.), *Urbanization in the World-Economy*, Orlando, FL. Academic Press.

Orozco-Pereira, R. A. and B. Derudder (2010) "Determinants of Dynamics in the World City Network, 2000–2004." *Urban Studies*, 47(9): 1949–1967.

Pirie, G. (2010) "Trajectories of North-South City Inter-Relations: Johannesburg and Cape Town, 1994–2007." *Urban Studies*, 47(9): 1985–2002.

Robinson, J. (2002) "Global and World Cities: A View from Off the Map." *International Journal of Urban and Regional Research*, 26(3): 531–54.

Ross, C. (1987) "Organizational Dimensions of Metropolitan Dominance: Prominence in the Network of Corporate Control, 1955–75." *American Sociological Review*, 52(2): 258–67.

Sassen, S. (1985) "Capital and Labor Migration: Their Expression in Core Cities." Chapter 11 in M. Timberlake (ed.), *Urbanization in the World Economy*. Orlando, FL: Academic Press.

Sassen, S. (1991) *The Global City: New York, London, Tokyo*. Princeton, NJ: Princeton University Press.

Sassen, S. (1995) "On Concentration and Centrality in the Global City." Chapter 4 in *World Cities in a World-System*. Cambridge: Cambridge University Press.

Sassen, S. (1998) *Globalization and its Discontents: Essays on the New Mobility of People and Money*. New York: New Press.

Sassen, S. (2002) *Global Networks, Linked Cities*. New York: Routledge.

Shin, K. H. and M. Timberlake (2000) "World Cities in Asia: Cliques, Centrality and Connectedness." *Urban Studies*, 37(12): 2257–85.

Smith, D. (1987) "Dependent Urbanization in Colonial America: The Case of Charleston, South Carolina." *Social Forces*, 66(1): 1–28.

Smith, D. and M. Timberlake (1995a) "Cities in Global Matrices: Toward Mapping the World-System's City-System." pp. 79–97 in P. L. Knox and P. J. Taylor (eds), *World Cities in a World-System*. New York: Cambridge University Press.

Smith, D. and M. Timberlake (1995b) "Conceptualizing and Mapping the Structure of the World-System's City System." *Urban Studies*, 32: 287–302.

Smith, D. and M. Timberlake (2001) "World City Networks and Hierarchies, 1977–97." *American Behavioral Scientist*, 44: 1656–78.

Taylor, P. (1995) "World Cities and Territorial States: The Rise and Fall of Their Mutuality." Chapter 3 in P. Knox and P. Taylor (eds), *World Cities in a World-System*. Cambridge: Cambridge University Press.

Taylor, P. (2004) *World City Network: A Global Urban Analysis*. London: Routledge.

Taylor, P. (2011) "Advanced Producer Service Centres in the World Economy." Chapter 3 in P. Taylor, P. Ni, B. Derudder, M. Hoyler, J. Huang and F. Witclox (eds), *Global Urban Analysis: A Survey of Cities in Globalization*. London: Earthscan.

Taylor, P., G. Catalano and D. R. F. Walker (2002) "Measurement of the World City Network." *Urban Studies*, 39: 2367–76.

Timberlake, M. (ed.) (1985) *Urbanization in the World-Economy*. Orlando, FL: Academic Press.

Timberlake, M. and J. Kentor (1983) "Economic Dependence, Overurbanization, and Economic Growth: A Study of Less Developed Countries." *The Sociological Quarterly*, 24 (Autumn): 489–507.
Vinciguerra, S., K. Frenken and M. Valente (2010) "The Geography of Internet Infrastructure: An Evolutionary Simulation Approach Based on Preferential Attachment." *Urban Studies*, 47(9): 1969–1984.
Walton, J. (1977) "Accumlation and Comparative Urban Systems: Theory and Some Tentative Contrasts of Latin America and Africa." *Comparative Urban Research*, 5(1): 5–18.
Walton, J. (1982) "The International Economy and Peripheral Urbanization." pp. 119–35 in N. Fainstein and S. Fainstein (eds), *Urban Policy Under Capitalism*. Newbury Park, CA: Sage.
Wasserstrom, J. (2008) *Global Shanghai, 1850–2010: A History in Fragments*. New York: Routledge.

7.4
How individuals shape global production

Frederick William Lee

The contemporary literature on the organization of transnational production is dominated by firm- and meso-level research such as the global commodity and value chains paradigm. In economic sociology the global value chains (GVCs) paradigm has characterized the contemporary literature and defines the extant knowledge. GVCs research emphasizes the economic motivations for participating in the world-economy and considers the unequal distribution of value in parallel with the distribution of labor within global production networks as sociogenic and representative of the unequal power relations within the world-economy.

In recent years, many Chinese production and manufacturing firms (e.g., OEMs) have achieved disintermediation in the way that they participate in the world-economy. The Chinese export manufacturing industry has become prominent for achieving success in industrial upgrading and disintermediation at the global level, and this is primarily achieved through industrial learning, such as the evolution from OEM to global original design manufacturing (ODM), and then brand development, such as the evolution from ODM to original brand manufacturing (OBM), for example, Haier and Hisense in the electronics sector. Chinese firms have also achieved industrial upgrading through the acquisition of foreign firms, for example, Lenovo's buyout of IBM's personal computing business. Many Chinese firms, including both large enterprises and small and medium sized enterprises (SMEs), have achieved end-to-end disintermediation of their export supply chains, and a prominent example is the Chinese computer firm Lenovo, a partially listed state-owned enterprise (i.e., a "red-chip" corporation), which has achieved global end-to-end disintermediation through retailing Chinese-made computers direct from its factories to global end consumers through internet retailing. Lenovo's case is typical of the large Chinese firms (typically red-chip) and industrial upgrading. However, what is more interesting is that small privately owned Chinese enterprises have also achieved global end-to-end disintermediation of their export supply chains, and one prominent way that this has been achieved is through the immigrant entrepreneurship of the Chinese.

A flaw of the extant knowledge is that sociological studies of GVCs have been firm-centric and have not considered the significance of individuals and their agency within the global organization of firms and production. In fact, GVCs analysis, which has been firm-centric, posits that firms seek out industrial upgrading to "move to more profitable and/or technologically sophisticated capital- and skill-intensive economic niches" (Gereffi 1999: 51–52). Hence, the research has premised economic explanations of industrial upgrading and disintermediation (which can be understood as a variant of industrial upgrading). This is a problem, because the way in which the world-economy is organized is as much sociogenic as it is economic, and Bair (2005) argues that the literature should focus more attention on the social nature of the GVCs within the contemporary world-economy through contextualizing chains within the wider social, historic, and political-economic conditions in which they exist. For instance, explanations of disintermediation may be enriched beyond the economic considerations of firms to incorporate the agency of entrepreneurs and the impact of social forces that shape the agency of entrepreneurs. This is important, as social forces have already been demonstrated to be significant for determining the participation of firms in the world-economy. For instance, dialectal and provincially derived ties have been found to impact upon the organization of the Chinese furniture export supply chain throughout the business networks of the Chinese Diaspora (Lee 2010).

As GVCs researchers posit that economic motivations, such as the retention of value and industrial learning, motivate firms to seek out industrial upgrading and disintermediation, they have largely excluded what motivates the individuals that control the firms and operate within GVCs to seek out disintermediation. This is inadequate for understanding how contemporary supply chains evolve, because the disintermediation of GVCs may result from non-economic forces such as immigrant entrepreneurship. The agency of individuals—that is, the volition of entrepreneurs—is highly pertinent to global supply chains, and the decisions they make concerning their personal lives can shape transnational business networks, the organization of global production and the distribution of labor and goods as well as profit. Lee (2010) demonstrated that non-economic motivations, such as personal preference for the "lifestyle" in Australia, may be the ultimate motivation of Chinese immigrant entrepreneurs in the Sydney furniture industry who have facilitated the disintermediation of the Chinese export industry (Lee 2010: 53). This means that firms are not the only agents that can shape the contemporary world-economy, and the agency of individuals shape the concatenation of economic actors and value inputs in the contemporary world-economy. Thus, the significance of individual agency needs greater recognition in the sociological literature on GVCs and transnational production.

Overall, the GVCs literature should explore the ways that non-firm actors and non-economic motivations operate in the world-economy as a way of enhancing sociological paradigms for understanding the global organization of production. The current GVCs literature does not elucidate the significance of non-firm actors, specifically individuals and their agency, and this is a barrier to fully developing the knowledge of the contemporary world-economy and society. Moving on, the GVCs research should encompass individual agency and the social dynamics that arise from the larger—often global—contexts as key concerns for GVCs scholarship, and this will allow researchers to provide better answers to existing questions and ask new questions about GVCs and their existence at the nexus of the global and the local and as the product of both individual agency and firm- and meso-level dynamics.

References

Bair, J. (2005). "Global Capitalism and Commodity Chains: Looking Back, Going Forward." *Competition and Change*, 9(2): 153–80.

Gereffi, G. (1999). "International Trade and Industrial Upgrading in the Apparel Commodity Chain." *Journal of International Economics*, 48: 37–70.

Lee, F. W. (2010). *The Disintermediation of the Furniture Supply Chain: Chinese Immigrant Entrepreneurs at the Global-Local Nexus.* Bachelor of Arts with Honours thesis, the University of Sydney, Sydney, New South Wales.

7.5
World cities in Asia

Kyoung-Ho Shin

The world city hypothesis formulated by John Friedmann (1986) opened a new strand of research that focuses on the positions of cities, including large Asian cities in the global network of cities (Knox and Taylor 1995). The Globalization and World Cities project (GaWC) and other researchers found that Asian super-cities such as Tokyo, Singapore, and Shanghai have relatively central positions in the global city system (Beaverstock et al 1999; Friedmann 1995; Shin and Timberlake 2000; Smith and Timberlake 2001). Asian world cities have increased their centrality and density of connectedness in the flows of trade, people and information, forming a discernable regional city system, which suggests three theoretical ramifications. First, as Sassen (1991) explains, the dispersion and spatial concentration of the global economic activities happen simultaneously. The rising Asian world cities play the roles of overseeing their globalizing economies, and reveal increased polarization within cities. Second, there are diverse modes of interactions between global force and the roles of nation-states in regional networks. Given that Asia became an important regional axis of world-economy, what Giovanni Arrighi (1998) calls "the rise of Asia," it is important to understand the roles and relations among Asian cities with various historical legacies and vast hinterlands through the contextual lens of rising Asia and the larger structure of the world-system (Taylor 2004). Third, Asian world cities have become spatial sites producing trans-local communities and identities with increased time-space compression that brings new types of political processes and cultural subjectivities (Harvey 1996; Sassen 2007).

Tokyo remains a global commanding center with strong trans-regional connections despite economic stagnation of recent decades (Jacobs 2006). Singapore is a key nodal city of Southeast Asia in intermediary trades and investment for high tech and finance. The Singaporean state's energetic social engineering and its highly visible championing of "Asian values" has had a significant impact on development within Singapore and as a model for other Asian polities (Salaff 2004). Seoul maintains a close link with Tokyo and, as China has emerged strongly in the global economy, it now has stronger connections with Chinese cities. Seoul has come to perform the role of "crucial intermediary" through which manufacturing orders of electronics and clothing from the United States and Japan flow to the low-wage manufacturing sites in free trade zones of Indonesia and China. As a destination of tourism, Bangkok competes with Singapore in inter-regional connectedness in Southeast Asia, and Mumbai (Bombay) has risen to become a center of film and software industries.

China's global cities (Beijing, Guangzhou, Hong Kong, and Shanghai) have rapidly developed despite the efforts of central planners to spread development toward smaller cities and other regions (Ma and Timberlake 2008). Beijing concentrates on political function of foreign policy, alliances and managing relations with the "Great Powers," while Shanghai attracts foreign investment and boasts the third largest stock market in Asia behind Hong Kong and Tokyo (World Federation of Exchange 2008). Shanghai, as China's globalizing gate city, evidences the shift of state policy away from manufacturing production with strong regional and domestic connections toward flexible producers' services of finance, insurance, and real estate (Chen and Zhou 2009). Hong Kong is the city of light manufacturing and financing in Asia, overshadowing Shanghai in its connections beyond the regional economy. The future of Hong Kong will depend in part on political decisions by the governments in Hong Kong and Beijing (Meyer 2000). Rapid integration of China's economy into the global economy has produced a significant increase in inequalities within Chinese cities and among regions.

Asian world cities do not simply stand in a world city system as intermediaries for New York, London, and Tokyo; rather, they compete and cooperate with each other. Research agenda for Asian world cities in the future include: trans-local communities and identities with new political processes and cultural subjectivities; positive and negative impacts of time-space compression on city (re)development from globalization and global restructuring; social consequences of deepening inequality within cities; and mapping out the emerging Asian city system with data on flows and interactions among the cities.

References

Arrighi, G. 1998. "Globalization and the rise of East Asia: Lessons from the past, prospects for the future," *International Sociology*, 13: 59–77.

Beaverstock, J. V., R. G. Smith and P. J. Taylor. 1999. "A roster of world cities," *Cities*, 16: 445–58.

Chen, X. and Z. Zhou (eds). 2009. *Shanghai Rising: State Power and Local Transformations in a Global City*. Minneapolis: University of Minnesota Press.

Friedmann, J. 1986. "The world city hypothesis," *Development and Change*, 17: 69–84.

Friedmann, J. 1995. "Where we stand: A decade of world city research" In P. Knox and P. Taylor (eds) *World Cities in a World-System*. pp. 21–47. Cambridge: Cambridge University Press.

Harvey, D. 1996. "Globalization and deindustrialization: A city abandoned," *International Journal of Politics, Culture & Society*, 10(1): 175–92.

Jacobs, A. 2006. "Embedded localities: Employment decline, inner city population growth, and declining place stratification among Japan's mid-size and large cities," *City & Community*, 5: 269–92.

Knox, P. and P. Taylor (eds) 1995. *World Cities in a World-System*. New York: Cambridge University Press.

Ma, X. and M. Timberlake. 2008. "Identifying China's leading world city: A network approach," *GeoJournal*, 71: 19–35.

Meyer, D. 2000. *Hong Kong as a Global Metropolis*. Cambridge: Cambridge University Press.

Salaff, J. 2004. "Singapore: forming the family for a world city." In J. Gugler (ed.) *World Cities beyond the West: Globalization, Development, and Inequality*, pp. 240–67. Cambridge: Cambridge University Press.

Sassen, S. 1991. *The Global City: New York, London, Tokyo*. Princeton, NJ: Princeton University Press.

Sassen, S. (ed.). 2007. *Deciphering the Global: Its Spaces, Scales, and Subjects*. New York: Routledge.

Shin, K. H. and M. Timberlake. 2000. "World cities in Asia: Cliques, centrality, and connectedness," *Urban Studies*, 37(12): 2257–85.

Smith, D. A. and M. Timberlake. 2001. "World city networks and hierarchies, 1977–97: An empirical analysis of air travel links," *American Behavioral Scientists*, 44(10): 1656–78.

Taylor, P. 2004. "Regionality in the world city networks," *International Social Science Journal*, 56(181): 361–72.

World Federation of Exchange. 2008. "*Latest statistics: YTD Monthly*." Available at: www.world-exchanges.org/statistics/ytd-monthly.

7.6

The Internet and the world-system(s)

Piotr Konieczny

The Internet promised to usher in many revolutions. The debate on what it has and will deliver is still ongoing and consensus is far from sight. Surprisingly, considering the prominent role that information and communication tools (ICTs) have played in delivering the current world-system, the world-system scholars are not among the most active participants of that discussion.

ICTs have accompanied the world-systems since the very beginnings of human history; *homo sapiens* were using mnemonic tools—painting caves, engraving sticks, and stacking stones—for tens of thousands years. Where scribes were recording business transactions in clay tablets filled with cuneiforms, or where printing presses were busily churning copies of the Bible and other works, the elites—not the scribes themselves, but rather, the priests, the nobility, and the kings who ordered them—reaped the benefits from the power given to them by their monopoly on the flow of information, propaganda, and censorship.

However, such monopoly was useful not only to individuals; it empowered the entire world-system. Commenting on the birth of the modern world-system, Wallerstein (1999) noted that the technology, notably advances in transportation and communication, was one of the major factors that allowed Europe—later, the West—to become the system's core region. As Pomeranz (2000) wrote, it was a time of Great Divergence; a time of improvements in maritime technology and the spread of printing presses throughout Europe. The former provided the tools needed to break the Middle Eastern lockdown of European trade. As European merchants and soldiers gained direct access to Americas and Asia, Europe was propelled from a periphery into a nearly unchallenged hegemon of the new world-system. The latter resulted in a growing deluge of steadily cheaper printed materials; the resulting growth in literacy transformed Europe into the one place in the world where the notion of progress became normal. The technological advancements, having found a receptive ground in a society, began a virtuous spiral: after printing came the telegraph, then the telephone, radio, television ... All those ICTs were first developed and deployed in the core, which had decades if not centuries of near monopoly on their use.

On the surface, the newest branch of ICTs, the Internet, is no different. It began as a US military project (ARPANET). It relies on expensive technology and skills that are hard to come by in the poor periphery, and it is most widely used in the developed core.

Yet while existence of a global digital divide cannot be denied (Chen and Wellman 2004), there are some interesting counter-trends underneath the surface. The Internet is no longer the bastion of the English-speaking West. According to Internet World Stats, in mid-2004, 26 percent of English-speakers had Internet connection, and constituted 36 percent of Internet users, way ahead of the Chinese, of whom only 8 percent had Internet access, constituting 13 percent of Internet users. In late 2009, Internet penetration among English-speakers was 39.5 percent and they accounted for 27.7 percent of the Internet population, while in China, the Internet has spread to almost 30 percent of the population, with Chinese accounting for 22.6 percent of the internauts worldwide. In 2007, Arabic became one of the 10 most widely used languages on the Internet, with the number of Arabic Internauts growing twice as fast as that of the Chinese, at a staggering rate of 2500 percent. According to Alexa, in 2009, two Chinese sites—Baidu, a search engine, and QQ, a communication portal, have broken into the Internet Top 10 most popular sites, previously an English-only list. At least 80 percent of articles in Wikipedia, the online encyclopedia, are in languages other than English.

The Internet is still evolving; sites such as Wikipedia, YouTube, and Facebook are just a few years old. Semiperiphery and periphery countries are catching up to the cores, and may be getting ahead by bypassing obsolete ICTs. In countries with weak infrastructure, landlines are rare—but mobile phones use has surged, as putting up cell towers is more efficient then laying cables. Where the Internet has been introduced only recently, people are beginning their cyber-experience not with email or static websites but with Web 2.0 applications like social networks.

The Internet is not limited to state or corporate actors. As it is becoming cheaper and more user-friendly, it is increasingly used by individuals and transnational NGOs focusing on the inefficiencies of the modern world-system. Traditional authorities are trying to retain control, and their efforts should not be underestimated (Lessig 2006), but if history is any guide, the growth of literacy and access to ICTs have been tied to empowerment of previously underprivileged actors. And when we consider the Internet's efficiency in asynchronous many-to-many communication, it could be argued that no ICT developed since the printing presses has been so versatile, and friendly to individuals and small groups, as the Internet.

History shows us that technology was one of the key forces that helped to usher in the current world-system. The global digital divide still tilts the balance to the core, but the gap is closing, as the Internet, still evolving, is rapidly penetrating the peripheries. When better ships and printing presses developed in a periphery, they propelled it toward centuries of core hegemony. Since then, the core has had almost absolute control over ICTs. The Internet is breaking that pattern. As the economies of regions such as China and India are churning on, and their populations are rapidly becoming digitally literate, could we be witnessing another periphery-core shift? We may find it beneficial to look closer at how this is shaping the current—and the future—world-system.

References

Chen, Wenhong and Wellman, Barry (2004). "The global digital divide-within and between countries". *IT & Society* V1, 17.

Lessig, Laurence (2006). *Code: Version 2.0*. New York: Basic Books.

Pomeranz, Kenneth (2000), *The Great Divergence: China, Europe, and the Making of the Modern World Economy*. Princeton, NJ: Princeton University Press.

Wallerstein, Immanuel Maurice (1999). *The End of the World as We Know It: Social Science for the Twenty-First Century*. Madison: University of Minnesota Press.

8. Globalization and distribution

8.1

Globalization: Theories of convergence and divergence in the world-system

Kelly F. Austin, Laura A. McKinney, and Edward L. Kick

Introduction

Theories of globalization and its causes and consequences are a dominant area of scholarship in macro-comparative sociology. A key focus is the degree to which expansion of the world-system causes the economic, military, political, and cultural conditions of nations and interrelations among them to converge or diverge. Some contend that there are greater international connections among states and as a result, there are increasing similarities in state attributes on the domestic level. For example, many scholars point out that as a result of converging international linkages, nation-states exhibit a great deal of isomorphism in their structures and policies (e.g., Boli and Thomas 1997; Hafner-Burton and Tsutsui 2005; Meyer et al 1997). Indeed, these approaches see globalization as a homogenizing force; escalating flows of people, information, and commodities increase various forms of integration, producing a single world society or global culture (Meyer et al 1997).

As an example of an alternative perspective, some emphasize that despite increasing integration, including economic realms such as international trade, international lending, and cross-boarder mergers and acquisitions, the degree of structural inequality between nations has only continued to increase over time (Robinson 2004). Indeed, despite rising interconnectedness and the adoption of common policy scripts, nation-states occupy similar hierarchical positions today as they did 40 years ago (see Kick et al 2011; Snyder and Kick 1979). Much of the world-system remains fragmented, as less-developed or periphery nations continue to have fewer and less beneficial international linkages than more-developed or core nations (e.g., Beckfield 2003, 2010). It is important to note that these perspectives address the degree of convergence or divergence across two dimensions: the interconnections among states and the domestic attributes of states. Although the distinction between these two dimensions is often taken for granted in the theorization, the convergence/divergence debate as it applies to international linkages *and* domestic attributes represents one of the central points of inquiry among globalization and world-system scholars, thus signifying an issue that deserves our further theoretical and analytical scrutiny in this essay.

Convergence in the world-system

Perspectives highlighting patterns of convergence in international linkages and nation-state attributes in the world-system include modernization theories of development and world-polity approaches to globalization. The modernization school is a historical product of several events in the post-World War II era, including the rise of the United States as an international superpower and the disintegration of European colonial empires in Asia, Africa, and Latin America. These dynamics led nation-states to search for optimal models for development (So 1999). With significant support from the US government, private foundations, and international organizations such as the World Bank, the development "movement" spread around the world (Stiglitz 2007).

Modernization theory adopted both evolutionary and functionalist approaches in an effort to illuminate the development experiences of poor countries. Functionalist perspectives center on interrelations and interdependencies in institutions in society (e.g., Parsons 1951), and evolutionary theories (e.g., Durkheim 1893; Tonnies 1887) see social change as unidirectional, moving societies from a primitive to an advanced state, asserting that this transition is fundamentally favorable. Thus, modernization perspectives argue that tradition is one of the greatest barriers to modernization; traditional ways block the development of scientific rationality, the employment of advanced technologies, and specialized integration into the economic world-system (Rostow 1960). In essence, poor nations can benefit by adopting the domestic conditions found in Western states, as modernization is perceived as a phased, homogenizing and Americanizing process (So 1999).

Modernization perspectives thus represent a convergence standpoint for nation-state attributes as they advocate Western cultural, political, and economic models for development, such as the spread of achievement motivation and policy scripts that call for increased economic integration, often in the form of free-trade or increased foreign investment (Inkeles and Smith 1974; McClelland 1964; Rostow 1960). Additionally, this approach, therefore, implies rising international linkages through the incorporation of domestic economies into global systems of exchange. A contemporary perspective that highlights these trends is the world-polity perspective. This strand of theorization points to increased interstate contact through the increased isomorphism of institutions, state structures, and policies across nations (Meyer 2000; Meyer et al 1997). Accordingly, worldwide structural convergence is occurring both across states and within them (e.g., Bell 1973; Meyer et al 1997). The diffusion and convergence of institutional forms has been observed for a number of outcomes, including scientific research (Schoefer et al 2000), worldwide environmental regimes (Frank et al 2000), and technology (Bell 1973).

Another cultural interpretation that adopts an alternative position is the world civilizations hypothesis. This line of thinking contends that there is no single global culture, but rather several distinct cultural groups that have a high degree of integration within each world-cultural region. For instance, Huntington (1996) argues that the world is divided into roughly nine different civilizations—Western, Latin American, African, Islamic, Chinese, Hindu, Orthodox, Buddhist, and Japanese—each with a distinct value system. Cultural differences are expressed here as civilizational divides; thus this perspective recognizes both convergence and divergence in the world-system, expecting convergence within civilizations and divergence across them (Huntington 1996).

Taken together, these comparative perspectives emphasize that the world-system is characterized by a trend toward convergence both in interstate contact and in domestic conditions (although for Huntington [1996] this would exist on a civilizational level), and this can be evidenced, for instance, in the growth of interstate relations and adoption of similar fiscal policies, institutional arrangements, and perspectives on individual freedoms and rights (Meyer et al 1997). Whether or not these structural adoptions result in actual changes is debatable, as some studies point to the purely symbolic character of nations' participation in global society (Hafner-Burton and Tsuitsui 2005). Nonetheless, world-polity theorists maintain that globalization leads to increasing cultural and political contact and isomorphism in institutional and organizational forms among nations, and insights from modernization theory contend that cultural and political convergence will lead to increased economic convergence as well. In fact, one recent assessment using a within-regions approach finds that economic development is indeed associated with shifts away from traditional norms and values toward increasingly rational and participatory values (Inglehart and Baker 2000). Interestingly, some theorists (e.g., Robinson 2004) see the homogenization of national identities and state structures as part and parcel of the globalizing trend of transnational capital accumulation. This approach views societal arrangements and state institutions as constructed in accordance with the drive of capitalism to become both more intensive and extensive in the production and accumulation of wealth. Thus, any observed similarity of structure across nations or of interstate contact is the direct result of the expansionary imperative of capital (Lenin 1917; Mezaros 1995).

Divergence in the world-system

While international exchanges of all sorts are prominent among nation-states in the world-system, scholars point out that this may be producing divergent rather than convergent patterns of interstate interaction and attendant outcomes. The "divergence" camp of theorists (e.g., Giddens 2000; Guillen 2001; Held et al 1999) critique the claim that nation-states converge as part of the globalization process in the modern era. These scholars view globalization as a dialectical process that can result in opposing forms and increased fragmentation (Giddens 2000; Held et al 1999). For instance, Giddens (2000: 31) argues the events of globalization, "often produce divergent or even contrary occurrences" among nation-states.

World-systems theory emphasizes divergent development dynamics among nation-states, which are attenuated and enabled by uneven interstate interactions. The strata of the world-system—the core, semiperiphery, and periphery—represent functionally distinct positions in the global division of labor and world-economy that are the direct result of cumulative power advantages in an unequal capitalist world-system. This hierarchy allows the more-developed nations of the core to use their advantaged political-economic power to maintain both their privileged status and the inferior positions of non-core nations (Chase-Dunn 1998; Wallerstein 1974). Despite some relative gains, structural inequality among nation-states has remained fairly constant over time, and in the most recent decades, international inequality has actually increased (Robinson 2004). Additionally, this perspective emphasizes that states have been interacting with one-another since at least the colonial era, but that these patterns in interaction are not even. Rather, core groups have always maintained higher levels of international contact, often through political-military conquest and the securing of power relations. Thus, the world-system perspective emphasizes patterns of divergence across and among states as central to the operation of the capitalist world-system.

From this perspective, a world-polity or global culture may still be in evidence, but it develops within the existing hierarchical organization of nation-states (Boswell and Chase-Dunn 2000). A number of world-system analysts argue that world orders are established by core hegemons to benefit elite classes and core states, and that political organizations simply become "boards of directors for ruling states" (Boswell and Chase-Dunn 2000: 238). Core nations have powerful international ambitions for the accumulation of wealth, and establish political and cultural representation in every corner of the world in order to monitor local investments and labor conditions, and expand their sphere of influence. Structurally, this implies a dominance of the core over the polities of states, with core nations being highly integrated, and semiperiphery and periphery nations engaging in limited participation with each other.

Although world-system scholars identify patterns of divergence, this perspective is often criticized for focusing exclusively on economic factors. A body of literature has emerged that combines economic and non-economic data to assess structural interrelations among states and related domestic outcomes, such as economic growth (e.g., Snyder and Kick 1979; Kick et al 2011; Van Rossem 1996). As relationships of divergence or convergence can be evidenced through patterns of nation-state interaction, scholars often employ network analysis across multiple dimensions to assess the structure of relational ties among states and resultant changes to domestic conditions.

Some studies utilizing network techniques examine patterns in nation-state interaction across multiple indicators; as a landmark study, Snyder and Kick (1979) utilize network data for trade flows, military interventions, diplomatic relations, and conjoint treaty memberships for the 1960–65 period. Their results indicate significant levels of hierarchy and inequality across all four dimensions, suggesting that the world-system features a high degree of divergence across contacts in economic, military, cultural, and political realms. Additionally, this research finds that states are also distinguished by divergent levels of economic growth. A replication of this study for the contemporary period (Kick et al 2011) supports the claim that these patterns of divergence are maintained in the world-system over time, as the structural positions of nation-states in the world-system today are largely consistent with the arrangements identified by Snyder and Kick (1979) over 30 years ago. Thus, this body of research emphasizes patterns of continuing divergence in nation-state interaction and domestic attributes, especially in economic and military spheres.

Convergence and divergence? Embassies and intergovernmental organizations as illustrative networks

Given the emerging importance of non-economic international linkages, we consider the role of embassies and intergovernmental organizations in transmitting political and cultural information between nation-states. An intergovernmental organization (IGO) is an organization composed primarily of sovereign states (referred to as member states), or of other intergovernmental organizations. Intergovernmental organizations differ in function, membership, and membership criteria. Some IGOs are developed to fulfill a need for a neutral forum for debate or negotiation to resolve disputes, such as the United Nations. Others are developed to carry out mutual interests in a unified form. The network of IGOs has been shown to influence patterns of democratization, international trade, neoliberal restructuring, international conflict, and social movements.

Embassies also encompass an important aspect of global ties, especially in an era of globalization, insofar as they are established to represent governments in foreign locales. In general, embassies convey their countries' policies to international organizations and states abroad, while negotiating agreements and treaties on a variety of issues, ranging from trade to weaponry transfers (Feigenbaum 2001; Gills 2002). Embassies also perform a diversity of other important functions, such as coordinating and supporting international activities of other government agencies, hosting official visits, promoting and assisting their countries' business interests, protecting their citizens abroad, managing international educational exchanges, accepting and adjudicating requests for visas, and managing the allocation of resources for foreign relations (Feigenbaum 2001).

Embassies and IGOs as institutions shape the actions of foreign states and their economic interests, thus exerting influence in cultural, political, military, and economic spheres (Beckfield 2010; Gills 2002). Political-cultural institutions such as these do more than orient action; however, they also constitute or legitimate actors. Embassies and IGOs, therefore, represent informational ties between nations, transmit influence to receiving nations, and signify the legitimacy of sending nations. The breadth of these activities has grown exponentially over the last century (Beckfield 2010; Feigenbaum 2001). In particular, there was a dramatic rise in the number of embassies and IGOs in the post-World War II era, not only due to an increase of "watch dog" efforts during this time, but also widespread decolonization led to the development of institutions that could exert long-distance political influence in foreign places. Indeed, whether or not the dramatic rise in these institutions signifies increased interstate contact is up for debate; some argue that nations were more connected in various dimensions in the 1800s and earlier 1900s, as this historical period is marked by direct colonial ties which could have linked states more significantly than the political-cultural linkages that we see today (e.g., Waltz 1999).

The recent study by Kick and colleagues (2011) utilizes network data for both intergovernmental organizations and embassies. Their results reveal that the embassy network is relatively sparse, however, with the strongest connections amongst core and Western European nations. While almost every nation has ties to the core/Western European groups, the periphery groups lack strong embassy relations among each other. In fact, only six nations (the isolates) of a sample of 166 are not linked to the international embassy network. Thus, there is significant potential for convergence in domestic attributes, as the majority of nation-states in the world-system are integrated into the global field of embassy relations. Yet, the embassy density matrix demonstrates uneven levels of embassy integration, with core/Western European groups having the highest levels of international participation and the periphery groups having much lower levels of participation.

Considering the network of IGO ties, Kick and colleagues (2011) find that although this system is much denser than the embassy network, the IGO density matrix still illustrates a high degree of fragmentation. The results of the blockmodeling analysis also reveal noticeable clustering along regional lines (Kick et al 2011). In order to fully assess the degree of convergence or divergence in the world-system, it is useful to consider changes in political-cultural networks over time. Although many studies report growth in the number of IGOs and embassies during the twentieth century, only a few studies have been able to examine changes in the structures of these networks over time.

One recent longitudinal study by Beckfield (2010), which utilizes IGO and state data dating back to 1820, finds that states have indeed become more interconnected through common memberships in IGOs; nearly every state holds at least one IGO membership in common with every other state in the world-system. Additionally, inequalities in the level of state involvement in IGOs are shown to be deceasing substantially over time. However, Beckfield (2003, 2010) also finds that the network of IGOs has become more dense and centralized, where, while states are growing more similar in the number of IGOs they belong to, they increasingly belong to different IGOs. Overall, there is evidence of growing disintegration, uneveness, and regionalization over time in the global IGO system. Thus, while states are increasing in levels of participation and interaction with one-another, the organizations that link them are becoming more fragmented over time. Additionally, Beckfield (2010) analyzes the different types of IGOs separately (general purpose, social/cultural, economic, and military/political), and finds that the economic IGO network is especially fragmented, largely along regional lines.

Taken together, these examples illustrate that while the amount of world-polity activity as measured by embassies and IGOs has increased dramatically over time, these world-polity networks are largely hierarchical and divided along regional lines. Much of this growing fragmentation is a newer trend, as political-cultural networks identified here were largely static in the late nineteenth and early twentieth centuries. After 1945, the world-polity grows more disintegrated and structurally uneven. Both of the current embassy and IGO networks described by Kick and colleagues (2011) and Beckfield (2010) resemble a star-shape, with Western European core groups sharing many political-cultural ties to each other and more peripheral groups, but with peripheral groups exhibiting few ties to other peripheral groups. Additionally, these studies emphasize the regionalization of cultural-political ties as clustering for both types of networks is evidenced within major geographical spaces. In particular, Kick's findings reveal that sharing a geographic region often predicts the presence of embassy ties among states, and Beckfield reveals that connections among states in the IGO network are strengthened significantly within regions. Not surprisingly, many of the more recently developed IGOs are regionally exclusive in nature, such as the European Union and the Association of Southeast Asian Nations.

It has been argued among world-polity scholars that political-cultural networks tend to be more evenly distributed, as there is a significant degree of political and cultural convergence among nation-state embassy and IGO ties in the world-system. Thus, it is possible to argue that there is evidence for mounting convergence in the world-system as the size of these networks has grown over time. Yet, the organization of international embassy relations and IGOs reveals that political-cultural networks exhibit a significant amount of divergence, which is therefore likely to produce uneven state attributes. If convergence were the dominant trend, we would also see a proliferation of embassy and IGO ties across the periphery groups. A relative lack of connections across non-core groups signifies the relative maintenance of structural divergence in the world-system.

Further, there is strong evidence for the civilizations hypothesis as this body of research finds considerable fragmentation in embassy and IGO networks along regional lines. It is also important to recognize that historical patterns in economic development are strongly regional, for example, with Western Europe/North American regions continuing to exhibit the highest degree of economic power and Sub-Saharan African nations exhibiting enduring legacies of deep impoverishment. Additionally, health outcomes research indicates that despite some increased convergence in physical well-being measures over time, such as life expectancy, the majority of the world-system remains marked by extreme disparities in quality of life, with 2.5 billion people lacking improved sanitation, 1.1 billion of which are still defecating in the open (WHO 2010). Thus, hierarchy and inequality most prominently characterize the structure of interstate relations and resultant national-level conditions. Economically powerful core groups, largely in the North Atlantic region, dominate the global embassy and IGO networks, and periphery groups fail to converge across each other. Thus, core nations are able to use political and cultural influence to maintain their advantaged status in the world-system. For example, the majority of embassies in the world-system are either hosted in or established abroad by core nations, therefore a rise in the number of embassies signals increasing global dominance by powerful states, rather than convergence which evenly integrates and affects the rest of the system.

Despite this, the proliferation of similar institutional forms and the propagation of common policies do signal the possibility of world-system convergence toward Western cultural and developmental models. But to what end? The world-system is a capitalist world-system, and many of these shared policies and institutional forms make it easier for states to interact economically, and function to draw more states into the networks of production and exchange. Capitalism depends on the increasing accumulation of capital, and the expansionary imperative demands that more states are integrated into international networks and exhibit similar state structures (Bunker and Ciccantell 2005; Mezaros 1995). Thus, capitalism and globalization are seemingly twin dimensions, where common policies and similar cultural perspectives across states make the world "safer" for core capital accumulation.

While many emphasize the fundamentally new components of globalization (e.g., Robinson 2004; Sklair 2001), classical world-system theory reminds us that states have had political, cultural, military, and economic exchanges on a global level since at least before the fifteenth century (Wallerstein 1974). States are integrated into global networks and exhibit some similar institutional attributes in order to increase the possibilities for capital growth for the powerful and privileged core (Lenin 1917). Hence, while nation-states and supporting institutions may have become more similar in many political and cultural forms with increasing contact, these elements of world society are largely transmitted from a small group of powerful actors that seek to maintain their structural advantages. Additionally, institutional pressures are increasingly likely to be channeled through regional polities, which might signal local convergence and global divergence. Many of these regional polities could also be a direct response to globalizing pressures; nation-states may seek protection from international capital or attempt to increase their autonomy and relative power through regional alliances.

The world-system is thus marked by patterns of divergence *and* convergence embodied in globalization processes. However, the greatest evidence for convergence across and among states constitutes only a small group of Western nations. More states are being pulled into international networks and changing some of their domestic attributes, such as institutional structures, but this largely occurs through and by core group influence. Periphery groups have little contact among each other, and economic inequality among nation-states has only widened over time. Most prominently, the world-system is systematically and historically typified by structural divergence and regional clustering. However, core nations would be unable to maintain their privileged status without a high degree of cultural and political assimilation among dependent semiperiphery and periphery states; thus patterns of convergence and divergence are both central and functional to the operation of the capitalist world-economy. Indeed, capitalist development is characterized by dialectical processes, as the expansionary imperative of capital demands both competition and coordination among nations in the world-system.

Advancing theoretical synthesis on globalization

We began by recounting the view that globalization dynamics are resulting in convergent patterns of interstate linkages and institutional isomorphism among nations in the world-system. This line of reasoning appears to be most consistent with modernization theory and world-polity perspectives. Yet, other theorists such as Huntington (1996) argue these patterns of convergence are most pronounced within civilizational divides, with greater divergence across civilizations resulting from conflicting worldviews, religions, and cultures. World-system theorists posit continuing divergence across and among states, due to unequal power, currency, and status differentials that uphold core nations in superior positions of power, while relegating the bulk of the world into structurally disadvantaged semiperiphery or periphery positions. Finally, a new group of globalization scholars, echoing themes of early theories of imperialism, conclude that any detectible degree of convergence across nations is an agent of capital accumulation and exists exclusively to facilitate its expansion (e.g., Robinson 2004). That is, greater similarity of domestic institutions, market economies, and state structures among nations, as well as increasing interstate contact, assists ever-growing trends of capital accumulation in the core.

Certainly these analytical models can be encapsulated by a world-system theoretical framework that articulates the logic of production, consumption, and division of labor in the global system, and the nation-states that are constituent parts of it. In addition, world-systems theory and its extensions in particular explain how the forces of the system may produce dynamics of convergence and divergence across hierarchical categories of the world-system. Indeed, our empirical examples of the global networks of embassies and IGOs demonstrate both patterns of convergence and divergence, as the majority of nations have adopted these forms of political-cultural integration and resultant state structures, but the organization of these networks remain highly fragmented and hierarchical. Further, it is unclear if like institutional forms prompt homogenization of domestic conditions, or whether participation in world society is purely symbolic (Hafner-Burton and Tsuitsui 2005). We hope that future analyses will build on this argument by continuing to draw from the full range of possible theoretical explanations, viewing them for their synergies instead of their incompatibilities. Thus, we advocate a synthetic theoretical approach and companion modeling. We recognize that a common disciplinary approach is to view strands of theoretical approaches as competing, rather than complementary, paradigms. But we are far less certain that this approach to theorization is optimal for achieving maximum explanatory power.

Theoretical synthesis as opposed to a disparate approach to theory construction and refutation/validation has been advocated in a number of sociological treatments over the last two decades. Although our list is incomplete, we draw attention to the work of Ritzer (1991), Collins (1989), Giddens (1996), and Habermas (1984). Aiding in theoretical synthesis, we contend that multiple-network analysis is especially well-suited to examining the nature of global linkages and hierarchical domains of different scopes, and structural equation modeling is a methodology that can assist in the blending of multiple perspectives. As elaborated by Bollen (1989), structural equation modeling techniques allow researchers to specify direct and indirect effects, include latent (unmeasured) concepts, achieve unbiased estimates even when independent variables are intercorrelated, and judge the "fit" of the model specified to the data provided, all of which promote theory synthesis rather than refutation/validation of specific hypotheses. In concert, we suggest a wedding of the central themes examined in the present essay, as theoretical synthesis and compatible modeling techniques provides a far more complete picture of globalization dynamics than does treating perspectives as competitor theories.

References

Beckfield, J. (2003) "Inequality in the world polity: the structure of international organization", *American Sociological Review*, 68: 401–24.

Beckfield, J. (2010) "The social structure of the world polity", *American Journal of Sociology*, 15: 1018–68.

Bell, D. (1973) *The Coming of Post-Industrial Society*, New York: Basic Books.

Boli, J. and Thomas, G. M. (1997) "World culture in the world polity: a century of international non-governmental organization", *American Sociological Review*, 62: 171–90.

Bollen, K. (1989) *Structural Equations with Latent Variables*, New York: John Wiley & Sons.

Boswell, T. and Chase-Dunn, C. (2000) *The Spiral of Capitalism and Socialism: Toward Global Democracy*, Boulder, CO: Lynne Rienner.

Bunker, S. G. and Ciccantell, P. S. (2005) *Globalization and the Race for Resources*, Baltimore, MD: Johns Hopkins University Press.

Chase-Dunn, C. (1998) *Global Formation*, Oxford: Rowman & Littlefield.

Collins, R. (1989) "Sociology: prescience or antiscience?", *American Sociological Review*, 54: 124–39.

Durkheim, E. (1893) *The Division of Labor in Society*, New York: The Free Press.

Feigenbaum, H. B. (2001) "Globalization and cultural democracy", *Center for Arts and Culture: Art Culture and the National Agenda*, Washington, DC: The George Washington University.

Frank, D. J., Hironaka, A. and Schofer, E. (2000) "The nation-state and the natural environment over the twentieth century", *American Sociological Review*, 65: 96–116.

Giddens, A. (1996) *In Defense of Sociology*, Cambridge: Polity.

Giddens, A. (2000) *Runaway World: How Globalization Is Reshaping Our Lives*, New York: Routledge

Gills, B. K. (2002) "Democratizing globalization and global democracy", *The Annuals of the American Academy*, 581: 158–71.

Guillen, M. F. (2001) *The Limits of Convergence: Globalization and Organizational Change in Argentina, South Korea, and Spain*, Princeton, NJ: Princeton University Press.

Habermas, J. (1984) *The Theory of Communicative Action, Vol. 1*, Boston, MA: Beacon Press.

Hafner-Burton, E. M. and Tsutsui, K. (2005) "Human rights in a globalizing world: The paradox of empty promises", *American Journal of Sociology*, 110: 1373–1411.

Held, D., McGrew, A., Goldblatt, D. and Perraton, J. (1999) *Global Transformations*, Stanford, CA: Stanford University Press.

Huntington, S. P. (1996) *The Clash of Civilizations and the Remaking World Order*, New York: Simon & Schuster Press.

Inglehart, R. and Baker, S. (2000) "Modernization, cultural change, and the persistence of traditional vales", *American Sociological Review*, 65: 19–51.

Inkeles, A. and Smith, D. H. (1974) *Becoming Modern: Industrial Change in Six Developing Countries*, Cambridge, MA: Harvard University Press.

Kick, E. L., McKinney, L., McDonald, S. and Jorgenson, A. K. (2011) "World-System Position: A Network Analysis of Nations?" in J. Scott and P. Carrington (eds.) *Sage Handbook of Social Network Analysis*, Thousand Oaks, CA: Sage Publications.

Lenin, V. I. (1917) *Imperialism, the Highest Stage of Capitalism: A Popular Outline*, Peking: Foreign Language Press.

McClelland, D. (1964) "Business Drive and National Achievement", in A. Etzioni and E. Etzioni (eds) *Social Change*, New York: Basic Books.

Meyer, J.W. (2000) "Globalization: sources and effects on national states and societies", *International Sociology*, 15: 233–48.

Meyer, J. W., Boli, J., Thomas, G. M. and Ramirez, F. O. (1997) "World society and the nation-state", *American Journal of Sociology*, 103: 144–81.

Mezaros, I. (1995) *Beyond Capital: Towards a Theory of Transition*, New York: Monthly Review Press.

Parsons, T. (1951) *The Social System*, Glencoe, IL: Free Press.

Ritzer, G. (1991) *Metatheorizing in Sociology*, Lexington, MA: Lexington Books.

Robinson, W. I. (2004) *A Theory of Global Capitalism: Production, Class, and State in a Transnational World*, Baltimore, MD: Johns Hopkins University Press.

Rostow, W. W. (1960) *The Stages of Economic Growth: A Non-Communist Manifesto*, 3rd edn., London: Cambridge University Press.

Schofer, E., Ramirez, F. O. and Meyer, J. W. (2000) "The effects of science on national economic development, 1970 to 1990", *American Sociological Review*, 65: 866–87.

Sklair, L. (2001) *The Transnational Capitalist Class*, Oxford: Blackwell Publishers Inc.

Snyder, D. and Kick, E. L. (1979) "Structural position in the world system and economic growth, 1955–70: a multiple-network analysis of transnational interactions", *American Journal of Sociology* 84: 1096–1126.

So, A. Y. (1999) *Social Change and Development: Modernization, Dependency, and World-Systems Theories*, Newbury Park, CA: Sage Publications.

Stiglitz, J. (2007) *Making Globalization Work*, New York: Norton Press.

Tonnies, F. (1887) *Community and Society*, Cambridge: Cambridge University Press.

Van Rossem, R. (1996) "The world-system paradigm as general theory of development: a cross-national test", *American Sociological Review*, 61: 508–27.

Wallerstein, I. (1974) *The Modern World-System, Volume I: Capitalist Agriculture and the Origins of the European World-Economy in the 16th Century*, New York: Academic Press.

Waltz, K. N. (1999) "Globalization and governance", *Political Science and Politics*, 32: 693–700.

World Health Organization (WHO) (2010) *World Health Statistics 2010*, Available at: http://www.who.int/whosis/whostat/EN_WHS10_Full.pdf

8.2

Social stratification and mobility

National and global dimensions

Timothy Patrick Moran

The degree to which a society is described by rigid categories of stratification and/or by more open routes of social mobility, and the extent to which these characteristics are changing over time, has been of persistent interest to social scientists since the beginning of the twentieth century. Although notable earlier studies existed, after the seminal publication of *The American Occupational Structure* (Blau and Duncan 1967), stratification and mobility became a burgeoning subfield in North America and Western Europe, developing into a prolific centerpiece of the sociological discipline; a bibliography of work on class structures, status attainment, and the intergenerational transmission of socioeconomic advantage would form a book-length manuscript.

From its inception, however, the study of stratification and mobility has been built on the fundamental premise that discreet, bounded "national" territories, or nation-states, constitute the key unit of analysis. Social forces and interactions that lead to categorical inequality, and to differential abilities for people to engage in social mobility, are assumed (sometimes explicitly, but most often implicitly) to take place primarily or even exclusively within a "national society" demarcated by its national borders. Furthermore, the mainstream social science literature that developed concentrated almost exclusively on the wealthy national societies of North America and Western Europe. With such a focus, much of the existing literature then proceeded to establish as universal the trends and patterns characteristic of these nation-states, even though their combined population constitutes a small fraction of the world's people.

But what does stratification look like from a global perspective? What constitutes social mobility when the unit of analysis is the world as a whole? In this chapter, I first review the study of stratification and mobility as it has existed, that is as a national phenomenon. This literature served as a key linchpin in what came to constitute the "modernization paradigm" that became orthodoxy across the post-World War II social sciences. Yet, while world-systems analysis was born as a theoretical and methodological intervention into this paradigm, studies of stratification and mobility that take a world-historical perspective are still rare. Aside from a few notable exceptions discussed below, the nation-state still predominates both the collection of relevant data, and the scope of its analysis. The second section of the chapter outlines what a world-historical perspective might reveal in terms of global dimensions of stratification and mobility.

National stratification, mobility, and the modernization paradigm

The first studies of stratification and social mobility appeared at the beginning of the twentieth century, and in particular with Sorokin's (1927) influential work emphasizing the mobility of modern/Western societies vis-à-vis "immobile" societies of past history. The field moved into one of centrality in the middle of the century, specifically the pioneering work of Glass (1954) in England, Carlsson (1958) in Sweden, and Rogoff (1953) in the United States, followed by the seminal volume by Blau and Duncan (1967). Blau and Duncan in particular set future agendas for the study of social mobility as they developed the first rigorous statistical models to study these processes empirically, while simultaneously recasting the theoretical problem as well. By focusing on describing the way occupational roles and statuses are allocated within societies, they in fact moved the field away from studying differential levels of mobility as characteristics of social systems, that is away from describing macro-level social mobility itself, and toward the more specific problem of individual status attainment They also ushered in the methodological transformation of mainstream sociology into the quantitative (dominated) social science it is today.

From the beginning, and in the field's subsequent development, the sociological study of stratification and mobility concerned itself with the wealthy countries of North America and Western Europe and was subsumed under what Erikson and Goldthorpe (1992) call the "liberal theory of industrialization." In this sense, the field became sociology's contribution to the broader modernization paradigm that directed the social sciences more broadly. As within the paradigm itself, social mobility research focused on demonstrating how individual mobility is simultaneously a function of industrialization and/or political democracy (or egalitarian ideology more broadly). By implication, the paradigm was also resisting Marxist conclusions of the inevitability of class polarization. Using the first large survey data on status and mobility as it became available in the 1960s, the literature sought to demonstrate empirically that class antagonisms and the "prolitarianization" of the middle classes were exaggerated claims. By contrast, the argument went, modernization leads to high rates of (upward) social mobility and fluid systems of stratification whereby endpoints are not preordained by characteristics at birth. As Tyree et al (1979: 413) explain, "Either industrialization is seen to require the allocation of individuals to roles on criteria at least partially inconsistent with ascription, or political democracy has been seen as requiring widely perceived mobility opportunities for its maintenance."

Crucial to the modernization paradigm, and underlying the literature on social mobility from the start, is the claim that modernity transforms the process through which individuals are allocated within the division of labor. This transition is seen as singular, linear, and national in scope. A more complex division of labor (e.g., as accompanying industrialization) requires a shift away from ascription and toward performance and achievement as universalistic criteria shaping stratification. Indeed, this became the key finding of *The American Occupational Structure* (Blau and Duncan 1967): in mid-century United States, achievement processes (and above all, educational attainment) are more important than ascriptive processes (i.e., family background) in determining one's station in life. In societies with expanding educational opportunities, the association between one's social origin and eventual destination weakens, and society becomes increasingly "open" as "meritocracy" comes to form the basis for social selection.

Concomitant with this overall theoretical approach, an empirical consensus formed around the consistency of these transitions across the wealthy, developed nations. Within the consensus, any cross-national variation in stratification criteria and rates of social mobility can be primarily attributed to the historical phasing of economic development. As summarized by Kurz et al

(1987: 418), "Industrial nations share a large amount of cross-national similarity, over-time constancy (especially in rates of relative mobility), and between-group invariability."

When the analytical gaze is the institutional changes occurring within the wealthy national societies of the twentieth century, then, what do you see? With stratification, the apparent withering away of inequality based on ascribed, birth characteristics as the predominant basis of stratification criteria, replaced by the universalistic criteria of performance and achievement. You see the decline of categorical inequality and the rise of a meritocratic society. For social mobility, only one path seems available: the accumulation of human capital, particularly education and skill. Since ascription-based inequality is increasingly less relevant to stratification processes, access to this mobility path is further assumed to be widely shared.

World-historical dimensions of stratification and mobility

The growing importance of flows of people, capital, and goods across borders is challenging mainstream perspectives within the social sciences that understand social inequality, stratification, and mobility as the outcome of interactions that take place solely within national boundaries. However, a few observers discussed the idea of "transnational" social structure as far back as the 1960s. Robinson (2004: 36), for example, quotes Goldfrank arguing in 1977 that "the study of class structure or stratification on a world level is in its infancy." Nonetheless, this field of inquiry has not developed since Goldfrank's prediction, and almost all discussion of people being integrated into global stratification structures and social life—be it nascent global class formations, or emergent global civil society networks—still largely operates from nation-state centered approaches. The global becomes an exercise in adding up colluding national elements (classes, labor unions, NGOs, etc.) at the level of an international system.

Theorizing stratification from a global conceptualization of class, a small but growing literature has developed around the idea of a "transnational capitalist class," which is most developed in the work of Robinson (2004) and Sklair (2001), but see also recent articles by Kentor and Jang (2004, 2006). Although key distinctions exist between the approaches, the central idea is that neoliberal globalization has created the material conditions for the rise of a world ruling class whose "coordinates are no longer national" (Robinson and Harris 2000: 13). This transnational integration of capitalists and elites provides objective existence, agency, and identity to an emergent class of people that manage global production and accumulation and are now the primary agents of the global economy. As Robinson and Harris (2000: 17) explain, the transnational capitalist class is "class conscious ...; and has been pursuing a class project of capitalist globalization, as reflected in its global decision making and the rise of a transnational state apparatus under the auspices of this fraction."

Yet other aspects of global class formation, for example, labor solidarity in the context of transnational social movements, or the supranational identity of other social groups, are discussed more in their theoretical (and political) potential than their empirical reality (a few notable examples of cross-border collaboration aside, see Evans 2005). Robinson and Harris (2000) argue that a transnational working class is increasingly a class "in-itself," employing the old Marxist distinction, but not yet a class "for-itself." But in Robinson (2004: 28), transnational working class formation is relegated to a footnote, a process described as being "underway" and a "matter to be taken up elsewhere."

Apart from this focused interest in a global ruling class, and from the general mandate of world-systems scholars to constitute a fundamental analytical protest that reflects deep global inequalities, little other work exists that tries to understand stratification and mobility as world-historical phenomena. How different do comparative levels of income, or the relative social positions of people, households, and occupations look as seen from a national versus a global analytical frame? What constitutes social mobility when seen from the world as whole? Who has the relative ability to access the different routes to social mobility, and to what extent over time?

In a recent book, *Unveiling Inequality: A World-Historical Perspective* (Korzeniewicz and Moran 2009), we attempt to theorize these concepts, guided by a key theoretical premise: processes of social inequality, stratification, and mobility have always been global in character. Methodologically, this means that effectively capturing the social interactions that are relevant to understand how the patterns at hand have changed over time requires observations aggregated for the world as a whole.

Table 8.2.1 outlines the key epistemological differences between the national, modernization approach to stratification and mobility that has dominated mainstream sociology, compared to a global, world-historical approach. When taking the world as a whole as the relevant unit of analysis, attention is redirected to the changing processes of institutionalized social interaction that result in inequality and/or mobility. The social attributes used in mainstream social science to "explain" inequality and individual attainment—such as education, skill, gender, race and ethnicity, and so forth—become instead the changing expression of processes of differentiation that entail both within- and between-country inequality. Definitions and standards for determining statuses, and for who constitutes an "elite," "middle-class," or "poor" are drastically redefined when such relative social positions are established using global criteria. Scholars and observers often use such descriptive terms without considering that such categories might be referring to very different income levels on a world scale, and that relative social positions when viewed from a national level look very different when assessed within global stratification.

Global stratification does not equal adding up the "classes" from different countries, as if they all occupied the same objective position in terms of interests. Social locations in global stratification are determined, not primarily by the relative position attained (or achieved) within a given country, but by the position occupied by a given country relative to others. Social inequality and stratification, when seen globally, is shaped by multiple and overlapping (as opposed to singular) transitions from the central importance of certain ascribed criteria in particular historical eras to other systems of categorical exclusion in other eras. Over the course of the last two centuries, we argue, the ascribed criteria that underlies current global inequality is the nation-state itself. That is, your location in global stratification is determined, above all other preordained characteristics of birth, primarily and above all else by the country in which you are born.

Table 8.2.1 Epistemological perspectives in the study of stratification and mobility

	Modernization	*World-Historical*
Unit of Analysis	Wealthy nation-states	World as a whole
Institutional Scope	National	Simultaneously (and in relation) national and global
Basis of Stratification	Singular transitions from ascribed criteria to achievement	Multiple and overlapping transitions from the importance of certain ascribed criteria to others
Paths to Social Mobility (accessibly)	Human capital accumulation (equal in theory)	Human capital accumulation (limited to elite few) National development (limited to certain populations) "Categorical mobility" (via migration)

From such a perspective, nationhood and citizenship have been as much a part of the constitution of the identity of social forces, and/or of the demarcation of their interests, than "class." As Wallerstein (1982: 99) argues "[i]f you were to sit down and do a content analysis of books of history, to see if political conflicts were between or within classes, I think you would find on any rapid checklist that about 95 percent of the conflicts within the world are in fact within classes." From a global perspective, the constitution of class interests often had the (intended or unintended) consequence of constructing nation-states as the crucial criterion of stratification, that is, "the unequal distribution of people across social categories that are characterized by differential access to scarce resources".

To illustrate a more global perspective on social stratification, we conduct an empirical exercise in the book combining within-country distributional information for countries with between-country measures of economic performance (see Korzeniewicz and Moran 2009: Ch. 5). More specifically, the share of income accruing to each country's national population deciles (beginning with the richest 10 percent, and moving through each decile until reaching the poorest 10 percent) is combined with each nation's gross national income to produce an estimate of the per capita income of each country decile. Hence, for every country in the dataset, the procedure yields ten "country deciles," each representing a tenth of the country's population, and each with an estimated per capita income based on the country's income distribution. USA10, for example, represents the wealthiest 10 percent of the population of the United States, and USA1 the poorest 10 percent. The dataset includes country decile information for 85 nations, which we used to calculate the distribution of income for 850 country deciles.

Once the 850 country deciles were resorted from richest to poorest, global income deciles were calculated, creating a global map inequality. Now each household in the sample is located in two income deciles: a country decile (its location within the national income distribution) and a global decile. Comparing these two locations yields interesting results. If global stratification was not defined primarily by national identity—and the high degree of inequality between-nations—then the country deciles should match up fairly closely with the global deciles. The richest 10 percent of each country would more or less form the richest global decile, and so forth on down the scale. Yet these figures are in weak correlation. The populations of most wealthy countries are contained wholly within the two top global income deciles: in the case of the United States, USA10 through USA4 are contained in the top global decile, while USA3 through USA1 are contained in the second wealthiest global decile. At the opposite end of the spectrum, the population of a country like Zimbabwe is almost wholly contained within the poorest global deciles, but high within-country inequality in Zimbabwe allows the wealthiest decile to occupy a significantly higher position within the global hierarchy.

This type of exercise quite dramatically shows the importance of national boundaries in determining your relative place in the world. Such a global map of inequality, albeit limited in sophistication by the availability of relevant data, shows where each household or individual stands vis-à-vis everyone else in the world. And the relative standing of each household or individual continues to be shaped most heavily by whether they inhabited a poor or wealthy nation (i.e., by ascriptive criteria relating to where you are born). This is a consequence of, as well as evidence for, the crucial impact of the high level of inequality between-nations on global patterns of stratification.

A world-historical perspective also carries implications for the ways we conceptualize paths to social mobility, and for the relative accessibility of these paths to different populations. As discussed earlier, the modernization perspective points to the accumulation of human capital as the principle path to upward social mobility and that relative access to this route is more or less shared equally, as ascribed inequality is weakened over time. Indeed, when the analytical gaze is limited to the national border, this is the only visible path to social mobility. But from a world-historical perspective, social mobility is recast: access to human capital accumulation is itself a condition of elite status. Seen from the world as a whole, access to education and skill is quite limited and available only to a relative few. Thus while well over a quarter of adults in the United States have college degrees, in a country like Brazil this number is in the low single digits. In fact, a possible measure of privileged social status within a stratification system is precisely the extent to which one has the ability to engage in this path to social mobility.

But there are at least two other paths to social mobility in global stratification. Another potential path entails national economic growth. The relative standing of each individual in the world can potentially be improved (or worsened) by the relative standing of the county in which you live in the global wealth hierarchy.

The pursuit of national economic growth, however, is often portrayed as a willingness of people to allow for greater inequality in their own country in exchange for the growth of the overall wealth available for distribution. Leaving aside the fact that not all strategies of economic growth entail rising within-country inequality, even the existence of such a tradeoff would not be indicative of a lack of concern with inequality. The pursuit of economic growth involves a recognition of the crucial role of between-country inequality in shaping world stratification. When people in South Korea and/or China come to endorse policies designed to generate economic growth, rather than leaving concerns for inequality behind, they are recognizing the potential significance of such a path for engaging in upward social mobility within a global system of stratification.

However, national economic growth as a path to social mobility has not been historically accessible to vast parts of the world, and success stories have been the exception rather than the rule for most of the world's population. For most of the past two centuries, the path of social mobility through national economic growth, the development project itself, failed to deliver its promise. From a world-historical perspective, this describes a situation where the goal posts are constantly being moved forward. For the past two centuries, to paraphrase Giovanni Arrighi, most developing countries have been "running very fast to stay in place." The development and implementation of successful growth panaceas (Japan in the 1970s, South Korea in the 1980s) have seldom provided a replicable model for success, and in fact have been part and parcel of the constant creation of obsolescence.

More recently, China, and to a lesser extent India, have been characterized by extraordinary rates of growth (albeit the benefits of this growth have not been equally distributed among all households and regions) and this has been driving an overall decline in between-country inequality (although the magnitude of this decline is a matter of academic dispute, see Korzeniewicz and Moran 2007). China and India today embody much of the optimism on the potential rewards of economic growth as a route toward social mobility. The much larger population sizes of these two countries, accounting for about 40 percent of the world, renders the economic growth path to social mobility accessible, for the first time in history, to significant amounts of people.

We thus arrive at the single most immediate and effective means of global social mobility for populations in most countries of the world: migration. Given the crucial role of nationality in shaping global stratification, "jumping" ascribed categories by moving from a poorer country to a wealthier one is a highly effective strategy of mobility.

The empirical exercise on income deciles discussed above easily demonstrates this reality, illustrating how global stratification produces strong incentives for migration for individuals and/or groups of people in relatively poorer countries. In the case of Guatemala, for example, anyone belonging to the poorest seven deciles would be engaging in upward mobility by gaining access to the average income of the second poorest decile in Mexico. In the case of Mexico, the incentives are even more striking, as all but the wealthiest decile would find upward mobility in gaining access to the average income of the second poorest decile in the United States.

In another type of global comparison, Table 8.2.2 depicts comparable, net hourly incomes for fourteen common occupations, expressed in 2009 $US in the cities of Sao Paulo and Miami. Here again, the notion of simply "adding up" the different classes or occupational classifications across places is dispelled. Three empirical points on global stratification are made in the table: (1) the spread in hourly income from top to bottom is much higher in Sao Paulo, a characteristic correlating with Brazil's very high levels of inequality; (2) Engineers in Sao Paulo (a seemingly "skilled" occupation) earn less than several seemingly "unskilled" occupations in Miami such as bus drivers and building laborers; and (3) the lower the occupational wage, the bigger the difference between the two cities. Such disparities illustrate why economic migrants often abandon a professional status in their country of origin to work in relatively more menial positions in their country of destination—underlining the contingent meaning of "skill" and "human capital" in global terms.

Of course, migration is not merely the product of differentials in income incentives. Engaging in migration requires access to manifold resources—from those needed to meet the costs of

Table 8.2.2 Net hourly income for selected occupations in 2009 $US

Sao Paulo			*Miami*
		39.47	Product Managers
		27.45	Engineer
		24.52	Department Heads
		21.11	Skilled Ind. Workers
		18.39	Primary School Teachers
		18.22	Personal Assistants
Department Heads	17.68		
Product Managers	16.42		
		16.15	Cooks
		15.29	Call Center Agents
		13.94	Bank Credit Officers
		13.56	Car Mechanics
		11.39	Building Laborers
		11.01	Female Sales Assistants
		10.81	Bus Drivers
Engineer	10.55		
		10.38	Female Factory Worker
Skilled Ind. Workers	8.22		
Cooks	7.69		
Personal Assistants	6.99		
Bank Credit Officers	6.26		
Car Mechanics	4.32		
Primary School Teachers	3.79		
Female Sales Assistants	3.61		
Bus Drivers	3.13		
Call Center Agents	2.21		
Building Laborers	2.12		
Female Factory Worker	1.97		

Source: UBS (2009).

transportation and entry into a foreign country, to those involving social networks that can facilitate access to housing and jobs—and these resources are not equally available to all populations. And even in the presence (or absence) of strong income incentives, decisions to migrate are also based upon broader considerations regarding security, safety, well-being and personal attachments.

The two previous paths of mobility, the upgrading of human capital and the pursuit of national economic growth, both require a long time to generate their intended returns, and are fraught by a high degree of uncertainty over whether such returns will be available as expected. By contrast, the third path of mobility in global stratification, migration, while often requiring a high level of courageous determination, tends to offer much more immediate and certain returns (although a different type of uncertainty might be precisely what calls for high doses of courageous determination, particularly for undocumented migrants). Thus, while academics might remain convinced that national borders provide the appropriate boundaries for understanding social mobility, migrants, in their crossing of such borders, reveal the boundaries of stratification to be global.

Conclusion

Current trends in global inequality, and their implications for the three paths to social mobility identified above, are generating a sense that we are in the midst of systemic change. For some, this is cause for celebration: economic growth in countries such as China and India allows previously poor populations to upgrade their human capital and take advantage of the opportunities afforded by globalization to reduce the gap separating them from wealthy nations. Globalization works, and convergence has finally arrived. For others, there is little to celebrate: for workers facing unemployment in declining industrial towns, globalization has brought hardship, growing competition from cheap manufactures, and the growing presence of migrants. Globalization only brings vulnerability, and must be constrained.

Changes in the global economy are restructuring labor markets and welfare states, increasing a sense of exposure among many who benefited from the arrangements of the past (e.g., the "deskilled" manufacturing workers in rich nations). But this same restructuring has provided a sense of opportunity for others that are doing better (e.g., the "newly-skilled" service workers in middle-income nations), or populations that in the past were excluded from markets (e.g., those in low-income nations with high growth rates). High levels of world inequality becomes a driving force for migration, holding the promise of providing a quick path for overcoming the gap between wealth and poverty. In this sense, migration embodies social mobility on a global scale. A truly free flow of people across the world, in fact, would provide the fastest means to thoroughly transform the equilibria, centered around remarkably high and unchanging levels of between-country inequality, that have characterized global stratification for the past 200 years.

References

Arrighi, Giovanni. 1991. "World Income Inequalities and the Future of Socialism." *New Left Review*, 189: 39–65.

Blau, Peter M. and Otis Dudley Duncan. 1967. *The American Occupational Structure*. New York: John Wiley & Sons.

Carlsson, Gosta. 1958. *Social Mobility and Class Structure*. Lund: Gleerup.

Erikson, Robert and John Goldthorpe. 1992. *The Constant Flux: A Study of Class Mobility in Industrial Societies*. Oxford: Oxford University Press.

Evans, Peter. 2005. "Counterhegemonic Globalization: Transnational Social Movements in he Contemporary Global Political Economy." In Thomas Janoski et al (eds) *The Handbook of Political Sociology*. Cambridge: Cambridge University Press.

Glass, David V. (ed.). 1954. *Social Mobility in Britain*. London: Routledge & Kegan Paul.

Goldfrank, Walter. 1977. "Who Rules the World? Class Formation at the International Level." *Quarterly Journal of Ideology*, 1: 32–37.

Kentor, Jeffrey and Yong Suk Jang. 2004. "Yes, There is a (Growing) Transnational Business Community: A Study of Global Interlocking Directorates 1983–98." *International Sociology*, 19: 355–68.

Kentor, Jeffrey and Yong Suk Jang. 2006. "Different Questions, Different Answers: A Rejoinder to Carroll and Fennema." *International Sociology*, 21: 602–06.

Korzeniewicz, Roberto P. and Timothy P. Moran. 2007. "World Inequality in the Twenty-First Century: Patterns and Tendencies." pp. 565–92 in George Ritzer (ed.) *The Blackwell Companion to Globalization*. Oxford: Blackwell Publishing.

Korzeniewicz, Roberto P. and Timothy P. Moran. 2009. *Unveiling Inequality*. New York: Russell Sage Foundation.

Robinson, William I. 2004. *A Theory of Global Capitalism: Production, Class and State in a Transnational World*. Baltimore, MD: Johns Hopkins Press.

Robinson, William I. and Jerry Harris. 2000. "Towards A Global Ruling Class? Globalization and the Transnational Capitalist Class." *Science and Society* 64: 11–54.

Rogoff, Natalie. 1953. *Recent Trends in Occupational Mobility*. Glencoe, IL: Free Press.

Sklair, Leslie. 2001. *The Transnational Capitalist Class*. Oxford: Blackwell.

Sorokin, Pitirim. 1927. *Social Mobility*. New York: Harper & Brothers.

Tyree, Andrea, Moshe Semyonov, and Robert W. Hodge. 1979. "Gaps and Glissandos: Inequality, Economic Development, and Social Mobility in 24 Countries." *American Sociological Review*, 44: 410–24.

UBS. 2009. *Prices and Earnings*. New York: UBS.

Wallerstein, Immanuel. 1982. "World-Systems Analysis: Theoretical and Interpretative Issues." pp. 91–103 in T. K. Hopkins and I. Wallerstein (eds) *World-Systems Analysis: Theory and Methodology*. Beverly Hills, CA: Sage Publications.

8.3

Income inequality in the world

Looking back and ahead

Volker Bornschier

For thousands of years during the evolution of human society, material inequality has been on the rise (see Lenski 1966). A distinct turning point in this history occurred two centuries ago when average income levels started to diverge markedly *between* societies (Bornschier 2010: Figure 2). Between 1750 and 1800, when comparing broad regions such as Europe and China, inequality was mainly due to class differences in income and regional difference in fertility of the soil. From then onward, diverging average income levels between societies increasingly became the main source of differences in the world distribution of income.

It would be tempting to relate the emerging divergence in income levels to the industrial revolution. This is the story that conventional modernization theory is telling us. Yet the size and pattern of this uneven development was not a consequence of industrialization as such. Instead, it was the consequence of two linked processes—industrialization at the core and colonial rule over the periphery. This double peripherization caused de-industrialization at the periphery (Bornschier 2010: Figure 1). This specific economic and political power relation functioned at the advantage of the core and at the disadvantage of the periphery.

Since then, material inequality in the world has continued to grow as it did before, but at a faster pace. Between 1820 and 1992, total inequality almost doubled. Furthermore, it changed its structure. Whereas in 1820, 87 percent of total inequality was due to the "within" society component, this dropped markedly, while the between component increased from 13 percent to almost two-thirds, reaching a peak in the late 1980s and early 1990s.

Average income differences between countries increased between 1980 and 1997 (Bornschier 2002: 125). These findings revealed the decisive effect of weighing cases by population size, in which case the difference between rising and falling global inequality comes down to only one single influential case: China. Countries like China and India have enormous populations—together about 37 percent of the world population. To the extent that the economic growth rates of such influential cases deviate from the general pattern, this will have a considerable impact on the evolution of the world stratification of incomes.

Forecasting the future is a hard venture, especially if it pertains to the future—a wise saying attributed to Keynes. Nevertheless, informed by theory and latest empirical evidence, some predictions about the future trajectory of inequlaity can be made, at least for the next two decades or so. I offer five forecasts.

Forecast 1: Increasing polarization

Both "within" as well as "between" country inequalities (unweighed) are likely to increase in the coming years. Two main theoretical reasons are: sector transition to the knowledge-based new technological style and marginalization. Both processes make for more income inequality within countries *and* at the same time polarize income between the three income layers.

Forecast 2: Changes in distribution

Although polarization "within" and "between" countries will be the prevalent trend in coming years, total world income inequality may not increase but remain rather stable at a historically very high level. This is because world stratification of income will change its shape, driven in large part by rapid economic growth in the population rich countries China and India. Although this rapid growth is unlikely to last forever, it seems likely to continue for the near future.

Forecast 3: Changes between layers

If the recent slow growth in rich countries continues, then the middle and upper income layers will come closer together. The shrinking relative income privileges of lower classes in rich countries contrasts with the relative income gains of the population in the middle income layer. This seems relevant for people's perceptions even if the relative shares of lower classes in the middle income layer is shrinking. It may mitigate or even overcompensate their shrinking relative share as long as world economic growth resumes after the recent slump.

Forecast 4: Collective action

There exists a long history of thought in social philosophy and science to look at conflict and rebellion as reactions to inequality. The Marxian model is that inequality→conflict→reduction of inequality. There are, however, reasons to doubt this. First, objective inequality has little effect on collective conflict aimed at changing the structure of power and rewards. Second, subjective evaluation of income distribution as unjust is hardly related to objective income inequality, probably because considerable absolute income gains due to rapid growth may compensate. Third, subjective evaluation of income distribution as unjust is by far the most important predictor of subjective awareness of conflict. Therefore collective reactions towards inequality are unlikely to have a significant feedback effect on inequality.

Forecast 5: The politics of inequality

Given the mentioned trends, larger parts of lower classes in the world will—after a long time—come closer together in terms of objective life conditions. In principle, this would work into the direction of world classes as such—much more than in the past. It seems, however, unlikely that such a "class in itself" will result in politics of a "class for themselves" to use Marx' terms. The necessary solidarity is too hard to achieve.

For the years to come, conflict within the lower classes will probably absorb much of the latent class conflict. Yet as long as world economic growth resumes after the present crisis, this is unlikely to cause severe disturbances in the near future. But the world economic boom associated with the diffusion of new technologies will eventually run out of steam. At this point, such diverted potential conflicts may cause severe trouble for world peace.

References

Bornschier, Volker. 2002. "Changing Income Inequality in the Second Half of the 20th Century—Preliminary Findings and Propositions for Explanations." *Journal of World-Systems Research* 8: 100–127.

Bornschier, Volker. 2010. "On the Evolution of Inequality in the World System." pp. 39–64 in Christian Suter (ed.), *Inequality Beyond Globalization: Economic Changes, Social Transformations, and the Dynamics of Inequality*. Zurich: World Society Publications and LIT Verlag.

Lenski, Gerhard E. 1966. *Power and Privilege. A Theory of Stratification*. New York: McGraw-Hill.

8.4

Billionaires and global inequality

Does an increase in one indicate an increase in the other?

Jenny Chesters

Inequality is an inherent feature within the capitalist world-economy as capital is accumulated through the process of exploitation, both between nations and within nations. Global inequality is the sum of these two types of exploitation: within-nation inequality and between-nation inequality. Wealthy nations (the core) exploit poorer nations (the periphery). The bourgeoisie (those with enough capital to employ waged labour) exploit the proletariat (those wholly dependent upon waged labour for their existence) (Terlouw 1993). Although exploited within their own state, the proletariat in wealthy nations benefit from the exploitation of the proletariat in poorer nations due to the creation of redistributive welfare regimes and the availability of cheaper goods and services, which allow them to enjoy a higher material standard of living than would otherwise be possible (Chase-Dunn 1989; Terlouw 1993).

Between 1992 and 2007, the world-economy experienced a prolonged period of growth during which the economies of several semiperipheral countries, notably China and India, became more closely integrated into the world-economy. China and India account for more than one third of the world's population, therefore, the dramatic growth in the GDP of China (700 percent) and India (380 percent) placed downward pressure on the between-nation component of global inequality. On the other hand, the marketization of the previously planned economy of China and the liberalization of the economy of India resulted in dramatic increases in within-nation inequality. For example, between 1991 and 2007, the Gini coefficient increased from 0.300 to 0.410 in China (OECD 2010). The increase in within-nation inequality that accompanied the closer integration of these two large nations into the world-economy highlights the process of exploitation inherent in capitalist economies. The exploitation of the proletariat in India and China benefited both the wealthy elites in these two nations and the proletariat in core nations (through the availability of cheaper manufactured goods and cheaper services).

The examination of trends over time in income inequality on a global scale is particularly difficult due to the paucity of suitable data. Although income data has been collected on a systematic basis in wealthy core countries for decades, income data from peripheral and semi-peripheral countries is less consistent. For example, income data allows for the comparison of the P90:P10 ratio (the income earned by those in the top 10 percent of the income distribution compared to the income earned by those in the bottom 10 percent of the income distribution) over time. The P90:P10 ratio allows researchers to determine whether increases in national income have been shared more or less evenly or have contributed to increasing polarization of incomes as those in the top decile take a disproportionate share.

One indicator of changes in the within-nation and between-nation components of global inequality that has not yet been explored is the distribution of billionaires. Billionaires are arguably an indicator of the concentration of wealth within the global economy. They represent the global elite: the winners in the capitalist world-economy, an economy which gives "priority to the endless accumulation of capital" (Wallerstein 2004). It would appear that the extended period of growth and the integration of post-communists states into the world-economy has been accompanied by a rapid increase in the number of billionaires.

Between 1996 and 2010, the number of billionaires in the world-economy increased from 422 to 1011, with much of this growth being concentrated in the United States, United Kingdom, India, Russia, and China. Increases in the number of billionaires per capita could provide an indicator of increasing polarization of incomes within particular nations. For example, increased exploitation of the proletariat in these economies may have resulted in the wages share of national income declining and the profit share of national income increasing. Increasing profits translate into increased wealth for the owners of capital, giving rise to an increase in the number of billionaires.

Changes in the distribution of world billionaires could also provide an indicator of changes in between-nation inequality. In 1996, there were no billionaires residing in either Russia or China, however, in 2010 there were 70 billionaires in Russia and 72 in China. By 2010, the number of billionaires per capita in Russia exceeded that of Germany, France, Italy, and Japan. The declining proportion of world billionaires residing in core economies and the increasing proportion of world billionaires residing in emerging economies located in the semiperiphery, may signal a shift in wealth from the core to the semiperiphery as the wealthy elite in the emerging economies become less reliant on capital investment from the core and are thus able to retain a greater proportion of the economic surplus they extract from their proletariat. It may also signal that these emerging economies are moving closer to the core as their wealthy elites look for stable investment opportunities for their billions, or even, perhaps, the development of a new core.

References

Chase-Dunn, C. (1989) *Global Formation: Structures of the World-Economy*. Cambridge, MA: Basil Blackwell.

OECD (2010) *OECD Economic Surveys: China 2010*. Available at www.oecd.org, accessed 20 October 2010.

Terlouw, C. P. (1993) "The elusive semiperiphery: A critical examination of the concept semiperiphery", *International Journal of Comparative Sociology*, 34(1/2): 87–102.

Wallerstein, I. (2004) *World-Systems Analysis: An Introduction*. Durham, NC, and London: Duke University Press.

8.5

The pervasiveness of ICT in our present modern world-system

Melsome Nelson-Richards, with the assistance of Kandu E. Agbimson

When Emmanuel Wallerstein embarked upon his world-system perspective, ICT was seldom in use, although by the time he completed the third of his three volumes, the technology had become a household name. Today, the driving force behind the capitalist global economy and hence the modern world-system is ICT. In the USA alone, the total related computer occupation is about 65 percent, and that of the EU is about 48 percent. Toward the end of the twentieth century, the world's ICT market was worth about USD 1.8 trillion contrasted with USD 1.3 trillion five years earlier (OECD). This illustrates the tsunamic force of ICT. For the global economy to sustain itself in its present form, the present world-system must engage in a robust and continuous international competitiveness in the core, semi periphery, and the periphery on new methods of production and absorption of information. Active research and development are key to further growth, development, and productivity in the present world-system. The core engages in competitive programs and has comparative advantage. This comparative advantage is crucial to the rise and/or fall of a country or a region's present and future use and innovation of ICT in the world-system.

Since ICT's contribution to the world-economy can be observed in its effectiveness in digital arrangements and structures, it is more likely to enhance productivity. Currently, ICT-related industries have a higher productivity growth rate than non-ICT related industries. In 2008 the world ICT market was USD 3.1 trillion (www.eito.com/reposi.FreeDataSheets/ict-marketOverview-world). This can be retained only with on-going research, and specifically with R&D. No core country in the present world-system can afford to reduce its budget on R&D on ICT because it can be safely assumed that R&D will result in increased economic growth. So effective has ICT become in the world-system, that even the reduction of cost of its productions has not dampened the volume of investment in this technology; in fact, there have been significant improvements. The refined products, their outgrowth and high quality have in turn enabled companies to dramatically improve productivity.

In the OECD countries, the core of the world-system, this growth has resulted in massive investment in ICT (Colecchia and Schreyer 2002). The efficacy of ICT in the world-system is indispensable in macro-economics, thus enhancing the world-system perspective. Its impact on business performance is tangible; effective investment in ICT translates into sound labour productivity. Rapid improvements in ICT goods and services results in more rapid and intense improvements, which further results in more rapid multiplier productivity in the ICT-producing sector. In addition, more comprehensive use of ICT results in better and more efficient productivity. Thus ICT is a *sine qua non* in the global capitalist economy, and by inference, in the world-system.

As we examine the mechanics of the capitalist global economy, we must conclude that it is exploitative. Analyses of the present modern world-system illustrate that in its *modus operandi*, it is exploitative; advocacy on concepts and practices of equality, social justice, and inclusion reduces the exploitation. These are the issues that led Blau (2000), Wieviorka (2005), and others to question the sincerity of the perspective in its present form. Mohmmad Yunus, after lending his own money to female petty commodity producers in Jorba, Bangladesh, at very reasonable rates of interest, realized that they no longer found themselves in perpetual debt traps. Wieviorka's views of injecting humanitarian principles into the system were not welcomed by the perspective's original thinkers because such could result in altering the scientific discourse of the perspective and lead to a possible ideological platform. This may not necessarily be so.

After all, the modern world-system perspective was founded on Marxist principles, which are based on the rigors of scientific procedures in which sociological analyses are combined with political action reform. The fact that the founding father's scholarship of the world-system perspective is anchored in the inequality, marginalization, and exploitative capitalism that he witnessed first in Africa (Wallerstein 2000) directs our attention to the notion that social sciences and humanities can combine forces to improve the lot of the underprivileged. The World Summit on Social Development Conference that took place in Copenhagen in 1995 eventually led to the Millennium Development Goals, which, if achieved, are intended to reduce the hardship of the people in the periphery. If there is no determined ideological commitment that leads to equality, inclusion, and good governance, the modern world-system perspective will have failed dismally in its original objectives and will have no standing in improving the lot of mankind, as the perspective intended.

The intransigence of class situation in the world-system, the many politico-economic groupings, the mushrooming of identity politics, all of which have manifested themselves through political, economic, regional, and tribal groupings have posed ever more complex problems for the class situation. These new variables have been facilitated by the effectiveness and rapidity of ICT. They add to the complex nature of class.

Some medium-income economies such as India, whose structure is both class and caste, act as mini world-systems and have all the bearings of the perspective. Furthermore, the political and economic groupings, for example, the EU in Europe, NAFTA in North America, the AU in Africa, and APEC in Asia constitute the new structure of the perspective. They all fall under our world-system perspective and function there because of the effectiveness of ICT in the capitalist global economy where the *de facto* means of production is also ICT.

References

Blau, J. (2000) "Inequalities", *Contemporary Sociology*, 29(6): 819–20.

Colecchia, A. and P. Schreyer (2002) "Measuring the Information Economy", *International Research and Development: Funds and Technology* (OECD Discussion Papers).

Wallerstein, I. (2000) *The Essential Wallerstein*. New York: The New Press.

Wieviorka, M. (2005) "A Symposium on Emerging Issues in the Twenty-first Century World-System", *Contemporary Sociology*, 34(1): 107.

Part IV

Development and underdevelopment

9. Indigeneity and incorporation

9.1

Early capitalist inauguration and the formation of a colonial shatter zone

Robbie Ethridge

Scholars now understand that the time of early European colonialism in the late sixteenth, seventeenth, and early eighteenth centuries were a time of a profound social transformation among the Native people of the southern United States. The Native people who stood on either side of this great transformational divide were organized into quite different kinds of societies. The Indians of the eighteenth-century South are the ones familiar to most people and whose descendants are recognized today as the Creeks, the Cherokees, the Chickasaws, the Choctaws, the Catawbas, and so on. We now know that these societies formed out of survivors of the polities of the sixteenth-century precontact world—Coosa, Mabila, Pacaha, Chicaza, Cofitachequi, and others—as they broke apart in the late sixteenth and seventeenth centuries. We still do not have an adequate vocabulary to describe the Native societies of the eighteenth century. They have been called "confederacies," "tribes," "nations," and so on. We now generally call them "coalescent societies" because they were all, to varying degrees, coalescences of people from different societies, cultures, and languages who relocated and banded together after the fall of their polities (Hudson and Ethridge 1998).

On the other side of this transformational divide, the ancestors of the eighteenth-century Southern Indians were organized into what archaeologists call "chiefdoms." A chiefdom is a particular kind of political and social type characterized by a ranked social order of ruling elite lineages and nonelite lineages. Chiefdoms were the prevailing political unit in much of the South during the time known as the Mississippi Period (900 CE to 1700 CE). It was during the Mississippi Period that people of the chiefdoms built the flat-topped earthen temple mounds that one still finds throughout the American South. The chiefdoms in a region were tied together in myriad ways—ecologically, politically, socially, economically, and religiously. Although the exact nature of these ties has yet to be fully explained, they served to integrate these ancient chiefdoms of the South into a single world which has come to be known as the "Mississippian world."

The disturbances in the American South in the first 175 years of European colonization, between 1540 and 1715, have not been difficult to identify. The archaeological and documentary evidence attests to the disappearance of Native chiefdoms, movements of people into tightly compacted and heavily fortified towns, a dramatic loss of life, multiple migrations and splintering of groups, the coalescence of some groups, the disappearance of many others, and an overall decline in Native artistic life (Ethridge 2010; Ethridge and Hudson 2002; Ethridge and Shuck-Hall 2009; Galloway 1995; Hudson and Tesser 1994). We are beginning to understand that the

Mississippian world did not fall asunder in one fell swoop. Rather, there was a more piecemeal disintegration and reorganization, as chiefdoms fell and people regrouped—and sometimes regrouped again—over the full span of almost 200 years. The full transformation of the Mississippian world, then, must be placed within an interpretive framework against which each instance of collapse and restructuring can be placed.

I have begun constructing such a framework, one that has come to be known as the "Mississippian shatter zone" (Ethridge 2006, 2009a). The Mississippian shatter zone was a large region of instability in eastern North America that existed from the late sixteenth century through the early eighteenth century. It was created by the combined conditions of: (1) the inherent structural instabilities of the Mississippian chiefdoms; (2) the introduction of Old World pathogens and the subsequent serial disease episodes and loss of life; (3) the inauguration of a nascent capitalist economic system by Europeans through a commercial trade in Indian slaves and animal skins; and (4) the intensification and spread of violence and warfare through the Indian slave trade and especially through the emergence of militaristic Native slaving societies, who sought a larger share of the European trade (Ethridge 2006: 208–09, 2009a: 2).

Because the fall of the Mississippian world is at the core of the Mississippian shatter zone concept, the history of its collapse sets the chronological parameters. Some chiefdoms, such as Apalachee, Natchez, and Caddo functioned into the eighteenth century, although how closely these resembled Mississippian chiefdoms is in question. Others, such as the paramount chiefdom of Coosa, fell by 1600. Such a broad temporal span is necessary to cover the complete collapse of the Mississippian world and the full reorganization of the geopolitical landscape that followed. The Mississippian shatter zone then is defined here by the chronological and spatial parameters of the Mississippian world, distinguishing it as a particular historical case. This is not to say, however, that the Mississippian shatter zone was the only shatter zone created in the early years of colonialism. Rather, I am attempting to distinguish this as one particular case of instability. Future comparative work with other places of colonial instability may find some similarities between them and the Mississippian shatter zone.

Structural instabilities of Mississippian chiefdoms

Mississippian chiefdoms were not uniformly alike across space and time. Archaeologists understand the classic Mississippian chiefdoms such as Cahokia and Moundville to have been quite large and complex. A "complex chiefdom" is a political designation for chiefdoms in which one large chiefdom exercised some sort of control or influence over smaller chiefdoms within a defined area. Archaeologists call the smaller chiefdoms, "simple chiefdoms," in that they were polities in which the elites only controlled the towns connected to that chiefdom (Anderson 1994: 4–9; Steponaitis 1978).

In the Late Mississippi Period, complex and simple chiefdoms existed side by side, in a few cases single and especially charismatic leaders forged an alliance of several complex and simple chiefdoms into "paramount chiefdoms." Archaeologists are not in agreement as to the specific organizational mechanisms that held a paramount chiefdom together. Whereas one can think of a simple chiefdom as a "hands-on, workaday administrative unit under the aegis of a particular chief, the paramount chiefdom may have been little more than a kind of non-aggression pact, and the power of a paramount chief may have been little more than that of first among equals"(Hudson et al 2008: 467). Paramount chiefdoms were political entities that could have ranged from strongly to weakly integrated and paramount chiefs possessed power or influence. The concept of the paramount chiefdom implies a less centralized administration than a large, statelike organization, but possessing a larger area of geographic influence than that of simple or even complex chiefdoms, sometimes spanning several hundred miles (Beck 2003; Hally 2006: 30–39; Hally et al 1990).

Archaeological evidence indicates the occurrence of a pattern of cycling in complex and simple chiefdoms—a seeming endemic rise and fall of chiefdoms through time (Anderson 1994: 1–52; Blitz 1999: 583–90). How and why Mississippian chiefdoms rose and fell is still poorly understood. Certain stresses, such as soil exhaustion, drought, depletion of core resources, military defeats, and contested claims to the chieftainship may have constituted proximate causes for collapse. However, archaeologists generally agree that within the structure of a chiefdom, there also were structural instabilities, most likely those associated with chiefly power, authority, and political succession (Anderson 1994: 1–52; Blitz 1993: 12–13, 1999: 583–90; Hally 1993, 1996, 2006: 33–37; Scarry 1996; Williams 1990; Worth 2003).

When Europeans came into the American South, they penetrated the Mississippian world. In the late fifteenth and early sixteenth centuries, European sailors explored portions of the Atlantic and Gulf of Mexico coastlines, but European colonial efforts to settle North America did not begin until the early 1500s, with the largest effort being that of Hernando de Soto, who launched his expedition in 1539. Landing in present-day Tampa, Florida, and moving north, Soto's expedition explored the Southern heartland but failed in its objective of establishing a colony. At the time of Soto's entrada there were dozens of chiefdoms in the South, many of which Soto and his army encountered (Hudson 1990, 1997). Soto's expedition came as a conquering army, and in the cases of those chiefdoms that saw intense action, the direct military assault of the Spanish may have precipitated their collapse (Ethridge 2010: 60–88; Hudson 1997: 110–15, 238–49, 266–74, 336–38).

Soto's prolonged stay and ransacking of the region for food would also have had quite serious repercussions. When Soto and his army of 600-plus men trekked through the Southern interior, they depended on the food stores of Native people. Soto's presence in any polity resulted in depleted local stores as well as depletion of any emergency surpluses. Although Mississippian people knew much about utilizing wild plant and animal foods, such a shortage of stored cultivated crops would have meant hardship for all and starvation for some. In addition, the leadership of these chiefdoms partly derived from being able to procure and secure stores of food for just such emergencies, and if the leadership failed on this count, these polities must also have been subjected to political unrest (Blitz 1993: 69–97, 123–25, 176–78, 180–81; Ethridge 2010: 60–88; Wesson 1999: 157–58).

It is also likely that Soto's presence upset the balance of power in some complex and paramount chiefdoms. Records from the expedition describe several instances in which the micos (or rulers) of simple chiefdoms challenged and defied the authority of a chief under whose power they had fallen. In these political struggles, the more powerful mico often enlisted Soto's influence and military aid to prevail against the rebellion. In some, a savvy lesser mico used Soto to bolster his authority in challenging the overarching mico. In either case, given the fragile nature of the ties binding complex and paramount chiefdoms, the presence of a new and powerful ally could have easily upset the balance of power, resulting in a reshuffling of authority among these chiefdoms (Ethridge 2010: 60–88; Hudson 1997: 179–80, 228–29).

The introduction of old world diseases

After the Soto entrada, Spain managed to plant a small colony in North America—St. Augustine—and to institute their mission strategy for colonization whereby missionaries labored to convert indigenous people into Spanish peasants. The Spanish presence in present-day Florida inaugurated many exchanges with Native peoples, culturally, socially, and materially. It also intensified the

French and English scouting of North America. With increasing contact between Indians and Europeans, Old World diseases began to circulate through the American South. Until recently, scholars attributed the collapse of Native societies across the Americas at the time of contact to the introduction of Old World diseases into virgin soil populations and a resultant demographic collapse on the order of 90 percent. They also understood this demographic collapse to have occurred before sustained European contact. Today, scholars still accept a 90 percent loss of life, but this loss now appears to have occurred over about 200 years in a gradual, not steep, demographic decline and only after sustained European contact (Alchon 2002). In the American South, for example, the Indians of Spanish Florida had the most prolonged and direct contact with Europeans from about 1550 to 1715. Although there is plenty of documentary evidence for epidemic disease episodes in Spanish Florida, there is no evidence for a catastrophic loss of life across the region. Rather, epidemics were localized occurrences, albeit with high mortality rates (Ethridge 2009a: 10–13). In addition, there was no sweeping cultural upheaval between 1550 and 1700 in La Florida. The chiefdoms continued to function into the eighteenth century and adjusted to the presence of the Spanish in various ways. The chiefdoms of Spanish Florida fell not because of disease or because of the Spanish presence, but because of the late seventeenth- and early eighteenth-century commercial and English-sponsored slaving campaigns of Indian groups such as the Westos, Yamasees, Apalachicolas, and others. Today, scholars agree that disease was but one factor in the collapse of Native societies and the demographic decline, and they now point to contributing factors such as slaving, internecine warfare, dropping fertility rates, violent colonial strategies such as genocide, and general cultural and social malaise from colonial oppression (Alchon 2002; Dye 2002: 137–38; Ethridge 2010; Ethridge and Shuck-Hall 2009; Galloway 1995: 134–41, 159–63; Kelton 2007, 2009; Worth 1998, v. 2: 1–37, 2009).

Inauguration of the capitalist economic system

Certainly the structural instabilities of the Mississippian chiefdoms, the introduction of Old World disease epidemics, and cultural exchanges with early Spaniards had profound effects on Native life. But these instabilities, epidemics, and exchanges in and of themselves were not transforming. Not until Native peoples became engaged in the world-economy through contact with the English, French, and Dutch did they revamp their social, political, and economic orders to adapt to this system (Boswell and Chase-Dunn 2000: 17–50; Hollis 2005: 121; Hudson and Ethridge 1998).

Spain and Portugal were core areas during the late fifteenth and early to mid-sixteenth centuries, but their prominence in the global economy was eclipsed by the Netherlands by the late sixteenth century. Dutch interests lay in India and Asia rather than in the Atlantic, and by the time Dutch attention turned to North America in the early seventeenth century, they had to vie with England and France for economic hegemony (Boswell and Chase-Dunn 2000: 39, Table 1.3). By the late seventeenth century, France and England were in a contest for the Atlantic arena, with England finally emerging as the core area by the early eighteenth century.

Who were these Europeans and what sorts of ties to their homelands did they bring with them? This has been a central question in American colonial history almost from its inception. In the context of world-systems analysis, during the seventeenth century, key European core countries (primarily the Netherlands and England in competition with declining Spain and Portugal) were knitting together a global economy by articulating economic circuits, such as the Atlantic trade circuit with the Asian trade circuit. Although the American colonies were in the Atlantic circuit, they were still a subset of the larger global economy (Coclanis 2005; Gallay 2009). In this light, by the seventeenth century, the American colonies were strategic commercial outposts situated in a global periphery through which European core countries extracted labor and resources.

The North American overseas ventures were largely one set of commercial endeavors among a host of many others that crisscrossed the globe. The North American settlements served mostly as extractive trade factories, at least in the first 100 years or so. These settlements may have been small and overwhelmed by Indian populations, but they were one conduit through which the economic power of the core countries flowed. Hence, one should not look on the American colonies and colonists themselves as wielding extraordinary transforming power over Native life, but rather it was the system in which they served that was so transformative. It was not Quebec, New Amsterdam, Jamestown, or Charles Town that created the shatter zone; it was the global commercial power of the Netherlands, England, and France as funneled through these settlements (Ethridge 2009a: 18–19; Gallay 2009).

Many of the European men and women who first came to eastern North America understood themselves to be embarking on commercial enterprises through which they hoped to gain wealth and prestige. Most of them had much prior experience working for international military and commercial enterprises and with European trade companies in other areas of the world, and they brought these experiences to North America. They accepted that this business was not for the faint of heart, and they knew that real dangers awaited them. Nor were they all novices in dealing with people different from themselves, and the most enterprising, such as John Smith of Virginia and the South Carolina slaver Henry Woodward, had experience in other cultures and good linguistic and negotiating skills. They also had some ideas about how to engage the locals in trade, especially through the sale of armaments, and they had a good sense for commodities, in this case animal skins and Indian slaves (Ethridge 2009a: 19).

Slavery was not completely new to North American Indians at contact, as most Native groups practiced an indigenous form of slavery in which war captives were sometimes put into bondage. Large-scale captive-taking, as occurred during the seventeenth and early eighteenth centuries, however, was most likely not conducted during the Mississippi Period but came about with the colonial commercial slave trade. Hence, the forms of bondage during the Historic Period that Europeans observed did not exist in the Mississippi Period (Snyder 2010).

A few decades into the colonial Indian slave trade, many Southern Indian groups were responding to a powerful dynamic by which the commercial trade in Indian slaves worked. In this dynamic, Indian men incurred debts when they accepted guns, ammunition, metal tools, and other items from European traders. European traders in turn requested that the debt be paid in slaves and furs and skins. Armed with European-manufactured weapons, Indian groups raided rival groups for slaves. Members of the unarmed group would then have to acquire guns with which to defend themselves, and they in turn would become slave raiders. The result was a vicious cycle of slaving, trading, and weapons escalation (Ethridge 2010; Ethridge and Shuck-Hall 2009; Gallay 2002: 40–52; Hudson and Ethridge 1998: 41–42).

The spread of violence and warfare

The Mississippian shatter zone, as the name implies, takes violence to be attendant to imperialism and the birth of the early modern world. It also takes violence to be part and parcel of capitalist incorporation as theorized by Brian Ferguson and Neil Whitehead in *War in the Tribal Zone* (1992a). Ferguson and Whitehead (1992a) have begun to devise a theoretical framework for understanding the historical context of indigenous violence. They argue that one general result of state expansion and capitalist incorporation is the militarization of indigenous peoples through an increase, intensification, and transformation of indigenous warfare on the edges of empire, or the "tribal zone." Ferguson and Whitehead, however, do not understand indigenous people to be victims of expanding states. They insist that in addition to the colonizers' violence, indigenous people who lived and participated in the colonial world also contributed to and sometimes created colonial violence. They point out that this is especially true when indigenous people became involved in commercial slave trading.

Of all the types of colonial commerce, the commercial trade in slaves requires a high level of force as a necessary accompaniment to trade (Ferguson and Whitehead 1992b: 23). In other words, whereas trading in furs or deerskins does not necessarily involve warfare, trading in slaves does because it requires force. Certainly, warfare penetrated Mississippian life, so much so that war, not peace, may have been the accepted state of affairs. Once Europeans arrived, however, one can see that Native war efforts became entwined with market interests and international commerce (Dye 2002). As warfare became tied to commercial interests, the Mississippian chiefdoms' mechanisms for mitigating war and brokering peace broke down, meaning that as Indian commercial interests in the colonial Indian slave trade intensified, so did warfare and the militarization of those Native groups who sought to control the trade (Ethridge 2009a: 20–21).

Ferguson and Whitehead (1992b) argue that once indigenous people became involved in European trade, colonial politics, and military agendas, indigenous warfare became directed by elements outside of a purely Native world. Access to European trade goods, control over the supply and distribution of Indian slaves and trade goods, the use of indigenous mercenaries by the colonizers, as well as mutual military, political, and economic manipulation of each other by Europeans and indigenous inhabitants, all began to figure into indigenous motivations for and strategies of war. In the American South, one can see that the trade in Indian slaves intensified warfare and folded precontact motivations for war such as revenge and retaliation into burgeoning commercial interests (Dye 2002). The result in some cases was the emergence of what I have termed militaristic slaving societies (Ethridge 2006: 108–9, 2009a: 24–25, 2009b; Ferguson and Whitehead 1992b: 18–28). Militaristic slaving societies were Indian societies that gained control of the trade and "through their slave raiding, spread internecine warfare and created widespread dislocation, migration, amalgamation, and, in some cases, extinction of Native peoples" (Ethridge 2006: 208–09). Most of these militaristic slaving societies were short lived, existing for only about 100 years—from 1620 to 1720. During this brief window of time, they were key agents in the creation and expansion of the Mississippian shatter zone through a relentless raiding of their Indian neighbors for slaves.

The Mississippian shatter zone takes the Indian slave trade to be primarily a commercial venture with violence and internecine warfare as part and parcel of the slaving business. In other words, in the Mississippian shatter zone, the commercial trade in Indian slaves was not a continuation and adaptation of preexisting captivity patterns; it was a new kind of slaving, requiring a new kind of occupation created on the edge of the modern world-system—that of organized militaristic slavers. The capture and sale of Indian slaves did not serve to weld a world together; rather, Indian slaving was one of the hammers that shattered the world in which they themselves existed (Ethridge 2009a: 24–26, 2010).

Conclusion

By 1715, with almost 90 percent of the indigenous peoples of the American South either enslaved, killed by disease or in slave raids, and with the survivors living close to or among Europeans or grouped into formidable coalescent societies such as the Creeks, Cherokees, Chickasaws, Catawbas, and Choctaws, the supply of Indian slaves dwindled and the trade in Indian slaves faltered. Three conflagrations—the Tuscarora War of 1712, the Yamasee War of 1715, and the Natchez Revolt of 1729 signaled not only the end of the slave trade era but also the collapse of the last chiefdom that retained vestiges of its Mississippian roots.

By 1715 a new colonial South was emerging. Gone were all the many polities that Soto saw, replaced by large Indian "nations." In this new South, Indians were fully engaged in the global trade system. Cloth, guns, metal tools, and other items were as common and necessary to Indians as they were to Europeans. Only now, the Indian trade was not in slaves but in deerskins, and instead of slave raiders, most of the men became commercial hunters. The new South of 1715 was born out of the Mississippian shatter zone. The Mississippian chiefdoms, structured to rise and fall, fell spectacularly with European contact, and they failed to rise again because of the turmoil and wide regional instability created by the introduction of Old World diseases and the increased violence and disruptions from the commercial trade in Indian slaves. Faced with the collapse of their chiefdoms, people invented new ways of putting their social lives back together, ways that sought to adapt to a field of play with nation states and a global capitalist economy.

In 1540 Indian life was lived out in a purely Indian world; by 1715 it was lived out on the edge of an expanding and conflict-ridden European world and in a new social landscape that included not only Indians but also Europeans and Africans. The first 200 years of European colonization of the American South, then, is a story about the collision of two worlds—the emerging modern world of Europe and its American colonies and the centuries-old Mississippian world. The meeting of these two worlds was not peaceful or orderly, and it was marked by warfare, violence, struggle, disease, and hardship for all involved. There can be little doubt that eventually the European world prevailed over the Mississippian world. But the meeting also opened new opportunities, new possibilities, and new ways of doing things for both Natives and newcomers. The result was a transformation of both worlds and a melding of both into the modern world-system.

References

Alchon, S. A. (2002) *A Pest in the Land: New World Epidemics in a Global Perspective*, Albuquerque, NM: University of New Mexico Press.

Anderson, D. G. (1994) *Savannah River Chiefdoms: Political Change in the Late Prehistoric Southeast*, Tuscaloosa: University of Alabama Press.

Beck, R. A. (2003) "Consolidation and Hierarchy: Chiefdom Variability in the Mississippian Southeast," *American Antiquity*, 68, no. 4: 641–61.

Blitz, J. (1993) *Ancient Chiefdoms of the Tombigbee*, Tuscaloosa: University of Alabama Press.

Blitz, J. (1999) "Mississippian Chiefdoms and the Fusion-Fission Process," *American Antiquity*, 64, no. 4: 577–92.

Boswell, T. and C. Chase-Dunn (2000) *A Spiral of Capitalism and Socialism: Toward Global Democracy*, Boulder, CO: Lynne Rienner.

Coclanis, P. (ed.) (2005) *The Atlantic Economy during the Seventeenth and Eighteenth Centuries: Organization, Operation, Practice, and Personnel*, Columbia: University of South Carolina Press.

Dye, D. H. (2002) "Warfare in the Protohistoric Southeast," in C. Wesson and M. A. Rees (eds) *Between Contact and Colonies: Archaeological Perspectives on the Protohistoric Southeast*, 126–41, Tuscaloosa: University of Alabama Press.

Ethridge, R. (2006) "Creating the Shatter Zone: The Indian Slave Trader and the Collapse of the Mississippian World," in T. J. Pluckhahn and R. Ethridge (eds) *Light on the Path: The Anthropology and History of the Southeastern Indians*, 207–18, Tuscaloosa: University of Alabama Press.

Ethridge, R. (2009a) "Introduction: Mapping the Mississippian Shatter Zone," in R. Ethridge and S. M. Shuck-Hall (eds) *Mapping the Mississippian Shatter Zone: The Colonial Indian Slave Trade and Regional Instability in the American South*, 1–62, Lincoln: University of Nebraska Press.

Ethridge, R. (2009b) "The Making of a Militaristic Slaving Society: The Chickasaws and the Colonial Indian Slave Trade," in A. Gallay (ed.) *Indian Slavery in Colonial America*, 251–76, Lincoln: University of Nebraska Press.

Ethridge, R. (2010) *From Chicaza to Chickasaw: The European Invasion and the Transformation of the Mississippian World, 1540–1715*, Chapel Hill: University of North Carolina Press.

Ethridge, R. and C. Hudson (eds) (2002) *The Transformation of the Southeastern Indians*, 1540–1760, Jackson: University Press of Mississippi.

Ethridge, R. and S. M. Shuck-Hall (eds) (2009) *Mapping the Mississippian Shatter Zone: The Colonial Indian Slave Trade and Regional Instability in the American South*, Lincoln: University of Nebraska Press.

Ferguson, R. B. and N. L. Whitehead (eds) (1992a; 2nd edn 1999) *War in the Tribal Zone: Expanding States and Indigenous Warfare*, Santa Fe: School of American Research Press.

Ferguson, R. B. and N. L. Whitehead (1992b; 2nd edn 1999) "The Violent Edge of Empire," in R. B. Ferguson, and N. L. Whitehead (eds) *War in the Tribal Zone: Expanding States and Indigenous Warfare*, 1–30, Santa Fe, NM: School of American Research.

Gallay, A. (2002) *The Indian Slave Trade: The Rise of the English Empire in the American South*, 1670–1717, New Haven, CT: Yale University Press.

Gallay, A. (2009) "Introduction," pp. 1–32 in A. Gallay (ed.) *Indian Slavery in Colonial America*, Lincoln: University of Nebraska Press.

Galloway, P. K. (1995) *Choctaw Genesis, 1500–1700*, Lincoln: University of Nebraska.

Hally, D. J. (1993) "The Territorial Size of Mississippian Chiefdoms," in pp. 143–68 J. A. Stoltman (ed.) *Archaeology of Eastern North America: Papers in Honor of Stephen Williams*, Archeological Report, no. 25, Jackson: Mississippi Department of Archives and History.

Hally, D. J. (1996) "Platform Mound Construction and the Instability of Mississippian Chiefdoms," pp. 92–127, in J. Scarry (ed.) *Political Structure and Change in the Prehistoric Southeastern United States*, Gainesville: University Press of Florida.

Hally, D. J. (2006) "The Nature of Mississippian Regional Systems," pp. 26–42 in T. H. Pluckhahn and R. Ethridge (eds) *Light on the Path: The Anthropology and History of the Southeastern Indians*, Tuscaloosa: University of Alabama Press.

Hally, D. J., M. T. Smith and J. B. Langford, Jr. (1990) "The Archaeological Reality of De Soto's Coosa," pp. 121–38 in D. H. Thomas (ed.) Columbian Consequences, Vol. 2, *Archaeological and Historical Perspectives on the Spanish Borderlands East*, Washington, DC: Smithsonian Institution Press.

Hollis, S. A. (2005) "Contact, Incorporation, and the North American Southeast," *Journal of World Systems Research*, 11, no. 1: 95–130.

Hudson, C. (1990; 2nd edn. 2005) *The Juan Pardo Expeditions: Exploration of the Carolinas and Tennessee*, 1566–1568, Tuscaloosa: University of Alabama Press,.

Hudson, C. (1997) *Knights of Spain, Warriors of the Sun: Hernando de Soto and the South's Ancient Chiefdoms*, Athens: University of Georgia Press.

Hudson, C., R. A. Beck, Jr., C. DePratter, R. Ethridge and J. E. Worth (2008) "On Cofitachequi," *Ethnohistory*, 55: 465–90.

Hudson, C. and R. Ethridge (1998) "The Early Historic Transformation of the Southeastern Indians," pp. 34–50 in C. E. Hill and P. D. Beaver (eds) *Cultural Diversity in the U.S. South: Anthropological Contributions to a Region in Transition*, Athens: University of Georgia Press.

Hudson, C. and C. C. Tesser (1994) *The Forgotten Centuries: Indians and Europeans in the American South* 1521–1704, Athens: University of Georgia Press.

Kelton, P. (2007) *Epidemics and Enslavement: Biological Catastrophe in the Native Southeast, 1492–1715*, Lincoln: University of Nebraska Press.

Kelton, P. (2009) "Shattered and Infected: Epidemics and the Origins of the Yamasee War, 1696–1715," pp. 312–32 in R. Ethridge and S. M. Shuck-Hall (eds) *Mapping the Mississippian Shatter Zone: The Colonial Indian Slave Trade and Regional Instability in the American South*, Lincoln: University of Nebraska Press.

Scarry, J. F. (ed.) (1996) *Political Structure and Change in the Prehistoric Southeastern United States*, Gainesville: University Press of Florida.

Snyder, C. (2010) *Slavery in Indian Country: The Changing Face of Captivity in Early America*, Cambridge, MA: Harvard University Press

Steponaitis, V. P. (1978) "Location Theory and Complex Chiefdoms: A Mississippian Example," pp. 417–53 in B. Smith (ed.) *Mississippian Settlement Patterns*, New York: Academic Press.

Wesson, C. B. (1999) "Chiefly Power and Food Storage in Southeastern North America," *World Archaeology*, 31, no. 1: 145–64.

Williams M. (1990) "Paired Towns," pp. 163–74 in M. Williams and G. Shapiro (eds) *Lamar Archaeology: Mississippian Chiefdoms in the Deep South*, Tuscaloosa: University of Alabama Press.

Worth, J. E. (1998) *The Timucuan Chiefdoms of Spanish Florida*, 2 vols, Gainesville: University Press of Florida.

Worth, J. E. (2003) "An Ethnohistorical Synthesis of Southeastern Chiefdoms: How Does Coosa Compare?", paper presented at the Annual Meeting of the Southeastern Archaeological Conference, Charlotte, North Carolina.

Worth, J. E. (2009) "Razing Florida: The Indian Slave Trade and the Devastation of Spanish Florida, 1659–1715," pp. 295–311 in R. Ethridge and S. M. Shuck-Hall (eds) *Mapping the Mississippian Shatter Zone: The Colonial Indian Slave Trade and Regional Instability in the American South*, Lincoln: University of Nebraska Press.

9.2

Indigenous peoples, globalization and autonomy in world-systems analysis

James V. Fenelon

Introduction

The study of indigenous peoples in all their diversity offers a unique lens into contemporary global processes and deeper insights into how people strive to retain some degree of local autonomy in the face of the onslaught of global, neoliberal capitalism. International area studies and anthropological histories differ markedly whether they are truly global or maintain focus on North American Indians, with more difference in how Native Nations view their own struggles, observed in a contrast of the late John Mohawk titling his books—*Utopian Legacies* (2000); and/or the late Vine Deloria Jr. titles—*God Is Red* (1994) and *Red Earth, White Lies* (1997). Today, indigenous survival and resistance continues against the forces of colonizing capitalism, found in the title of the book—*Paradigm Wars, Indigenous People's Resistance to Globalization* (Mander and Tauli-Corpuz 2006).

Globally, indigenous peoples number over 350 million, and possibly many more depending on how one defines "indigenous," comprising more than the population of the United States or the equivalent of the European Union. They include on the order of five thousand different cultures. Thus, indigenous peoples constitute about 4 percent of the world's population, but more significantly, account for as much as 95 percent of the world's cultural diversity (Coates 2004). While many writers and observers have thought Native Americans would disappear, such as Thomas Jefferson Morgan, the nineteenth-century Commissioner of Indian Affairs who claimed, "The great body of Indians will become merged in the indistinguishable mass of our population" (quoted in Iverson 1999: 16–17), Indian Nations are "still here" and one of the fastest growing segments of the population of the United States (Snipp 1992), as is true for indigenous peoples in much of the rest of the world.

Nevertheless, confrontations and conflicts between states and indigenous peoples are as old as states themselves (Chase-Dunn and Hall 1997). While states have been successful in displacing, absorbing, incorporating, assimilating, or destroying indigenous peoples over the last 5000 years, indigenous peoples have resisted and survived attempts at total annihilation (Thornton 1987), surviving despite dire predictions of their disappearance.

Some of the ways in which indigenous peoples have survived illustrate a need to use precise concepts when discussing indigenous survival, and to re-examine theoretical issues of indigenous peoples, their origins, and processes of globalization (Hall and Fenelon 2009). Many scholars use a common gloss for globalization: intense speeding up of communication, travel, culture, commerce, and interconnections across planet Earth in the late twentieth and early twenty-first centuries, primarily built around capitalist systems in rich countries that exploit market structures. For indigenous peoples, it is instructive to think of globalization as divided into three general categories of economic, political, and cultural patterns that are involved in a close and complex process of interaction. Distinctions between a world-economy and a global economy, and economic patterns of production and commodification associated with an "extensive and intensive enlargement" (Robinson 2004) help to identify the social relations of domination.

Within recent political patterns of globalization, no single event stands out more than the election of Evo Morales in 2005. As the first indigenous president of Bolivia to assume office, Morales' victory exemplifies changing socio-political scenes in countries where indigenous movements have taken hold. It not only demonstrates the global importance of these successes, but adds to a pattern of resistance that opposes neoliberal hegemony and demonstrates the establishment of close ties with other Latin American leaders.

Globalization accompanies this expansion of the modern world-system. Andre Gunder-Frank (1966), Immanuel Wallerstein (2004), and others have identified much of this as emerging at the same time and in tandem with colonialism, and have identified what others have called underdevelopment as a way of constructing and controlling world markets. Frank (1966) uses the term "the development of underdevelopment," others (e.g., Wallerstein 2004) call this neo-imperialism by core states. This understanding is essential in any analysis of indigenous peoples for two reasons. First, American Indian Nations and indigenous peoples in the western hemisphere—especially North America—are subsumed in this process of "development" as conquest, coercive assimilation, and nation-building. Second, the surviving nations are further oppressed and marginalized through forms of internal colonialism. These populations are relatively small in core states, and relatively large although diffuse in Latin American states, typically less "developed" than the United States or Canada. Recent conflicts in Mexico, Nicaragua, Guatemala, Ecuador, Chile, Peru, and especially Bolivia underscore these relationships of indigenous peoples, resistance, and larger globalization processes.

Definitions and timeframes

The category "indigenous peoples" is a gross simplification of thousands of distinct cultures (Champagne 2005). A UN definition of indigenous peoples by José Martinez Cobo (1986) is:

> Indigenous communities, peoples and nations are those which, having a historical continuity with pre-invasion and pre-colonial societies that developed on their territories, consider themselves distinct from other sectors of the societies now prevailing in those territories, or parts of them. They form at present non-dominant sectors of society and are determined to preserve, develop and transmit to future generations their ancestral territories, and their ethnic identity, as the basis of their continued existence as peoples, in accordance with their own cultural patterns, social institutions and legal systems.

Cobo further notes that several additional factors are relevant to this definition: a historical continuity … for an extended period reaching into the present … of the following factors: (1) Occupation of ancestral lands … (2) Common ancestry with the original occupants of these lands; (3) Culture in general … (4) Language; (5) Residence in certain parts of the country, or regions … (Cobo 1986).

Terms like "indigenous peoples," "non-state societies," or "first peoples" lump these highly diverse groups into overly broad categories emphasizing their important differences from states. First, these are *not* state-based organizations, since there were indigenous states in North and South America prior to European contact—Aztecs, Maya, Inka, and so on (Chase-Dunn and Hall 1997). Furthermore, some indigenous societies took on state-like qualities, including the Cahokia (Forbes 1998) and the Haudenosaunee peoples (Snow 1996). While differences in social organization are crucial, they are *not* assertions about claims to rights, or of international status. All these forms of social organization are non-capitalist, glossed as "pre-capitalist," with unfortunate connotations that such organizations are precursors of capitalism. In this view they are seen as at best "primitive" forms of social organization, and at worst as outmoded. Instead, these societies are rooted in modes of organization, production, and accumulation that have little to do with capitalist accumulation of capital, and thus resist assimilation (Bodley 2003).

It is critical to recognize, as Eric Wolf argued (1982), that these peoples have histories separate and distinct from those of Europeans states, and, indeed, all states (Chase-Dunn and Hall 1997). Furthermore, indigenous peoples have been forced to deal with waves of European expansion and the increasing globalization of capitalism over the last 500 years. Many peoples have been incorporated into the capitalist world-system, but far from completely. Many have resisted incorporation heroically, and untold numbers have died doing so (Dunaway 2000). Culture and identity politics have become contested issues, with names of indigenous peoples particularly contested (Fenelon 2007). On one hand, to label chiefdoms "nations" confounds a profoundly modern form of social organization with older, very different forms of social organization. On the other hand, variations among "nations" are sufficiently distinct from the concept of state or "nation-state" (Deloria and Wilkins 1999).

It is useful to re-articulate basic terms. First is the *state*, which means a formal, bureaucratically organized form of government, first developed some 5000 years ago with the formation of Ur, in Mesopotamia. Second is *nation*, which means a body of people with a shared identity that recognize themselves as a more-or-less united group. Third is *nation-state*, which means a state characterized by a population with a shared identity. True nation-states are rare, and most are multiethnic, with the United States not particularly exceptional in diversity, and the notion of why people ever came to hold the nation-state as ideal is problematic (Hall 1998). Fourth is *country*, meaning the territory of a state. Fifth are entities such as tribes, chiefdoms, and bands. These are terms for various types of non-state organization that may be nations, but are **not** states. These terms are politicized and often used for political ends to deny treaty rights (Fenelon 2002).

These social structures have evolved over time, so that existing indigenous peoples are *not* "living fossils." They have evolved, adapting to contexts when surrounded by one or more typically hostile states. One powerful insight from world-system theory is that the fundamental entity evolving is the system itself, and any component of it—state, nation, nation-state, or indigenous group—must be understood as system evolution, and not as a rigid sequence moving from "primitive" [bad] to "modern" [good]. Feedback within the world-system limits the possibilities for its components within a capitalist world-system that is continuing to change, and at times indigenous peoples have played integral parts in the system.

One important set of social formations during the rise of the modern world-system that is critical to understanding American Indians and later indigenous peoples are those of race and racism.

The initial formation of "race" for indigenous Americans became salient after 1492, with the first usage by the Catholic Church and European nation-states interested in colonial expansion in the "New World." Taino-Arawak peoples on Hispaniola are representative (Fenelon 2007), with a few taken back by Columbus to Spain, shown to monarchs to prove the existence of *los Indios* and then sold in the slave markets of Seville. These indigenous Natives were racially identified. On Hispaniola there were almost complete genocides of Arawakan peoples by 1542; this, coupled with Cortez's conquest of the Aztecs and subsequent reduction of all the indigenous peoples in Mexico to the lowered status of "Indian," lead to permanent recognition by "race" stratification (Berkhofer 1978; Fenelon 2002). Racial hierarchies would continue to be utilized for three centuries by the Spanish, with Columbus starting the encomienda system, the French building race-based plantations throughout Haiti, and all leading to either genocidal destruction or racial subordination of Indians, who experienced loss of sovereignty and freedom. The "modern world" began with European expansion into the western hemisphere over indigenous Americans called *Indios*, or "Indians" that racially aggregated millions of diverse peoples under one race term (Mohawk 2000). This expansion unleashed socioeconomic changes globally, within the totality of global development and conquest with a centrality of exploitation and profiteering critical to these social formations.

Wallerstein (2004) has pointed to a strong reliance by global capitalism on the interstate system, and a multitude of connections to military-political networks. The United States has a most well-developed and codified relationship with indigenous peoples and Canada follows closely, having surpassed the United States by recognizing the oral tradition of their "First Nations" (Perry 1996). Most colonial systems conducted genocide but did not develop treaty-based legal systems. Central and South America incorporated Native peoples into systems of racial subordination, segregation and partial assimilation. States have used two concepts to justify this expansion: the Doctrine of Discovery and the Prince's Rights to Conquest (Wilkins 2006), denying strong Native Nations predicated on treaties and various "non-intercourse" acts, intended to contain indigenous peoples within one state in the expanding world-system. In the United States, such relationships became known as "tribal sovereignty" for surviving indigenous peoples. This relationship evolved into complex "dual sovereignty" relationships (Fenelon 2002). Western hemisphere states assumed they could extinguish claims to tribal sovereignty at a later date (Champagne 2007). Generally, that did not happen. Similar patterns, albeit on autonomy, occurred with indigenous peoples around the world (Hall and Fenelon 2009).

Indigeneity, global stratification, resistance, and revitalization

Indigenous social systems of governance, community, local economy, and land tenure became arenas of conflict within individual states and sites of resistance to capitalist domination, along with imposition of private property and profit-driven value systems. Globalization, besides developing new consumptive patterns, attempts to privatize and corporatize all social systems. Examples of resistance include the remote areas of Chiapas, which gave rise to the Zapatistas; Aymara coca leaf growers in the Bolivian highlands collectivizing and organizing with unions; Inuit in Nunavut, Mohawk in Quebec, Navajo in southwestern United States, and Lakota in the northern plains states in the United States re-organizing under collectivities striving to maintain traditional life.

Globalization often produces "dependency" and "underdevelopment" in developing economies common in Latin America where corporate profits go to other countries, usually in core areas (Dello Buono and Lara 2007). For indigenous peoples, this means socioeconomic depression and continuing cultural repression. Indigenous resistance to global capitalism is worldwide, diverse, and interconnected, with some forms covert, in the form of weak weapons that allow co-existence. Other forms defy easy description. Events in Chiapas are sometimes seen as part of a regional, peasant (hence class), or a *caudillo* driven rebellion. They are rarely discussed as an indigenous Mayan rebellion (Ramirez and Munoz 2003). Movements in the United States, such as the American Indian Movement (AIM), are often seen solely in terms of localized ethnic, urban, or racial rebellions. In other cases, traditional culture and organization itself is a resource that facilitates resistance and survival (Champagne 2007). Indigenous resistance struggles are occurring all over the world: in Europe among the Saami, with Kurdish actions in the Middle East, others in West Asia, and the Miskito resistance in Nicaragua, long noted as indigenous movements (Harff and Gurr 2004). Many researchers are now developing catalogues of such movements, and Linda Smith's (1999) *Decolonizing Methodologies* lists resistance modalities.

The most fundamental challenge to capitalism, though, comes from communal ownership of resources, because it denies the over-arching dominance of private property rights. Contrary to what many early colonizers thought, indigenous people do understand individual ownership, except they have long recognized what environmental movements are forcing capitalists to accept: most resources are partially, if not wholly, "public goods" and are thereby sites of political, economic, or social contestation, including environmental alliances (Gedicks 2001).

Culture-building is another form of resistance. There are more than 33 tribal colleges in the United States. In some cases, language programs have aimed at reviving or reinvigorating a language, with Tribal colleges as a means of preserving and enhancing "traditional cultures." Inroads have recently been made in universities around the world, although these institutions are purveyors of hegemonic ideologies and history (Champagne 2007). Resistance can also take the form of building localized institutions that conform to traditional cultural values, as the Diné (Navajo) have done, including tribal police culturally sensitive to Navajo traditions, "peacemaker courts" avoiding adversarial techniques of Anglo courts by pursuing Navajo concepts of harmony. New social forms arise as resistance, with the "junta del buen gobierno" justice of the Zapatistas, coca leaf growers in Bolivia with Evo Morales, or the autonomous yet Islamic justice of the Kurds.

Other forms of resistance are less institutionalized, with many analysts observing how women have adapted conventional alcohol abuse or spouse abuse programs to indigenous culture, such as the northern Cheyenne. Typically, indigenous feminists focus on issues of identity and cultural preservation. Rigobertu Menchu speaks of the oppression and resistance by Mayan women and families in Guatemala, as Gloria Ramirez (2003) does for Zapatista women.

Religion is another form of resistance (Champagne 2003), especially over lands considered sacred and necessary for religious ceremonies. This leads to conflicts over use of the land for sacred functions versus "productive" and/or "recreational" use. Native scholars see these intrusions into "traditional culture" as a last arena of conflict, where the holistic knowledge of indigenous peoples can be exploited by western markets that stress profits over people and land, and of knowledge itself. The revival of older traditions, such as the Sun Dance (Fenelon 1998: 114, 288–94), can be another form of religious resistance. These revivals hark back to revitalization movements: the Longhouse religion of the Iroquois, the Ghost Dance movement (Thornton 1987), and the Native American Church.

Many indigenous religions oppose privatization and support collective ownership of goods—land and livestock—that are typically privately owned commodities in the capitalist world-system. As Champagne (2007) suggests, many indigenous peoples see their roles primarily as stewards who are a part of nature, not as its controllers or owners. One dramatic example of resistance is the continuing effort of Lakota peoples to regain control of the Black Hills after court decisions determined that the region was taken from them illegally (Fenelon 1998). Capitalist US jurisprudence over settlement of this claim has been monetary. Lakota peoples have steadfastly refused commodified settlements and insisted on the return of the land that they consider sacred. This commitment is underscored by the poverty on the Pine Ridge (Lakota) reservation. Despite the temptation to take a cash settlement, the Lakota have continued to reject it, preferring instead to struggle for the return of their land.

All of these discussions for indigenous peoples in the United States revolve around issues of sovereignty with the US federal government (Deloria and Wilkins 1999). Indeed, sovereignty issues are often the basis of challenges to states around the world and cut to the heart of the interstate system built on the 1648 Peace of Westphalia. Native leaders worry about how much they give up to win victories, accepting premises of the European civilization turf they have been fighting on. Thomas Biolsi argues that the law is "a fundamental constituting axis of modern social life—not just a political resource or an institution but a constituent of all social relations of domination" (Biolsi 1995: 543). Courts have been an institutional means of commodifying everything; especially land (Biolsi 1995). Still, indigenous peoples continue to use legal systems to resist incorporation and global capitalism, noting that important differences between indigenous struggles in the core, "first" world or the global north and those in "third world" or peripheral areas or the global south, exist, especially concerning the rule of law.

Indigenous people's relationships in resistance to globalization and neo-liberal capitalism demonstrates struggles within four spheres, community as collectivity, economic redistribution, land tenure as collective responsibility, with consensual decision-making systems providing the conduit to the dominant society. These social practices, which are essentially non-capitalist, survive as alternatives to globalization, and are less destructive and less violent in each sphere of interaction—political, the economic, environmental, and community. As non-dominant options, they meet the criteria for alternatives to neoliberal capitalistic societies (Sklair 2002).

For instance, Lakota peoples had *Eyapaha* (traditional speakers), while Gond leaders had a "spokesperson" representing the elders, and a reconstituted *Ghotul* similar to the Sun Dance revitalizing the Lakota, all with a focus on community, kinship, stewardship of the environment, and group decision-making. Other examples include control over "patented" corn by the Mixtec and Zapotecs in Oaxaca, and Gond or Warli adevasi resisting bio-piracy in India. Struggles over forests management exemplify resistance by Gond Adevasi, Zapotec in Oaxaca, Mapuche resistance in Chile, coastal wet-lands of the Aquinnah Wampanoag, and the Foreshore Seabed fights by Māori citing the Waitangi treaty (Hall and Fenelon 2009: 27–30), as anti-commodification and de-colonization actions to preserve traditional life.

Contemporary indigenous political and social movements

In recent decades, indigenous movements have challenged globalizing capitalism (Champagne 2003), including those by INGOs and for indigenous organizations. The Zapatista movement (EZLN) centered in Chiapas has been dramatic, since the Zapatista ideology and practices contradict the logic of capitalism. They reject modernization's "development" (Ross 1995). Mignolo (2002) argues that the Zapatista movement constitutes an alternative to greco-roman legacies of state making, seeking to maintain traditional life in the face of overwhelming forces to assimilate to capitalist systems, in opposition to NAFTA (North American Free Trade Agreement) and FTAA (Free Trade in the Americas Agreement).

These movements arise from resistance and revitalization (Fenelon and Hall 2008), building on the arguments that globalization, especially capitalist forms, engenders resistance movements that are anti-systemic or are counter-hegemonic. Key factors are their levels of sovereignty, autonomy, or minority status, and revolve around legitimate rights to organize and live their lives in ways other than those permitted or favored by neoliberal capitalism (Robinson 2004; Sklair 2002), challenging forms of privatization, commodification, and conventional views of the fundamental relations between human beings and the environment. Nearly all seek to preserve some aspects of ancient ways of living while adapting to contemporary social conditions.

Examples of these movements include the Mohawk in Canada, a "First Nation" that experienced three hundred years of colonial domination and the exchange of gunfire with the military as late as 1990, with issues of transnational historical ethnicity. The Lakota (Sioux) represent about

200 years of conflicts, ranging from war and formal treaty-making with the United States over sovereignty and control of the land, to recent anti-hegemonic social movements from the 1960s until today. The Cherokee were removed under genocidal conditions, with all three forms of sovereignty—federal, tribal, state—remaining in contention over limiting Indian Country. The Puyallup started out as a treaty-tribe and made a stunning comeback in the twentieth century. The Pequot were thought to have been eliminated for 300 years and now receive federal recognition, having built up stunning economic success through Indian Gaming, based entirely on sovereignty. Wampanoag people went from once-great nations to winning a limited partial recognition through the court system, with nominal sovereignty. The Yaqui complete the US examples, straddling borders with Mexico, ultimately achieving recognition and growing substantially in the last two decades in terms of its territorial claim.

Mexico, historically an assimilative nation toward indigenous peoples, treated most of their "Indians" with discriminatory repression, existent today. Tarahumara peoples, with difficult relationships and an *ejido* system of land tenure, represent state control over bounded peoples. Those of Mayan descent in Chiapas exhibit both armed and socio-political resistance to US globalization, struggling for 500 years against transnational capitalism, hemispheric hegemony, and repression of peasant Indians, with revolutionary struggle linked to indigenous resistance. The Miskito in Nicaragua resisted a socialist revolutionary government that imposed tough conditions, boundaries, and forced removal (Harff and Gurr 2004). Yanomami people, in both Venezuela and Brazil, faced imposed state borders, were subject to genocidal internal colonialism in Brazil, and in Venezuela, faced limited protections but an invasive market economy with trading posts and timber companies. Quechuan people in Ecuador, and in Peru, maintain a sizable demographic presence and are involved in elections that are nearly always followed by military coups. Separatists movements in Peru, Venezuela, and Colombia attempted to make coalitions in similar fashion to dominant groups. Finally, Native Hawai'ians achieved limited sovereignty through political recognition.

Six levels of indigenous survival exist with respect to sovereignty, autonomy, and minority status, so that on the highest level, sovereignty is recognized or at least contested—offering opportunity during times of hegemonic decline, with caveats. The Mohawk, Lakota, Cherokee, Puyallup, Pequot, perhaps Yaqui, and Native Hawai'ian cases appear to be operating on all these levels. Autonomy struggles exist, with assets such as land or mineral rights, labor and trade determined by Latin American countries. The Mayan and the Miskito are examples of peoples with minority status who are dependent on dominant policies of the state, a common feature in the European expansion over the western hemisphere. The Yanomami and Quechuan exemplify such vulnerable positions.

Movements have related to the location in a declining hegemon in the last two centuries. By the time the United States realized that repression of sovereignty was not working, "popular" ideas of Indians emanated through publicity around the occupation of Alcatraz, the Trail of Broken Treaties, and the siege at Wounded Knee. Resistance activities of the late 1960s and early 1970s gained press coverage and support. The US government did mount serious efforts to suppress them, especially against AIM. A notorious example on Pine Ridge (Oglala) Lakota reservation in South Dakota became a veritable zone of "low intensity conflict," with unsolved murders and incarceration of leaders.

Conclusions

In September 2007, the United Nations General Assembly voted to adopt the Declaration on the Rights of Indigenous Peoples after three decades of struggle, acknowledging self-determination and autonomous relationships of indigenous peoples as separate communities with the right to organize their own governance, relationship to the land, economic practices, and sense of identity and membership. Only four countries worked against it, voting no—the United States, Canada, New Zealand, and Australia. Each of these countries has a treaty-basis with indigenous peoples, a strong and violent history of colonization by the English during the expansion out of Europe into the modern world-system, and is ensconced in hegemonic forces of globalization.

John Mohawk captured the essence of these struggles when observing, during that initial presentation to the United Nations in 1977, that the "Haudenosaunee position is derived from a philosophy which sees The People with historical roots which extend back tens of thousands of years ... And they are speaking, in this instance, to a world which dates its existence from a little over 500 years ago, and perhaps, in many cases, much more recently than that" (Mohawk 2000).

These events are the "tip of the iceberg" of indigenous struggles around the world. Great progress and duress for indigenous peoples will probably accompany a rise and demise of hegemonic forces, with recognition and continuing suppression of indigenous peoples by individual states continuing to shape, but not determine, their futures. Movements and "international" meetings of indigenous peoples react to and use forces and processes of globalization and the growing global civil society to aid their efforts, maintaining a strong focus on the lands of their grandmothers, the communities formed by their grandfathers, and on economies that support all their people.

References

Berkhofer, Robert. 1978. *The white man's Indian.* New York: Random House.

Biolsi, Thomas. 1995. "Bringing the Law Back In: Legal Rights and the Regulation of Indian-White Relations on the Rosebud Reservation." *Current Anthropology*, 36: (4): 543–71.

Bodley, John H. (ed.) 2003. *Power of Scale: A Global History Approach.* Armonk, NY: ME Sharpe.

Champagne, Duane. 2003. "Renewing American Indian Nations: Cosmic Communities and Spiritual Autonomy." pp. 167–81 in Philip Anderson (ed.) *Diversity and Community: A Critical Reader*, . Oxford: Blackwell. (republished pp. 9–24 in Duane Champagne (ed.) *Social Change and Cultural Continuity among Native Nations*, Lanham, MD: AltaMira Press.

Champagne, Duane. 2005. "Rethinking Native Relations with Contemporary Nation-States." pp. 1–33 in Duane Champagne, Karen Torjesen, and Susan Steiner (eds) *Indigenous People and the Modern State*, Walnut Creek, CA: AltaMira Press.

Champagne, Duane. 2007. *Social Change and Cultural Continuity among Native Nations.* Lanham, MD: AltaMira Press.

Chase-Dunn, Christopher and Thomas D. Hall. 1997. *Rise and Demise: Comparing World-Systems.* Boulder, CO: Westview Press.

Coates, Ken S. 2004. *A Global History of Indigenous Peoples: Struggle and Survival.* New York: Palgrave.

Cobo, Jose R. Martinez. 1986. "Problem of Discrimination against Indigenous Populations." United Nations, Doc. E/CN:4/Sub.2/1986/7 & Adds.4, PARAS. 380.

Dello Buono, Richard A. and José Bell Lara, (eds) 2007. *Imperialism, Neoliberalism and Social Struggles in Latin America.* Boston, MA: Brill

Deloria, Vine Jr. 1994. *God is Red: A Native View of Religion.* Golden, CO: North American Press.

Deloria, Vine Jr. 1997. *Red Earth, White Lies: Native Americans and the Myth of Scientific Fact.* New York: Scribner.

Deloria, Vine Jr. and David E. Wilkins. 1999. *Tribes, Treaties, and Constitutional Tribulations.* Austin: University of Texas Press.

Dunaway, Wilma A. 2000. "The International Fur Trade and Disempowerment of Cherokee Women, 1680–1775." pp. 195–210 in Thomas D. Hall (ed.) *A World-Systems Reader: New Pespectives on Gender, Urbanism, Cultures, Indigenous Peoples, and Ecology*, Lanham, MD: Rowman & Littlefield.

Fenelon, James V. 1998. *Culturicide, Resistance, and Survival of the Lakota (Sioux Nation).* New York: Garland Publishing (Routledge).

Fenelon, James V. 2002. "Dual Sovereignty of Native Nations, the United States, & Traditionalists." *Humboldt Journal of Social Relations*, 27(1): 106–45.

Fenelon, James V. 2007. "The Struggle of Indigenous Americans: A Socio-Historical View." pp. 15–38 in Hernan Vera and Joseph Feagin (eds) *Handbook of the Sociology of Racial and Ethnic Relations.* New York: Springer Press.

Fenelon, James V. and Thomas D. Hall. 2008. "Revitalization and Indigenous Resistance to Globalization and Neoliberalism." *American Behavioral Scientist* 51(12): 1867–1901.

Forbes, Jack. 1998. "The Urban Tradition among Native Americans." *American Indian Culture and Research Journal* 22(4): 15–41.

Frank, Andre Gunder. 1966. "The Development of Underdevelopment." *Monthly Review* September: 17–31 [Reprinted in many collections]

Gedicks, Al. 2001. *Resource Rebels: Native Challenges to Mining and Oil Corporations.* Cambridge, MA: South End Press.

Hall, Thomas D. 1998. "The Rio de La Plata and the Greater Southwest: A View from World-System Theory." pp. 150–66 in Donna Guy and Thomas Sheridan (eds) *Contested Ground: Comparative Frontiers on the Northern and Southern Edges of the Spanish Empire*, Tucson: University of Arizona Press.

Hall, Thomas D. and James V. Fenelon. 2009. *Indigenous Peoples & Globalization: Resistance and Revitalization.* Boulder, CO: Paradigm Publishers.

Harff, Barbara and Ted Robert Gurr. 2004. *Ethnic Conflict in World Politics*, 2nd edn., Boulder, CO: Westview Press.

Mander, Jerry and Victoria Tauli-Corpuz (eds) 2006. *Paradigm Wars, Indigenous People's Resistance to Globalization.* Berkeley: University of California Press.

Mignolo, Walter D. 2002. "The Zapatistas Theoretical Revolution: Its Historical, Ethical, and Political Consequences." *Review* 25(3): 245–75.

Mohawk, John C. 2000. *Utopian Legacies: A History of Conquest and Oppression in the Western World.* Santa Fe, NM: Clear Light Publishers.

Mohawk, John C. 2007. *Indian Country Today* (indiancountry.com), September 20, Original "From the Natural People to the Western World," in *A Basic Call to Consciousness: the Hau de no sau nee Address to the Western World*, edited by Six Nations. Mohawk Nation, Rooseveltown, NY: Akwesasne Notes.

Ramírez, Gloria Munoz. 2003. *EZLN: 20 y 10, el fuego y la palabra.* Mexico, DF: *La Jornada Ediciones.*

Perry, Richard J. 1996. *From Time Immemorial: Indigenous Peoples and State Systems.* Austin: University of Texas Press.

Robinson, William I. 2004. *A Theory of Global Capitalism: Production, Class, and State in Transnational World.* Baltimore, MD: Johns Hopkins University Press.

Ross, John. 1995. *Rebellion from the Roots: Indian Uprising in Chiapas.* Monroe, ME: Common Courage Press.

Sklair, Lelsie. 2002. *Globalization: Capitalism and Its Alternatives*, 3rd edn., Oxford: Oxford University Press.

Smith, Linda Tuwahi. 1999. *Decolonizing Methodologies Research and Indigenous Peoples.* Dunedin: University of Otago Press, Zed Books.

Snipp, C. Matthew. 1992. "Sociological Perspectives on American Indians." *Annual Review of Sociology*, 18: 351–70.

Snow, Dean R. 1996. *The Iroquois.* New York: Blackwell.

Thornton, Russell. 1987. *American Indian Holocaust and Survival.* Norman: University of Oklahoma Press.

Wallerstein, Immanuel. 2004. *World-Systems Analysis: An Introduction.* Durham, NC: Duke University Press.

Wilkins, David E. 2006. *American Indian Politics and the American Political System*, 2nd edn., Lanham, MD: Rowman & Littlefield.

Wolf, Eric R. 1982. *Europe and the People without History.* Berkeley: University of California Press.

9.3

Peasants, peasantries and (de)peasantization in the capitalist world-system

Eric Vanhaute

The survival and persistence of peasantries in a globalizing and ever more commodified world has been puzzling social scientists for a long time now. Time and again, the demise of the peasant was announced by capitalists, intellectuals, national, and development planners; "indeed, by virtually everyone but the peasants themselves" (Desmarais 2007: 195). However, as Wallerstein reminds us: "What is surprising is not that there has been so much proletarianization, but there has been so little. Four hundred years at least into the existence of a historical system, the amount of fully proletarianized labor in the capitalist world-economy today cannot be said to total even fifty percent" (Wallerstein 2003: 23). The very notion of peasants and peasantries confronts us with the flaws of traditional/mainstream economic development theory. Understanding old and new "agrarian questions" requires new historical knowledge about the role of peasantries within the long-term transformations in the capitalist world-system.

The mainstream image of the peasant and of peasantries is still deformed by a twofold myopia. First, the much praised English Road to capitalist agriculture, built on rapid depeasantization, has never been the standard road to development. The quasi permanent transformation of peasantries and small-scale agriculture within the expanding modern world-economy was much more diverse. Secondly, the European Experience, the dissolution of peasant societies within industrial and post-industrial economies, is neither inevitable nor desirable for most of the non-Western world. Because it was at the top of the modern world-system, nineteenth- and twentieth-century Europe could rather easily and cheaply dismantle its rural economies by importing the basic products it needed and exporting surplus labor to its old and new colonies. For most of the world, this is currently a very different story.

When we look beyond the old premises of westernized development, we see a different picture. We still see worlds of family- and village-based agricultural societies that combine diversified production chains and multiple strategies of risk minimization with locally and regionally anchored income and exchange systems. These include local markets, access to land and rights of use of common goods such as water and natural resources (Altieri and Nicholls 2005; McMichael 2006; Vanhaute 2008). These worlds are under increasing stress. This essay tries to understand the survival of the peasantry as a social process within historical capitalism.

Peasants as a social category

Social categories shape and reshape our social knowledge. They are constructs, ever redefined within changing social contexts. Social categories can also shape and reshape reality. When institutionalized, social categories become "bounded," they create boundaries and categorical differences. When these differences become durable, as Charles Tilly has argued, they create categorical inequality (Tilly 1998). Due to local, historical, and organizational contingencies, different sorts of categorical pairs referring to gender, class, and race occupy distinct positions in social life (Tilly 1998: 240). Stressing the exclusive character leads to essentialism that can create unilinear, often teleological and mostly biased explanatory stories. In social sciences, many of these categories became framed and institutionalized; a topic of intense academic discussions. This process of framing and eventual deconstruction is closely linked with the everyday struggle with social reality. "Since all groups are socially created, they are socially created for some purpose. And the purpose is to advance the rights (and privileges) of the group. (...) How local, regional, or transregional we wish to define the location of a group ... is function of the political alliances we are creating and recreating constantly" (Wallerstein 2007: 5–6). This also applies to the story of the peasant as a social category. The "essence" of the peasant is revealed in its most basic definition as "a countryman working on the land" and "a member of the class of farm laborers and small farmers" (*Oxford Advanced Learned Dictionary*). The dualistic view of the rural versus non-rural worlds, mostly with a negative connotation, can also be found in the French equivalent *paysan/ paysanne*. Disdain toward the "louts and oafs" has been part of the discourse of the wealthy, the powerful and the literate in Europe for a long time (Freedman 1999). Anette Desmarais has often repeated that the Anglo-Saxon concept of peasant continues to keep its narrow meaning, basically related to the European era of feudalism. The words *paysan* or *campesino* have *in se* a broader meaning but they still often refer to a social group from the (far away) past (Desmarais 2007, 2008).

In nineteenth and twentieth century modernization thinking, the peasant represented the left (starting) point on the axis of evolution, the traditional community and the opposite of modernity. In this "stationary" society, the economy was still dominated by agricultural subsistence activity; its output was consumed by the producers rather than traded. Production was labor intensive, using only limited quantities of capital, and social mobility was low. Western-based historiography has long developed and described the "anti-modern" model of a "familistic," family-based society, as a relatively undifferentiated economy of family farms and rural crafts and services, structured by internal agencies such as family, kinship, and village. A "peasant set of values" opposed the development of a new, open, mobile, individualistic, and market-based society. The success of modernity depended on "the degree to which the prevailing ideology of social relations was predicated on familistic, or individualistic, principles" (Schofield 1989: 304). The peasant's ambivalent relationship with the outside world is the main reason for the often schizophrenic scientific interpretation of this social group. Markets and exchange systems are the most visible, but also the most difficult relationship to grasp; see for example the famous quote of Fernand Braudel: "The peasant himself, when he regularly sells a part of his harvest and buys tools and clothing, is already a part of the market. But if he comes to the market town to sell a few items-eggs or a chicken-in order to obtain a few coins with which to pay his taxes or buy a plowshare, he is merely pressing his nose against the shop window of the marketplace" (Braudel 1977: 19). Market versus non-market, economic versus cultural forms of exchange, a long tradition of rural sociology is grafted upon these dichotomies.

The 1966 publication of the English translation of two texts by the Russian agrarian economist and rural sociologist Alexander V. Chayanov (1888–1937) triggered a new wave of peasant studies, and more importantly, a new debate about the nature of peasant societies (Thorner et al

1966). The two works, "Peasant farm organisation" and "On the theory of non-capitalist systems," written around 1925, compile Chayanov's main ideas. First, to explain the economic behavior of peasants, traditional concepts such as wages, rents, and profits do not apply. The absence of wage labor (and a labor market) and the predominance of a separate logic of household consumption-labor balance differentiates the peasant farm from capitalistic units of production. Secondly, a peasant economy is a distinct system (mode of production) within the national economy that is based on fundamentally non-capitalistic principles. Chayanov's definition of a peasant focuses, therefore, on the family as a production/consumption unit, or the "economic unit of a peasant family that does not employ paid workers" (Chayanov 1966: 1). Intense debates about Chayanov's work delegitimized former, ethnographical perceptions of peasant societies as undifferentiated, primitive, and static. In the 1970s and 1980s, a series of "local," "micro," and "village" studies attempted to understand the internal logics of survival within past and contemporary peasantries. Nevertheless, a formal demarcation line between peasant-based non-capitalist and capitalist economies frequently constituted the underlying macro-story.

Around the same time, the anthropologist Eric Wolf published his groundbreaking booklet "Peasants" (Wolf 1966). By framing the peasantry within an evolutionary time frame he rejected a binary, a-historical interpretation model: "This book is concerned with those large segments of mankind which stand midway between the primitive tribe and the industrial society" (Wolf 1966: VII). Moreover, he stressed the necessity to analyze peasant societies not outside, but within broader societal contexts: "Neither primitive nor modern," the story of peasant villagers "cannot be explained in terms of that village alone; the explanation must include consideration both of the outside forces impinging on these villages and of the reactions of villagers to these forces" (Wolf 1966: 1). He defined peasants as "rural cultivators whose surpluses are transformed to a dominant group of rulers that uses the surpluses both to underwrite its own standard of living and to distribute the remainder to groups in society that do not farm but must be fed for their specific goods and services in return" (Wolf 1966: 3–4). Tom Scott argued that "Eric Wolf's *Peasants* is something of a summary of, as well as a new departure from, these debates. (…) Wolf moves the debate beyond whether peasants were naturally conservative, values-rational, safety-oriented investors of their land and labor or whether they tended to be risk-taking, market-oriented maximizers, by showing that the coordinate strategies for balancing their private familial with their communal needs they had to be both—and that they engaged in a special 'peasant rationality' only in so far as this appeared in terms of agricultural and village contingencies that could vary greatly in proportion to the manner and complexity of their internal and external articulations with both local and wider markets" (Scott 1998: 197).

The wide and rich oeuvre of rural sociologist Teodor Shanin is a quasi permanent struggle with the difficult integration of internal and external analyses. His definition of peasants tries to reconcile the insights of Chayanov and Wolf: "Peasantry consists of small agricultural producers who, with the help of simple equipment, and the labour of their family, produce mainly for their own consumption and for the fulfillment of the holders of political and economic power" (Shanin 1990: 5; first published in 1971). Central concepts are the farm (the pursuit of an agricultural livelihood combining subsistence and commodity production), the family (internal social organization based on the family as the primary unit of production, consumption, reproduction, socialization, welfare, and risk-spreading), and class (external subordination to state authorities and regional or international markets which involve surplus extraction and class differentiation). What is largely missing is the community, the village society. He justifies his integrated view by arguing that "measuring peasant capitalism lies at the heart of the major concerns of contemporary social science. It has to do with capitalism as a process; it relates the understanding of the origins of our time to the characterization of the essential tenets of the global system we live in" (Shanin 1980: 89).

He criticized both classical and Marxist political economies that explained capitalism "outside peasant economies and societies" with the assumption "that capitalism equals de-peasantation" (Shanin 1980: 89). Peasants are neither remnants of the past nor victims of the present. Ethnographical research and modernization theory chained the peasant in static, a-historical narratives. The search for "other," "backward," "non-capitalist" characteristics and for separate modes of production has burdened peasant studies for a long time. This is especially true of their relationship with capitalism: "Ultimately peasantry is considered as a class whose significance will necessarily diminish with the further development of capitalism, as occurred in Europe a century ago" (Owen 2005: 369–71). On the other hand, the picture of the peasant as a (eternal) victim, part of dependency thinking that originated in the 1970s, gave birth to what Shanin has called a new essentialism, peasantism or peasantology (Shanin 1986: 6, 1990: 3).

Peasantry as a social process

Post-modern and globalization studies have amplified the thesis of "the end of peasantries" while dismissing the concept of the peasant altogether. This deconstruction and "hybridization" dispossesses history of its ability to shape contexts. The alternative, according to John Owen, is to construct "articulated" social concepts: "The question is thus a matter of reintroducing a localized concept of peasantry whilst acknowledging the extent of changing capitalist relations in places of articulation" (Owen 2005: 379). This peasant is a set of social relationships. The household is the basic economic unit and the gateway to the wider world. It is engaged in economic transactions for the main purpose of securing a level of subsistence, within the framework of a broader market economy. That is why the concept of the peasant needs to be contextually redefined in order to be sensitive to local situations and not to obscure non-capitalist entities into essentialist or dualistic frameworks such as agency-structure, west-rest, self-other, capitalist-non-capitalist (Owen 2005: 382).

In his manifold publications, Henry Bernstein challenges the view of peasants as a separate social category within the contemporary deruralizing world. He denies that contemporary peasantries constitute a general (and generic) social "type" or group, determined by a set of distinct qualities, from household subsistence over village solidarity to social/ecological harmony, as opposed to other social groups such as rural proletarians and market-oriented farmers (Bernstein 2010: 2–4). This so-called "peasant essentialism" is apparent in both historical (peasants as pre-capitalist remnants) and contemporary (agrarian populism) analyses. The rejection of a contemporary sui generis peasantry is supported by the "classic" view of class formation in the countryside, following the emergence of agrarian capital and wage labor. However, this "differentiation of the peasantry" also involved the transition to petty commodity production, with its varying scales of reproduction costs. Peasants, according to Bernstein, become petty commodity producers "when they are unable to reproduce themselves outside the relations and processes of capitalist commodity production, when those relations and processes become conditions of existence of peasant farming and are internalized in its organization and activity" (Bernstein 2003: 4). This model of peasant differentiation supplements the binary Marx/Lenin model, not by suppressing the peasantries as a social reality but by incorporating them (gradually) in a polarizing capitalist world-economy as producers of export crops, of food staples for domestic markets, and of labor power via (free or indentured) migrant labor systems. This created a large variety of systems of land tenure and differential forms of access to markets of land, labor, and credit (Bernstein 2003: 10, 2010: 101–12).

Within this framework, peasantry is an open concept that interacts within multiple forms and scales of action and conflict and leaves room for different levels of autonomy. (De-)peasantization is an ongoing process, both of adaptation and of resistance. Moreover, "like every social entity, peasantry exists in fact only as a process" (Shanin 1987: 6). Contemporary peasant studies since the 1990s have shown time and again how useless binary, static concepts are for understanding the fate of the rural and agrarian populations: "Peasantries are best understood as the historical outcome of an agrarian labour process which is constantly adjusting to surrounding conditions, be it fluctuations of climate, markets, state exactions, political regimes, as well as technical innovations, demographic trends, and environmental changes. These rural populations become peasants by degree and relinquish their peasant status only gradually over time" (Bryceson et al 2000: 2–3).

Peasantization and depeasantization within the capitalist world-system

The capitalist world-system has historically expanded and transformed in coexistence with frontier-zones or zones of contact (Hall 2000). These zones, where non-, semi- and fully-integrated actors and structures meet, are vital to the inherent expansive drive of historical capitalism. The processes of interaction that emanate from these contacts are challenged by pressures for incorporation into the modern world-system. These pressures contribute to the homogenization of the world-system by reducing its frontiers, but they simultaneously lead to heterogenization because they are answered by the (re)formulation of (new) frontiers. Throughout history, peasant societies and rural zones represent geographically dispersed frontier-zones. Rural communities are not and have never been able to escape the pressures of incorporation since coming into contact with the modern world-system (Wallerstein 1974: 400). In response, they have been developing strategies for survival and resistance, articulated toward expanding state power, expanding market relations, class struggle, ethno-cultural identity. Over the long-term, the scales on which these social power relations are expressed have not only been widening and multiplying, they have also become increasingly interdependent. This is translated in the interconnected processes of de- and re-peasantization.

For more than a century, debates about the "peasant question" have been dominated by two groups of protagonists (Araghi 1995: 338–43). On the one hand, the "disappearance thesis" defends that the expansion of capitalism will lead to the extermination of the peasantry. Following Lenin and Kautsky, the former, more or less undifferentiated class of peasants is transformed into new, distinct groups: capital owners (capitalist farmers) and wage laborers. On the other hand, advocates of the "permanence thesis" argue that, according to Chayanov's peasant mode of production, peasant societies have a distinct development logic that supports the survival of the peasantry within capitalism. Araghi labels the first option as teleological and the second as essentialist, both suffering from a-historical and often functionalistic presumptions. According to Araghi, "depeasantization has been neither a unilinear process, nor has it taken the historically particular form of differentiation in the countryside within each and every nation-state" (Araghi 1995: 359).

Over time, the combined process of overburdening, restricting, and reducing peasant spaces has considerably weakened the material basis of this once so dominant economic system. From this point of view, the twenty-first century seems to become the era of "the end of peasantries." In 2007, the United Nations declared that for the first time in human history, more than half the world's population was living in cities and towns (United Nations 2007). Less developed regions will hit the halfway point later, but more than likely before 2020. However urbanization, as one of the major indices of the process of "modernization," is a very narrow and often misinterpreted index. The concept of depeasantization has to be understood as a multi-layered process of erosion of an agrarian way of life, the increasing difficulty to combine subsistence and commodity agricultural production with an internal social organization based on family labor and village community settlement (Bryceson 1999a: 175). As Heather Johnson has stressed, the biggest problem with the concept of depeasantization is its (mostly inherent and often not explicated) links with urbanization, industrialization, development, and marginalization. Measuring this process is risky and difficult, not only because of the diffuse strategies of labor and income pooling within households, but also because seemingly opposite processes such as urbanization and migration can become part of rural income strategies. This also includes a diversification of rural coping mechanisms, such as petty commodity production, rural wage labor, seasonal migration, subcontracting to (multinational) corporations, self-employment, remittances, and income transitions. Rural-urban migration patterns are often part of rural household strategies as in the form of two-way remittances (Johnson 2004: 56, 61). What is often regarded as "depeasantization" is, in essence, part of more diversified labor and income strategies of the peasantry. Due to the marginalization of a growing part of the world's population, these mixed survival strategies become more important than ever. In his recent works, Jan Douwe van der Ploeg coined these revived multi-level strategies of survival, autonomy and resistance a "recreation of a peasant strategy" (van der Ploeg 2010: 20–23). This century would even witness a new turning point, via a re-emergence of the peasantry. One of the signs, according to van der Ploeg, is that in response to the agrarian crisis of the last decades in many regions, farming is increasingly being restructured in a peasant-like way.

Because of these complex transformations, depeasantization (the erosion of an agrarian way of life) is supplemented with the concepts of deruralization (as a synonym of urbanization, or the decline of rural areas) and deagrarianization (Bryceson 1996). Deagrarianization refers to the process of income differentiation, resulting in the long-term decline of agrarian-based activities and a shrinking self-sufficiency. Deagrarianization (a decline of reliance on agriculture within the diversification of livelihood) does not necessarily imply depeasantization (the erosion of the family basis of their livelihoods) (Ellis 2006: 387). Diversification has always been part of peasant survival strategies. The accelerated process of erosion is a sign of the growth of highly vulnerable peasant populations in the last two decades. Depeasantization can be seen as a specific form of deagrarianization in which peasantries lose their economic capacity and social coherence, and shrink in size. Conversely, peasantization can be regarded as an answer to deagrarianization, when the loss of (an exclusive) agrarian income is supplemented by other forms of income pooled by the rural household. On a global scale, processes of deagrarianization in the core zones often created new peasantries in the periphery. For example, twentieth-century colonialism engendered processes of peasantization that facilitated the colonial government's agricultural commodity export aims. Spurred by colonial taxation, African agrarian producers increasingly produced agricultural commodities in conjunction with their subsistence production, or alternatively exported male labor on the basis of circular migration. Recent forces of deagrarianization are triggered by the enforcement of neoliberal policies and Structural Adjustment Plans. This often stimulated rural producers to reallocate land and labor to smaller residential "garden" plots whose output is oriented to domestic production and gift-giving rather than commercial sale (Bryceson 1999b: 2–7).

Conclusion: New peasant strategies

In a retrospective of their 1977 pamphlet "Theses on Peasantry," Johnson, Wisner, and O'Keefe list what they see as the most important research questions regarding the twenty-first century peasantry (Johnson et al 2005: 951–52). These include peasant production and knowledge systems, peasant land holding (access to land, land rights, land use), peasant food production and food systems (food security, food sovereignty), rural migration and remittances, and peasant movements and forms of resistance. This implies, in their words, a mental and ideological repeasantization, the resurrection of a peasant movement. What might this look like?

The early twenty-first century has put peasants back on the global agenda of governmental and non-governmental institutions alike. In recent reports, the World Bank has revalued smallholder farming as "a powerful path out of poverty" (World Development Report 2008) while still propagating the imperial road of "commodification." After five centuries of capitalism, two centuries of industrialization, and three decades of neo-liberal globalization, self-provisioning family farming continues to be a major mode of livelihood in the twenty-first century world. A large part of world food production remains in the hands of small-scale sustainable farmers, outside the control of large agribusiness companies or supermarket chains. Millions of small farmers in the South still produce the majority of staple crops needed to feed the planet's rural and urban populations. Small increases in yields on these small farms that produce most of the world's staple crops will have far more impact on food availability at the local and regional levels than the uncertain increases predicted for large corporate-controlled monocultures (Altieri and Nicholls 2005). In this context, a strategy of "peasantization" can be a powerful answer to real marginalization. Massive declines in the reliance on agriculture (deagrarianization), the erosion of the family basis of peasant livelihoods (depeasantization), and an exodus from the countryside (deruralization and "slumpization") are quickly redefining the place and the nature of the remaining peasantries.

Vulnerability, the link between risk and the precariousness of people's livelihood, has always been part of the peasant's existence; a diversification of income and coping strategies (individual, in the household and in the village) has always been the main answer. However, a continuing erosion of the family basis of livelihoods has created new and more massive forms of vulnerability. According to Frank Ellis, vulnerability has switched from a temporary to a structural state of being (Ellis 2006: 393). This is countered by the intensification of old and the introduction of new forms of livelihood diversification such as taking up non-farm activities and relying on non-farm income transfers. Rural household income becomes less based on farm activities and on the exploitation of own assets. This erodes former household and village security mechanisms and affects their ability to overcome short-term economic stress, such as harvest shortages or variations in income or food prices from one year to the next or even within shorter time spans (Bengsston 2004: 33–35; Vanhaute 2011). Three decades of economic liberalization and institutional restructuring, resulting in multiple and intensified involvement in markets—for commodities, credit, technology, land, and all kinds of services—have created growing and interconnected vulnerabilities and new risks. New forms of organized peasant reactions, such as *Via Campensina*, try to formulate an answer to the predominantly neoliberal mode of food production (Patel 2006: 84–85). Food sovereignty, control over one's own food production and food markets, is put forward as an alternative for food security, a concept agnostic about food production systems. A call for localizing food power implies support for domestic food production and promotion of the return to smallholder farming (Holt-Giménez 2008: 13–14). At the same time, peasant's rights are now defined as a set of "transgressive rights," challenging the primacy of the nation-state and calling for international (international business) and even universal (human rights) spaces (Edelman 2005; Patel 2007). This clarifies how the present material and ideological struggles for "peasant spaces" (McMichael 2008) put the peasantry in the hearth of the twenty-first century's systemic crisis.

References

Altieri, Miguel and Nicholls, Clara (2005) *Agroecology and the search for a truly sustainable agriculture*, Berkeley: University of California.

Araghi, Farshad (1995) "Global depeasantization, 1945–90", *The Sociological Quarterly*, 36: 601–32.
Bengsston, Tommy (2004) "Living standards and economic stress", pp. 27–60 in T. Bengston, C. Campbell, J. Z. Lee et al (eds), *Life under pressure. Mortality and living standards in Europe and Asia, 1700–1900*, Cambridge, MA: MIT Press.
Bernstein, Henry (2003) "Farewells to the peasantry", *Transformation*, 52: 1–19.
Bernstein, Henry (2010) *Class dynamics of agrarian change*, Halifax and Winnipeg: Fernwood Books.
Braudel, Fernand (1977) *Afterthoughts on material civilisation and capitalism*, Baltimore, MD: Johns Hopkins University Press.
Bryceson, Deborah Fahey (1996) "Deagrarianization and rural employment in sub-Sahara Africa. A sectoral perspective", *World Development*, 24, 1: 97–111.
Bryceson, Deborah Fahy (1999a) *African rural labour, income diversification and livelihood approaches: A long-term development perspective*, Leiden: African Studies Center.
Bryceson, Deborah Fahy (1999b) *Sub-Sahara Africa betwixt and between. Rural livelihood practices and policies*, Leiden: African Studies Center.
Bryceson, Deborah Fahey, Kay C. and Mooij J. (eds) (2000) *Disappearing peasantries? Rural labour in Africa, Asia and Latin America*, London: Intermediate Technology Publications.
Chayanov, Alexander (1966) "On the theory of non-capitalist economic systems", in D. Thorner, B. Kerblay and R. Smith (eds), *The Theory of Peasant Economy*, Homewood, IL: Irwin.
Desmarais, Annette A. (2007) *Globalization and the power of peasants. La Vía Campesina*, Halifax: Fernwood Books.
Desmarais, Annette A. (2008) "The power of peasants: reflections on the meanings of La Via Campesina", *Journal of Rural Studies*, 24, 2: 138–49.
Edelman, Marc (2005) "Bringing the moral economy back in … to the study of 21st-century transnational peasant movements", *American Anthropologist*, 107, 3: 331–97.
Ellis, Frank (2006) "Agrarian change and rising vulnerability in rural Sub-Saharan Africa", *New Political Economy*, 11, 3: 387–97.
Freedman, Paul (1999) *Images of the medieval peasant*, Stanford, CA: Stanford University Press.
Hall, Thomas D. (2000) "Frontiers, ethnogenesis and world-systems: rethinking the theories", pp. 237–70 in Thomas D. Hall (ed.), *A world-systems reader. New perspectives on gender, urbanism, cultures, indigenous peoples, and ecology*, Oxford: Rowman & Littlefield Publishers.
Holt-Giménez, Eric (2008) "The World food crisis. What's behind it and what can we do about it?", *Food First. Institute for Food and Development Policy (Policy Brief nr 16)*.
Johnson, Heather (2004) "Subsistence and control: The persistence of the peasantry in the developing world", *Undercurrent*, 1, 1: 55–65.
Johnson, Kirsten, Wisner, Ben and O'Keefe, Phil (2005) "'Theses on peasantry' revisited", *Antipode*, 37, 5: 944–55.
McMichael, Philip (2006) "Peasant prospects in the neoliberal age", *New Political Economy*, 11, 3: 407–18.
McMichael, Philip (2008) "Peasants make their own history, but not just as they please …", *Journal of Agrarian Change*, 8, 2–3: 205–28.
Owen, John R. (2005) "In defence of the 'peasant'", *Journal of Contemporary Asia*, 35, 3: 368–85.
Patel, Rajeev (2006) "International agrarian restructuring and the practical ethics of peasant movement solidarity", *Journal of Asian and African Studies*, 41, 1–2: 71–93.
Patel, Rajeev (2007) "Transgressing rights. La Via Campensina's call for food sovereignty", *Feminist Economics*, 13, 1: 87–93.
Schofield, Roger (1989) "Family structure, demographic behaviour, and economic growth", pp. 279–304 in J. Walter and R. Schofield (eds), *Famine, Disease and the Social Order in Early Modern Society*, Cambridge: Cambridge University Press.
Scott, Tom (ed.) (1998) *The peasantries of Europe from the fourteenth to the twentieth centuries*, London and New York: Longman.
Shanin, Teodor (1980) "Measuring peasant capitalism", pp. 89–104 in E. J. Hobsbawm et al (eds), *Peasants in history. Essays in honour of Daniel Thorner*, Oxford: Oxford University Press.
Shanin, Teodor (1986) "Chayanov's message. Illuminations, miscomprehensions, and the contemporary 'Development Theory'", pp. 1–24 in A.V. Chayanov (ed.), *The Theory of Peasant Economy*, Madison: The University of Wisconsin Press.
Shanin, Teodor (1987) "Introduction. Peasantry as a concept", in T. Shanin (ed.), *Peasants and Peasant Societies. Selected Readings*, Oxford and New York: Basil Blackwell.
Shanin, Teodor (1990) "Agenda's of peasant studies and the perception of parallel realities", in T. Shanin (ed.), *Defining peasants*, Oxford and New York: Basil Blackwell.

United Nations (2007) *State of World Population 2007*, Report from the United Nations. Data from *World Urbanization Prospects. The 2007 Revision Population Database* (http://esa.un.org/unup).

Thorner, Daniel, Kerblay, Basile and Smith, R. E. F. (eds) (1966) *The theory of peasant economy*, Homewood, IL: Irwin.

Tilly, Charles (1998) *Durable inequality*, Berkeley and Los Angeles: University of California Press.

Van der Ploeg, Jan Douwe (2010) "The peasantries of the twenty-first century: the commoditisation debate revisited", *The Journal of Peasant Studies*, 37, 1: 1–30.

Vanhaute, Eric (2008) "The end of peasantries? Rethinking the role of peasantries in a world-historical view", *Review of the Fernand Braudel Center*, 31, 1: 39–59.

Vanhaute, Eric (2011) "From famine to food crisis: what history can teach us about local and global subsistence crises", *Journal of Peasant Studies*, 38, 1: 47–65.

Wallerstein, Immanuel (1974) "The rise and future demise of the of the world-capitalist system. Concepts for comparative analysis", *Comparative Studies in Society and History*, 16, 4: 387–415.

Wallerstein, Immanuel (2003) *Historical capitalism with capitalist civilization*, London: Verso.

Wallerstein, Immanuel (2007) "Naming groups. The politics of categorizing and identities", *Review of the Fernand Braudel Center*, 30, 1: 1–15.

Wolf, Eric (1966) *Peasants*, Englewood Cliffs, NJ: Prentice-Hall.

World Development Report (2008) Agriculture for development. www.worldbank.org.

9.4

Chiefdom world-systems (with a focus on Hawaii, 1390–1790)

Elena Ermolaeva

It is extremely difficult to study chiefdom world-systems. Chiefdoms are developmentally located in between the kin-based and state societies, having features of both. The chiefdom level should be viewed on a scale which requires a precise empirical documentation of its positioning. For these reasons, the affirmative models (usually expected in sciences) are problematic for chiefdoms and can be hardly generalized.

The researchers of a specific chiefdom (e.g., Hawaii before 1779) are in disagreement about the evolutionary stage reached by pre-contact society; the trajectory of the developmental process; the mechanisms which contributed or undercut hierarchy formation; the significance of regional variations, and the role of culture in the political evolution.

In case of the ancient Hawaiian world-system, the focus is on five major islands—The Big Island of Hawai'i, Maui, Oahu, Molokai, and Kauai. The time frame is from 1390 CE (the beginning of Early Traditional period) to 1790 (when Kamehameha fought with Kalanikupule and for the first time he used guns which were provided by Europeans).

Until recently, many Western observers had assumed that Pacific Island societies were geographically isolated and culturally homogeneous. The historical and ethnographic material presented contradicts such easy generalizations (Earle 1977; Hommon 1976; Kirch 1984; Sahlins 1985). The research on Hawaii documented more than 40 archipelago-wide differences in material culture, occupations, traditions, religion, and political behavior (Ermolaeva 1997).

The evidence allows us to speak about intersocietal differentiation and hierarchy (Chase-Dunn and Hall 1991). The islands differed in population density, degrees of stratification, size of polities, and technological intensification. During circa 400 years, the core/periphery relations within the archipelago were changing. In the Hawaiian context, the process of material accumulation is "at risk" as the result of pursuing either a wife-taking or wife-giving strategy. As explained by Valrio Valeri (1972, 1985), ruling chiefs seek to take wives only, but if they do so they risk rebellion from their dissatisfied juniors; if they give, they predispose themselves to eventual decline because ceding a sister to a junior relative carries the potential for subverting the senior line's authority in the next generation. The history of the older cores—Kauai and Maui—fits nicely into this model. Only the Big Island of Hawai'i had a protracted competitive advantage over the others in the system. The junior chiefs were pursuing a wife-taking strategy and they counterbalanced wife-giving by incestuous marriages.

The Hawaiian narrative on warfare presents rich evidence of the transition from a kin-based to a state order (in spite assertions that it is not possible to document the transition empirically). The strengthening of internal structures of power is associated with the formation of hierarchical archipelago-wide power configurations. First, circa 1580–1600, Kauai became a great power, but shortly after that, in 1600–620, Maui had reached the same status in the system. This bipolarity continued for more than 100 years when, circa 1720–40, Oahu became a superpower. Oahu did not exhibit core-like characteristics based on marriage data, it pursued relatively egalitarian marriage relations with its neighbors. However, the alliances which Oahu built around intermarriages contributed to its rise in a politico-military sense. Oahu's dominance did not last long; in 1740–60 Maui took the firm status of a superpower up until 1775–80. At this point, we witness the rise of a new core—the Big Island of Hawai'i.

The analysis of forms of interactions as a basis of social power would be incomplete without a study of symbolic or ideational interactions. The case here is an information exchange through the medium of travelers and messengers. The continuous measure of centrality pointed to Maui, Oahu, and Molokai as the most central in the interisland information network. Interisland news exchange can be viewed in the context of geopolitical alliances—bipolarity versus unipolarity—because of the changing number of group members in the interisland structure of news flow. The presence of six group members before 1700 (bipolarity) as opposed to four group members after 1700 (unipolarity)—with the previous group members being relegated to the position of attached isolates—indicates the process of polarization and hierarchy formation in the system.

Pololu valley on the Big Island was a particular kind of semiperiphery, a buffer zone, and it was an integral element in the Hawaiian inter-chiefdom political economy. The search for independent resources led the chiefs to gain control over the lands that had not already been allocated and, as a correlate, were spatially distant from immediate subalterns. Pololu buffer zone assisted the great powers and local usurpers in arms supply and military support.

Civilizationists (Friedman 1994; Wilkinson 1985) argue that certain conditions favor state systems, other conditions may favor empire formation. Specifically, ethnocultural heterogeneity is a correlate with state systems, homogeneity—with empires. Two tendencies were at work in Hawaii: cultural uniformity created pre-conditions for an empire formation; cultural variations supported an inter-chiefdom system. The net result was an unstable balance between the two and four hundred years of oscillation between a chiefdom social formation and a state.

References

Chase-Dunn, C. and Hall, T. (eds) (1991) *Core/periphery Relations in Precapitalist Worlds*, Boulder, CO: Westview Press.

Earle, T. (1977) "A reappraisal of redistribution: complex Hawaiian chiefdoms", in T. K. Earle and J. F. Ericson (eds) *Exchange Systems in Prehistory*, New York: Academic Press.

Ermolaeva, E. (1997) "Chiefdom to state: cultural identity and hierarchy formation in the ancient Hawaiian world-system", unpublished Ph.D. dissertation, The Johns Hopkins University.

Friedman, J. (1994) *Cultural Identity and Global Process*, London: Thousand Oaks; CA: Sage Publications.

Hommon, R. (1976) "The formation of primitive states in pre-contact Hawaii", unpublished Ph.D. dissertation, University of Arizona.

Kirch, P. V. (1984) *The Evolution of the Polynesian Chiefdom*, Cambridge: Cambridge University Press.

Sahlins, M. (1985) *Islands of History*, Chicago, IL: University of Chicago Press.

Valeri, V. (1972) "Le fonctionnement du systeme des rangs a Hawaii", *L'Homme*, 12: 29–66.

Valeri, V. (1985) *Kingship and Sacrifice. Ritual and Society in Ancient Hawaii*, Chicago, IL, and London: The University of Chicago Press.

Wilkinson, D. (1985) "States systems: ethos and pathos", paper presented at the meeting of the International Society for the Comparative Study of Civilizations, Yellow Springs, Ohio.

10. Models of growth and stagnation

10.1

Position and mobility in the contemporary world-economy

A structuralist perspective

Salvatore J. Babones

Introduction

Viewed from a very broad perspective, the distribution of rewards in the modern world-economy has been incredibly stable over time. Since the full incorporation of all areas of the world into the Europe-dominated world-economy in the late nineteenth century there has been very little change in relative incomes. The richest areas then (Europe and North America) are still rich today and the poorest areas then (Africa, indigenous central America, south and southeast Asia) are still poor today. Coastal China is still somewhere in the middle. Parts of east Asia have become very rich, but the richest part of all—Japan—was already middle-income in the late nineteenth century. In fact, according to data from Maddison (2010) the correlation of logged national income per capita in 1870 with logged national income per capita in 2008 for the 10 major world regions is a remarkable $r = 0.93$. There has been virtually no change in relative incomes in over a century of otherwise massive change.

Neoclassical economists have been predicting the convergence of poor countries with rich countries ever since the 1950s. Neoclassical economics predicts that capital will flow from rich countries to poor countries, labor will flow from poor countries to rich countries, and technology will equalize everywhere, leading to one great global economic convergence. Since most neoclassical economists have been based in the United States, their assumption has generally been that the rest of the world will converge to US levels of economic productivity; they have universally ignored the obvious implication of their arguments, which is that all countries would be expected to converge to the global mean (i.e., Mexico), not to the global maximum. In any case, the simple fact is that neoclassical unconditional convergence has absolutely failed to occur (Babones 2009).

Sociologists at least would not be surprised to find out that there is a long-lasting structure to the world-economy. Wallerstein (1974a) seminally recognized the existence (and persistence) of a three-tier structure in the modern world-economy. He adapted the dependency school approach of Frank (1967) and other Latin Americanists that divided locations in the world-economy into a metropolitan "core" and a marginal "periphery," inserting it within a more traditionally sociological framework of class analysis. Wallerstein went further, however, to distinguish economic and political systems, embedding one inside the other. Where an economic system includes within it multiple polities, a capitalist world-economy results. Where a political system includes within it multiple economies, a military world-empire results. World-economies and world-empires together comprise the two historical types of world-systems.

Wallerstein went on to identify three broad structural locations in both kinds of world-system: the core and periphery, certainly, but in addition to these an intermediate position he called the semiperiphery. In world-empires, Wallerstein identified long-distance luxury trading networks as occupying a semiperipheral position. In world-economies, he identified a group of widely-disparate sovereign states that were both "exploited and exploiter" (Wallerstein 1974b: 405). Wallerstein (1974b) argued that the existence of the semiperiphery is not just an empirical fact, but a necessary factor in ensuring the stability of the world-system. He explicitly argued that in the absence of a semiperiphery, the world's periphery would ultimately unite and overwhelm the (numerically much inferior) core. The existence of semiperiphery polities prevents the unification of the exploited areas of the world-system by setting their immediate interests against each other.

Wallerstein (1974a) observed that world-systems throughout history have always had three-tier structures, whether they were world-economies like today's world-system or world-empires like ancient Rome. Given the very long time frames over which world-systems can exist (Hall and Chase-Dunn (2006) trace the direct lineage of today's world-system back to 1500 BCE), this implies that the specific identities of the core, semiperipheral, and peripheral areas can change over time. In fact, Wallerstein argued throughout his early work that though the structure of core, semiperiphery, and periphery was always present in world-systems, the specific identities of the particular countries and areas that made up the core, semiperiphery, and periphery change as an ordinary feature of world-system dynamics. He argued that even in the modern world-economy, countries and areas regularly move up and down in the systemic hierarchy.

Surely in the *longue durée* they do. The question is: how *longue* is *longue*? In this essay I argue that countries' locations in the contemporary world-economy are relatively fixed. Certainly there has been mobility since the era of the ancient Hittites—or the era of the Hapsburgs—but that is hardly relevant to most people and policymakers today. Over the *moyenne durée* in which the overall character of the system remains stable, countries' locations in the core, semiperiphery, and periphery generally do not change. Considering the contemporary world-system to date from the point in the late nineteenth when the modern world-system came to circumscribe the entire geographical world, there has been virtually no mobility at all. This leads me to propose a new, structuralist theory of world-system stratification that may explain the observed patters of stability (and mobility) better than the canonical approach derived from Wallerstein.

A structuralist perspective on position and mobility

In the contemporary world-economy, a single system-wide division of labor cuts across many different countries. Though individuals occupy economic locations in this unified division of labor, countries occupy political locations. Core, semiperiphery, and periphery thus describe differing political (not economic) locations in the world-system. Countries—or at least the state apparatuses that govern countries—are influenced by at least a portion of their citizens and serve as vehicles through which these citizens attempt to gain advantages in the super-national world market. Countries generally do not, however, participate directly in that market. What then makes a specific country a core, semiperipheral, or peripheral country?

The answer to this question is not well-defined in Wallerstein, who takes a historicist approach to the emergence of world-system structure. Arrighi and Drangel (1986) identified this as a major gap in our understanding of world-systems and offered an economistic answer based on locations in commodity chains. Arrighi and Drangel divided commodity chains into "core-type," high value-added nodes and "peripheral-type," low value-added nodes. In their view, a core country is one in which a large proportion of the population participates in economic activities that correspond to core-type nodes, while a peripheral country is one in which a large proportion of the population participates in peripheral-type nodes. Countries are continually working through their state machineries to upgrade the activities within their borders into core nodes (or prevent their sliding into peripheral nodes). A semiperipheral country is simply one in which there is a "more or less even mix" (p. 26) of core and peripheral nodes.

In contrast with Wallerstein's (rather teleological) argument that the world-system must have three tiers to remain stable, Arrighi and Drangel's explanation of world-system structure is at least theoretically-grounded. Nonetheless, in its details it strains credulity. Arrighi and Drangel claimed that semiperipheral states are active in resisting the slide of their countries into the periphery through careful management of their economies, using an ever-changing mix of selective protection of their home markets and competitive exploitation of their low-cost advantages to maintain their precarious middle positions. This seems more just-so story than empirically-grounded reality. If there is little empirical justification for Wallerstein's (1974b) claim that semiperipheral countries today are some kind of conduit for the exploitation of peripheral countries by core countries, there is even less for Arrighi and Drangel's (1986) account of canny semiperipheral states outsmarting the system to maintain (somewhat) privileged economic positions for their citizens.

A more realistic theory to explain the three-tier structure of countries in the contemporary world-system can be derived from a crumb dropped from Wallerstein's grand table. Wallerstein (1979: 72) observed that "a clear and distinctive feature of a semiperipheral state" is the fact that "[t]he direct and immediate interest of the state as a political machinery in the control of the market (internal and international) is greater than in either the core [or] the peripheral states." Arrighi and Drangel (1986) similarly observed the very large role played by semiperipheral states in their own countries' economies. Given the nature of core, semiperiphery, and periphery as the locations of countries—not individuals—in the modern world-economy, it would make sense to derive the categories from the attributes of countries, not as some kind of average of the attributes of the individuals who live within them. In other words, individuals participate in "core-type" or "peripheral-type" activities because of the systemic positions of the countries in which they live, not the other way around.

In this structure-first or "structuralist" perspective, core countries are seen as those that are governed by state machineries that are strong enough to secure major advantages in the world market for their own citizens, especially through the exploitation of populations in peripheral countries. So far, so Wallerstein. Peripheral countries are those that do not possess strong enough states to prevent the exploitation of their populations by economic actors located in countries and operating with the support of core states. So far, so Frank. The main difference between the structuralist perspective and the canonical approach in world-systems analysis is in its understanding of the semiperiphery. In the structuralist approach, the semiperiphery is made up of those countries in which the state is strong enough to retard major exploitation from the core but not strong enough to project its own power externally. Instead of focusing on helping its citizens engage in external exploitation, semiperipheral states mainly focus on helping internal elites exploit their own populations.

Today's (unambiguously) core countries are—without exception—either (1) the imperial countries that conquered major colonial empires during the rapid expansion of the Europe-centered world-economy through the end of the nineteenth century or (2) countries that are both geographically and culturally contiguous with those imperial countries (and which participated in imperial colonialism indirectly through negotiating access rights for their citizens). All of these countries gained their current positions through dispossession of one form or another, directly or indirectly. The formerly imperial countries that are not current core countries—China, Russia, Turkey—acquired their empires before being incorporated into the modern world-economy and were incorporated (empires included) as semiperipheries. Japan, it should be noted, embarked on its program of colonialism only after incorporation and in an explicit policy effort to avoid peripheralization.

In contrast, today's (unambiguously) peripheral countries have, for the most part, never been allowed to develop strong state machineries. Throughout sub-Saharan Africa (with the exception of South Africa) the currently existing countries have culturally and geographically arbitrary borders and have, for the most part, never in all of history been governed as countries by unified state machineries. There once were strong African polities, but these were long ago shattered Humpty-Dumpty style and have never been put back together again. The same is true in Central America and the Andean region. Throughout Latin America, as in Africa, only the European settler states have avoided peripheralization. In Asia the most peripheral states are those in which the state machinery was imposed and then withdrawn (French southeast Asia) or in which there had never existed any state machinery to speak of (Afghanistan, Bangladesh).

The semiperiphery, on the other hand, is composed of countries that are governed by states that are strong enough to police their borders, collect taxes and customs, and maintain effective rule of law. In other words, semiperipheral countries are those that possess states that are strong enough to enforce "order"—including exploitative property rights—throughout their territories. In this view, semiperipheral status doesn't guarantee that a country will achieve a medium income level, but medium income levels are only achievable in semiperipheral countries. In Arrighi and Drangel's terms, semiperipheral countries (unlike peripheral countries) can support core-type nodes. On the other hand, it is very difficult for a semiperipheral country to break into the core of the world-economy, especially in an increasingly open global economy.

The differences between the structuralist perspective presented here and the received wisdom in world-systems analysis may seem minor, but they have major implications for understanding economic growth trajectories outside the core of the world-economy. Wallerstein and Arrighi find development opportunities for the periphery and semiperiphery at times when the larger world-economy is in crisis, particularly during what they see as Kondratieff "B" phases. The structuralist perspective would suggest no such opportunities. Instead, the structuralist perspective would suggest that the main route to improved living standards in peripheral countries is through the development of strong, effective government, while the main route in semiperipheral countries is through the transition from government in the interests of elites to government in the interests of all citizens.

The empirical literature on structural income mobility

What are the empirical facts on mobility in the contemporary world-economy? There have been a dozen or so major empirical studies of the structure of the world-economy. Most adopt a network approach, analyzing networks of international trade (and sometimes investment) flows to arrive at measures of world-economic centrality. A recent example that reviews most of the prior literature is Clark and Beckfield (2009). Network-based studies have generally shown a three-tier structure in the international trading system that corresponds reasonably well to qualitative categorization of countries into core, peripheral, and semiperipheral locations. Though not based on income, the results of network-analytic studies show that the world's trading system is patterned similarly to its income distribution.

Another stream of research on world-system structure has focused specifically on the distribution of national income per capita. These studies have shown an empirical stratification of the world-economy by national income into core, semiperipheral, and peripheral bands. Very roughly speaking, core countries have been identified as having national income levels over $20,000 per person per year, peripheral countries under $2000, and semiperipheral countries in between $2000 and $20,000. Separating each income category is a range of national income levels (around $2000 and around $20,000) that are relatively under-populated, suggesting that the core, semiperipheral, and peripheral bands represent distinct hierarchical positions, not just locations on a continuum. The results of income-based studies generally agree with the results of network-based studies. A third approach, based on the discriminant analysis of development indicators (Dezzani 2001) similarly identifies three distinct bands.

A major focus of these studies of system structure has been on quantifying the degree of mobility in the world-economy. Focusing on income-based studies as the most relevant here, Arrighi and Drangel (1986) found that 95 percent of the world's countries occupied substantially the same position in 1975/1983 as compared to their starting points in 1938/1950. Using a much wider panel of countries, Babones (2005) similarly reports that 83 percent of countries occupied the same position continuously between the years 1975–2002. Out of 103 countries studied, only 17 changed position, and 15 of these 17 changes were upward and downward mobility between the semiperiphery and the periphery. Dezzani (2002) also found stratum persistence rates of between 72 percent and 100 percent over the period 1960–90. Given the r = 0.93 long-term correlation of national income levels cited at the beginning of this essay, such high levels of positional persistence are not surprising.

Though there have been many cases of mobility over the *moyenne durée* covered in these studies, they are mostly cases of small countries that were already close to the boundary between two structural locations. Moreover, as Arrighi and Drangel (1986: 47–49) were at pains to point out, in the majority of cases the observed mobility is temporary: countries revert to their starting positions after some fleeting run of good or bad macroeconomic or geopolitical luck. I would go further than Arrighi and Drangel and argue that income level is just an indicator of world-economic position, not world-economic position itself. In the structuralist understanding of world-economic position, structural positions are political, not economic. Countries' income levels (and trading positions) serve as reasonably good indicators of those positions, but the positions themselves are much less volatile.

The tendency of mobile countries to return to their starting points is strong (though indirect) evidence in support of this view. Fundamentally the issue is this: when a country is affected by a random economic shock (e.g., the discovery of vast oil reserves), does that place the country in a new position (and consequently a new development trajectory)? Or does it provide a temporary one-off boost that will slowly disappear over time as old patterns, based on internal politics, reassert themselves? In the next section I consider this question in light of actual examples of income mobility in the contemporary world-economy.

Specific cases of high income mobility

Since the late nineteenth century, there have been three kinds of mobility in the world-economy that together encompass almost all cases of countries making substantial moves up or down the national income hierarchy of the world-economy. These are (1) mobility due to swings in commodities prices, (2) mobility due to the growing economic primacy of cities, and (3) mobility due to returns to equilibrium. All three types are more consistent with the structuralist approach than with the canonical approach to world-system structure. The only major case of country mobility in the contemporary world-system not covered by these three types is Japan, which provides perhaps an example for other states wishing to escape the dictum that structure is destiny.

(1) Mobility due to commodity price swings

An unsurprising feature of Babones' (2005) study of position and mobility in the contemporary world-economy is that whenever oil prices were high, Saudi Arabia advanced to core-type income levels, but whenever oil prices were low, Saudi Arabia receded to a national income level more characteristic of the semiperiphery. Obviously, Saudi Arabia's economy is dependent on oil. Dezanni (2001, 2002) confirms as much when his discriminant analyses show that the OPEC countries form a distinct grouping in the world-economy that do not obey the general structural rules that apply to the rest of his panel. This is not surprising. The lessons of OPEC, however, can be applied beyond oil. For example, the national income per capita of the small Pacific island of Nauru peaked at core levels at the height of its phosphate boom, only to drop back again when the boom was over.

Structural contrarians are fond of pointing to the formerly high living standards of the southern cone countries of Argentina, Chile, and Uruguay as evidence that countries can and do fall out of the core of the world-system. This is, frankly, nonsense. In 1900, the southern cone countries were commodity economies, pure, and simple. They were never able—never tried—to turn their commodity booms into society-wide economic development (Cypher 2010). The real world prices of non-fuel commodities rose by a factor of four or more between 1820 and 1910 and stayed high until the early 1950s (Ocampo and Parra-Lancourt 2010: 17), and national income rose and stayed high accordingly. From the 1950s to the 1990s, non-fuel commodities prices fell back by half, and the southern cone returned to semiperipheral income levels.

From a structuralist standpoint, neither in the OPEC case nor in the southern cone case did the world-systemic position of the relevant countries change. All of these countries have been continuously semiperipheral since the times of their decolonization. They all decolonized with working government administrations, reasonably rational geographies, and reasonably secure borders. Their economic roller-coaster rides are emblematic of the semiperipheral character of their economies. Highly diversified, modern core economies don't experience massive economic swings due to commodity price movements, even when they are highly commodity dependent (witness Australia and Norway). These massive economic swings occur in the semiperiphery because semiperipheral economies are built on a narrow base of exploitation by elites, not a broad base of wide economic participation by the masses.

(2) Mobility due to the global rise of cities

The most striking example of mobility in the contemporary world-economy has certainly been the rise of east Asia. The four "Asian tiger" economies of Hong Kong, Singapore, Taiwan, and Korea moved from low to high incomes in little more than a generation. None of these states, however, is in the full sense a country. Arrighi himself recognized this, saying of Korea and Taiwan that:

> neither of them is a nation-state in the full sense—South Korea living in constant hope or fear of being reunited with its northern half, and Taiwan in constant hope or fear of becoming the master or the servant of Mainland China. Finally, the two smallest but by no means least important states, the semisovereign Hong Kong and Singapore, are not nation-states at all but city-states.
>
> *(Arrighi 1996: 8–9)*

I would go further to claim that Taiwan and Korea are also not much more than city-states. In Taiwan, the extended Taipei metropolitan area makes up 38 percent of the population of national population. In Korea, metropolitan Seoul makes up 42 percent of the population. Moreover, in both cases the capital is the richest part of the country.

This is relevant because a worldwide phenomenon of the post-1950 period has been the rise of cities. For whatever reasons, cities everywhere—from New York to Nairobi—are now much richer than their hinterlands than they were previously. Where Arrighi sees Hong Kong and Singapore as classic pre-modern semiperipheral trading states, I think it makes more sense simply to see them as cities shorn of their hinterlands. Hong Kong is rich, but were its data submerged in the data for China, it would be merely a rich city in a poor country, like Shanghai and Beijing. Singapore is rich, but were it still attached to Malaysia (as it was until 1965) it would be merely a rich city in a middle-income country. Both Hong Kong and Singapore are statistical oddities. They are not core countries, any more than Moscow (a similarly rich city) would become a core country were it to secede from Russia.

Similarly, Taipei is a rich Chinese city more than it is the capital of a core country. Taipei's population was swelled by relatively privileged refugees from China, and much of Taipei's income today is earned in its economic hinterland on the mainland. The government of Taiwan does not even officially recognize its own country as a distinct national entity. Of the four tigers, Korea comes closest to being an unambiguous country, but it is a country that is dominated by its two largest cities. It is also a country with a restricted hinterland (on account of the isolationism of North Korea). Perhaps not coincidentally, Korea is by far the poorest of the four Asian tigers, only about two-thirds as rich as Hong Kong. It is the closest to being a full country, and thus the farthest from wealthy core-type income levels.

The economic success of the four Asian tigers is not in doubt. This economic success, however, should not be equated with full participation in the core of the world-economy. The four Asian tigers are all relatively successful semiperipheral city-states.

(3) Mobility due to return to equilibrium

Today, of course, all the talk is of China and India, but mostly of China. I have recently treated the Chinese case in detail (and the Indian case in passing) elsewhere (Babones 2011). China is and has been since the end of its civil war a clearly semiperipheral country. Wallerstein consistently treated contemporary China as absolutely semiperipheral. On my criteria, it certainly has a state that is strong enough to police its borders, collect taxes and customs, and maintain effective rule of law. What kept China so poor for so long after 1950 was its disastrous economic policies. It was a poorly administered semiperipheral country that as a result attained income levels that were more characteristic of the periphery of the world-economy. Freed from the shackles of Maoism, China is fast becoming a classic semiperipheral country, with a few rich cities presiding over the mass exploitation of the population for the benefit of a state-connected elite.

Japan

That leaves Japan. Japan is the one unambiguous country to have risen to unambiguous core status only after incorporation into the contemporary world-economy. Japan, of course, is exceptional in many ways. It is the one formerly semiperipheral country to have conquered and exploited a large colonial empire after its own incorporation. In the post-war period, Japan had imposed on it what Arrighi (2003) called "the global new deal:" a US program to remake Japan in its own 1940s image. Japan implemented state-directed economic planning for the broad development of the entire country, urban and rural. In a wrenching fiscal transition, the Japanese state went directly from financing itself through indirect consumption taxes that fell mainly on the poor to a highly progressive income tax with a top rate of 75 percent. It also pushed interest rates on savings artificially low (while preventing people from moving their money out of the country), a further, indirect tax on the wealthy.

The Japanese state invested all these newfound takings from the elite into an intensive internal development program. It brought roads, electricity, and bullet trains to previously isolated areas. It also pushed investment funds onto sometimes reluctant industrial groups, again preventing them from moving these funds out of the country. As a result, Japan's industrial groups invested inside the country. Whatever the social cost, the overall economic effect of these policies was to develop the country as a whole rather than to enrich a few elite enclaves. Japan bootstrapped its way into the core.

Implications and conclusions

It may be going too far to say that structure is destiny, but from the time that the modern world-system came to encompass the entire world in the late nineteenth century, there has been only one unambiguous case of structural mobility from the semiperiphery into the core (Japan). Status mobility between the periphery and the semiperiphery has occurred on and off as powerful states have been able to establish full, effective control over their territories or (more rarely) have lost it. Income mobility has been rather more common, especially between peripheral and semiperipheral levels, as semiperipheral states have pursued more or less disastrous economic policies. Absolute self-interest in almost all cases has prevented semiperipheral elites from allowing their states to pursue development strategies that might have a chance of lifting their countries into the core of the world-economy.

The structural approach suggests that most economic mobility in the world-economy derives from one of four root causes: (1) the establishment of effective states in peripheral countries (turning them into semiperipheral countries); (2) the disintegration of effective states in semiperipheral countries (turning them into peripheral countries); (3) global price shocks that temporarily raise or lower countries' levels of national income; (4) returns to equilibrium after price shocks. This is not to say that real mobility is impossible. It is just to say that real mobility—especially mobility into the core—is rare. The system is fixed not just in its overall contours but also in its specific details.

What of the future? The best hope for the future is that peripheral countries will gain effective government and that semiperipheral countries will wrest their state apparatuses away from the control of national elites. The former is a tough nut, but history and reason suggest that strong states will emerge more often than they will disintegrate. Over time, more and more of the world's peripheral countries should transform themselves into semiperipheral countries. The latter—the economic democratization of semiperipheral states—seems less certain. National elites have strong incentives to use their states to further their own parochial interests. The 2009–10 almost-revolution in Iran and the 2011 revolutions in Tunisia and Egypt suggest that perhaps the Internet is finally living up to its transformative potential to break this cycle, but internet-crazy China shows no signs of an online revolution. For most countries, the semiperiphery may be as good as it gets.

References

Arrighi, Giovanni. 1996. "The Rise of East Asia and the Withering Away of the Interstate System." *Journal of World-Systems Research,* 2(15): 1–34.

Arrighi, Giovanni. 2003. "The Social and Political Economy of Global Turbulence." *New Left Review,* 20: 5–71.

Arrighi, Giovanni, and Jessica Drangel. 1986. "The Stratification of the World-Economy: An Exploration of the Semiperipheral Zone." *Review,* 10: 9–74.

Babones, Salvatore J. 2005. "The Country-Level Income Structure of the World-Economy." *Journal of World-Systems Research,* 11: 29–55.

Babones, Salvatore J. 2009. *The International Structure of Income: Its Implications for Economic Growth.* Saarbruecken: VDM Publishing.

Babones, Salvatore J. 2011. "The Middling Kingdom: The Hype and the Reality of China's Rise." *Foreign Affairs,* 90(5): 79–88.

Clark, Rob, and Jason Beckfield. 2009. "A New Trichotomous Measure of World-system Position Using the International Trade Network." *International Journal of Comparative Sociology,* 50: 5–38.

Cypher, James Martin. 2010. "South America's Commodities Boom: Developmental Opportunity or Path Dependent Reversion?" *Canadian Journal of Development Studies* 30: 565–638.

Dezzani, Raymond J. 2001. "Classification Analysis of World Economic Regions." *Geographical Analysis,* 33: 330–52.

Dezzani, Raymond J. 2002. "Stability Analysis of World Economic Regions." *Journal of Regional Science,* 42: 595–625.

Frank, Andre Gunder. 1967. *Capitalism and Underdevelopment in Latin America.* New York: Monthly Review Press.

Hall, Thomas D., and Christopher K. Chase-Dunn. 2006. "Global Social Change in the Long Run." pp. 33–58 in Christopher K. Chase-Dunn and Salvatore J. Babones (eds) *Global Social Change: Historical and Comparative Perspectives,* Baltimore, MD: The Johns Hopkins University Press.

Maddison, Angus. 2010. Statistics on World Population, GDP and Per Capita GDP, 1–2008 AD. Groningen: Groningen Growth and Development Centre, University of Groningen.

Ocampo, Jose Antonio, and Mariangela Parra-Lancourt. 2010. "The Terms of Trade for Commodities since the Mid-19th Century." *Revista de Historia Economia, Journal of Iberian and Latin American Economic History,* 28: 11–43.

Wallerstein, Immanuel M. 1974a. *The Modern World-System, Vol. I: Capitalist Agriculture and the Origins of the European World-Economy in the Sixteenth Century.* New York: Academic Press.

Wallerstein, Immanuel M. 1974b. "The Rise and Future Demise of the World Capitalist System: Concepts for Comparative Analysis." *Comparative Studies in Society and History,* 16: 387–415.

Wallerstein, Immanuel M. 1979. *The Capitalist World-Economy.* Cambridge: Cambridge University Press.

10.2

O'Connorian models of peripheral development—or how third world states resist world-systemic pressures by cloning the policies of states in the core

Samuel Cohn

The affinity between world-systems theory and James O'Connor

The great contribution of world-systems theory to development theory was breaking the assumption that there was one trajectory of development that characterized both core and peripheral nations. Modernization theory (McClelland 1966; Rostow 1960) treated economically developed and underdeveloped nations as if they were both manifestations of the same phenomenon—social evolution. Thus, the formulae for economic growth were the same for both countries—and involved principles such as rationalization, universalism and demographic transition.

World-Systems Theory in its pre- and post-Wallerstein forms, was a healthy antidote to such simplistic thinking. Frank (1967) and the Comisión Económica para América Latina y el Caribe (CEPAL) school (Prebisch 1950) rightly caused such thinking to be discarded by showing that the global south is poor *because* of what rich countries do to it. More specifically, market forces (in particular unequal terms of exchange), and foreign institutional actors (in particular, colonialists, multinational corporations, and international creditors) systematically restrict capital accumulation in the periphery.

As a result, the ability of the periphery and semiperiphery to develop is utterly dependent on the capacity of states within these disadvantaged regions to resist this systematic imperialism. As a result, world-systems theory has emphasized two central forms of resistance: hard bargaining with multinational firms and core governments (the prime focus of Celso Furtado [1983] and Peter Evans [1979]), and restriction of the international capital market through either import substitution (the CEPAL school) or through fully fledged state-led development (advocated in its fullest form by the students of the developmentalist states of East Asia [Amsden 1989; Wade 1990]). Both models argued that the state was critical to development in the semiperiphery and periphery—and that the role of the state was to buffer the nation from adverse market forces by either restricting the actions of predatory foreign corporations (thus correcting for sins of commission) or by using state investment to compensate for the failures of international capital markets (thus correcting for sins of omission).

This has become a not-uncommon claim in world-systems theory. The subsequent decade, with its preoccupation with globalization, has changed little of this intellectual framework. Globalizers argue that third world states have lost the capacity to hard bargain with foreign capital or to engage in state centered heavy industrialization (Sassen 1996; Yergin and Stanislaw 2002). If world-systems theorists were to follow this logic strictly, the next set of questions is whether the state can regain its capacity to resist, making Porto Alegre, Venezuela, and Bolivia critical cases.

All of these developments have been constructive. However, there has been some intellectual loss by the narrow focus on the ability of third world states to resist the forces of international capital. The absolute insistence on the non-parallelism of core and periphery has distracted from a lot of useful neomarxist, macrosocial theory development that would stem from the presupposition that core and peripheral governments use similar principles in promoting capitalism within their nations. Few serious world-systems theorists would really want to argue that state provision of education had no effect on economic growth in either the core or periphery. Few serious world-systems theorists would really want to argue that state funded population control programs were not important in promoting development. Comparable cases can be made for the provision of infrastructure, the funding of research and development or the Keynesian management of demand. Within the world-systems literature, this type of "garden variety" government program is generally treated with benign neglect.

This benign neglect would be fully justifiable under the assumption that peripheral states are undergoing fiscal crises, and that they can not afford to implement the "garden variety" programs of state-led stimulation of economic growth that characterize routine public administration in the core. The literature on the most recent Third World debt crisis made this argument, and made it compellingly well (Chossudovsky 1997; Potter 2000). Such neglect is less justifiable in the face of peripheral states that are fully financially solvent, or who under conditions of partial duress, can somehow fund a subset of the "garden variety" core development package.

Empirically, both regimes really exist, making the existence of effective routine state development policy a contingent rather than absolute circumstance. Christian Suter's work (1992) on long-term waves of international debt crisis documents compellingly that at least since 1800, the periphery has oscillated between periods of ample state financial resources and periods of acute fiscal crisis. There are regular waves of debt crises that produce sustained budgetary crises in third world nations. However, in each period, there are individual nations that avoid the worst of the financial strictures that affect the rest of the periphery (such as Taiwan in the 1980s and 1990s). There is also the flush period *before* the debt crisis (such as the 1970s in the most recent case), when non-core states have substantial resources to spend on development (or on other goals).

For world-systems theory to be able to meaningfully address this issue, a theoretical structure is required that

(1) Grounds capitalist development in materialist class-based dynamics—a key strength of world-systems theory.
(2) Addresses the precise mechanisms by which routine state expenditures produce capital accumulation.
(3) Incorporates fiscal crisis as a fundamental mediating variable in any link between state policy and development.

The obvious theorist of choice for this problem is James O'Connor.

The James O'Connor model

James O'Connor's 1973 *Fiscal Crisis of the State* laid out a general model of the role of the state in capital accumulation. The argument was very simple.

(1) States produce economic growth by constructing physical infrastructure that capitalists would never construct themselves.
(2) States produce economic growth by investing in human capital that capitalists would never finance themselves.
(3) States legitimize capitalism by providing welfare.
(4) States maintain aggregate demand via Military Keynesianism.
(5) Limits on the ability to collect taxes produce a fiscal crisis that inhibits the state's ability to perform (1) through (4).
(6) Monopoly capital expropriates most of the benefits of state spending, since government programs disproportionately favor large corporations.
(7) Monopoly capital effectively avoids taxation due to its influence on the tax code.
(8) The tax burden thus effectively falls on petty capital and on individuals.
(9) Once petty capital and individuals become fully class conscious regarding this process, they will mobilize against the state by refusing to pay taxes.
(10) This will produce a fiscal crisis of the state—which will inhibit state capacity to provide the programs required for the reproduction of capitalism. Capitalism will fall not from catastrophic collapse but from slow decay.

Contemporary readers will note the eerie and unsettling resemblance between this model and the current Tea Party mobilizations in the United States.

For the purposes of development theory, what matters is items (1) through (4), the specification of the routine "garden variety" government programs that generate development. O'Connor is all about building roads, building airports, building schools, supporting science, and running pension programs. There is little that O'Connor advocates that would actually be in opposition to the World Bank's 2002 report on the appropriate role of the government in responsible pro-capitalist development—or to Joseph Stiglitz's heterodox prescription for state led economic growth in *Making Globalization Work* (2006).

When this theory is applied to peripheral states, what is remarkable is the absence of the following argument:

(0) States directly invest in strategic industries that will improve the long term international competitiveness of the economy.

O'Connor is completely confident that monopoly capital will provide whatever investments are needed to fund any industry that would be intrinsically viable. Argument 0 is a key position of Celso Furtado, Peter Evans, and the developmental state theorists, but not of O'Connor. O'Connor's implicit dismissal of interventionist state investment and proactive industrial policy is probably too extreme. It would be hard to dismiss the substantial empirical success rate of interventionist states, notably in Europe and in East Asia (Chang 2007; Haggard 2004).

However, O'Connor's formulation has an important advantage over "active state investment" models. Infrastructure provision is easier to execute than is developmental statism. The most recent wave of sociological scholarship on the state and development has emphasized unsuccessful activist states, where local class structures and absence of endogenous bureaucratic capacity have undercut governmental attempts to promote development (Chibber 2006; Lange and Rueschemeyer 2005). Capture of the state apparatus by the local bourgeoisie can be enough to undercut technocratic government planning. Unsuccessful developmentalist governments can turn into a neoliberal's stereotype of the rent-seeking state. Dani Rodrik (2007) argues that developmentalist states have to make strategic choices. Because seemingly similar countries can often have dissimilar problems, it is easy for governments acting in good faith to "guess wrong." The structural impediments to developmentalism identified by Chibber and Lange and Rueschemeyer, and the tactical impediments to developmentalism identified by Rodrik help explain why highly successful developmentalist states such as Singapore exist, but are relatively rare.

In contrast, it is not difficult to improve ports, pave roads, hire rural nurses, or open more high schools. Infrastructure provision generally provides some benefit, unless the country has become so developed that the public good in question is now in surfeit. Most developing nations are nowhere near the point where surfeit can be an issue. O'Connorian models provide a slow safe model for relatively incrementalist growth, while the more traditional world-system models suggest formulas for high risk but relatively spectacular growth. Naturally, some of the most successful developing nations, such as those in East Asia, use both strategies. However, when only one strategy can be implemented, the cumulative gains from steady incremental progress can be significant.

Empirical support

What empirical support is there for the O'Connorian model? For most world-systems theorists, the aspect of O'Connor that has received the least attention, and thus the empirical material with which they are likely to be the least familiar, is the argument about the importance of infrastructure. Therefore, the review of the data puts a disproportionate emphasis on that component of the model.

Very few sociologists of any theoretical persuasion have written extensively on physical infrastructure provision. The significant exception is John Kasarda (Irwin and Kasarda 1991; Kasarda and Sullivan 2006). Kasarda essentially correlates employment growth with the volume of air traffic in and out of metropoli. In theory, this ought to be a completely artifactual correlation, since there is an obvious relation between volume of economic activity and the volume of air traffic that is caused by that economic activity. What makes Kasarda's findings important, and what rescues them from spuriousness, is the use of the hub system by American and foreign airlines. Airlines do not fly point to point between all centers of economic activity; air traffic is channeled to strategic central airports, where both cargo and passengers change flights in order to get to their final destination. Therefore, cities with hub airports receive volumes of air traffic far in excess of that volume needed to support their own locally generated arrivals and departures. It is this excess traffic that is correlated with job growth and GNP growth in the future. Excess traffic is implicitly a measure of infrastructure. In order to have a hub airport, a city has to invest in substantial runway space, passenger terminal space, and cargo terminal space to support the transshipment function. By the standards of physical infrastructure projects, airports are relatively expensive, since they tend to require large amounts of contiguous urban or suburban space, most of which has to be obtained at a premium price. Those cities that have made such an investment have tended to prosper.

Intuitive support for Kasarda's argument comes from the rise of Atlanta at the expense of New Orleans. In the nineteenth century, both cities were equally important centers of southern commerce and service. Neither city has a locational advantage, such as Pittsburgh or Houston, that puts them near stocks of physical raw material used in manufacture. While some of Atlanta's early advantages came from its status as a rail junction, its gigantic take-off in the late twentieth century was centered around the construction and dramatic expansion of Hartsfield Airport, with Delta Airlines using Atlanta as its main hub. New Orleans' airport development was desultory, and the city is not associated with any major airline.

Work by the Marxist geographer, Doreen Massey (1974), provides further support for Kasarda's argument. Late twentieth-century Britain was marked by a movement of employment away from traditional manufacturing centers in northern England (such as Tyneside, Liverpool, Manchester, and Birmingham) and lower Scotland (such as Clydeside) to southeastern Britain in the area around London. This could not be accounted for by any objective economic advantage of the London area—such as access to population or labor, access to raw materials, transportation costs relevant to continental export, or access to an educated labor pool. The key factor was proximity to Heathrow Airport. In the postwar era, there was substantial multinational penetration of the British industrial structure, with American firms buying up a large percentage of important UK employers. The American executives wanted easy physical access to their offices and production facilities and did not want to travel extensively once they arrived in Britain from the United States. They tended, therefore, to relocate offices and production centers to facilities within easy commuting distance of Heathrow Airport in west London. This created a substantial secondary wave of British-owned firms relocating to southeastern England in order to be able to service and work with these key American owned companies—with the end result being the dramatic de-industrialization of northern England.

Cohn (2010) provides further support for Kasarda, on a case in which publically funded airports were correlated with growth—without a transition from local to multinational ownership being required to produce the effect. Cohn contrasts the growth rates of paired sets of nearly identical Brazilian states—with one member of the pair but not the other receiving a significant airport expansion. He combines this with contrasting growth rates in states with airport expansions before and after the year of the actual expansion. Both types of contrasts show dramatically higher growth rates in the states which have received an airport expansion. What is particularly noteworthy is how rapidly the improvement in regional GDP occurs. Most development projects involve a substantial lag between project completion and the ultimate amelioration of growth rates. Airport improvements seem to take effect almost immediately. The GDP jumps the very year the terminal is completed, and the favorable contrasts between airport-expanding states and their non-airport-expanding peers are marked in the immediate years after the completion of the project. Commercial agriculture and tourism seem to be particularly responsive to enlargements of airport capacity. In some rural northeastern locations, the expansion of a nearby airport led to the rapid construction of refrigerated warehouses next to the airport, followed by the conversion of nearby farms into internationally oriented fruit and vegetable agribusinesses oriented towards sending perishable products to American and Asian markets. Airports near attractive beaches were often accompanied by the quick construction of new hotels and coastal resorts along with the creation of "vacation packages" in encourage Brazilian tourists to visit the new leisure complex.

Naturally, airports are not the only form of publicly provided infrastructure that produces economic growth. Nineteenth-century governments constructed railways, and substantially stimulated industrialization by doing so (Dobbin 1997; Hawke 1970; Reed 1969). Fogel's (1964) famous contrarian position that econometrically railroads contributed nothing to US economic growth is based on the argument that the United States already had a fully functional set of canals; canal construction is as O'Connorian as rail construction. Dobbin's argument that Victorian states milked railway companies as cash cows does not gainsay the fact that the state was key to getting these railways constructed—even if their treatment of the companies afterwards was somewhat rough.

The largest body of systematic evidence on the role of infrastructure in economic growth comes from the work of development economists. Fedderke et al (2006) did a long time-series analysis of economic growth in South Africa to show the independent positive effects on GDP of railway lines, volume of rolling stock, telephone lines, paved roads, electricity provision, and air passenger volume. Since reverse causation is a serious potential problem in such an analysis, they carefully rule this out by contrasting forward effects of external shocks in both variables. Even in this constrained test, all six infrastructure variables perform well.

Pereira (2000) did a similar time-series for the United States. His analytical categories are odd, but the findings are basically positive. His dependent variable—private sector productivity—was related to aggregate public investment, energy and transportation systems (aggregated together for some reason), sewage and water supply systems, public buildings (a category that is dominated by schools and public hospitals), and a none-of-the-above infrastructure category.

Aschauer (2000), in a cross-national sample of less developed nations, found that private investment, secondary school enrolment, and public investment had virtually identical positive effects on growth. World-systems theorists can raise reasonable questions about the interpretations of his findings. In countries such as South Korea, some developmentalist state policies, such as state-run enterprise, would have undoubtedly found its way into the public investment measure. Likewise, we know from Bornschier and Chase-Dunn (1985) to differentiate between the short-term and long-term effects of private investment, at least in the case of foreign direct investment. That said, in less developed nations which are not developmentalist states—which is most of them, public investment data would be dominated by infrastructure investments such as those discussed by O'Connor. An effect of public investment as least as large as that of private investment merits serious attention.

Esfahani and Ramirez (2003) find large effects on GDP growth within a large cross-sectional sample of countries of all income levels for increases in telephone availability and power generating capacity. Mitra et al (2002), in an analysis of physical infrastructure and social infrastructure combined found that an index of roads, electricity, railroads, postal services, education and health and banking correlated with differences in technical efficiency of manufacturing establishments in Indian states. Cohn (2010) shows positive employment effects in Brazil for water, sewage, airport, and road projects. Easterly and Serven (2003) document a dramatic difference in infrastructure provision between East Asia and Latin America. They argue that this difference explains about 30 percent of the productivity gap between these regions. Telephone lines, power generation, and roads all contribute equally to this gap. Shioji (2001) estimates the effect of infrastructure growth and education on GDP growth in the United States and Japan and argues that the effect of infrastructure improvements were substantially greater than those of education. Sanchez-Robles (1998) finds that measuring actual infrastructure constructed produces higher GDP effects than measuring expenditure on infrastructure; her methodology may implicitly correct for the corruption (and lost funds) that can be associated with infrastructure products, thus showing that actual service delivery of public goods makes a difference.

Fernald (1999) suggests a reasonable limit to the effectiveness of infrastructure. At some point, wealthy nations get "enough" of a certain type of infrastructure, after which further productivity gains are less likely to occur. He documents that the building of the US Interstate Highway system brought dramatic increases in productivity, but that road building after the completion of the highway system produced substantially smaller increases.

There is a small contrarian body of economists who argue that public spending on infrastructure is ineffective (Holtz-Eakin 1994; Hulten and Schwab 2000). They present equations with low coefficients for the infrastructure variable that either have a huge number of region and time dummies, or include private investment as a control. Any variable can be shrunk to insignificance by the inclusion of enough time and space dummies. Including private investment as a "control" ignores the fact that state investment in infrastructure provides favorable conditions for investors, which increases their willingness to provide capital.

Overall, the body of evidence showing that state infrastructure raises development is sufficiently large to support O'Connor. Education, research and development, and Keynesian stimulus are not reviewed here; however, these claims, while not entirely undisputed, are not highly controversial.

Conclusion: Keeping development alive in the face of fiscal crisis

The other important feature of O'Connor's model is the role of fiscal crisis. O'Connor's model was designed to explain growth trajectories in the core. However, accumulation crises induced by states facing critical budget shortages are only recently affecting the growth of core nations. In contrast, this has been an enduring and fundamental feature of the periphery. The tax crises of the core have taken the form—as O'Connor predicted—of petty bourgeois anti-tax political movements. The tax crises of the periphery come from widespread tax avoidance, a very large informal sector, and relatively weak state capacity to enforce tax laws either through audit or through effective criminal sanctions. (For a review of third world tax issues, including problems of enforcement and compliance, see Ahmad and Stern [1989], Bird [1992] or Prest [1962].) The persistent debt crises in the periphery, which have been well documented by Suter (1992), are partially caused by the inability of many of those states to obtain financial resources from their own populations, notably from their own elites. The anti-tax voter of present-day Atlanta has an ancestor in the 1950s Rio de Janeiro grocer with two sets of books; American multinationals who pay very few taxes have ancestors in the tax-dodging Brazilian latifundistas and coroneis. The result was a series of debt problems in Brazil, just as the United States is becoming increasingly indebted.

The Brazilian experience also provides significant lessons on how to avoid the social damage associated with O'Connorian fiscal crises. Albert Fishlow (1990), the great Latin American economist who specialized in studying the Brazilian case, wrote a surprising essay on the role of the state in economic development. He was equally comfortable with a Marxist, heavily interventionist, dirigiste regime, and a neoliberal Washington Consensus mini-state. For him, the question was not how large the state was, but whether the government could pay its bills. As long as a government could maintain its financial sovereignty and not become dependent on predatory foreign creditors, it could manage its development policy any way it saw fit.

Fernando Henrique Cardoso (2009), world-systems theorist and former Finance Minister and President of Brazil, wrote an essay reflecting on the theoretical lessons of his experience in actually administering a semiperipheral economy. His essay absolutely echoes Fishlow's sentiments. In his opinion, external debt is anathema because it compromises the capacity of the peripheral state to act. Cardoso has more faith than do many American "globalization" writers, on the capacity of third world states to autonomously reduce their exposure to foreign creditors. The Cardoso administration dramatically increased the administrative capacity of the Brazilian treasury to collect taxes—and obtained an immediate increase in overall tax revenues (Font 2003).

Cardoso also had a second tactic. He followed the precepts of the heterodox economic theorist, Luiz Carlos Bresser-Perreira (who became Cardoso's finance minister). Bresser-Perreira (1992) argued that semiperipheral states could avoid the worst consequences of post-debt fiscal crisis if they redesigned their programs to be cheap. Traditional state-led industrialization with import substitution approaches were simply too capital intensive and expensive to be viable in a world of international dependency. However, many of the benefits of such regimes could be simulated simply by designing lower cost versions that offered high economic yields relative to their price.

This is exactly what Cardoso did. He sold off expensive, loss making government firms, notably in steel, energy, and telecommunications. However, he vastly increased his investment in research (notably agricultural research), and primary and secondary education, and vocational education. Furthermore, he went on an infrastructure binge, which included building new airports around the country, and pouring road, water, and sewerage money into the impoverished Brazilian northeast (Cohn 2010; Font 2003). The result was a resurgence of economic growth in what had previously been viewed as a stagnant economy—and the laying of the groundwork for Brazil to enter its status in the Lula regime as being a promising and rapidly growing member of BRIC.

The Cardoso-Bresser-Perreira formula of increasing tax collections while promoting cheap effective pro-growth government expenditure is an interesting response of a semiperipheral state to the financial constraints imposed on it by a global financial system. This approach may be useful to the core as well as the periphery and may represent a template for a new and effective strategy for economic nationalism. World-systems theorists have generally not paid very much attention to taxation and they certainly have not paid very much attention to routine government programs. However, these may be a key towards mediating the adverse effects of world-systemic financial dynamics and producing locally based balanced alternatives to neoliberal globalization. O'Connor may not only have been prophetic about the decline of late capitalism, but he may have implied solutions for how a humanistic late capitalism can be maintained in core and periphery alike.

References

Ahmad, E. and N. Stern (1989) "Taxation for Developing Countries" pp. 1006–92 in H. Cherery and T.N. Srinisvan (eds), *Handbook of Development Economics*, New York: Holland.

Amsden, A. (1989) *Asia's Next Giant: South Korea and Late Industrialization*, New York: Oxford.

Aschauer, D. (2000) "Public Capital and Economic Growth: Issues of Quantity, Finance and Efficiency." *Economic Development and Cultural Change,* 48: 391–406.

Bird, R. (1992) *Tax Policy and Economic Development*, Baltimore, MD: Johns Hopkins University Press.

Bornschier, V. and C. Chase-Dunn (1985) *Transnational Corporations and Underdevelopment*, New York: Praeger.

Bresser-Perreira, L.C. (1992) *Crise do Estado: Ensaios Sobre a Economia Brasileira* (Trans: The Crisis of the State: Essays on the Brazilian Economy) Sao Paulo: Nobel.

Cardoso, F. (2009) "New Paths: Globalization in Historical Perspective." *Studies in Comparative International Development* 44: 296–317.

Chang, H. (2007) *Bad Samaritans: Myth of Free Trade and the Secret History of Capitalism*, New York: Bloonsbury.

Chibber, V. (2006) *Locked in Place: State Building and Late Industrialization in India*, Princeton, NJ: Princeton University Press.

Chossudovsky, M. (1997) *Globalization of Poverty: Impacts of IMF and World Bank Reforms*, New Brunswick, NJ: Zed Books.

Cohn, S. (2010) "What State Policies Actually Raise Employment in Brazil or Why James O'Connor Can Be As Important as Peter Evans." Paper Presented at Annual Meetings of American Sociological Association.

Dobbin, F. (1997) *Forging Industrial Policy: United States, Britain and France in the Railway Age*, New York: Cambridge University Press.

Easterly, W. and L. Serven (eds) (2003) *Limits of Stabilization: Infrastructure, Public Deficits and Growth in Latin America*, Washington, DC: World Bank.

Esfahani, H and M. Ramirez (2003) "Institutions, Infrastructure and Economic Growth", *Journal of Development Economics,* 70: 443–77.

Evans, P. (1979) *Dependent Development*, Princeton, NJ: Princeton University Press.

Fedderke, J.W., P. Perkins and J.M. Luiz (2006) "Infrastructural Development and Long Term Economic Growth: South Africa 1875–2001." *World Development*, 34: 1037–59.

Fernald, J. (1999) "Roads to Prosperity: Assessing the Links Between Public Capital and Prosperity." *American Economic Review*, 89: 619–38.

Fishlow, A. (1990) "Latin American State." *Journal of Economic Perspectives*, 4: 61–75.

Fogel, R. (1964) *Railroads and American Economic Growth: Essays in Econometric History*, Baltimore, MD: Johns Hopkins University Press.

Font, M. (2003) *Transforming Brazil: Reform Era in Perspective*, New York: Rowman & Littlefield.

Frank, A.G. (1967) *Capitalism and Underdevelopment in Latin America*, New York: Monthly Review Press.

Furtado, C. (1983) *Accumulation and Development: Logic of Industrial Civilization*, New York: Saint Martins.

Haggard, S. (2004) "Institutions and Growth in East Asia." *Studies in Comparative International Development*, 38: 33–51.

Hawke, G. (1970) *Railways and Economic Growth in England and Wales 1840–1870*, Oxford: Oxford University Press.

Holtz-Eakin, D. (1994) "Public Sector Capital and the Productivity Puzzle." *Review of Economics and Statistics*, 76: 12–21.

Hulten, C. and R. Schwab (2000) "Does Infrastructure Investment Increase the Productivity of Manufacturing Industry in the U.S.?" pp. 81–104 in L. Lau (ed.), *Econometrics and the Cost of Capital*, Cambridge, MA: MIT Press.

Irwin, M. and J. Kasarda (1991) "Air Passenger Linkages and Employment Growth in U.S. Metropolitan Areas." *American Sociological Review*, 56: 524–37.

Kasarda, J. and D. Sullivan (2006) "Air Cargo, Liberalization, and Economic Development." *Annals of Air and Space Law*, 31: 214–30.

Lange, M. and D. Rueschemeyer (eds) (2005) *States and Development: Historical Antecedents of Stagnation and Advance*, New York: Palgrave Macmillan.

Massey, D. (1974) *Towards a Critique of Industrial Location Theory*, London: Center for Environmental Studies.

McClelland, D. (1966) "Does Education Accelerate Economic Growth?" *Economic Development and Cultural Change*, 14: 257–78.

Mitra, A., A. Varoudakis and M. Veganzones-Varoudakis (2002) "Productivity and Efficiency in Indian States Manufacturing: Role of Infrastructure." *Economic Development and Cultural Change*, 50: 395–426.

O'Connor, J. (1973) *Fiscal Crisis of the State*. New York: Saint Martins.

Pereira, A. (2000) "Is All Public Capital Created Equal?" *Review of Economics and Statistics*, 82: 513–18.

Potter, G. (2000) *Deeper Than Debt: Economic Globalization and the Poor*, Sterling, VA: Kumarian.

Prebisch, R. (1950) *The Economic Development of Latin America and Its Principal Problems*, New York: United Nations.

Prest, A.R. (1962) *Public Finance in Developing Countries*. London: Weidenfeld and Nicholson.

Reed, M.C. (ed.) (1969) *Railways in the Victorian Economy: Studies in Finance and Development*, Newton Abbott: David and Charles.

Rodrik, D. (2007) *One Economics: Many Recipes*. Princeton, NJ: Princeton University Press.

Rostow, W. (1960) *States of Economic Growth*, New York: Cambridge University Press.

Sanchez-Robles, B. (1998) "Infrastructure Investment and Growth: Some Empirical Evidence", *Contemporary Economic Policy*, 16: 98–108.

Sassen, S. (1996) *Losing Control: Sovereignty in an Age of Globalization*, New York: Columbia.

Shioji, E. (2001) "Public Capital and Economic Growth: A Convergence Approach." *Journal of Economic Growth* 6: 205–27.

Stiglitz, G. (2006) *Making Globalization Work*. New York: Norton.

Suter, C. (1992) *Debt Cycles in the World Economy: Foreign Loans, Financial Crises and Debt Settlements*. Boulder, CO: Westview.

Wade, R. (1990) *Governing the Market: Economic Theory and the Role of Government in East Asian Industrialization*. Princeton, NJ: Princeton University Press.

World Bank (2002) *Globalization, Growth and Poverty*. Washington, DC: World Bank.

Yergin, D. and J. Stanislaw (2002) *Commanding Heights; Battle for the World Economy*, New York: Touchstone.

10.3

The embedded periphery

Slums, favelas, shantytowns and a new regime of spatial inequality in the modern world-system

Delario Lindsey

The future is urban. More specifically, the future of the Capitalist World-Economy will be firmly situated in the expansive terrain of the global urban. Even now, it can be said that we live in an era where cities and urban spaces have become the most vital and important components of the global economy. According to a 2008 UN-HABITAT report, fully half of the human beings on the planet live and work in and around a city or urban area. Growth in the global urban has been consistent over the past 30–40 years and wide-spread, with growth in urban populations occurring in developed and to a greater extent, developing regions of the world-system. Growth in world cities (or globally significant urban centers predominantly located in the developed nations of world [Friedmann and Wolff 1982]) has mirrored the expansion and intensification of the Capitalist World Economy since the end of World War II. As a consequence of this growth, a fundamental tension has developed between the city as economically essential space (particularly in the case of world cities, which are planned and built to be spaces of commerce and consumption), and the city as lived-environment where people can work and make lives for themselves within viable communities. As those cities around the world that make up the global urban have become increasingly important to the stable functioning and operation of the Capitalist World-Economy and the broader Inter-State System, they are becoming necessary spaces of economic survival for poorer populations around the world. Traditional rural economic structures (typically centered on agriculture) have all but collapsed, leaving the cities (particularly those located in the global south) as one of the few economic havens for those seeking to earn a reasonable livelihood. As a consequence of these ongoing population shifts, urban centers all over the world are experiencing extreme forms of spatial inequality that impact the lived-environment of poorer urban residents as an increasing number of these people find themselves relegated to favelas, slums, and shantytowns.

According the UN-HABITAT 2003 working paper on the Slums of the World, a slum can be defined as: "a wide range of low-income settlements and/or poor human living conditions and note that these inadequate housing conditions exemplify the variety of manifestations of poverty ... " Slums can include deteriorated housing settlements that were at one time "desirable," as well as improvised informal or squatter settlements that in many instances lack legal recognition by city and state officials. Slums, favelas, and shantytowns go by different names, and have different topographies (depending on location), but share several crucial features: (1) these are spaces that are typically underserved (economically and politically) by the local administrations, (2) these spaces tend to be plagued by violence and crime, and are vulnerable to all manner of environmental or ecological hazards, (3) they are almost always inhabited by the working poor and working classes of the city, (4) they represent on the community level, the reality of unequal core-periphery social, economic, and political relations. Due to the rapid expansion of cities in the developing world, slum areas there are particularly vulnerable to various health crises as municipal governments may not have built out vital infrastructure and services to incorporate newly established informal settlements on the urban periphery. In cities where they are present, slums, favelas, and shantytowns clearly occupy the political, economic, social, and spatial periphery. An understanding of the formation and structure of these spaces may also shed important analytical light on the future of the global urban and the Modern World-System in general.

The global urban and the modern world-system

Over the last 30 years, a compelling case has been made for the historical and contemporary significance of the global urban to the Modern World-System. Much of the analysis of global urban has been concerned with the prominent role cities have played in the Capitalist World-Economy in the years since World War II. Cities (especially world cities) have served to articulate regional, national, and international economies (Sassen 2001; Taylor 2002), and during the global economic transition (sometimes referred to as a "crisis" [Lash and Urry 1994]) that has led to a reorganization of the management structures of the world-economy, cities have become vitally important to the maintenance and growth of the global economy. Before this transition in global capitalism, nation-states played the most important role in the organization and management of global capitalism. This is not to say that nation-states no longer play an important role in global economic processes, but the place national governments have occupied in the mediation of transnational economic processes has definitely declined in significance relative to the era of state manage capitalism (also known as Fordism, or Keynesism ca. 1932–72 [Harvey 2009]).

A major component of the reorganization of global capitalism, processes of production in the wider Capitalist World-Economy have transcended traditional notions of space and territoriality or have become "hyper-spatialized" (Jameson 1991), and are no longer centralized within any one particular zone, region, or nation-state. Capitalist production processes have become globalized and disintegrated, and as a consequence, urban spaces such as world cities have become key nodal points for the command and control of global capitalism (Friedmann and Wolff 1982). The world of late modern capitalism is one where capital and capitalists float above it all (particularly the boundaries of nation-states), and when capital and the global elite that manage it return to earth, they typically return to a world city. The increasing significance of the global urban has led to interesting forms of spatial inequality including the re-articulation of world-systemic core-periphery relations between and within those cities that comprise the global urban.

Since John Friedmann and Goetz Wolff (1982) first proposed the rigorous study of the world city phenomenon, there has been the development of two different approaches to addressing the question of how world cities are incorporated into the capitalist world-economy: the world city system as hierarchy, and the world city system as network. John Friedmann is closely associated with the hierarchical model while Peter Taylor stands as an advocate of the network approach. Both models use many of the assumptions of world-system theory (which include features such as a capitalist world-economy, a global division of labor, core-periphery relationships in an inter-state system) as a basis for their understanding of world city development and functioning.

Both models view the acquisition and maintenance of command and control functions associated with the everyday maintenance of the capitalist world-economy as a key determiner of world city-ness, but the hierarchal model states that world cities tended to be incorporated into the capitalist world-economy in a way that mirrors the incorporation of the nation-states that contain them. This means that the higher order world cities tend to be located in those nation-states that make up the core of the capitalist world-economy. This is the basis of the hierarchy as world cities like New York, London, and Tokyo form the top tier of the world city system because it is from these cities that the most advanced command and control functions (those associated with the greatest returns in profit, like advanced services, biotechnology, or high finance) tend to be located (Friedmann and Wolff 1982, 1995). Cities lower down the hierarchy have fewer of these command and control functions. Though there are many large urban centers in the peripheral nation-states, there are very few world cities among them. Cities like Sao Paolo in Brazil, Mumbai in India, and Johannesburg in South Africa are examples of world cities in the periphery.

The network model tends to focus less on the concentration of all important command and control functions in the richer core states and more on the actual functions required by global capital. World cities are argued to produce and reproduce important functions of (economic, political, social, and cultural) globalization (Taylor 2005). All world cities in a networked world city system function as productive nodes in the larger capitalist world-economy as such, some nodal points have a greater concentration of (for example) high technology firms while others are more important to the system for heavy manufacturing. In a networked world city system, the extent of a city's integration into the capitalist world-economy is determined by functions allocated to a city by market forces (Taylor 2002). Inequality and difference between various world cities is a function of the profitably of the sectors associated with the various cities in the system.

Another important feature of world cities (or almost any major metropolis around the globe) is its social, political, and (what is more important) economic gravity. Cities have always had the tendency to draw in resources and people from outlying areas. This gravity becomes a major factor in the persistence and growth of slums, favelas, and shantytowns as the steady decline in rural economies and the threat of systemic redundancy pushes individuals and communities from the countryside and toward the city. Many of those living in Rio de Janeiro's 600 favelas have been internal migrants subject to the push-pull forces affecting the north and northwest portion of the state of Brazil over the past 50 years (Neuwirth 2006). We can almost think of emigration (particularly in the global south and developing regions of the world) as an implicitly urban phenomenon; as the poor move from rural to urban, they almost invariably finds themselves living in a slums, favelas, or shantytowns.

The embedded periphery of the global urban

The situation for slums, favelas, and shantytowns in the developing world will be particularly harsh as most of the growth of cities will happen in the poorer parts of the global economy. Though much of the discussion of future changes in the structure of the Capitalist World-Economy tends to focus on realignments that could take place in the core of the Modern World-System (perhaps leading to the hegemonic dominance of the nation of China [Gunder Frank 1998]), the reality may be that it is in the periphery of the World-System, which is increasingly subject to redundancies of space and people, that these changes could be most devastating. On the level of the global urban and within the space of individual cities, the core-periphery relationship and its various consequences are readily identifiable. Though urban studies has adopted the phrasing "core" and "periphery" to delineate zones within a city (read: urban core as the city center, urban periphery as the suburban areas), the core-periphery relationship is traditionally associated with World-System Analysis (Wallerstein 1974; Hopkins and Wallerstein 1982), and it describes not only the central systemic division between the developed and developing nations of the world, but also references the idea that developed core spaces and developing peripheral spaces are implicitly and inexorably bound to one another within the globe spanning totality that is the Capitalist World-Economy. The core-periphery relationship is at essence an economic relationship (indicated by the realities of economic dependency, and unequal-exchange), but it is also a political and social relationship, structured by historical concentrations of power and influence in core areas of the global system. The core-periphery concept is scalable, and can be used to not only describe relations of inequality between nations in an Inter-State System, or between cities in a World City system, but also as a means of describing an ongoing reality of spatial inequality within cities. As the favela Rocinha can be viewed as the periphery to Central Rio de Janiero, or as the Kibera slum is the periphery to Central Nairobi, or as Dharavi slum is the periphery to Central Mumbai, the city will become the new frontier for spatial inequality and the articulation of the core-periphery relationship.

The utility of the core-periphery concept for analyzing the problem of spatial inequality within the cities that comprise the global urban (in both developed and developing regions) rests on the most basic and persistent of social realities that can exist between the affluence one may encounter in a thriving urban center, and the abject poverty typically associated with the far-flung slum areas of a city, and this is the reality of separation. I am of course referring to the social, economic, and political distance or separation between the urban core and the urban periphery that prescribes the spatial relationship between the city center and the slum. As the twenty-first century progresses, we will see the expansion and intensification of the seemingly contrary processes of urban development and urban peripheralization in major cities throughout the world. To understand the nature of contemporary forms of urban peripheralization in the face of expanding urban development, I have come to see the idea of "embedded-ness" as a useful way of describing the significant place slums, favelas, and shanties occupy in the growing cities of the developing world. I have deployed the idea of embedded periphery to describe two very important features of the expanding global urban which are: (1) the spatial embedding of a growing urban underclass in internally peripheralized slums, favelas, and shantytowns; (2) the embodied periphery which refers to the processes and potential consequences for individuals and communities when being located within peripheralized spaces (like slums, favelas, or shantytowns) is perceived (by those in the urban core) as an ascribed component of their identity (as with usage of the expression "Favelado" to disparagingly describe those who live in the favelas of Brazil's major cities).

Recent studies (such as Davis 2007) discuss the consequences of an expanding global economy for many of the large cities of the developing world in the twenty-first century. Global economic processes that were once located only in the core spaces of the world-economy are today becoming increasingly associated with cities of the developing world. Add to this the UN-HABITAT (2003) wherein it was estimated that nearly one billion people live in slums or favelas around the world. In addition, as the portion of the world's population living in urban areas is estimated to increase from just over three billion (est. 2008) to five billion by 2030, with 85 percent of this population living in the cities of the developing world, there is the likelihood that this growth will led to the deepening of the social, economic, political, and spatial inequality already persistent in many of these developing areas (Harvey 2009).

"Favelização" and the future of the world-system

The space of the slum, favela, or shantytown has long been thought as an outsider space, a space of lawlessness, a space of the undesirable poor. These are spaces that not only point to large-scale systemic problems (like economic redundancy or over-population), but slums, favelas, and shantytowns also shed light on the way state and local authorities create and exacerbate the conditions of separation that impact the lives of those who inhabit the embedded periphery. Urban scholars like Anthony D. King (1980) used the concept of the built-environment to refer to the influence political, social, and cultural forces in the city can have over every aspect of planning related to the urban landscape. The city, from the location of parks and to schools, to the width of its avenues and boulevards, to the availability of public restrooms, can be seen as the continually unfolding creation of ongoing political, social, and cultural discussions and debates about the space of the city. In this same way, we can think of the embedded periphery (in the form of the slum, favela, and shantytown) as a type of built-environment that reflects the social and political responses of influential stakeholders (that can include politicians and city administrators, security officials, and a corporate/business elite) to the large-scale structural shifts in economic processes. The slum can also represent a kind of built-environment that has been structured with long-standing prejudices and historical inequities.

A fairly prominent illustration of this idea of the slum as a built-environment can be seen in the city of Rio de Janeiro, Brazil, which has long struggled with its own local version of the core-periphery relationship as it tries to manage life in and around its estimated 600 favelas. As a means of combating what some call the "favelização" (Editorial, Jornal do Brasil, 2009) or the favelization of the city, there are plans in the works to construct walls around some of the city's larger favelas (such as Rocinha) as a means of halting the growth of these improvised housing settlements. Among the various forms of social/spatial control developed to combat favelization in Rio De Janeiro are the paramilitary pacification programs, and the euphemistically named "eco-barrier," which is intended to protect the natural beauty of rolling green hills that surround the city from the unauthorized development of ever-growing favelas. An ancillary effect of the barrier or wall is to isolate and conceal the manifest signs of economic inequality that are associated with Rio's growing, yet long embedded favela communities.

The idea of "favelização" is an interesting one, because it not only refers to the large-scale world-systemic processes that help to create and exacerbate the situation of slums, favelas, and shantytowns, but it can also be used to describe the ways in which local responses to the presence of slums, favelas, and shantytowns (and the associated social problems such as crime, blight, and disease) can actually intensify the broader problem of urban inequality and peripheralization.

The responses of local administrators in some cases reflect a privileging of an exclusive, elite vision of the global urban (Davis 2007). Favelização is the persistent spatial manifestation of separation, alienation, and social difference. More specifically, slums, favelas, and shantytowns as urban peripheries are the products of municipal governance and policy creation that favors the positions and presence of the urban elite; they are the product of advocacy and activism that consistently promotes the city as a space intended for those who occupy the privileged urban core. The Black African Townships (also known as "Bantustans") were developed during the Apartheid era as a type of spatial enforcement of the racist policies of the South African government (Paul Maylam 1990). In 2005, the city of Harare saw the burning and bulldozing of homes and businesses in the Porta Farm area of the city. As part of a program intended to combat crime and blight, government officials forcibly evicted nearly 700,000 residents of the shantytown, who were also believed to be supporters of the political rivals of President Robert Mugabi (Tibaijuka 2005).

In December of 2004, India's largest city, Mumbai, began a campaign to disrupt favelization there by demolishing portions of India's largest slum, Dharavi. Mumbai has a metro area population of almost 18 million, and the city itself has a population of over 12 million people and as of 2005, 7.5 million of these people live in the city's huge slum areas. The typical inhabitant of a Mumbai slum is a worker who has moved to the city from the countryside in search of greater economic opportunity. The slums themselves are not mere squatter communities, as most residents have paid as much as $6000. Part of the reason so many people are forced to live in the slums is due to the relatively extreme cost of housing in Mumbai. Mumbai is a very desirable place to live for many Indians, as the city serves as the financial capital and is also the center of India's "Bollywood" film industry. The desire behind the removal of the slum involve the ambition of administrators in the city of Mumbai to follow the development model of presumptive rival Shanghai China, which has seen tremendous growth and global integration in the past 20 years. The development of Mumbai is not only viewed as a matter of national pride for India, but is also viably viewed as a national development strategy. Having a "world-class" city in a developing nation like India may be an essential component of national development and integration into the capitalist world-economy. A first-world type city with all of the structural and infrastructural cues of the developed world could spur the economy (national, regional, local) by encouraging diversification of economic processes, thus helping India to move the economy from a developing economy to a developed economy that commands a greater portion of advanced sector processes (such as high finance and banking). In the case of Mumbai, the control of favelização is a development issue, as projecting the image of a prosperous world city (such as those found in the core areas of the World-System), meant the effective erasure of the reality of an embedded periphery that existed in the city center. In the end, controlling favelização meant deepening the peripherization of the poor and working classes by limiting the access and visual presence of the slum-dwelling population, most of whom contributed greatly to the city's economic vibrancy (Seabrook 1996).

As our world becomes increasingly urban, favelização, and other localized forms of peripheralization will become everyday realities for many people around the globe. The growth and expansion of slums, favelas, and shantytowns will owe just as much to negligence and indifference on the local level as to large-scale structure shifts on a systemic level. Nonetheless, the consequences of favelização for those who live and work in these spaces will be serious. These urban peripheries will be subject to continued and increasingly harsh forms of control and exclusion. I use the term "de-located" to describe how peripheralized communities (like some slums, favelas, and shantytowns) are effectively written out of the broader narrative of the rest of the city (Lindsey 2009). As cities like Mumbai plan for future growth and development, communities like Dhavari are excluded from any image or narrative of what the city will become through economic development. The successful "writing" out of undesirable subjects and the spaces they inhabit allows for the deployment of novel forms of social control. After peripheralized communities have been excluded discursively, physical control and exclusion becomes a logical next step. De-location serves as the discursive mechanism by which actual social and economic exclusion is made politically reasonable.

Conclusions: External arenas reborn?

Toward the end of his groundbreaking work, Modern World-System, Vol. I, Immanuel Wallerstein, says " … the line between periphery and external arena is fluid, both in the sense that it is hard for an analyst to fix it and in the sense that it shifts easily." In this particular instance he was referring to the relationship between Europe and Russia in the sixteenth century, but this comment has some contemporary relevance when applied to urban spaces in the developing world. Extreme manifestations of the slum, or favela can be seen as actual examples of peripheral communities falling out of the organized Modern world-system into a form of contemporary external arena. I use the term "systemic disincorporation" as a way of describing the particular form of spatial inequality associated with the slum areas of major metropolitan cities where the separation and social distance between the urban core and the urban periphery may be so great that the peripheral space may no longer be considered periphery in the conventional sense, and begins to behave more like an unincorporated external area. These new external areas can exist within the core of the world-system, but may be so far removed (socially, politically, and economically) that there may be no viable relationship between the core and the (urban) periphery. In the urban future, favelização might well serve as the one-word description for new levels of social exclusion and deprivation, and for the inability (or unwillingness) of those with power and influence to work to fully integrate into the Capitalist World-Economy all those who may be subject to its systemic forces. Favelização could also serve as a byword for a completely reorganized Capitalist World-Economy, one where the primary organizing social relationship of the system is not that which exists between the core and the periphery, but rather the relationship will be structured by a fundamental tension between the urban, the urban periphery, and the disincorporated multitudes living on the fringes of the worlds cities in slums, favelas, and shantytowns.

References

Anonymous. (2009) "Slums Grow without Restraint, Editorial", *Jornal do Brasil*, 21: 30, April, 23 2009.

Davis, Mike. (2007) *Planet of Slums*, New York: Verso.

Gunder Frank, André. (1998) *ReOrient: Global Economy in the Asian Age*. Berkeley: University of California Press.

Friedmann, John and Wolff, Goetz. (1982) "World City Formation: An Agenda for Research and Action", *International Journal of Urban & Regional Research*, 6: 309–36.

Friedmann, John and Wolff, Goetz. (1995) "Where We Stand: A Decade of World City Research", in Paul L. Knox and Peter J. Taylor (eds) *World Cities in a World-System*, Cambridge: Cambridge University Press.

Harvey, David. (2009) *Social Justice and the City*. Athens: University of Georgia Press.

Hopkins, Terrence, and Immanuel Wallerstein. (1982) *Processes of the World-System*. Beverly Hills, CA: Sage Publications.

Jameson, Frederick. (1991) *Postmodernism, or, The Cultural Logic of Late Capitalism*. New York: Verso.

King, Anthony. (1980) *Buildings in Society: Essays on the Social Development of the Built Environment*. London: Routledge.

Lash, Scott and Urry, John. (1994) *Economies of Signs and Space*. London: Sage Publications.

Lindsey, Delario. (2009) "Spectacular Urbanism: The Image, Development and the (De)Location of the Urban 'Other'", paper presented at International Sociological Association-Research Committee 21, Sao Paulo Conference Inequality, Inclusion and the Sense of Belonging, August 2009.

Maylam, Paul. (1990) "The Rise and Decline of Urban Apartheid in South Africa". *African Affairs*, 89: 57–84.

Neuwirth, Robert. (2006) *Shadow Cities: A Billion Squatters, A New Urban World*. New York: Routledge.

Sassen, Sakia. (2001) *Global City: New York, London, Tokyo*. Princeton, NJ: Princeton University Press.

Seabrook, Jeremy. (1996) *In the Cities of the South: Scenes from the Developing World*. New York: Verso.

Taylor, Peter J. (2002) "Firms and Their Global Service Networks", in Saskia Sassen (ed.) *Global Networks, Linked Cities*, New York: Routledge.

Taylor, Peter J. (2005) "Leading World Cities: Empirical Evaluations of Urban Nodes in Multiple Networks." *Urban Studies*, 42: 1593–1608.

Tibaijuka, Anna Kajumulo. (2005) Report of the Fact-Finding Mission to Zimbabwe to assess the Scope and Impact of Operation Murambatsvina by the UN Special Envoy on Human Settlements Issues in Zimbabwe. UN-HABITAT.

UN-HABITAT. (2003) *Slums of the World: The Face of Urban Poverty in the New Millennium*. Earthscan.

UN-HABITAT. (2008) *State of the World's Cities 2008/2009*. Earthscan.

Wallerstein, Immanuel. (1974) *The Modern World-System I: Capitalist Agriculture and the Origins of the European World-Economy in the Sixteenth Century*. New York: Academic Press.

10.4

Urbanization and poverty in the global "South"

Shahadat Hossain

The rural population operated subsistence economies maintaining only limited external contacts until the nineteenth century. The expansions of capitalist systems, however, underway for a half a millennium and accelerated by the Industrial Revolution, incorporated even more outlying regions into the emerging world-economy (Wallerstein 1989). Existing cities were integrated into the new system, their functions transformed. New cities were established to exercise political control and to channel resources to the metropolitan centers, and rural populations all over the world have been drawn into the urban nexus. This process of incorporation into the world-system has spread across the entire globe. The self-centered society that had only limited contact with the outside world has virtually disappeared. Rural populations have become subject to political control exerted from urban centers (Gugler 1992, 2004).

Massive changes are taking place in patterns of urbanization on a global scale and the South is urbanizing very rapidly. According to the UN Population Division (2004) the global urban population is set to double from 2.6 billion in 1995 to 5.1 billion in 2030. The South's share of the world's urban population has risen roughly in line with its total population share. Davis (2006) argues that the most celebrated result of urban transformation is the burgeoning of new megacities. In 1995, only one city (Tokyo) in the world had reached that threshold. By 2025, Asia alone could have ten or eleven conurbations as large, including Mumbai (33 million), Shanghai (27 million), Jakarta (24.9 million), Dhaka (25 million), and Karachi (26.5 million).

The issue of rapid urbanization is interlinked with massive poverty and the growth of informal sectors in the global "South." On the basis of research in low-income urban neighborhoods in Ghana, Hart (1973) emphasized the great variety of both legitimate and illegitimate income opportunities available to the urban poor. Subsequently, McGee (1976) explored various approaches toward what he called the "proto-proletariat." A great deal of research on the informal economy focused on a workforce that is typically under-enumerated and commonly characterized as unproductive in the 1970s (De Soto 1989; Castells and Portes 1989; Roberts 1995). These studies reveal that the informal economy is based on household labor and is therefore closely linked with household strategies.

Exploring social organizations of the urban poor is essential for understanding urban poverty and marginality. The traditional economic functions of marriage are not as important to poor communities as they are for other sectors of society. This is largely due to the poor economic prospects of the male and the corresponding fact that the male is not economically indispensable in the family. According to Roberts (1995), under conditions of poverty and job instability, men may be unwilling to commit themselves to relatively permanent obligations; likewise, women may be unwilling to take on the liability of a permanent attachment when the man proves to be a drunkard, unable to earn a living, or prone to violence. This explanation for marital instability in cities of developing countries emphasizes the elements of rational calculation entering into arrangements that to the casual observer, may seem to be result of individual or social disorganization (Eames and Goode 1973).

The final type of marginality to consider is political marginality, or the extent to which low-income families are disconnected from urban or national political issues and are unable to organize themselves to influence decision-making. Perlman (1976) documents the complex political organization of the urban poor in Brazil and points out that the level of political awareness among the poor is higher than that found in rural areas. Their level of direct participation in politics, through demonstrations or political meetings, is comparable with local and national administrative agencies. Roberts (1995) also found that the poor are not politically marginal in the sense of not participating in or affecting urban politics. When the political structure permits, the poor enter readily into the electoral game, organizing on behalf of middle-culture candidates and extracting what benefits they can for themselves or their neighborhoods.

In recent times, most of the urban poverty research focuses on economics and policy issues that lack theoretical frameworks. Although huge numbers of studies were conducted in Latin America in relation to the "culture of poverty" debate during the 1960s and 1970s, very few were conducted in the context of Asian and African cities. A number of quality studies were conducted on urbanization in the Third World from political economic perspectives, which analyze inequality and poverty through macro-level data and often lack empiricism. Therefore, the existing gaps in literature justify more theoretically and empirically balanced research on urban poverty in the near future.

References

Castells, M. and Portes, A. (1989). "World underneath: the origins, dynamics, and effects of the informal economy", in A. Portes, M. Castells and L. A. Benton (eds), *The Informal Economy: Studies in Advanced and Less Developed Countries* (pp. 11–37), Baltimore, MD: The Johns Hopkins University Press.

Davis, M. (2006). *Planet of Slums*, London: Verso.

De Soto H. (1989). *The Other Path: The Invisible Revolution in the Third World*, London: I.B.Tauris.

Eames, E. and Goode, J. G. (1973). *Urban Poverty in a Cross-Cultural Context*, New York: The Free Press.

Gugler, J. (1992). "The urban-rural interface and migration", in A. Gilbert and J. Gugler (eds), *Cities, Poverty and Development: Urbanization in the Third World* (pp. 62–86), Oxford: Oxford University Press.

Gugler, J. (2004). *World Cities Beyond the West: Globalization, Development and Inequality*, Cambridge: Cambridge University Press.

Hart, K. (1973). "Informal income opportunities and urban employment in Ghana", *Journal of Modern African Studies*, Vol. 11, pp. 61–89.

McGee, T. G. (1976). "The persistence of the proto-proletariat: occupational structures and planning for the future of Third World cities", *Progress in Geography*, Vol. 9, pp. 3–38.

Perlman, J. E. (1976). *The Myth of Marginality*, Berkeley: University of California Press.

Roberts, B. (1995). *The Making of Citizens: Cities of Peasants Revisited*, London: Arnold.

United Nations Population Division (2004). *World Urbanization Prospects: The 2003 Revision*, New York: United Nations.

Wallerstein, I. (1989). *The Modern World System* (Vol. 3), New York: Academic Press.

11. Food and agriculture

11.1

Global environmental governance, competition, and sustainability in global agriculture

Brian J. Gareau and John Borrego

World-system scholars are well aware that no matter how the elusive term "sustainable" might be defined, conventional capitalist agriculture is anything but (Foster and Magdoff 2000). While for millennia social groups had indeed found ways to overproduce, to overuse, to ignore (often at their peril) the intimate links between agriculture and surrounding ecosystems (Chew 2007; Moore 2003), the capitalist conversion of agriculture revolutionized the scale of the flows of food, and other natural resources from the countryside to cities (Foster 1994, 1999). This process has led to various "crises of the soil" throughout modern history (Foster and Burkett 2000: 420). Indeed, agriculture is the classic example of how indiscriminately applying technology solely to maximize yields can cause socio-ecological destruction. For in agriculture, all the usages of the term "yield" are brought into play: yields in the sense that technologies are designed to expropriate from the soil (and from farm workers) all its vitality; yields in the sense that the goal is to generate as much product as possible; yields in the sense that the product is a harvest, a living crop for consumption; and yields in the sense that the desired outcome is maximum economic gains.

In capitalist agriculture, these yields can not be maintained without the constant application of dangerous inputs. Today these are pesticides, herbicides, fungicides, and fertilizers that degrade the immediate and surrounding socio-ecological landscape. The unsustainable nature of capitalist agriculture was something recognized by political economists at its naissance:

> In modern agriculture, as in urban industry, the increase in the productivity and the mobility of labor is purchased at the cost of laying waste and debilitating labour-power itself. Moreover, all progress in capitalist agriculture is a progress in the art, not only of robbing the worker, but of robbing the soil; all progress in increasing the fertility of the soil for a given time is a progress towards ruining the more long-lasting sources of that fertility. The more a country proceeds from large-scale industry as the background of its development, as in the case of the United States, the more rapid is this process of destruction.
>
> *(Marx 1977 [1867]: 638)*

From the Industrial Revolution onward, agriculture (especially, but not exclusively, in grains) became increasingly global whilst continuing the deeper-rooted process of distancing practitioners from the ecological fabric of the landscape (Polanyi 2001). Intensification generated the need for entire markets for inputs to replace nutrient loss, and to destroy pests that thrived in the mono-crop systems that came to dominate industrial agriculture (Friedmann 2000).

From the 1940s to the late 1970s, Green Revolution technologies disseminated worldwide provided new chemical-intensive means of production that were far more damaging than simply snatching "more produce from the soil by robbing it of its fertility" (Marx 1977 [1867]: 376). Agricultural practices became equipped with fossil fuel-dependent technologies to run machinery, provide chemical inputs and pump irrigation water, with huge energy increases occurring in peripheral zones. For example, from 1955–2000, China's fossil energy use in agriculture increased 100-fold (Goodland and Pimentel 2000).

True to Marx's prediction, the United States proved to be a hotspot for such developments (Carson 1994; Kloppenburg 2004), but the effects were shared with the world entire. Starting in the early 1940s, high-yielding plants dependent on pesticides, fungicides, fertilizers, and irrigation were bred by the Rockefeller and Ford Foundations and integrated into the world-economy (McMichael 2003). "Petro-farming" reconfigured agriculture to depend entirely on fossil fuels, from pre-plant soil treatment to commodity transport (Walker 2004). Consequently, global grain production, as well as chemical input use, increased from 1950–90 worldwide. However, the increases were unevenly distributed; for example, whereas Latin America experienced increased per capita food production, Sub-Saharan Africa experienced a decrease (Jorgenson and Kuykendall 2008). As core-based transnational corporations further integrate farming systems into the global economy, we can expect chemical and water-intensive crop use to increase, with consequential negative effects on socio-ecological systems (Altieri 2000; Longo and York 2008). At present, some of those consequences include 10,000 annual pesticide poisoning deaths in peripheral countries, 400,000 pesticide poisonings worldwide (Jorgenson and Kuykendall 2008), and the creation of 200 dead zones in oceans due to nitrate (fertilizer) contamination (Speth 2008). Eighty percent of agricultural land in dry regions suffers from desertification—typically caused by overuse and mismanaged irrigation, but also by population growth and poverty—affecting mostly Africa (70 percent dry land), but also India and other parts of Asia, northern Mexico, and the southwestern United States (Speth and Haas 2006).

Global environmental governance and global agriculture

Global environmental governance is a world order led by the core countries, albeit with constant contestation by various institutions and social groups. In practice, it amounts to reinforcing what McMichael posits "development" represents: "the political relations of global capitalism" (2007: 170). Over the years, the global governance of both agriculture *and* the environment has been systematically designed as "ideological expression[s] of capitalist development" (McMichael 2007: 170). Unfortunately, attempts to make more equitable global agricultural and environmental sustainability have been, on the whole, rather dismal failures.

The idea of global environmental governance arose out of the recognition of global social and environmental interdependence in the late nineteenth century (Frank 1997). As countries became more strongly tied to the world-system, core states pressured others to participate, thus linking state participation in global environmental governance with state legitimacy (Frank et al 2000). Roberts (1996) shows that peripheral and semiperipheral countries have participated less in environmental treaties than core countries for all manner of reasons, including the fact that poor economic conditions dissuade any commitment that would reduce natural resource-extraction, and that repressive regimes are less willing to listen to demands from civil society. Nevertheless, participation in global environmental governance spread worldwide, likely augmented by environmental movements in peripheral countries in the 1980s (Schaeffer 1999), culminating most clearly in the 1992 Rio Earth Summit.

Efforts to govern food distribution via global governance emerged in the mid-1940s with the establishment of the UN Food and Agriculture Organization (FAO). The FAO's mandate was to direct international trade in order to "stabilize world agriculture and establish food security" (McMichael 2007: 172). Simultaneously, the FAO served to disseminate the wide-spread use of petro-farming and legitimize the penetration of core zone firms and states into peripheral markets. In its efforts to contain the Soviet Union, the United States established bilateral food programs with key peripheral and semiperipheral states (such as Mexico, Brazil, and India), which served to subsidize industries in the periphery whilst helping the US agro-industrial complex to manage its domestic over-production problem (Bartley and Bergesen 1997; Friedmann 1982).

The Green Revolution served as the basis for a new phase of capitalism—neoliberalism. Peasant agro-ecosystems worldwide were subjected to brutal shifts to make room for higher yielding hybrid crops. Laboratory-produced privatized seeds linked bio-technology to the global food market as never before. Peripheral zones became interlocked into the petro-farming treadmill that increased global food production at the cheapest world market prices in history (from 1980–2000) by combining agro-technologies with cheap labor, expropriated nutrient-rich land, and minimal water costs (Moore 2009). Recognizing agriculture as "the stepchild of development," World Bank President Robert McNamara standardized the use of "the transnational development agency network" to link global governance to rural farmers (Goldman 2005: 71–71). This energy-intensive system reached its highpoint in the 1980s, when, for example, global grain production peaked (IMF 2008). Via Structural Adjustment Programs, the World Bank and IMF virtually coerced peripheral countries into opening their markets to the core and shifting emphasis to agro-exportation and away from domestic foodstuffs:

> The resulting expanded, and energy-intensive, circulation of food on a global scale was institutionalized in the [World Trade Organization]'s 1995 Agreement on Agriculture, which prescribes a model of agriculture that has basically only one dimension: increasing agricultural production for exports, importing what cannot be produced without tariff protection or subsidies … .
>
> *(McMichael 2007: 174)*

Thus, neoliberalism became embedded in the governance of global agriculture, which has dispossessed up to 30 million people who lost their land due to trade liberalization, not to mention exacerbating the spread of environmental degradation associated with capitalist agriculture, as mentioned above (McMichael 2007).

While the neoliberal shift in agriculture has been explained by world-systems scholars in political economic terms (Cf. McMichael 2007; Moore 2009), it is useful to illustrate the concomitant *ideological shift in global environmental governance*. Such an exposition highlights the extreme difficulty that social movements will face when attempting to change state tactics in the global arena in the neoliberal phase of capitalism. Social movements and other factions of global civil society will find it increasingly difficult to force global governing bodies to consider the needs of the poor, the food-insecure, and those suffering from environmental degradation via demands that the Rousseauian general will or the global environment be considered. In global environmental governance today, the will supreme is that of competition and protectionism of core industries.

From Stockholm to Copenhagen, from global environmental welfare to free market ideology

The 1972 UN Conference on the Human Environment (UNCHE) held in Stockholm took place when environmental concern among core citizens was extremely high, as was their trust in "big government" (Frank 1997). Command-and-control regulation of industry by the state and demand for multilateral governing regimes became mantras of the early core zone environmental movements. Regulation of the economy and of citizen consumption rates was not, as they are today, seen as being contradictory to democratic freedoms. Early environmental movement scholar William Ophuls expressed what was commonly felt among the growingly concerned public: "Man is a passionate being," he stated, and requires "checks on will and appetite; if these checks are not self-imposed, they must be applied externally as fetters by a sovereign power" (Ophuls 2004: 60). In Stockholm, states were urged by core zone citizenry to develop international environmental law in response to environmental destruction caused by industry (Speth and Haas 2006). Core citizens supported the state's role in intervening in the market. This approach to alleviating environmental destruction typified the political culture of the welfare state, which worked to de-commodify areas of economic and social life deemed too valuable to be subjected to market forces.

After Stockholm international environmental treaties grew in number, the United Nations Environmental Programme (UNEP) was created. Some examples include the 1973 International Convention for the Prevention of Pollution from Ships, the 1973 Convention on the International Trade in Endangered Species, the 1974 Paris Convention for the Prevention of Marine Pollution from Land-based Sources, the 1979 Convention on the Conservation of Migratory Species of Wild Animals, and the 1979 Convention on Long-range Transboundary Air Pollution. The 1980s witnessed global agreements on more complicated environmental issues, including the 1985 Vienna Convention on the Protection of the Ozone Layer, the 1987 Montreal Protocol on Substances that Deplete the Ozone Layer, the 1982 UN Convention on the Law of the Sea, the 1989 Basel Convention on Control of Transboundary Movements of Hazardous Wastes and their Disposal, and, in the next decade, the 1992 Framework Convention on Climate Change. This surge in global environmental treaties signified a rise in conviction that strong state intervention combined with new multilateral agreements could resolve global ecological issues (Hunter et al 2006: 173–74). Studies on issues such as the CFC-ozone link by Rowland and Molina (1975) and the National Academy of Science's Charney Report on climate change (NAS 1979) played major roles in encouraging core countries to take precautionary measures to control pollution on national and global scales.

After Stockholm, the next milestone was naturally the 1987 Montreal Protocol, which made use of the precautionary principle to phase out chlorofluorocarbons (CFCs) before understanding of the human impact of ozone layer loss was conclusive. In harmony with attitudes on environmental regulation expressed in Stockholm, the Montreal Protocol utilized command-and-control measures to compel industry to develop CFC substitutes. The results were successful: today over 95 percent of CFCs are phased out (Gareau 2010). For two decades, the impetus to create stronger global governance, begun in Stockholm, continued. After the fall of the Berlin Wall, cries for a world-government were not uncommon. International environmental law scholars called for centralized, powerful global governing bodies to manage the ecology of the world-system, just as peripheral states had acquiesced control to the WTO regarding international trade (Gareau and DuPuis 2009). Global environmental problems, it was argued, required the creation of "a proper international environmental agency within the United Nations system that has real power and authority" (Palmer 1992: 262).

Similar to the effects of the 1995 WTO Agreement on Agriculture, the 1992 UN Conference on Environment and Development (the Rio Earth Summit) marked the moment in which market-based approaches became a central focus of global environmental governance. With the promotion of the 1987 Brundtland Report (1987) at the Earth Summit, "sustainable development" discourse gained prominence, linking environmental degradation to poverty (Lélé 1991). However, Rio Summit deliberations de-emphasized command-and-control regulations, questioned erstwhile accepted "limits to growth" arguments (Meadows 1972), and were exposed to a rash of business/government partnership solutions partly through participation by the Business Council for Sustainable Development. Nevertheless, the Rio Summit still contained the attitude that global environmental governance was the appropriate arena in which to tack global environmental problems, with advocates pointing to the Montreal Protocol as evidence. Therefore, in the Agenda 21 language of the Rio Summit, the role of the state remained clearly central in each of its 23 principles.

Thenceforth, however, core zones intensified the political/environmental focus on the broader neoliberal agenda. With the 1980s global financial crisis, the rise of Reaganomics and Thatcherism among conservatives, and the surfacing of the post-Soviet era, the US position on global politics increasingly reflected a "global managerial state of mind" which "reformulated the post-1989 question of democratization and governance, and the green revolution era concerns with redistribution and equity, into the neoliberal question of the *freedom and sovereignty of capital*" (Goldman 2005: 91). Here, the United States promoted the privatization of state institutions, opening areas of society and the environment previously protected from market forces through elimination of command-and-control regulatory instruments (Harvey 2005), and the promotion of stakeholder politics over the usage of society-based measures (DuPuis and Gareau 2008). The core countries promoted the application of these principles to environmental issues by emphasizing free-marketeering and decentralized governance (Tickell and Peck 2003: 167). This "Washington Consensus" was "increasingly negative on multilateralism, environmental regulation, the United Nations, foreign aid, and treaties and similar agreements, and, indeed, *government itself*" (Speth and Haas 2006: 80, our emphasis).

The 2002 World Summit on Sustainable Development in Johannesburg marked the wholesale move toward the Washington Consensus in global environmental governance. Business coalitions took a central role, calling for business-government partnerships and expressing "a serious doubt about whether a multilateral system is able to effectively address our sustainable development challenges" (Stigson 2004: 267). Rather, they argued it was necessary for governments, "*in consultation with business*, to create the conditions that allow business to contribute fully to sustainable development.... If financial markets do not understand and reward sustainable behavior, progress will be slow" (Stigson 2004: 266, our emphasis). At the World Summit, Meena Rahman, Chairperson of Friends of the Earth, exclaimed that "nothing happened since Rio in 1992 because governments embraced a neoliberal agenda, which represents the interest of big powers" (UNEP 2004). Indeed, the principal outcome of the World Summit was a lack of commitment from core zone governments to provide aid, and instead, to promote voluntary business/government partnerships (Roberts and Parks 2007).

Outcomes of the 2009 UNFCCC 15th Conference of the Parties (COP 15) held in Copenhagen further suggest that global environmental governance is experiencing a devolution as markets and market-based principles are employed to defend environmental (in)action. Since first refusing to sign the Kyoto Protocol in 1998, the United States has argued that a global trading scheme would be unfair to core zones because the global South is exempt from any immediate carbon reductions (although the United States led the adoption of the Montreal Protocol, which also did not bind peripheral countries to immediate cuts. See Nordhaus and Boyer [2000]; Gareau [2010]). At

COP 15, the Copenhagen Accord, which was "taken note of" but not adopted (UNFCCC 2009: 19), was drafted by the United States, China, India, Brazil, and South Africa behind closed doors. Although the Yale Climate Project found that a vast majority of Americans—regardless of political affiliation—support the passage of federal climate and energy policies, including over 70 percent supporting the regulation of CO_2 as a pollutant, US leadership at the global scale remains tenuous. Some critics blamed COP 15 failures on the US Senate and President Obama. They argued that by negotiating the Copenhagen Accord with only a select group of nations, most UN member states were excluded. Others observed how the British and American governments both blamed China for the failure of the talks, but others still noted that Obama placed China in an impossible position by demanding concessions while offering none. These arguments are all valid, yet it is important to also remember that current environmental crises require an historically unprecedented conversion in frame of mind for citizens in the core countries. As George Monbiot laments, "It is a campaign not for abundance but for austerity, … not for more freedom but for less, … not just against other people, but against ourselves" (2006: 215).

Monbiot's sentiments echo arguments made by Ophuls and others in the 1970s, and they appear grossly outdated when considering the neoliberal turn in global environmental affairs. Recognizing world-system theoretical understandings of the ability of global competition to stall progress in global governance, we are left pessimistic about the role that global environmental regimes will play in alleviating present and future global environmental problems, including those associated with agriculture. The answer to a more "sustainable" agriculture will likely come from political economic conditions and institutional reforms pressured by social groups in the semiperiphery, probably in Asia, and perhaps centered in organic production in China.

Conclusion

With 80 percent of the 1$US billion spent daily on agricultural subsidies going to farmers in the EU, United States, and Japan (Pinstrup-Andersen 2002), it is clear that neoliberal agro-policies are unevenly shouldered by the peripheral and semiperipheral zones (Bello 2009). However, as Arrighi (1994, 2007) points out, amid the expansion and decline of Western-dominated neoliberal globalization, China has been able to create a quasi-neoliberal platform of its own. With stringent state-controlled economic growth, China has succeeded in developing an export-oriented capitalist mode of production, a project that it began in the early post-Mao period (Borrego 1990). In the agro-sector in particular, this process does not rely on "market forces" alone to attract investment, but importantly it involves state intervention in the management of local and regional resources and governance "links" in the global commodity network (Peine and McMichael 2005). The organic sector is emerging as a mainstream network that may help create a more sustainable agro-industrial complex. Yet the export-oriented nature of China's growing agro-market threatens to limit the socio-ecological sustainability of its organic model.

Worldwide, organic production systems are increasingly challenged by commercial conventions rooted in economic competition (Raynolds 2004). They are increasingly taking the shape of conventional commodity chains that have existed for centuries, as world-system analysts have understood them (Hopkins and Wallerstein 1986). Today, China is arguably molding the structure of the future of organic agriculture, one that makes use of interregional commodity networks that may soon become global. Unlike the EU and the US's historical rootedness in a local, bottom-up, ideologically-driven organic sector, China's model is export-oriented, top-down, and price-driven. It regards the organic farm firstly as an economic enterprise, and it uses core zone standards as its national guide in order to potentially access those markets. As Paull relates, "there is potential for China to redefine the standard for internationally traded food, as certified organic" (Paull 2008: 272).

While there is no denying that China has accelerated its energy-intensive and ecologically destructive agro-industrial production platform (Sachs 1999; Yang 2007; Ye et al 2002), here we also find astounding acceleration of its organic sector. Re-initiating its ancient agricultural practices in the 1980s with government-funded research, by the 1990s, China had already churned out 7 million hectares devoted to organic agriculture projects, which has led to economic growth in the region in some cases (Marsden 2006: 193). In 2002, it was estimated that certified organic production in China exceeded 40,000 hectares (Marsden 2006), and by 2005 estimates stated that acreage reached roughly 700,000 hectares, generating 3.7 billion Yuan in revenue and close to $150US million in exports (IFAD 2005). Moreover, organic processing companies were established in international supply chains (Yang 2007). By 2005–06 China's global ranking in number of hectares devoted to organic moved from forty-fifth to second, to 3.5 million hectares. Now, China is the global leader in terms of organically cultivated acreage. In 2005–06 alone, China accounted for over 60 percent of the world's annual increase in organic land (Paull 2008).

While China regulates what are probably "the most stringent of production and conformity assessment requirements" (Willer and Klicher 2009), the EU and United States have yet to recognize their organic produce domestically. This is significant because the core zones contain more organic demand than they can produce, and their citizenry willingly pay the higher prices (Raynolds 2004). Meanwhile, Southeast Asian countries (where 70 percent of agricultural trade is still interregional) have collaborated on strengthening their organic network and creating an inspection/ certification standard for the region's growers who seek international recognition (Willer and Klicher 2009). While the declining Western hegemonic powers will continue to struggle to protect themselves from the immersion of China into their organic agro-markets, it is likely that this organic sector will continue to gain legitimacy and grow, first regionally in Southeast Asia (as others have in Latin America), and then globally.

It is clear that organic is not coterminous with sustainable agriculture. China's focus on export-oriented growth in organic threatens to continue reorganizing traditional productive relations and decreasing the food security of rural communities whilst increasing tolls on water supplies and other socio-ecological resources. Growing labor shortages and civil unrest, however, might force China to consider food sovereignty, small farms, and diversity (Altieri 2009) in organic production, which could lead to the development of a state-led regional production platform worthy of replication. As we see it, the potentiality of sustainable agricultural networks will probably not grow out of global governance regimes, nor will it grow out of an altruistic counter-cultural movement, or pastoral ideals. Rather, it will likely grow out of state-regulated investment in opportunities to capture market-share in a globalizing sector driven by demand from the core and emerging core zones of the world-economy.

References

Altieri, M.A. (2000) "Ecological impacts of industrial agriculture and the possibilities for truly sustainable farming", in F. Magdoff, J.B. Foster and F. Buttel (eds) *Hungry for Profit*, New York: Monthly Review Press.

Altieri, M.A. (2009) "Agroecology, small farms, and food sovereignty", *Monthly Review*, 61(3): 102–113.

Arrighi, G. (1994) *The Long Twentieth Century*, New York: Verso.

Arrighi, G. (2007) *Adam Smith in Beijing*, New York: Verso.

Bartley, T. and Bergesen, A. (1997) "World-system studies of the environment", *Journal of World Systems Research*, 3: 369–80.

Bello, W.F. (2009) *The Food Wars*, New York: Verso.

Borrego, J. (1990) "La reintegracion de China: un analisis de las zonas economicas especiales de China en el contexto del capitalism global", in A.A. Bejar and J. Borrego (eds) *La Insercion de Mexico en la Cuenca del Pacifico, Volumen II*, Mexico, D.F.: Universidad Nacional Autonoma de Mexico.
Carson, R. (1994) *Silent Spring*, Boston, MA: Houghton Mifflin.
Chew, S.C. (2007) *The Recurring Dark Ages*, Lanham, MN: Altamira Press.
Dupuis, E.M. and Gareau, B.J. (2008) "Neoliberal knowledge: the decline of technocracy and the weakening of the Montreal Protocol", *Social Science Quarterly*, 89: 1212–29.
Foster, J.B. (1994) *The Vulnerable Planet*, New York: Monthly Review Press.
Foster, J.B. (1999) "Marx's theory of the metabolic rift: classical foundations for environmental sociology", *American Journal of Sociology*, 105: 366–405.
Foster, J.B. and Burkett, P. (2000) "The dialectic of organic/inorganic relations: Marx and the hegelian philosophy of nature", *Organization & Environment*, 13: 403–25.
Foster, J.B. and Magdoff, F. (2000) "Liebig, Marx, and the depletion of soil fertility: relevance for today's agriculture", in F. Magdoff, J.B. Foster and F. Buttel (eds) *Hungry for Profit*, New York: Monthly Review Press.
Frank, D., Hironaka, A. and Schofer, E. (2000) "The nation-state and the natural environment of the twentieth century", *American Sociological Review*, 65: 96–116.
Frank, D.J. (1997) "Science, nature, and the globalization of the environment, 1870–1990", *Social Forces*, 76: 409–35.
Friedmann, H. (1982) "The political economy of food: the rise and fall of the postwar international food order", in M. Burawoy and T. Skocpol (eds) *Marxist Inquiries*, Chicago, IL: University of Chicago Press.
Friedmann, H. (2000) "What on earth is the modern world-system? foodgetting and territory in the modern era and beyond", *Journal of World Systems Research*, XI: 480–515.
Gareau, B.J. (2010) "A critical review of the successful CFC phase-out versus the delayed methyl bromide phase-out in the Montreal Protocol", *International Environmental Agreements: Politics, Law, and Economics*, 10(3): 209–31.
Gareau, B.J. and Dupuis, E.M. (2009) "From public to private global environmental governance: lessons from the Montreal Protocol's stalled methyl bromide phase-out", *Environment & Planning A*, 41: 2305–23.
Goldman, M. (2005) *Imperial Nature*, New Haven, CT: Yale University Press.
Goodland, R. and Pimentel, D. (2000) "Environmental sustainability and integrity in the agriculture sector", in D. Pimentel, L. Westra and R.F. Noss (eds) *Ecological Integrity*, Washington, DC: Island Press, 121–37.
Harvey, D. (2005) *A Brief History of Neoliberalism*, New York: Oxford University Press.
Hopkins, T. and Wallerstein, I. (1986) "Commodity chains in the world-economy prior to 1800", *Review*, 10: 157–70.
Hunter, D., Salzman, J. and Zaelke, D. (2006) *International Environmental Law and Policy*, 3rd edn., New York: Foundation Press.
IFAD (2005) *Organic Agriculture and Poverty Reduction in Asia*, Rome: International Fund for Agricultural Development.
IMF (2008) *World Economic Outlook*, Washington, DC: International Monetary Fund.
Jorgenson, A.K. and Kuykendall, K.A. (2008) "Globalization, foreign investment dependence and agriculture production: pesticide and fertilizer use in less-developed countries, 1990–2000", *Social Forces*, 87: 529–60.
Kloppenburg, J.R. (2004) *First the Seed*, Madison: University of Wisconsin Press.
Lélé, S.M. (1991) "Sustainable development: a critical review", *World Development*, 19: 607–21.
Longo, S. and York, R. (2008) "Agricultural exports and the environment: a cross-national study of fertilizer and pesticide consumption", *Rural Sociology*, 73: 82–104.
Marsden, T. (2006) "Denial or diversity? creating new spaces for sustainable development", *Journal of Environmental Policy & Planning*, 8: 183–98.
Marx, K. (1867) *Das Capital*;trans. Ben Fowkes (1977) *Capital*, New York: Vintage.
McMichael, P. (2003) *Development and Social Change*, London: Pine Forge Press.
McMichael, P. (2007) "Feeding the world: agriculture, development and ecology", in L. Panitch and C. Leys (eds) *Coming to Terms with Nature*, New York: Monthly Review Press.
Meadows, D.H. (1972) *The Limits to Growth*, New York: Universe Books.
Monbiot, G. (2006) *Heat*, Essex: Penguin Press.
Moore, J.W. (2003) "Nature and the transition from feudalism to capitalism", *Review*, XXVI: 97–172.

Moore, J.W. (2009) "Ecology and the accumulation of capital: a brief environmental history of neoliberalism", paper presented at the Workshop on Food, Energy, Environment: Crisis in the Modern World-System, Fernand Braudel Center, Binghamton University, October 9–10.

NAS (1979) *Carbon Dioxide and Climate*, Washington, DC: National Academy of Sciences.

Nordhaus, W.D. and Boyer, J. (2000) *Warming the World*, Cambridge: MIT Press.

Ophuls, W. (2004) "The scarcity society", in K. Conca and D. Geoffrey (eds) *Green Planet Blues*, Oxford: Westview Press.

Palmer, G. (1992) "New ways to make international environmental law", *American Journal of International Law*, 86: 259–83.

Paull, J. (2008) "China's organic revolution", in S. Bhaskaran and S. Mohanty (eds) *Marketing of Organic Products*, Hyderabad, India: The ICFAI University Press, 260–75.

Peine, E. and McMichael, P. (2005) "Globalisation and global governance", in V. Higgins and G. Lawrence (eds) *Agricultural Governance and the New Politics of Regulation*, London: Routledge.

Pinstrup-Andersen, P. (2002) "Towards a sustainable global food system: what will it take?" Keynote presentation for the Annual John Pesek Colloquium in Sustainable Agriculture. Iowa State University, March 26–27.

Polanyi, K. (2001) *The Great Transformation*, Boston, MA: Beacon Press.

Raynolds, L.T. (2004) "The globalization of organic agro-food networks", *World Development*, 32: 725–43.

Roberts, J.T. (1996) "Predicting participation in environemntal treaties: a world-system analysis", *Sociological Inquiry*, 66: 38–57.

Roberts, J.T. and Parks, B. (2007) *A Climate of Injustice*, Boston, MA: MIT Press.

Rowland, F.S. and Molina, M.J. (1975) "Chlorofluoromethanes in the environment", *Reviews of Geophysics and Space Physics*, 13: 1–35.

Sachs, W. (1999) *Planet Dialectics*, New York: Zed Books.

Schaeffer, R.K. (1999) "Success and impasse: the environmental movement in the United States and around the world", in W.L. Goldfrank, D. Goodman and A. Szasz (eds) *Ecology and the World-system*, Westport, CT: Greenwood Press.

Speth, J.G. (2008) *The Bridge at the End of the World*, New Haven, CT: Yale University Press.

Speth, J.G. and Haas, P.M. (2006) *Global Environmental Governance*, Washington, DC: Island Press.

Stigson, B. (2004) "Walking the talk: the business case for sustainable development", in K. Conca and G.D. Dabelko (eds) *Green Planet Blues*, Oxford: Westview Press.

Tickell, A. and Peck, J. (2003) "Making global rules: globalization or neoliberalization?", in J. Peck and H.W.-C. Yeung (eds) *Remaking the Global Economy*, Thousand Oaks, CA: Sage Publications.

UNEP (2004) "Earth report (TVE): global environmental governance in 2050", *The Environment in the News*, November 30, Nairobi: United Nations Environmental Programme.

UNFCCC (2009) *Report of the Conference of the Parties on its Fifteenth Session, Held in Copenhagen from 7 to 19 December 2009*, Bonn: UNFCCC.

Walker, R. (2004) *The Conquest of Bread*, New York: The New Press.

Willer, H. and Klicher, L. (eds) (2009) *The World of Organic Agricutlure*, IFOM, Bonn, FIBL, Frick, ITC, Geneva.

Yang, Y. (2007) *China Environmental Health Project Factsheet*, China Environmental Forum in partnership with Western Kentucky University.

Ye, X.J., Wang, Z.Q. and Li, Q.S. (2002) "The ecological agriculture movement in modern China", *Agriculture, Eco-systems and Environment*, 92: 261–81.

11.2

Hunger and the political economy of the world food system

Stephen J. Scanlan

Introduction

Walking in Boston I happened upon the Irish Famine Memorial dedicated in 1998 to commemorate the 150th anniversary of an Gorta Mór—the Great Hunger. I was moved, not only by the imagery the memorial's sculptures invoked, but also by the narrative accompanying them. In particular were words included from Irish poet Lady Jane Wilde noting how "the stranger reaps our harvest, the alien owns our soil" in reference to British colonial rule at the time, which in effect starved or displaced millions of Irish people.

These words capture the essence of hunger in the world food system—not only then but today as well. They reflect the reality of a system of haves and have-nots in a world where there is hunger for some and plenty for others, where inequality, politics, and power enable sustenance at the expense of the exploited. In this essay I discuss these dynamics, laying out the relevance of food insecurity for world-systems analysis.

2007 and 2008 witnessed an explosion of discontent as a global food crisis led to protests and food riots in dozens of countries (Patel and McMichael 2009). Demonstrations erupted in Indonesia over soybean shortages, for example, and China put price controls on staple foods to quell potential violence among the poor and working classes increasingly vulnerable to hunger. The accompanying economic collapse and soaring food prices due to rising energy costs, growing demand in biofuels, and production concerns resulted in more people than ever in human history experiencing hunger at 1.02 billion in 2009, a total decreasing to 925 million people at the end of 2010.

The typical argument is that food insecurity exists primarily because of production shortfalls, natural disasters, or population pressure. Such "scarcity arguments" can be compelling, but from a critical world-systems perspective, miss the reality that food is plenteously available and that the core problem of hunger is rooted in inequality in food access associated with the socioeconomic and political dynamics of the world-system. In what follows, I elaborate on sociological insights and research regarding food insecurity and important areas of inquiry regarding the global political economy.

Food insecurity and famine: Conceptualization and theory

Although often presented by the media and relief agencies in the wake of crisis situations such as the conflict in Darfur or the earthquake in Haiti as such, food insecurity and its connection to the world-system is much more than what occurs in emergencies—it is an emergency in and of itself.

Academic and policy discussions use the multi-dimensional concept of "food insecurity" to account for hunger's complexities. Hunger and famine have existed for millennia but food insecurity as a concept came from the 1974 World Food Conference creating the UN Food and Agriculture Organization (FAO) Committee on World Food Security. Definitions have evolved since then, with one commonly cited being that of Reutlinger (1986: 1) who defines food security as "access by all people at all times to enough food for an active healthy life"; food *insecurity* is individuals lacking this access and living without enough food in their day-to-day lives. DeRose et al (1998) further elaborate, noting that food insecurity entails food shortage (the food supply), food poverty (lacking access to food), and food deprivation (the nutritional value of available and accessible food). Multiple dimensions of food insecurity are essential to consider because too often the focus is on supply, neglecting the food poverty and deprivation resulting from power, politics, and the commoditized nature of food.

The conceptualization of food security has recently taken on additional components: stability and sovereignty. Food stability refers to the ability of individuals or nations to handle sudden shocks such as a natural disaster or war, or disruptions in cyclical events such as growing seasons or markets that impact food security (FAO 2006). An important world-system consideration to food stability is who is most vulnerable to such shocks and how have they been exacerbated by the characteristics of "the global food order." McMichael defines food sovereignty as "the social right of a community or country to determine its own policies regarding food security (adequate supply and appropriate cuisine) and the cultural, social and ecological conditions under which it is sustained" (2004: 347). Here the emphasis is on controlling food shortage, poverty, and deprivation where individuals and entire countries are not empowered to feed themselves. The goal is maintaining the integrity of self-sustaining, local food systems—an important consideration for world-systems analysis. Taken together, the above components provide a well-rounded approach to what is typically described as *hunger.*

Like hunger, *famine* is a more familiar term than food insecurity. Famine implies widespread displacement, starvation, and death from shocks to and lack of sovereignty within the food system, be it in supply, distribution, or the nutritional value of the available food. Although complex emergencies derived from conflict or disasters are typically believed to be the root cause of famines, a critical world-systems perspective would emphasize macro-structural determinants tied to politics and inequality within the world-system.

When examining famine, one should contrast "outsider" versus "insider" perspectives. According to Devereux (1993: 18), outsider definitions of famine are preoccupied with three characteristics: a critical food shortage, starvation, and excess mortality. As Devereux (1993) notes and Ó Gráda (2009) confirms, one can look to the history of famine and see that hunger and starvation has occurred in many places where food was plenteously available, meaning that numerous famines fail to meet the assumed criteria. Therefore a different perspective and conceptualization is needed. "The contrast between 'insider' and 'outsider' definitions is starkest when it is recognized that victims see widespread death as the end result of a famine process, while outsiders see it as the beginning" (Devereux 1993: 18). Those looking at famine from the outside need to consider that the real experience for hundreds of millions of people around the world is the life of destitution that puts them in a vulnerable position. Famine is thus an ongoing condition tied to the constant dynamics of the political economy of the world-system, not simply an "event" that "happens"

Famines have primarily human-induced causes as opposed to being strictly "natural" events. Devereux (2007: 10) accentuates this in a paradigm shift that moves thinking on famine from asking the question "*what* caused the famine?" to instead considering "*who* caused the famine?" thus emphasizing the centrality of politics and calling for accountability connected to human actions and decision-making. The capacity exists to feed the planet but malevolent intent in some cases and limited political will in others have failed to eliminate famine (Devereux 2007: 10).

The most important body of theoretical work on food insecurity has derived from the "entitlement perspective" of Sen (1981; see also Drèze and Sen 1989; Drèze et al 1995). Referencing hunger and malnutrition, Ellmann speaks to this approach noting "it is not the lack of food but the inability to purchase it that causes such catastrophes. People starve because they *have* no food, not because there *is* no food" (1993: 5). Entitlement theory therefore emphasizes that hunger stems from differential access to the capital and purchasing power to acquire available food in the world-system.

Numerous other theories exist for explaining hunger, including discussions of modernization, natural disasters, political economy, and social disruptions, among others (see Buttel 2000), but the most attention has been given to the perils of population. Population pressure and ecological concerns connect to dire Malthusian predictions regarding the dangers of unchecked population growth for human well-being. At the risk of oversimplification, Malthus and his modern-day advocates emphasize that limited resources and the inability to meet the demands of overpopulation will contribute to widespread suffering and famine. Malthusian ideas have long been at odds with Marxist interpretations of "the population problem" and its connection to hunger, which are at the roots of a more radical political economic perspective. Although population problems exist and environmental degradation is also an essential consideration, from a world-system perspective such issues are a result of capitalism (Araghi 2009; Foster 2009) and are products of power and inequality in their relationship with food insecurity. As Cereseto (1977: 34) notes, this system is driven by dependency that generates underdevelopment and poverty that actually *cause* population growth in poor countries as families struggle to survive displacement from the land, low wages, and limited safety nets (see Bennett 1986; George 1991). Population concerns thus matter mostly in that they exacerbate the problems of the poor who are more adversely affected by such dynamics.

Globalization and the world food order

Building on Sen, World-systems analysis emphasizes the centrality of access to food insecurity and the need to move beyond a focus on production and supply problems. In this regard the perspective looks to politics and power in determining Sen's entitlements and Devereux's accountability for hunger. However, world-systems analysis offers a more critical perspective on the political economy of the world-system that questions the very nature of capitalism itself and the inequality and exploitative dynamics driving it. Global inequality is ultimately what matters most—and hence the emphasis of this essay.

Hunger and the political economy of the world food system have been examined in a number of fundamental works that speak to the centrality of global capitalism. Of most importance has been work within sociology including Bello (2009), Bonanno et al (1994), Buttel (2000), de Janvry (1976), Friedmann (1982, 1993), Magdoff et al (2000), McMichael (1995, 1997, 1998, 2009), Wimberley (1991), and Wimberley and Bello (1992) in addition to numerous others taking critical cultural, environmental, geographical, historical, and political perspectives (Bennett 1986; George 1991; Holt-Jiménez and Patel 2009; Lappé et al 1998; Patel 2007; Shiva 2000, 2007, 2008; Thurow and Kilman 2009; Weis 2007) on the global food economy.

An essential starting point of any discussion of world-systems dynamics is inequality inherent to global capitalism, and as a symptom of that inequality, hunger is no exception. As McMichael notes, famine and related issues are a product of the historical legacy of empire and contemporary neoliberalism in which, through global capitalism, "agriculture has been converted into a branch of industry" (2009: 1; also see Magdoff et al 2000). Poverty and hunger are closely interwoven with globalization and the presence of a world food system that dictates not only supply but also food poverty, deprivation, stability, and sovereignty. Trade in and acquisition of foodstuffs have been a part of an expanding world-system for centuries, but over the last few decades the world has become an increasingly interdependent cultural, economic, political, and social system fueled by advances in technology. Everything that is a part of our daily lives—including basic needs such as food—is dependent on borderless processes. Hunger results when food is commoditized in a global economy that has successfully increased the overall food supply with advances in agriculture, marketing, storage, and transportation but done little to guarantee access to that food for a global citizenry increasingly unable to acquire it. The roots of hunger have long been considered to stem from the inequality in the world-system and the significance of globalization should therefore not be underestimated (Cereseto 1977; George 1991; Magdoff et al 2000).

Agribusiness is at the center of globalization as the world shifted to large-scale agricultural production and increases in food imports and exports worldwide. The system is characterized by highly specialized, industrial food production, market-focused distribution, and a food surplus that seeks to keep the price of food cheap as demanded by growing urban populations that become increasingly distant from the sources of their sustenance (Beardsworth and Keil 1997; Magdoff et al 2000; Patel 2007). Such dynamics have enormous consequences regarding the power one has over what is eaten, where it comes from, how it is produced, and how much it is going to cost. Hunger remains a large concern because of this.

Globalization in its modern form has created interdependency in which problems and needs in one corner of the world affect another and the global food order amplifies this. There is an unprecedented global connection to the food on our plates as it travels thousands of miles from field to table (Petrini 2006). Developing countries have moved from feeding themselves to producing and exporting cash crops for the global marketplace and becoming net food importers through purchasing food for consumption. In this system, the United States has sought to become the breadbasket to the world through geopolitics and the institutionalization of "food power" under the banner of neoliberalism and free trade (McMichael 1998). To effectively participate in the system, less industrialized countries have been pushed by development experts and international organizations such as the World Bank and International Monetary Fund to pursue niche markets and comparative advantage policies that have left many countries and their citizens dependent and poor (Jarosz 2009; Madeley 2000). Structural adjustment programs intending to boost agricultural productivity and alleviate poverty, for example, "probably wrought greater havoc on peasant agriculture globally than any other force, social or natural" (Bello 2009: 31) and this is worsened with deepening debt, environmental degradation, gender inequality, immigration, and weak wages among other connections (Chattopadhyay 2000; Mukherjee 1994; Samatar 1993). This often makes the IMF and World Bank targets of protest and great discontent in less industrialized countries (Ryan 1998; Walton and Seddon 1994). Needless to say, food is central to human well-being and when people are powerless to meet their needs, desperate actions are certain to result.

The world food system is about satisfying market shares and profits (Magdoff et al 2000) more than feeding the hungry, and without sufficient entitlements to food, access remains problematic. Consider a government push for commercial prawn aquaculture in the Philippines, for example (Dunaway and Macabuac 2007). Here, the drive for export-oriented fishing ventures in the global supply-chain has made the country one of the world's top shrimp exporters, but only at the expense of subsistence fishing households, who suffer from increased food insecurity, the destruction of local livelihoods, the loss of access to ecological resources, and the loss of social services connected to structural adjustment agreements with the policy. The end result was a development agenda, which among other impacts, "privileged a small elite of export-oriented capitalists … (and) left fishing households with fewer survival options than they had before … " such outward-focused policies (Dunaway and Macabuac 2007: 324).

Building on this and turning to players in the global food order, the world's powerful agribusiness multinational corporations (MNCs) demand scrutiny. MNCs dominate the global food system in which "agribusiness imperialism is central to the coercive use of institutional mechanisms to monopolize control of world agriculture and flows of food" (McMichael 1998: 208; also see Patel 2007). Contributing to hunger, global agribusiness forms enormous conglomerates that operate at every level of the food commodity chain, establishing a presence in all corners of the world and capitalizing on humanity's need of food for survival. Although claiming themselves to be good corporate citizens as when Archer Daniels Midland touts itself as "the supermarket to the world," agribusiness practices can have adverse effects on hunger with links to land access, food sovereignty, biodiversity, loss and sustainability.

Often with cooperation from the state, global agribusiness practices hearken to colonialism and foreign dominance. Consider "land grabs" (Spieldoch and Murphy 2009; Zoomers 2010), for example, which shift productive land from local to global use with little benefit to the local citizenry dependent on that land. Such practices are touted as "win-win" situations by foreign governments and businesses that purchase or lease land for export agricultural production and other development purposes, claiming to introduce much-needed revenue into poor countries. Critics argue, however, that this is another form of exploitation creating dependency with food production parallel to other resource extraction practices in the global economy (Wilk 2004). There is also the practice of bioprospecting—or "biopiracy" as called by its critics—and its impact on indigenous communities. Although typically associated with global pharmaceutical corporations and their search for cures for disease, there are hunger implications as agribusiness seeks control over and patents on the natural world through biotechnology (Mendieta 2006; Shiva 2000) or privatization and control over natural resources such as water (Barlow 2003; Shiva 2002). Research has shown the detrimental effects of foreign investment on food security (Barkin 1982; Wimberly 1991; Wimberly and Bello 1992), while agribusiness sees itself as doing no harm. From a world-system perspective, then, the power of MNCs reflects a new form of imperialism in which exploitation and dominance govern who has access to the land, its products, and the benefits and basic needs that can be derived from it—all of which has implications for food insecurity.

The power of global agribusiness is exercised in the "proletarianization" of the agricultural labor force and its persistent poverty. The small landholder or peasant farmer most dependent on the land often has to work for wages in plantation agriculture or is displaced from the land altogether and forced to migrate to urban slums in search of alternative means for survival (Bello 2009). Those producing the world's food ironically go hungry (Barkin 1982; McMichael 2004) from their lack of purchasing power. Even the neoliberal World Bank agrees (2007: 135–37) acknowledging poverty and food insecurity problems from agribusiness concentration. The World Bank calls for greater competition and recognizes the significance of small- and medium-sized farms for development and food security, reinforcing the importance of agriculture not just for subsistence but sale in small-scale markets. Recognition of such issues is encouraging, but the system itself needs greater repair if empowerment from below will be possible. The "supermarket revolution" in which the global citizenry relies on large-scale operations to obtain their food can be devastating—particularly when the poor are shut out by rising global food prices and limited food sovereignty (Patel 2007).

Emphasizing marketing and scientific solutions to increasing the global supply, agribusiness has sought to address hunger by producing, packing, shipping, and selling food throughout the world. This does nothing for food poverty or deprivation, however, in that the global citizenry is not guaranteed access to expanding markets, and food stability and sovereignty become increasingly threatened. Global processes have local consequences and it is in understanding the vulnerabilities of these dynamics that a need to engage global versus local food systems more fully through a world-systems lens becomes evident (Lacy 2000).

Conclusions

World-systems research has offered much toward understanding the dynamics and complexities of the political economy of the world food system. Taking a critical perspective on global capitalism, conflict, power, environmental considerations, globalization, politics, and poverty and inequality have hunger consequences. Offering a much-needed alternative perspective to that typically espoused by agencies such as the FAO and the USDA (US Department of Agriculture) let alone the neoliberal leanings of academic research in development economics, world-systems ideas have emphasized stratification that generates food poverty and deprivation, while threatening food stability for those most vulnerable to shocks and the food sovereignty of the global citizenry.

The processes creating hunger are as political as they are economic. Additional issues connecting powerful and powerless in the world food system include the politics of food aid and food subsidies. Research often shows that the assistance countries receive has little to do with actual food insecurity needs and more to do with global geopolitics (George 1991). Murphy and McAfee (2005), for example, highlight the problems with the international food aid regime, noting that the primary winners of US food aid policy are not citizens in hungry countries desperate for survival but global agribusiness producing the food surplus. Critics further argue that aid creates a system of dependency on food imports in developing countries that historically had sustainable, self-sufficient food systems (Lappé et al 1998). Thus, as with marketing food globally, food assistance disrupts local markets and organizations such as Oxfam and others have argued that the best type of aid is depoliticized, direct cash payments that support livelihoods through an injection of much-needed money that will circulate locally. Similarly problematic for the world's poor and their ability to survive in the global food economy are government subsidies to agribusiness for the production of export crops in the developed world (Weis 2007). Tens of billions of dollars is given to farmers annually by governments in the European Union and the United States to produce food, keep food prices cheap, and agribusiness coffers flowing. Subsidies are an essential ingredient for corporate dominance of the world food system, indicative of a vastly uneven playing field that puts the world's poorest nations relying most on agriculture at a vast disadvantage—such state intervention on behalf of agribusiness is a far cry from the free trade mantra of the World Trade Organization. Analyzing the role of agribusiness in these and other issues is essential to understanding corporate power and influence in food systems and on agencies such as the USDA that develops programs and policy and ultimately what politics means for hunger (Nestle 2007).

There is much research to be done regarding hunger in the world-system. World-systems analysis reveals the interconnectedness and globalized nature of numerous important issues examined from this perspective. Food insecurity is no exception and hunger is an excellent example of how "human needs are shaped by human history" (Ellmann 1993: 5). Future research on food insecurity must foster these connections and seek to understand hunger in a way that is enlightened by and contributes to the larger body of work from political economy of the world-system scholars.

The world food system remains an important structural condition related to hunger, and analysis is needed regarding the institutions and organizations wielding power within. One area of particular importance is the politics of the global food aid regime and connections to poverty and power noted above. In addition, discussion with regard to environment and society and the hunger considerations that are central to these dynamics is increasingly important. Although hunger is not always directly discussed as an environmental issue, food availability starts with the land and must be produced using sustainable means. Next, with regard to the political economy of contention, there is a great need for new and ongoing research. This essay has not done justice to the significance of such concerns and research must build on the links between hunger and conflict in the world-system. Until the world finds a more peaceful means to resolve its disputes, militarization and war will remain a root cause of food insecurity. At the same time we must consider conflict and instability as a consequence of hunger and the revolutionary potential that poverty and want can generate. Be it a small-scale food riot or much larger social upheaval, the world-system roots to these are prevalent and in need of new analysis (see Natsios and Doley 2009; Patel and McMichael 2009). Similarly, food movements in a variety of forms such as the back to the land, food justice, food sovereignty, locavore, slow foods, and sustainable agriculture movements that seek "food security from below" (Wekerle 2004) are ripe with potential. Finally, these questions ultimately come down to democracy, human rights, and social justice in the global political economy (Kent 2005). Until food security is not only guaranteed but upheld as a human right as claimed in the Universal Declaration of Human Rights, the global food regime will continue to starve hundreds of millions throughout the planet (George 1986).

To conclude, hunger has been, is presently, and will remain important for world-systems analysis. Using the theoretical expertise and range of methodological skills incorporated by those doing research from this perspective, the causes and consequences of food insecurity in all of its forms can be better understood and research can ultimately contribute to much-needed solutions that lead to the goal of "a world without hunger" long-desired by policymakers and practitioners on the frontlines.

References

Araghi, F. (2009) "Accumulation by displacement: global enclosures, food crisis, and the ecological contradictions of capitalism", *Review*, 32: 113–46.

Barkin, D. (1982) "The impact of agribusiness on rural development", *Current Perspectives in Social Theory* 3: 1–25.

Barlow, M. (2003) *Blue Gold: The Fight to Stop the Corporate Theft of the World's Water*, New York: New Press.

Beardsworth, A. and T. Keil. (1997) *Sociology on the Menu: An Invitation to the Study of Food and Society*, New York: Routledge.

Bello, W. (2009) *The Food Wars*, London: Verso.

Bennett, J. (1986) *The Hunger Machine*, Cambridge: Polity.

Bonanno, A., L. Busch, W. Friedland, L. Gouveia, and E. Mingione. (1994) *From Columbus to ConAgra: The Globalization of Agriculture and Food*, Lawrence: University Press of Kansas.

Buttel, F. (2000) "Ending hunger in developing countries", *Contemporary Sociology*, 29: 13–26.

Cereseto, S. (1977) "On the causes and solution to the problem of world hunger and starvation: evidence from China, India, and other places", *Insurgent Sociologist*, 7: 33–52.

Chattopadhyay, R. (2000) "Zimbabwe: structural adjustment, destitution, and food insecurity", *Review of African Political Economy*, 84: 307–16.

De Janvry, A. (1976) "Material determinants of the world food crisis", *Berkeley Journal of Sociology*, 21: 3–26.

DeRose, L., E. Messer, and S. Millman. (eds) (1998) *Who's Hungry? And How Do We Know?* New York: United Nations University.

Devereux, Stephen. (1993) *Theories of Famine*, New York: Harvester Wheatsheaf.

Devereux, Stephen. (ed.) (2007) *The New Famines: Why Famines Persist in an Era of Globalization*, London: Routledge.

Drèze, J. and A. Sen. (1989) *Hunger and Public Action*, New York: Oxford.

Drèze, J., A. Sen, and A. Hussain. (eds) (1995) *The Political Economy of Hunger: Selected Essays*, New York: Oxford.

Dunaway, W. and M.C. Macabuac. (2007) "'The shrimp eat better than we do': Philippine subsistence fishing households sacrificed for the global food chain", *Review*, 30: 313–37.

Ellmann, M. (1993) *The Hunger Artists: Starving, Writing, and Imprisonment*, Cambridge, MA: Harvard University Press.
Food and Agriculture Organization of the United Nations. (2006) "Food security", *FAO Policy Brief*, Rome: FAO. Retrieved December 3, 2008 (ftp://ftp.fao.org/es/ESA/policybriefs/pb_02.pdf).
Foster, J.B. (2009) *The Ecological Revolution: Making Peace with the Planet*, New York: Monthly Review.
Friedmann, H. (1982) "The political economy of food: the rise and fall of the postwar international food order", *American Journal of Sociology* (Supplement) 88: S248–S286.
Friedmann, H. (1993) "The political economy of food: a global crisis", *New Left Review* 197: 29–37.
George, S. (1986) "The right to food and the politics of hunger", *Razvoj/Development International* 1: 121–37.
George, S. (1991) *How the Other Half Dies*, New York: Penguin.
Holt-Jiménez, E. and R. Patel (2009) *Food Rebellions!* Cape Town: Pambazuka Press.
Jarosz, L. (2009) "The political economy of global governance and the world food crisis", *Review*, 32: 37–60.
Kent, G. (2005) *Freedom from Want: the human right to adequate food*, Washington, DC: Georgetown University.
Lacy, W.B. (2000) "Empowering communities through public work, science, and local food systems: revisiting democracy and globalization", *Rural Sociology*, 65: 3–26.
Lappé, F.M., J. Collins, and P. Rosset. (1998). *World Hunger: Twelve Myths*, 2nd edn., New York: Grove.
Madeley, J. (2000) *Hungry for Trade: How the Poor Pay for Free Trade*, London: Zed.
Magdoff, F., J.B. Foster, and F.H. Buttel. (eds) (2000) *Hungry for Profit: The Agribusiness Threat to Farmers, Food, and the Environment*, New York: New York University Press.
McMichael, P. (ed.) (1995) *Food and Agrarian Orders in the World Economy*, Westport, CT: Praeger.
McMichael, P. (1997) "Rethinking globalization: the agrarian question revisited", *Review of International Political Economy*, 4: 630–62.
McMichael, P. (1998) "Global food politics", *Monthly Review* 50: 97–111; reprinted in C.R. Humphrey, T.L. Lewis, and F.H. Buttel (eds) (2003) *Environment, Energy, and Society: Exemplary Works*, Belmont, CA: Wadsworth.
McMichael, P. (2004) *Development and Social Change*, 3rd edn., Newbury Park, CA: SAGE.
McMichael, P. (2009) "Interpreting the world food crisis of 2007–8", *Review*, 32: 1–8.
Mendieta, E. (2006) "Biopiracy and bioterrorism: banana republics, NAFTA, and Taco Bell", *Peace and Change* 31: 80–89.
Mukherjee, A. (1994) *Structural Adjustment Programme and Food Security: hunger and poverty in India*, Brookfield, VT: Aldershot.
Murphy, S. and K. McAfee. (2005) *U.S. Food Aid: Time to Get it Right*. Minneapolis, MN: Institute for Agriculture and Trade Policy. Retrieved November 1, 2011 http://www.iatp.org/files/451_2_73512.pdf.
Natsios, A.S. and K.W. Doley. (2009) "The coming food coups", *The Washington Quarterly*, 32: 7–25.
Nestle, M. (2007) *Food Politics: How the Food Industry Influences Nutrition and Health*, Berkeley: University of California.
Ó Gráda, C. (2009) *Famine: A Short History*, Princeton, NJ: Princeton University Press.
Patel, R. (2007) *Stuffed and Starved: The Hidden Battle for the World Food System*, Brooklyn, NY: Melville House.
Patel, R. and P. McMichael. (2009) "A political economy of the food riot", *Review*, 32: 9–35.
Petrini, C. (2006) *Slow Food Revolution*, New York: Rizzoli.
Reutlinger, S. (1986) *Poverty and Hunger: Issues and Options for Food Security in Developing Countries*, Washington, DC: World Bank.
Ryan, C.R. (1998) "Peace, bread, and riots: Jordan and the International Monetary Fund", *Middle East Policy*, 6: 54–66.
Samatar, A.I. (1993) "Structural adjustment as development strategy? Bananas, boom, and poverty in Somalia", *Economic Geography*, 69: 25–43.
Sen, A. (1981) *Poverty and Famines: An Essay on Entitlement and Deprivation*, New York: Oxford.
Shiva, V. (2000) *Stolen Harvest: The Hijacking of the Global Food Supply*, Cambridge, MA: South End Press.
Shiva, V. (2002) *Water Wars: privatization, pollution, and profit*, Cambridge, MA: South End Press.
Shiva, V. (2007) *Manifestos on the Future of Food and Seed*, Cambridge, MA: South End Press.
Shiva, V. (2008) *Soil Not Oil: Environmental Justice in an Age of Climate Crisis*, Cambridge, MA: South End Press.
Spieldoch, A. and S. Murphy. (2009) "Agricultural land acquisitions: implications for food security and poverty alleviation", pp. 39–53 in M. Kugelman and S.L. Levenstein (eds) *Land Grab? The Race for the World's Farmland*, Washington, DC: Woodrow Wilson International Center for Scholars.
Thurow, R. and S. Kilman (2009) *Enough: Why the World's Poorest Starve in an Age of Plenty*, New York: PublicAffairs Books.

Walton, J. and D. Seddon. (1994) *Free Markets and Food Riots: The Politics of Global Adjustment*, Cambridge, MA: Blackwell.

Weis, T. (2007) *The Global Food Economy: The Battle for the Future of Farming*, London: Zed.

Wekerle, G.R. (2004) "Food justice movements: policy, planning, and networks", *Journal of Planning Education and Research*, 23: 378–86.

Wilk, R. (2004) "The extractive economy: an early phase of the globalization of diet", *Review* 27: 285–306.

Wimberley, D.W. (1991) "Transnational corporate investment and food consumption in the third world: a cross-national analysis", *Rural Sociology*, 56: 406–31.

Wimberley, D.W. and R. Bello. (1992) "Effects of foreign investments, exports and economic growth on third world food consumption", *Social Forces*, 70: 895–922.

World Bank. (2007) *World Development Report 2008: Agriculture for Development*. Washington, DC: World Bank.

Zoomers, A. (2010) "Globalisation and the foreignisation of space: seven processes driving the current global land grab", *Journal of Peasant Studies*, 37: 429–47.

11.3

Incorporating comparison

Sandra Curtis Comstock

In 1990, Philip McMichael coined the term incorporating comparison to draw attention to an alternative strategy for investigating how institutions, economic relations, regulatory practices, or cultural phenomena vary or converge over time and space (McMichael 1990). This strategy is best understood by how it relates to others in sociology's comparative historical tradition. In general, comparative historical approaches either emphasize locality-defined social processes or cross-place-defined social processes. Within these rubrics appear two sub-sets: locality-oriented traditions gravitate either toward scientistic or hermeneutic comparisons; while cross-place comparisons lean toward encompassing or incorporating comparison (Steinmetz 2004). Comparing the first three approaches alongside incorporating tactics will clarify how incorporating comparisons are specially equipped for indentifying and explaining the types of social processes that generate unprecedented, world-reorganizing change.

Within the locality-oriented tradition, scientistic approaches such as Mahoney (2001) assume that places compared are discrete entities and do not influence one another in any significant way. Scientistic comparisons view social forces of reproduction and change as those conditions and events that *repeatedly* produce a common social outcome across most cases (Mahoney et al 2009: 122). They therefore seek to identify the conditions and events that most consistently and uniformly occur across all cases. To do so, they select localities on the basis of their initial similarity with one another in the expectation that these are the places whose paths will diverge only in minimal ways.

In contrast, incorporating comparisons such as O'Hearn (2001) do not view localities as discretely functioning entities. Rather, they assume places are mutually influencing and inter-dependent. For this reason, they investigate regularities in terms of the historically specific, cross-place relations or adaptations that account for how similarities are relationally produced. Additionally, incorporating comparisons go beyond regularities to ask whether and how *unique* cross-place events sometimes cause the social practices and relations of different localities to "combine and redefine one another in ways that form something new and unpredictable" (O'Hearn 2001: 3).

Locality-oriented hermeneutic approaches such as George Steinmetz (2007) also understand social forces of reproduction and change as more than constantly repeating events and actions. For them, social outcomes in each locality are shaped by the culturally distinctive concepts and systems of signification with which actors interpret and respond to events and actions. They select localities they believe share underlying ways of generating symbolic relations that will deepen our understanding of how systems of signification operate in distinctive contexts. The goal of hermeneutic approaches is to show how the variability of social concepts and systems of signification from place to place explain differences in actions and outcomes across place (Steinmetz 2004: 377, 382, 392). However, emphasizing locally derived difference, hermeneutic approaches interpret case-by-case differences as propelled *solely* by case-bound actors' concepts and reactions to events and conditions.

Incorporating comparisons such as Friedmann (2006) also recognize the significance of case-by-case particularities. However, they do not view different localities' concepts or projects as *wholly* derived from local contexts. Rather, they presume that cross-place processes or events occasionally *bring distinctive social concepts, systems of signification, and practices together* in new ways which generate local and/or cross-place transformations. Additionally, incorporating approaches consider unequal relations of power between places as important determinants of how cross-place interactions unfold and transform the world. A goal of incorporating comparison is therefore to identify the critical junctures where *differentiated localities, social concepts, projects, and practices* uniquely confront and redefine one another, along with their broader, cross-place relations.

Encompassing comparisons such as Wallerstein (1979) share the incorporating perspective that asymmetric cross-place relations generate change. However, like their scientistic cousins, they define causally significant forces of change as the conditions and events that repeatedly accompany general world-systemic patterns of reproduction and change (Wallerstein 1979: 3, 6–7). They presume that relations between localities consistently shift in general ways according to the intrinsic tendencies of the overall system. With this logic, the encompassing method selects cases on the basis of their representativeness of common positions in the overall system. The goal of analysis is to identify patterns of difference in how hierarchically related localities respond to the same system-level dynamics and perpetuate systemic inequality.

Incorporating comparisons recognize the role of world-systemic contradictions as catalysts of change. However, studies such as Arrighi and Silver (1999) maintain that new relations between *particular* localities and groups are also key to dramatic world transformation. This view stems from the insight that when contradictions in the world-system generate problems for particular localities, local actors often resist or problem-solve by experimentally joining their own social practices and projects to those of others. This radically reconfigures them. For incorporating approaches, such novel connections produce the wedge innovations that break increasingly fraught world-systemic patterns and generate new arrays of possible world-reorganizing solutions and alliances. In sum, incorporating comparisons re-conceptualize critical turning points as more than "accidents" or pre-determined "systemic necessities" and offer unmatched tools for tracing and explaining where unprecedented, path-changing innovations come from (Mahoney 2000).

References

Arrighi, G., and Silver, B. (1999). *Chaos and Governance in the Modern World System*. Minneapolis: University of Minnesota Press.

Friedmann, H. (2006). "From Colonialism to Green Capitalism". *Research in Rural Sociology and Development*, 11: 227–64.

Mahoney, J. (2000). "Path Dependence in Historical Sociology". *Theory and Society*, 29: 501–48.

Mahoney, J. (2001). *The Legacies of Liberalism*. Baltimore, MD: Johns Hopkins University Press.

Mahoney, J., Kimball, E., & Koivu, K. (2009). "The Logic of Historical Explanation in the Social Sciences". *Comparative Political Studies*, 42(1), 114–46.

McMichael, P. (1990). "Incorporating Comparison within a World-Historical Perspective: An Alternative Comparative Method". *American Sociological Review*, 55 (June): 385–97.

O'Hearn, D. (2001). *The Atlantic Economy*. Manchester: Manchester University Press.

Steinmetz, G. (2004). "Odious Comparisons: Incommensurability, the Case Study, and 'Small N's' in Sociology". *Sociological Theory*, 22(3): 371–400.

Steinmetz, G. (2007). *The Devil's Handwriting*. Chicago, IL: University of Chicago Press.

Wallerstein, I. M. (1979). *The Capitalist World-Economy*. Cambridge: Cambridge University Press.

11.4

Equalizing exchange through voluntary certification?

The case of palm oil

Kristen Shorette

The Fair Trade (FT) system, based on the understanding that unequal exchange is one of several mechanisms in producing and reproducing global inequalities, seeks to counter underdevelopment via restructured trading relationships. Product and organization certifications indicate that goods are produced under favorable labor, economic, and environmental conditions.

As such, FT represents a deliberate attempt to equalize exchange through the shortening and redistributing of profits along commodity chains in order to improve the wellbeing of producers and promote economic and social development in the Global South. However, whether a voluntary certification system can substantially affect development outcomes from within the traditional market remains contested.

A growing body of literature indicates that FT offers positive but limited benefits to participant producers (and even non-participating producers in the vicinity of FT farms) for the production of coffee, bananas, and tea (Jaffee 2007; Ruben 2009). However, skepticism over the system's potential to fundamentally alter global development inequalities remains. Current literature focuses on two key limitations. First, the contradiction of undermining the traditional market while simultaneously located within it undermines the system's transformative power (Bacon 2005). Second, the system's dependence on exclusivity where FT is attractive to customers at least partly because it differentiates them from mainstream consumers, limits the bounds of expansion (Guthman 2009). However, some important limitations of the FT system have remained less scrutinized. For instance, the complexity of production processes and the visibility that component parts take in a final product affects their ability to be regulated via certification.

Tropical commodities generally lend themselves to informative GCC analysis because they have relatively few stages of production and are cultivated in the Global South and widely consumed in the Global North (Talbot 2009). While most certified FT products are tropical commodities, such as coffee, tea, and bananas, not all tropical commodities that are widely consumed in the Global North readily lend themselves to FT certification (or GCC analysis, for that matter). Palm oil highlights the limitations of the current FT system in regulating component products with somewhat complex commodity chains. Here, I briefly explore the importance of palm oil in terms of labor, environmental, and development consequences and the limitations of the current FT system to improve these conditions.

In today's marketplace, palm oil is both ubiquitous and largely hidden. Crude palm oil has at least five different stages of processing depending on the final product. After various stages of transformation, it is found in thousands of products ranging from cooking oils to cosmetics, plastics, textiles, explosives, and pharmaceuticals. However, it is rarely labeled as a separate ingredient even in food items, making its presence unknown to the vast majority of consumers.

Its production has harmful effects on producers, the natural environment and developing economies. Facilitated by the combination of natural, business and political climates, palm oil has become a boom industry across the developing world. The top 20 producing countries in 2008 are all located in South East Asia, Latin America, or Africa (FAO 2009). The clearing of tropical rainforests necessary to establish new palm oil plantations and the application of chemical pesticides to deter predators displaces local populations, further threatens endangered species, facilitates low-paid and child labor, and negatively affects the health of workers and the natural ecosystem (Rosenthal 2007). Current international trade laws do not include provisions for palm oil workers' health and safety despite their exposure to toxic pesticides. Further, export dependence on palm oil has not proven to facilitate economic development (WDI 2010).

Together, these factors make palm oil a key product for FT regulation in order to improve labor, environmental and economic conditions for producers. However, palm oil production does not lend itself to the current system of voluntary certification that distinguishes favorable practices. The many multiple commodity chains involving palm oil are much more complex than tropical commodities such as bananas, coffee, or tea. After harvest, palm kernels must be processed first in mills, then in refineries, transformed to various extents depending on the final product, and then distributed to local manufacturers (Vermeulen and Goad 2006). Because palm oil is largely sold as an ingredient after it has undergone some transformation, the economic, social, and environmental conditions of its production cannot be easily regulated with a single label on the final product.

Despite the limitations of the current FT system in regulating a product such as palm oil, it has adjusted in the past to meet the needs of commodities that challenged previous governance structures. The incorporation of agricultural products spurred the advent of certification systems that currently far outnumber the original system of direct sales networks. Perhaps the problem of regulating palm oil will spur a new form of alternative trade regulation to the (limited) benefit of palm oil producers throughout the developing world.

References

Bacon, C. (2005) "Confronting the Coffee Crisis: Can Fair Trade, Organic, and Specialty Coffee Reduce Small-Scale Farmer Vulnerability in Northern Nicaragua?" *World Development*, 33: 497–511.

FAO (2009) *FAO Statistical Yearbook*, Rome: Food and Agricultural Organization of the United Nations.

Guthman, J. (2009) "Unveiling the Unveiling: Commodity Chains, Commodity Fetishism, and the 'Value' of Voluntary, Ethical Food Labels", in J. Bair (ed.) *Frontiers of Commodity Chain Research*, Stanford, CA: Stanford University Press.

Jaffee, D. (2007) *Brewing Justice: Fair Trade Coffee, Sustainability, and Survival*, Berkeley: University of California Press.

Rosenthal, E. (2007) "Once a Dream Fuel, Palm Oil May be an Eco-Nightmare". *The New York Times*, January 31.

Ruben,R. (2009) *The Impact of Fair Trade*, Wagening, The Netherlands: Wagening Academic Publishers.

Talbot, J. (2009) "The Comparative Advantages of Tropical Commodity Chain Analysis", in J. Bair (ed.) *Frontiers of Commodity Chain Research*, Stanford, CA: Stanford University Press.

Vermeulen, S. and Goad, N. (2006) *Toward Better Practice in Smallholder Palm Oil Production*, London: International Institute for Environment and Development.

WDI (2010) *World Development Indicators*, Washington, DC: The World Bank.

Part V
Sustainability

12. Natural resources and constraints

12.1

New historical materialism, extractive economies, and socioeconomic and environmental change

Paul S. Ciccantell

Introduction: Raw materials and long term change in the capitalist world-economy

The relationship between raw materials and economic development has been a central concern of political economic analysis since the 1700s (Marx 1962; Ricardo 1983; Smith 1776). Despite more than two centuries of research, no clear understanding has emerged. Natural resource wealth aids economic development in some times and places, and hinders it in others. Capitalists, government officials, and analysts during the nineteenth and much of the twentieth century viewed raw materials as natural capital to be exploited for development as "gifts of nature." In the mid-twentieth century, analysts formalized this conceptualization in the growth pole (Perroux 1956) and linkage models of development. However, from the 1960s onward dependency theorists and world-systems analysts emphasized the dramatically contradictory relationship between raw materials wealth and development for colonized extractive peripheries (see e.g., Bunker and Ciccantell 2005; Cardoso and Faletto 1969; Chase-Dunn 1989; Frank 1967; Jalee 1968; Wallerstein 1974, 1979, among many others). More recently, this two-sided relationship between raw materials and economic development has been portrayed by some economists as the "resource curse" of peripheries where these natural resources are extracted (Auty 1995; Gelb 1988; Ross 1999; Sachs and Warner 1999).

However, at the same time it is equally obvious that as the classic political economists argued, raw materials are essential ingredients for economic development; the rapid economic ascent of the core economies that have risen to challenge for hegemony over the last five centuries relied critically on raw materials extraction, processing, and consumption (Bunker and Ciccantell 2005, 2007; Ciccantell 2009). How can these dramatically different effects of raw materials wealth on economic development in different areas and different eras of the capitalist world-economy over the last five centuries be explained? A recent reformulation of world-systems theory, new historical materialism, offers a theoretical model that focuses on the contradictory roles of raw materials extraction, processing, and consumption in shaping the long term evolution of the capitalist world-economy and its constituent national economies, as well as the socioeconomic and ecological impacts of this evolution.

In this chapter, I examine the dual character of the role of raw materials in the evolution of the capitalist world-economy. The following sections will outline the key elements of new historical materialism, a theoretical model that builds on the work of Wallerstein (1974, 1979), Chase-Dunn (1989), and Arrighi (1994, 2009) in order to understand the multiple roles of raw materials in shaping the past five centuries of the capitalist world-economy. New historical materialism highlights the need to examine economic ascent in the core, the exploitation of the periphery, the relationships between these processes, and the impacts of these relationships on long term socioeconomic and environmental change. This chapter concludes with an outline of directions for future research from the new historical materialist perspective.

New historical materialism

The initial formulation of our theoretical model in the late 1990s used the term "raw materialism," a term we eventually replaced with the less emphatic term "new historical materialism." This term reflects our intellectual foundation in world-systems theory and the classical political economy of Marx. More critically, it also reflects our goal of understanding social relations of power and inequality in the capitalist world-economy, rather than creating a new model of "environmental determinism" that claimed that the material characteristics of iron ore, coal, trees, and other natural resources determined social outcomes.

However, the key contribution of our theoretical model and research methodology lies precisely in its "raw materialism": the goal is to understand the material bases of power and inequality in the capitalist world-economy by beginning from first principles of chemistry, physics, geology, hydrology, and ecology, and understanding how the relationship between society and nature has been shaped by these processes and how human action has learned about, adapted to, and sometimes reshaped natural processes. This nature-society nexus is the central focus of our theoretical model and analytic method. Using this approach, we were able to identify and explain the key causal similarities across the five most spectacular cases of systemically transformative economic ascent over the past five centuries: the economic development of Holland, Great Britain, the United States, Japan, and China (Bunker and Ciccantell 2005, 2007).

The key elements of new historical materialism bring together natural and social processes in an integrated understanding of social change. The key locus of innovation and change in the capitalist world-economy at any point in time is found in those societies with rapidly growing economies that (from a post hoc perspective) have the potential to rise to challenge for core or even hegemonic status. These rapidly growing economies face a host of challenges at this nature-society nexus.

State and firm strategies for economic ascent must respond to and take advantage of contemporary technological, geopolitical, environmental, and market conditions in the capitalist world-economy of that time period and of the rising nation's position and location within that context. States and firms must also coordinate the physical characteristics and location in space and topography of the various raw material resources actually or potentially available with the physical characteristics and location in space and topography of the national territory. In the critical early years of economic ascent, states and firms face the challenge of coordinating domestic technological, organizational, and institutional innovations, particularly in heavy industry and transport, with the often external problem of access to cheap and secure sources of the raw materials used for heavy industry.

The raw materials used in the largest volumes present the greatest challenges to and best opportunities for achieving economies of scale, the reduction of the cost of each unit produced that results from producing more units more rapidly and with fewer inputs of matter, energy, capital, and labor. These economies of scale, however, drive a contradictory increase in transport costs. Societies deplete the closest, cheapest, and most secure reserves of raw materials rapidly as the scale of their industrial production increases, forcing states and firms to seek more distant resources to supply rapidly growing industries (see Bunker and Ciccantell [2005] for an extended presentation of this theoretical model).

The tension of this contradiction between the economies of scale and the cost of space creates tremendous incentives and opportunities for technological innovation in transport, in technologies that reduce material, energy, and labor inputs per unit of output, and in technologies for controlling heat, pressure, and the mix of chemicals to make each unit stronger and lighter while performing the same work. All of these technological fixes drive each other, and all of them tend to generate increases of scale, thus exacerbating over the long term this same contradiction between scale and space that they are designed to solve (Bunker and Ciccantell 2005).

The efforts undertaken to resolve this contradiction are more than simple technological innovations. These efforts involve organizational and institutional innovations that spread from heavy industry and transport to other economic sectors. The rising economies that have created the most effective technological and organizational solutions to this contradiction simultaneously generated their own rise to economic dominance, restructured the mechanisms and dynamics of accumulation across the capitalist world-economy, and expanded and intensified the commercial arena of raw materials trade and transport. The innovations that resolve this raw materials challenge require the coordination of multiple physical and social processes across geopolitical and physical space with domestic relations between firms, sectors, the state, labor, and new technologies. These innovations include not just new technologies, but also new organizational forms and institutions that become the framework for the rest of the national economy and that eventually transform international markets and institutions in support of the rising economy's continued development (Bunker and Ciccantell 2005).

The challenges and the opportunities presented by the basic raw materials industries and by the transport systems on which they depend foster generative sectors: sectors that, beyond creating the backward and forward linkages that underlie the concept of a leading sector, also stimulate a broad range of technical skills and learning along with formal institutions designed and funded to promote them, vast and diversified instrumental knowledge held by interdependent specialists about the rest of the world, financial institutions adapted to the requirements of large sunk costs in a variety of social and political contexts, specific formal and informal relations between firms, sectors, and states, and the form of legal distinctions between public and private and between different levels of public jurisdiction (Bunker and Ciccantell 2005).

Generative sectors are not necessarily high profit sectors, even though high profit sectors (whether gold and silver in the seventeenth century or computer technology in the twentieth century) are those that typically attract the most analytical attention (see e.g., Arrighi 1994; O'Hearn 2001). However, as has been shown elsewhere (Bunker and Ciccantell 2003a, 2003b), the highest profit sectors are not necessarily generative sectors, and generative sectors are not necessarily the highest profit sectors. Instead, generative sectors provide the material building blocks, cost reductions across many sectors to increase competitiveness, and patterns of state-sector-firm relations and other institutions that combine to drive economic ascent. Moreover, generative sectors rest on the creation of national and global commodity chains (Ciccantell and Smith 2009; Gereffi et al 1994) that tightly link the rising economy to its extractive peripheries and generate growing levels of economic and political inequality between the rising economy and its peripheries.

Raw materials acquisition strategies

How can rapidly growing economies acquire the raw materials essential to sustain these generative sectors, particularly in the face of domestic raw materials depletion and the resulting diseconomies of space as these raw materials must be brought from more distant areas outside the political control of the ascendant state? One critical strategy to accomplish this task has been to steal raw materials peripheries from earlier ascendant economies that have already undertaken the difficult and expensive tasks of building the necessary infrastructure, creating political, organizational, and legal forms that promote international trade and investment relations between a particular raw materials-producing state and the world-economy, and incorporating these peripheries economically and politically into the world-economy. Earlier processes of economic ascent progressively globalized the world-economy and brought new raw materials peripheries into the global trading system to supply the earlier ascendants' industries, so new ascendant economies and states have the opportunity initially to purchase raw materials from this established supply system (Ciccantell 2009).

The newer ascendants' rapid growth, however, means that their demand is increasing dramatically and necessitating a dramatic increase in supply if these growth rates are to be sustained. The combination of the existing social and material infrastructures in the raw materials peripheries established by earlier ascendants, rapid demand growth in the ascendant economy, and the willingness of the newer ascendant economy to pay higher prices for raw materials in order to sustain their domestic growth creates an opportunity that states and firms in the raw materials periphery find very attractive. Higher prices for rapidly increasing volumes of exports (in contrast to slower demand growth in the mature economies of earlier ascendants) motivate firms and domestic elites in the periphery and even from existing core powers with fewer opportunities for profitable investments to invest in production for export to the new ascendant. States in raw materials exporting regions typically support this investment with subsidies for transport and extraction, both in an effort to promote economic development and in hopes of gaining better returns and more political freedom from the power of the existing hegemon. This is particularly apparent in postcolonial situations in which newly independent states seek to break free from neocolonial ties and in situations of resource nationalism in which states seek greater control over and benefits from raw materials exports. Firms, elites, and states in raw materials peripheries come to see the new ascendant as a potential ally in their attempts to promote political independence and economic development (Ciccantell 2009).

From the perspective of the new ascendant, building these relationships with existing raw materials peripheries is much less expensive and difficult than creating their own new peripheries. One of the most important benefits is that most of the cost and the risk of expanding extraction and transport is borne by firms and states in the extractive periphery and often by firms from the earlier ascendant. At the same time, these investments in mines and transport systems also often create opportunities for exports of industrial products from the ascendant economy to the periphery to support the development of these extractive industries and for consumption by the owners of and workers in these industries. "Stealing" these peripheries from earlier ascendants thus further enhances the rapid growth of the new ascendant by reducing costs and risks while simultaneously creating significant new opportunities for profit from trade and investment (Ciccantell 2009).

Over the last 500 years, this process of stealing peripheries has been a key element of rapid ascent in each case of rapid, transformative ascent (Bunker and Ciccantell 2005, 2007). Holland progressively captured more and more of the Brazilian trade from the waning Portuguese empire in the 1600s, taking control of much of the sugar, precious metals, and other raw materials of trade and capturing the benefits of this trade for its own domestic development in shipbuilding, shipping, finance, and other industries. Great Britain rapidly displaced the Dutch from North America, the Caribbean, India, and Southeast Asia in the late 1600s and 1700s and took control over trade in timber, sugar, and a host of other raw materials. During the 1800s in Latin America, Great Britain similarly displaced the Spanish empire as the region's main trade and investment partner, inducing states to subsidize the construction of British-owned and manufactured railways to ensure high rates of profit and steady supplies of grain, beef, silver, tin, and other products to British consumers and industries. The rapidly growing United States did the same to support its ascent in the nineteenth and twentieth centuries, displacing Great Britain first in much of North America and later from Canada, Latin America, and the Caribbean to acquire the raw materials for US industrialization. The United States often used the opportunity to escape British hegemony as a key enticement for firms and states to redirect their exports of copper, bauxite, and other raw materials to the US market with its rapidly growing demand (Ciccantell 2009).

Moreover, this process has continued in the late twentieth and early twenty-first centuries. After World War II, the United States effectively took control over Australia's resources in order to establish a supply relationship with Japan, despite Australian objections, in order to support US Cold War efforts in Asia by rebuilding Japan. This US-led reorganization of Australia's raw materials trade relations created the pattern for Japan's progressive globalization of the iron and coal industries via long-term contracts and joint ventures. This raw materials access strategy came to rely on its seeming support for resource nationalism (often to escape the control of the US hegemon and its firms), but greatly benefited Japanese firms and the Japanese state by transferring much of the costs and risks of supplying Japan's industries to states and firms in Australia, Brazil, Canada, and other nations. The raw materials peripheries that Japanese strategies stole from the United States and Great Britain provided billions of dollars in subsidies to Japanese industrial development and global competitiveness in the form of lower raw materials costs, forming a key pillar of Japan's rapid economic ascent and ensuring that the steel and other raw materials processing and consuming industries that states and firms in Australia, Brazil, Canada, and other raw materials exporting regions wanted to develop using export revenues from raw materials exports to Japan were instead built in Japan (Bunker and Ciccantell 2007).

The relevance of raw materials in the "Post-Material" world

But why should raw materials continue to be a concern in the current era of high technology industries and supposed dematerialization (see e.g., Cleveland and Ruth 1998; Herman et al 1989)? Claims about dematerialization became possible as heavy industry in the United States and Western Europe became increasingly uncompetitive since the 1970s and closed. However, these industries were simply relocating to lower cost producing nations in the semiperiphery and periphery such as South Korea, Brazil, and especially China. The world-economy uses more material every year; it is simply produced increasingly outside the core nations. World steel production in 2007 was 1.34 billion metric tons, the largest volume of steel production in history, and more than 36 percent of this steel was made in China (Skillings 2008). Steel today is seen as an anachronistic relic in the United States because of the industry's collapse in the early 1970s in the face of first Japanese and then later other global competition. However, under conditions of rapid economic growth in China in recent years, huge and rapidly growing quantities of steel are needed to build office buildings, factories, roads, and other infrastructure, in addition to the products being exported. The Chinese steel industry simultaneously generates huge profits for steel-producing firms, many of them operating in partnership with Japanese, South Korean, and European steel firms. Steel remains a key ingredient in the process of rapid economic ascent in China (Ciccantell 2009), and China's ascent may prove to be another transformative case that reshapes the capitalist world-economy.

On the other side of this dual-sided process, how does raw materials extraction affect the peripheries where extraction takes place? Despite the optimism of the growth pole and linkage models of development (Hirschmann 1958; Perroux 1956), most efforts to use raw materials as the basis for economic development have had at best mixed success and, more commonly, partial or complete failure (see Ciccantell [1994] for a discussion of many examples; Bunker 1989). Extractive peripheries created by rising core powers or stolen from existing core powers typically confront growing domestic and international inequality, political powerlessness, environmental degradation, and a plethora of other negative consequences from raw materials extraction for export to industrial areas, whether these areas are within the same national boundaries or across thousands of miles of ocean. Even in recent years and even in relatively wealthy, highly developed nations with rich raw materials resources such as Australia and Canada, firms and states in extractive regions continue to subsidize the economic ascent first of Japan and, more recently, of China (Bunker and Ciccantell 2007); the cost in strictly economic terms, ignoring all other forms of social and ecological costs, were probably far larger for other extractive peripheries.

In such circumstances, the critiques advanced by Cardoso and Faletto (1969), (Frank 1967), Wallerstein (1974, 1979), and a host of other analysts remain just as salient today: for extractive peripheries, raw materials wealth is extremely unlikely to lead to sustained economic development, despite the essential role of raw materials in making rapid economic ascent possible in other places (Bunker and Ciccantell 2005, 2007). The areas of greatest importance as colonial raw materials peripheries are now among the poorest in the world, with Haiti and northeastern Brazil, once the world's most valuable sugar-producing colonies, among the world's poorest and most environmentally degraded areas.

Directions for future research

The new historical materialist approach opens a variety of avenues for research into new questions and topics, as well as a fresh approach to topics typically examined from other perspectives. Individual case studies and comparative case studies of particular natural resource-based industries and extractive regions remain critically important, particularly in this era of globalization, the rapid economic growth of China and India, and the growing potential for conflict over raw materials access. One strand of my current research focuses on coal mining in western Canada, while other research focuses on much less well-studied industries, including Brent Kaup's research on the natural gas industry in Bolivia (Kaup 2008). These case studies should examine a very important issue: are the consequences of extraction different for raw materials peripheries today in the postmodern, "post-material," transnationalized world, in comparison to the imperial phase prior to 1945 or the postcolonial phase from 1945 through the 1970s? The last three decades of "post-materialism" in the core, globalization, the information age, neoliberalism in most of the world (with the crucial exception of China), and the rise of China and India are generally regarded as a new phase in the evolution of the capitalist world-economy. Is it really a new phase, from the new historical materialist perspective? Or is it merely a continuation of the post-1945 era with a few new characteristics? The rise of China over the past three decades has strong parallels with the ascent of Japan after 1945 (Bunker and Ciccantell 2007; Ciccantell 2009), so this remains an open question.

Another key topic for future research focuses on comparative studies of partial or failed processes of economic ascent and of cases of failed challenges for hegemony. Spectacularly successful cases have transformed the capitalist world-economy and attracted a great deal of analytic attention, but even comparisons of successful cases need to be expanded to comparison with other cases in which rising economies failed to achieve hegemonic status. Historically, the cases of Germany and France are clearly two of the most important for this broader comparative effort, given their repeated efforts to challenge Great Britain and the United States for hegemonic positions over the last two centuries. Borocz' (2010) insightful examination of European economic development and the integral role of colonialism in shaping European nationalism and economic development provides an intriguing perspective on the role of heavy and transportation industries and of colonial raw materials access strategies, and I have been working on historical research on this topic.

Another key area for future analysis is bringing new historical materialist analysis of raw materials and transport industries into dialogue with the other major strain of sectoral analysis in world-systems research: the global commodity chains perspective. What are the major similarities and differences between raw materials and transport industries and other key sectors of the world-economy, such as textiles, automobiles, and information technology? Are such comparisons even possible in any meaningful way? What other sectors have served as generative sectors? David Smith and I believe they are and we are working on a research project aimed at making this sort of comparison (Ciccantell and Smith 2009), but expanding this comparative analysis to many different industrial sectors in different locations and different time periods will entail the work of a significant number of analysts.

One of the world's most important raw materials industries, petroleum, has not been explicitly examined from the new historical materialist perspective, although other world-systems analysts have focused on this industry and its critical roles in the world-economy (Podobnik 2005). Despite its immense profitability and importance, it is not clear that petroleum has served or could ever serve as a generative sector driving economic ascent. This industry shaped US economic ascent in some important ways during the twentieth century, but it seems very doubtful that it will play anything close to a transformative role anywhere else in the capitalist world-economy, despite being the most significant economic sector in a large number of countries. This industry warrants a much closer inspection from this perspective.

Remaining in the area of energy industries, the role of alternative energy and its impacts on core, semiperipheral, and peripheral regions of the capitalist world-economy and on the environment merit careful attention from a new historical materialist perspective. Wind, solar, tidal, and other forms of "clean" energy are the focus of government policies and corporate investment in many areas of the world in an effort to slow global climate change, but we still know very little about this industry and its consequences. Rigorous comparison with other raw materials-based industries is likely to inform our understanding and public policies in this area that is generally regarded as a "good thing" without being subjected to serious social scientific analysis.

Perhaps most fundamentally for the future of the capitalist world-economy, what is the potential for the emergence of a new era of intense conflict between core powers and rising economies over access to key raw materials? Some analysts have already taken up this issue (see e.g., Engdahl 2005; Kenny 2004; Mahbubani 2005; Nayar 2004; Zweig and Jianhai 2005), but a longer term, new historical materialist analysis could contribute significantly to our understanding of what could be the key to the next major global conflict. My current research focuses on the role of conflicts over raw materials during earlier cases of economic ascent and increasing rivalry between core powers, conflicts that helped lead to and then shaped the World War I and World War II. In the case of Germany prior to World War I, firms, political leaders, and government agencies sought access to agricultural land, iron ore, petroleum, and copper in Europe, Africa, and Asia, and emphasized the need for a navy capable of protecting long-distance transport routes, despite their awareness of the potential for these efforts to generate conflict with Great Britain,

France, and Russia and to force these three nations into an alliance against Germany. The choice was stark: without access to these resources, German economic ascent would be severely constrained, but in order to gain access, severe conflict was almost inevitable. After defeat in World War I, the same choice was recognized and the same decision made in the 1930s, with similar disastrous results for Germany and many other parts of the world. Concerns over access to food supplies, oil, iron ore, water, and other resources today confront rapidly growing economies in China and India; raw materials access strategies may once again shape geopolitical alliances and the contours of future conflicts.

References

Arrighi, G. (1994) *The Long Twentieth Century: Money, Power, and the Origins of Our Times*, London: Verso.

Arrighi, G. (2009) *Adam Smith in Beijing*, London: Verso.

Auty, R. (1995) "The Resource Curse Hypothesis", *Economic Geography*, 71 (3): 257–72.

Borocz, J. (2010) *The European Union and Global Social Change*, London: Routledge.

Bunker, S. (1989) "Staples, Links and Poles in the Construction of Regional Development Theories", *Sociological Forum*, 4 (4): 589–610.

Bunker, S. and P. Ciccantell (2003a) "Transporting Raw Materials and Shaping the World-System: Creating Hegemony via Raw Materials Access Strategies in Holland and Japan", *Review of the Fernand Braudel Center*, 26 (4): 339–80.

Bunker, S. and P. Ciccantell (2003b) "Generative Sectors and the New Historical Materialism: Economic Ascent and the Cumulatively Sequential Restructuring of the World Economy", *Studies in Comparative International Development*, 37 (4): 3–30.

Bunker, S. and P. Ciccantell (2005) *Globalization and the Race for Resources*, Baltimore, MD: Johns Hopkins University Press.

Bunker, S. and P. Ciccantell (2007) *An East Asian World Economy: Japan's Ascent, with Implications for China*, Baltimore, MD: Johns Hopkins University Press.

Cardoso, F. and E. Faletto (1969) *Dependency and Development in Latin America*, Berkeley: University of California Press.

Chase-Dunn, C. (1989) *Global Formation: Structures of the World-Economy*, Cambridge: Basil Blackwell.

Ciccantell, P. (1994) "Raw Materials, States and Firms in the Capitalist World Economy: Aluminum and Hydroelectricity in Brazil and Venezuela", unpublished Ph.D. dissertation, University of Wisconsin–Madison.

Ciccantell, P. (2009) "China's Economic Ascent Via Stealing Japan's Raw Materials Peripheries", Chapter 6 in Hung, Ho-fung (ed.) *China and the Transformation of Global Capitalism*, Baltimore, MD: Johns Hopkins University Press.

Ciccantell, P. and D. Smith (2009) "Rethinking Global Commodity Chains: Integrating Extraction, Transport, and Manufacturing", *International Journal of Comparative Sociology*, 50: 361–384.

Cleveland, C. and M. Ruth (1998) "Indicators of Dematerialization and the Materials Intensity of Use: A Critical Review with Suggestions for Future Research", *Journal of Industrial Ecology*, 2 (3): 15–50.

Engdahl, F. (2005) "China Lays Down Gauntlet in Energy War: The Geopolitics of Oil, Central Asia and the United States", *JapanFocus.org*, December 22, 2005.

Frank, A. (1967) *Capitalism and Underdevelopment in Latin America: Historical Studies of Chile and Brazil*, New York: Monthly Review Press.

Gelb, B. (1988) *Oil Windfalls: Blessing or Curse?*, New York: Oxford University Press.

Gereffi, G., M. Korzeniewicz and R. Korzeniewicz (1994) "Introduction: Global Commodity Chains", pp. 1–14 in G. Gereffi and M. Korzeniewicz (eds), *Commodity Chains and Global Capitalism*, Westport, CT: Praeger.

Herman, R. et al (1989) "Dematerialization", pp. 50–69 in J. Ausubel and H. Sladovich (eds), *Technology and Environment*, Washington, DC: National Academy Press.

Hirschmann, A. (1958) *The Strategy of Economic Development*, New Haven, CT: Yale University Press.

Jalee, P. (1968) *The Pillage of the Third World*, New York and London: Monthly Review Press.

Kaup, B. (2008) "Negotiating Through Nature: The Resistant Materiality and Materiality of Resistance to Bolivia's Natural Gas Sector", *Geoforum*, 39: 1734–42.

Kenny, H. (2004) "China and the Competition for Oil and Gas in Asia", *Asia-Pacific Review*, 11 (2): 36–47.

Mahbubani, K. (2005) "Understanding China", *Foreign Affairs*, 84 (5): 49–60, 51–56, 57–58, 59–60.
Marx, K. (1962) *Capital: A Critique of Political Economy*, Vol. I, New York: International Publishers.
Nayar, B. (2004) "The Geopolitics of China's Economic Miracle", *China Report*, 40(1): 19–47.
O'Hearn, D. (2001) *The Atlantic Economy: Britain, the U.S. and Ireland*, Manchester: Manchester University Press.
Perroux, F. (1956) "Note on the Concept of 'Growth Poles' ", in I. Livingstone (ed.) (1981) *Development Economics and Policy: Readings*, London: George Allen and Unwin.
Podobnik, B. (2005) *Global Energy Shifts*, Philadelphia, PA: Temple University Press.
Ricardo, D. (1983) *On the Principles of Political Economy and Taxation*, London: J. Murray.
Ross, M. (1999) "The Political Economy of the Resource Curse", *World Politics*, 51 (2): 297–322.
Sachs, J. and A. Warner (1999) "Natural Resource Abundance and Economic Growth", NBER Working Paper #5398.
Skillings. (2008) "Total 2007 World Steel Production." *Skillings Mining Review*, January 28, 2008.
Smith, A. (1776) *An Inquiry into the Nature and Causes of the Wealth of Nations*. Oxford: Clarendon Press (1976 edition).
Wallerstein, I. (1974) *The Modern World-System*, New York: Academic Press.
Wallerstein, I. (1979) *The Capitalist World-Economy*, Cambridge: Cambridge University Press.
Zweig, D. and B. Jianhai (2005) "China's Global Hunt for Energy", *Foreign Affairs*, 84 (5): 25–26, 27–30, 31032, 33–36, 37–38.

12.2

World-system structure, natural capital and environmental entropy

Edward L. Kick and Laura A. McKinney

Introduction

We discuss the foundational role of natural capital (Flora and Flora 2008) and entropic processes as crucial world-system dynamics. Low-entropy resources such as raw materials maintain the maximum capacity for re-arrangement and serve multiple purposes. Highly entropic materials (e.g., finished goods) and resulting waste, in contrast, create disorder. The history of human civilization is characterized by the seizure and manipulation of low-entropy natural resources for consumption and trade. However, the effects on sustainability of low-entropy natural resource extraction and high-entropic waste generation escalated substantially after the onset of the industrial revolution and the use of machinery to transform natural resource inputs. Low-entropy inputs are appropriated from the periphery and semiperiphery of the world-system, while the resulting high-entropy output and waste from the core are in part returned to those same locales. It is arguable that societal development in contemporary core settings would not have been possible without furthering the exploitative and entropic global hierarchy that benefits so few at the expense of so many. We examine long-term historical evidence of such raw material exchanges in substantiating our case, but first we turn to the theoretical foundations for our work.

Theoretical foundations and historical evidence

The modern, entropic processes of most interest to us were articulated by theories of imperialism. As Marx (1976: 579–80) put it, "A new and international division of labour springs up, one suited to the requirements of the main industrial countries, and it converts one part of the globe into a chiefly agricultural field of production for supplying the other part." V.I. Lenin's (1916) global interpretation after the turn of the nineteenth century recognized that imperialism in earlier epochs (e.g., Rome) differed in a number of fundamental ways from the imperialism of finance capital in the modern era. The most central feature of modern capitalism is that imperialism and monopoly capital become firmly established when raw materials are captured, by buying up, for example, iron ore and oil fields, thus making it impossible for rivals to compete. Thus, financial capital and supporting militaries try to seize territory and maximal control over raw materials, in a globally competitive struggle for access to natural capital.

Dependency theory borrows from these interpretations, arguing that in the post-World War II system, the former European core and the emerging global hegemon—the United States—expanded global capitalism to (former) colonial settings outside the center, controlling natural capital, and benefitting the center at the relative expense of the periphery. Dependency theory emerged in 1949 with works by Hans Singer and Raúl Prebisch, who focused on the periphery's loss of natural capital to the ever-enriched core. Dependency theorists Baran (1957), Amin (1974), and Galtung (1971), among many others, developed this theme in identifying a global division of labor tied to surplus extraction from plantation agriculture in the periphery and capital accumulation among the local elites and, largely, the multinationals of the core. Recent applications of dependency and related world-system theories (Wallerstein 1974) have treated the global distribution of environmental woes that accompany these processes (see e.g., Jorgenson and Kick 2009).

Dependency theorist Andre Gunder Frank's work (1978, 2005) is particularly important to the present effort. Frank concurs with others that the economies of poorer nations tend to produce a few commodities only, robbing from their natural capital, which they export to a handful of rich nations. The global North is far more diversified in its trade, and this asymmetry exacerbates already wide gaps in economic well-being between the global poor and rich, producing what Frank called the "development of underdevelopment." But Frank also identified, although apparently did not fully develop, a related approach to entropy, noting that entropic processes are fully a part of global power-dependency relationships. Frank (2005: 2) argues that dependency evolved throughout Afro-Eurasia over several thousand years, while it has been in place globally for the past five hundred years. Frank adds (2005: 2) "… the growth of those in positions of advantage generates entropy or disorder, which compromises the continuance of such growth—unless that entropy can be … dissipated to … other regions who are obliged to absorb this entropy … due to their unfavorable position in the system as a whole." It is precisely this type of argument we turn our attention to, after reviewing the historical periods of natural capital exchange that Frank introduces.

Frank and Gills (1993) maintain that the world has been one world-system since about 3000 BCE, when it was initiated by routinized Mesopotamian-Egyptian interactions based on the trading of the products of low-entropy resources. Gills (1995: 140) emphasizes that as far back as ancient Assyria (which formed Mesopotamia with Sumer, Akkadia, and Babylonia) there was evidence of a market/credit economy with private capitalists and wage laborers. Gills argues (1995: 141) the expansionary, territorial logic of conquest followed established trade routes of raw material or low-entropy exchanges. Gills is persuasive in contending that the confluence of major centers of raw material (and other) exchanges was the foundation for the later development of a full blown Eurasian world-economy (1995: 142).

A number of specifics about the production and trade of low-entropy natural capital are offered by Bernstein (2008), who inventories the inter-societal raw material exchanges of the ancient world. The earliest axis of trade "… covered the 3,000 miles from the hills of Anatolia, through Mesopotamia, out of the Persian Gulf, across the shores of the Indian ocean, and up the Indus river," with trading hubs at Ur, Akkad, Babylon, and Nieveh, all of which constitute modern Iraq (2008: 16). For Bernstein the likely beginnings of enduring low-entropy trades were the exchanges of grain from advanced farming communities in alluvial areas for metals, found in areas of lesser fertility (2008: 25). More specifically, he contends (2008: 28–29) the very survival of Mesopotamia depended "on the exchange of their surplus food, and likely fish and wool, for metals from Oman and the Sinai, granite and marble from Anatolia and Persia, and lumber from Lebanon." Other Mesopotamian imports included ivory, jewels, slaves, perfumes, and oils. Bernstein avers that trade along this vast network must certainly have involved scores of intermediaries and transactions, opening up the probability of capital accumulation from raw material exchanges at each step in the process.

Around 2500 BCE, Egypt and Babylon conducted trading voyages to modern day Yemen and Somalia for incense, especially frankincense and myrrh. Low-entropy incenses were at minimum valued for their ability to mask the brutal smells of human waste, slaughterhouses and the cemetery. Physicians used myrrh for medicinal purposes and it was an ingredient of choice for embalmers (Bernstein 2008: 60). Further, Chase-Dunn et al (2010) and Wilkinson (2002) conclude that the Mesopotamian and Egyptian interstate systems merged around 1500 BCE.

The Medo-Persian Empire also figures prominently in low-entropy exchanges among the territories of Iran, Pakistan, part of northern India, Central Asia, Asia Minor, Thrace and Macedonia, Iraq, Jordan, Palestine, Israel, Lebanon, Syria, parts of Saudi Arabia, and Egypt over to Libya. While a complete inventory of raw materials exchanges among the Empire's vast reaches is not possible, King Darius's records of the construction of his palace at Susa are instructive. Darius notes "The cedar timber from there (Lebanon) was brought, the yaka timber was brought from Gandara (Afghanistan) and from Carmania (Iran). The gold was brought from Sardis (Turkey) and from Bactria (Afghanistan) … the precious stone lapis-lazuli and carnelian … was brought from Sogdiana (Iran). The turquoise came from Chorasmia (Uzbekistan), the silver and ebony from Egypt, the ornamentation from Ionia (Turkey), and the ivory from Ethiopia and Sindh (Pakistan)," (Lendering 2010).

Bernstein (2008: 41–42) also substantiates the somewhat loosely coupled trade of low-entropy materials across three continents during the Roman Empire. Rome imported pepper, Chinese silk, and Indian wildlife. In return, it exported coral to India and glass to China. Bernstein also notes the low-entropy exchanges of the Greecian Empire as the Greeks imported cattle, wood, fish, and timber. Although there are variations across civilizations in the precise resources traded, the periods of these ancient empires are characterized by raw material exchanges, with rather low-entropic consequences.

The subsequent expansion of trade then was propelled by Islam, which would integrate trade through the Indian Ocean, across Asia, and to the extremities of Eurasia. The Islamic golden age ushered in early merchant capitalism and a money economy during the eighth to twelfth centuries in what some have referred to as the "Arab Agricultural Revolution" (Glick 1977). Indeed, Muslim traders linked the products of natural capital across parts of Eurasia, ranging from the sorghum of Africa, citrus from China, rice, cotton, and sugar cane from India, wheat from Yemen, and rice from Iraq (Watson 1974). While technological innovations and the spread of knowledge abounded, the Crusades from the West challenged the power of the Islamic world during the eleventh century, and by the early thirteenth century the Mongol Empire led by Genghis Khan conquered much of Eurasia opening overland trade (although competition and wars with the Islamic world continued for centuries). As the Mongols captured an ever expanding empire of territories, new trading routes were established and opportunities for taxation and tribute were multiplied.

Many see the Mongol Empire as effectively incorporating isolated imperial systems (Abu-Lughod 1989), reducing tribute gatherers along the Eurasia land mass and re-establishing the Silk Road as a viable means of transporting low-entropy resources. Under Mongol domination, caravans carried Chinese silk, pepper, ginger, cinnamon, and nutmeg to the West, along with gunpowder. Europe was also provisioned with Indian muslins, cottons, pearls, and weapons, as well as carpets and leather goods from Iran (Prawdin 2006). Moving West to East was silver, fine cloth, horses, and linen (Prawdin 2006). Chase-Dunn and Hall (1997: 163, 191) point out, however, that there was little the Chinese wanted other than gold, and then silver, American ginseng and Indian opium, giving them an exchange and accumulation advantage. Yet, Pax Mongolia was effectively disrupted in the mid-1300s by the transmission of fleas and the Black Death along the re-established trade routes (McNeill 1965). World-system theorist Abu-Lughod (1989) views this era as the early foundation for the world-system elaborated upon by Braudel (1966) and Wallerstein (1974). The era ushered in large cities as centers of commerce, with bills of exchange (Weatherford 2004), banking and credit systems (Bernstein 2008), and the initiation of greater levels of finished product (or, "high-entropy") trades.

The domination of modern world trade by the West circa 1500 is chronicled in a sizeable number of treatments (see, in particular, Wallerstein 1974, 1989), and we only briefly treat it here. Portugal's geographical location in Iberia placed it in a fortunate spatial position, and it rather quickly "controlled" a huge empire stretching from the Indian Ocean and its trade routes in spices to the slave trade in western Africa, while initiating sugar production in Brazil. Portugal parlayed the natural capital of the Western hemisphere in the form of low-entropy raw materials such as sugar, tobacco, turtle oil, spices, silver, and gold into the global trade of raw materials in Europe (albeit based in part in Asia, Bunker and Ciccantell 2005: 100). With respect to the slave trade, Bunker and Ciccantell (2005: 36) make the compelling observation that the Portuguese, "... conducted their slave raids as if they were extractive enterprises—that is, as if the Native Americans were forces of production that they could appropriate from nature."

Spain quickly followed suit from a comparable geographical location. Its enslavement of local populations in the Americas and the acquisition of silver from the Aztecs and Incas is so deeply documented that it requires no elaboration here (see e.g., among world-system treatments, Bunker and Ciccantell 2005). However different these conquests were from empires established in prior historical periods in their explorations and establishment of trans-Atlantic trades, they shared with them the collection and distribution of low-entropy natural capital, based on control over trade routes and wrested from enslaved peoples. Spain and Portugal relegated the transport of a range of cheap staple goods to the Dutch (Bunker and Ciccantell 2005: 102), and the Dutch successors established their colonial empire in Asia in the 1600s by taking it from the Portuguese, including key sea routes that were control points for the collection and exchange of pepper and spices. They also came to control the slave trade from Africa to American sugar plantations (Chirot 1986: 35). Bunker and Ciccantell (2005: 106) enumerate the breadth of Dutch commerce principally in low-entropy goods such as grain, salt, timber, brick, woolen cloth, silks, spices, slaves, bullion, cheap textiles, fur, fish, oil, rice, cereals, fruit, wines, sugar, tobacco, logs and lumber, tar, flax, crystal, glass lenses, paper, tea, and copper fittings (see also Chirot 1986: 36; Wallerstein 1974). The Dutch then were outdistanced by France and England, which warred throughout the 1700s for control of the Americas, Africa, and Asia. The eventual domination of what was to become the British Empire over the French coincided, approximately, with the age of industrialization.

The contemporary world-system and entropy: Natural capital, industrialization and finished products

Prior to the industrial revolution, manufacturing was home-based handwork done in "cottage industries," principally in rural areas. Rural producers made their own clothing, furniture, and tools from raw materials on their farms or in nearby forests. Thus, trade in the world-system was still intimately tied to low-entropy raw materials derived directly from nature. Entropic waste for the most part was dissipated locally, with little by way of far-reaching impacts resulting from their transfer. The Industrial Revolution, however, marked a pivotal change in the transfer of low-entropy natural resource inputs from, and entropic products and waste to, less developed areas.

Natural capital in the form of coal and iron deposits fueled the industrial revolution, which was founded upon power-driven machinery and the modern factory system of production. Great Britain's usage of natural capital in the textile and iron industries, for instance, was complemented by other raw materials Great Britain gained from trade with its American colonies (e.g., wood) and elsewhere in the world-system (Adams 1982). The colonies also provided an outlet for Great

Britain's highly entropic manufactured products, as elaborated in the theories of imperialism and dependency discussed earlier. This division of labor and pattern of exchange introduced a relatively recent period of highly entropic processes into the world-system. Indeed, we find Tilly's (1984: 62) approach to connectedness, addressed by Chase-Dunn and Hall (1997: 14), to be persuasive, " … (H)uman history has seen many world systems, often simultaneously dominating different parts of the globe. Only in the last few hundred years, by the criterion of a rapid, visible, and significant influences, could someone plausibly argue for all the world as a single system." Among those influences, and accompanying the expansion of full-scale industrialization, is the system-wide role of entropic processes, including their serious environmental consequences.

A number of theorists have emphasized world-system dynamics such as the global division of labor (e.g., Wallerstein 1974), geopolitics (e.g., Wilkinson 1987), financial and military imperialism (e.g., Lenin 1916), and technology (Bunker and Ciccantell 2005; Lenski 1966) as drivers of the world-system and its constituent societies. Key centers of attention have been systemic cycles, hegemonic shifts, and global expansions (Wallerstein 1974), as well as the rise and fall of empires, growth and stagnation, and exploitation of periphery areas (Frank and Gills 1993). The locus of our concern is different, however, in that we focus on asymmetries in environmental exchanges and differential ecological damage throughout the system that are most pronounced in the truly modern era. The switch to an industrialized mode of production becomes so pivotal when it is understood that it introduced an exponential increase of entropic exchanges made possible by machinery power, which is itself constructed out of low-entropy natural resources and fueled by low-entropy natural resources, respectively.

A reoccurring theme within the physical and natural sciences that is the tension between the "downward spiral" of physical processes, on the one hand, and the ascendancy of evolutionary phases of life systems, on the other hand. To illustrate, the first law of thermodynamics states that no energy can be created or destroyed. More consequential for this illustration, however, is the second law of thermodynamics, which explains the irreversibility in nature, that is, the decrease in usefulness that is referred to as "entropy" (Prigogine 1967). The manipulation of natural resources generates entropy as the resources lose the capacity to be rearranged, or become disordered. Consider the low-entropy natural resource of land. From its pristine form, infinite possibilities for land transformation exist, given the huge inventory of products, materials, and goods that can be created by industry and manufacturing sectors via the use of land to house such operations. Once the land is used, limitations are placed on other uses. The resulting stage is highly entropic (lowest capacity for arrangement for use), as, for example, industrial use renders it unable to serve as a habitat for other species, as a human dwelling, or for a range of other productive purposes. Similarly, coal has exponential possibilities for use as it has been a foundational source of the energy that powered the bulk of industrial processes. Yet, once the decision is made to combust and use coal for fuel, the resultant ash is in a highly entropic state with little to no demonstrable use. These final states of high entropy become, indeed, suitable only for disposal as waste and toxins that threaten low-entropy resources that now literally make the world run. These examples illustrate the impact of human actions on increasing the entropy (decreasing utility) of the physical world (Daly and Cobb 1989).

Empire expansion and imperialism directly relate to the quest for low-entropy natural resources through the acquisition of land area and extractive resource trades. The ascendancy of Great Britain and the United States to hegemonic positions in the world-system are obvious applications of this theorization. Our earlier accounts stopped with the British Empire, but we now point to the extensive arsenal of coal, natural gas, minerals, and land in the United States that helped to ensure its ascendancy to become the world's hegemon during the 1900s. But American dominance in agriculture and manufacturing in the twentieth century was possible due as well to its ability to draw on the storehouse of low-entropy resources (e.g., fossil fuels) from the periphery and semi-periphery of the world-system. Originally, America's landscape was cleared and exported as lumber and fiber for use in Great Britain especially. Specifically, the predominance of agriculture in the south (i.e., "Cotton is King") resulted from the capture and export of low-entropy natural resources from vast areas of land. Further, the manufacturing dominance in the Northern states (i.e., the Industrial Revolution) was possible due to the enormous amounts of coal, natural gas, and oil that fueled the expansion of industry generally in America. Taken together, this "perfect storm" of low-entropy natural resources propelled both Great Britain and the United States to their positions as global superpowers, a position each successively maintained. The US's ascendancy to hegemony coincided with the rise of industrialism, precipitating unprecedented wealth accumulation due to exponential increases in the use of low-entropy inputs to produce high-entropy finished goods and waste for export. In an important way, the maintenance of this hegemonic position is possible due to raw material imports from lesser developed countries in the periphery and semiperiphery, coupled with the transfer of high entropic exports (finished goods) and highly entropic waste (Pellow 2007) back to those same areas.

Implications

While a number of relatively recent treatments (Appelbaum and Robinson 2005; Sklair 1999) emphasize that globalization is a recent phenomenon, the axis of exchange and its consequences suggest a continuity of processes across thousands of years, with a sharp point of inflection in the dynamic that accompanies the comingling of empiric expansion of Great Britain and its industrialization (see also Frank and Gills 1993; Bunker and Ciccantell 2005). On this one point at least, the truly modern world-system emerged in the late 1700s and early 1800s, coincident with British hegemony and an entirely new and expansive system of production—industrial capitalism.

Our present effort complements a number of recent studies of environmental degradation, including the ecological footprint (Jorgenson and Clark 2009), greenhouse gas emissions (Grimes and Kentor 2003; Jorgenson 2006), biodiversity loss (McKinney et al 2010), deforestation (Burns et al 2003), toxic waste disposal (Pellow 2007), and general climate change (Clark and York 2005). Each of these rather thoroughly researched areas of environmental degradation relate precisely to entropic dissipation via the expansion of natural resource extraction and land use for consumption or finishing in core settings. The ecological footprint and its constituent parts (e.g., deforestation, carbon dioxide emissions) result from the acquisition of low-entropy natural resources, many of which are shown to be concentrated in less developed areas (Burns et al 2003; Jorgenson 2006). These dynamics reproduce global inequalities in the world-system as they rob from the natural environments of non-core settings, with pernicious effects on their sustainability and the sustainability of the world as a whole (see e.g., Sachs and Warner 2001).

Much like other dimensions of development and power, such as trade and industrialization, the existence of low-entropy natural resources favor core nations but jeopardize peripheral settings. The identification of low-entropy natural resources in peripheral settings is a liability, as core states seize those resources for their own gain; the same discovery of low-entropy resources in core settings buttresses their relative scope of wealth and power. Thus, in core settings, low-entropy natural resources are used to further their already advantaged positions; in the periphery they motivate further exploitation for those precious stocks. To sum, transfers of low-entropy resources and highly entropic waste figure prominently to the inner workings of the world-system.

In tandem, the resultant entropic waste, whether it be in the form of toxic waste, e-waste, or recycled materials, follows world-system dynamics that relegate offensive environmental outcomes to less developed areas; in this case it is the exportation of core waste to other locales. The introduction of entropic waste to less developed countries incites disorder (entropy) and further degrades their already frail environmental conditions. The application of dependency and world-systems theory to global entropic exchanges introduces a new, focal axis of imperialism that relates to current environmental degradation worldwide. As the scientific community reaches a near consensus on the certitude and perils of global climate change, attention to the axiomatic role of entropy will become ever more important.

Our treatment of entropic processes has implications for both theorizations and empirical examinations of world-system dynamics for the history of civilizations and today. We uncover the underlying physical component that is the foundational for market exchanges; that is, embodied entropy is linked to degradation of the physical environment and capital accumulation within states of the world-system. Specifically, accounting for the "unequal entropic exchanges" across world-systems is consequential for examinations of ecology and the world-system, as well as studies of economic, development, health, well-being, and sustainability outcomes in the world-system. Our approach adds to the existing body of literature on the world-system by combining the physical, natural, and social sciences to formulate a holistic approach to better understand its inner workings. Future research could apply this extension to theorizations of the world-system. We also advocate subsequent empirical examinations utilizing creative operationalizations for the entropic concepts offered here. Thus, noting the preliminary nature of the present effort, we are confident further explorations of those types will bring a fruitful level of precision to articulations of entropic dynamics in the world-system.

In conclusion, the contemporary global system depletes natural capital at an unsustainable pace, processing it in unsustainable ways and returning the perilous by-products to the earth; if these processes remain unchecked, they will lead to a collapse of the commons (Diamond 2005). While entropic exchanges in some fashion characterizes a succession of world empires, it is in their most modern industrial and global form that they pose the newest and gravest threat to the sustainability of all earthly life forms. It is possible in the short term for core countries to escape their entropic destiny by importing low-entropy natural resources in exchange for highly entropic goods and waste, but in the long run, there is but one planet that we all must depend on for sustenance and survival. The greater disorder or entropy we introduce into the environment will only haunt us at a later time when we realize the limits to our present unsustainable use of embodied entropy.

References

Abu-Lughod, J.L. (1989) *Before European Hegemony: The World System A.D. 1250–1350*, New York: Oxford University Press.

Adams, R. (1982) *Paradoxical Harvest: Energy and Explanation in British History, 1870–1914*, New York: Cambridge University Press.

Amin, S. (1974) *Accumulation on a World Scale*, New York: Monthly Review Press.

Appelbaum, R.P. and W.I. Robinson (2005) *Critical Globalization Studies*, New York: Routledge.

Baran, P.A. (1957) *The Political Economy of Growth*, New York: Monthly Review Press.

Bernstein, W.J. (2008) *A Splendid Exchange: How Trade Shaped the World*, New York: Grove Press.

Braudel, F. (1949; 2nd edn. 1966) *La Méditerranée et le monde méditerranéen à l'époque de Philippe II*; trans. Sian Reynolds (1999) *The Mediterranean and the Mediterranean World in the Age of Philip II*, London: Collins.

Bunker, S.G. and Ciccantell, P.S. (2005) *Globalization and the Race for Resources*, Baltimore, MD: Johns Hopkins University Press.

Burns, T.J., Kick, E.L., and Davis, B.L. (2003) "Theorizing and rethinking the linkages between the natural environment and the modern world-system: Deforestation in the late 20th century", *Journal of World Systems Research*, 9: 357–92.

Chase-Dunn, C. and Hall, T.D. (1997) *Rise and Demise: Comparing World-Systems*, Boulder, CO: Westview Press.

Chase-Dunn, C., Pasciuti, D., Alvarez, A., and Hall, T.D. (2010) "The ancient Mesopotamian and Egyptian world-systems", unpublished thesis, Institute for Research on World-Systems.

Chirot, D. (1986) *Social Change in the Modern Era*, San Diego, CA: Harcourt, Brace, Jovanovich.

Clark, B. and York, R. (2005) "Carbon metabolism: Global capitalism, climate change and the biospheric rift", *Theory and Society*, 34: 391–428.

Daly, H. and Cobb, J. Jr. (1989) *For the Common Good: Redirecting the Economy Toward Community, the Environment, and a Sustainable Future*, Boston, MA: Beacon Press.

Diamond, J. (2005) *Collapse: How Societies Choose to Fail or Succeed*, New York: Viking.

Flora, C.B. and Flora, J.L. (2008) *Rural Communities: Legacy and Change*, 3rd edn., Boulder, CO: Westview Press.

Frank, A.G. (1978) *Dependent Accumulation and Underdevelopment*, New York: Monthly Review Press.

Frank, A.G. (2005) "Urban location and dissipation of entropy (A programmatic statement)", unpublished thesis, http://rrojasdatabank.info/agfrank/research.html#dissipation.

Frank, A.G. and Gills, B.K. (1993) *The World System: Five Hundred Years or Five Thousand?* London: Routledge.

Galtung, J. (1971) "A structural theory of imperialism," *Journal of Peace Research*, 8: 81–117.

Gills, B.K. (1995) "Capital and power in the processes of world history", pp. 136–62 in S. Sanderson (ed.) *Civilizations and World Systems: Studying World Historical Change*, Walnut Creek, CA: AltaMira Press.

Glick, T.F. (1977) "Noria pots in Spain", *Technology and Culture*, 18: 644–50.

Grimes, P. and Kentor, J. (2003) "Exporting the greenhouse: Foreign capital penetration and CO_2 emissions 1980–96", *Journal of World-System Research*, 9: 261–75.

Jorgenson, A.K. (2006) "Global warming and the neglected greenhouse gas: A cross-national study of methane emissions intensity, 1995", *Social Forces*, 84: 1777–96.

Jorgenson, A. and Kick, E. (2009) *Globalization and the Environment*, 2nd edn., Chicago, IL: Haymarket.

Jorgenson, A.K. and Clark, B. (2009) "The economy, military, and ecologically unequal relationships in comparative perspective: A panel study of the ecological footprints of nations, 1975–2000", *Social Problems*, 56: 621–46.

Lendering, J. (2010) "Achaemenid Royal Inscriptions: DSf," unpublished thesis, http://www.livius.org/aa-ac/achaemenians/DSf.html.

Lenin, V.I. (1916 [1975]) *Imperialism, the Highest Stage of Capitalism: A Popular Outline*, Peking: Foreign Language Press.

Lenski, G. (1966) *Power and Privilege: A Theory of Social Stratification*, New York: McGraw-Hill Publishing.

Marx, K. (1976) *Capital*, Vol. 1, New York: Vintage.

McKinney, L.A., Kick, E.L., and Fulkerson, G.M. (2010) "World-system, anthropogenic, and ecological threats to bird and mammal species: A structural equation analysis of biodiversity loss", *Organization and Environment*, 23: 3–31.

McNeill, W.H. (1965) *The Rise of the West: A History of the Human Community*, New York: The New American Library.

Pellow, D.N. (2007) *Resisting Global Toxics: Transnational Movements for Environmental Justice*, Cambridge, MA: MIT Press.

Prawdin, M. (2006) *The Mongol Empire: Its Rise and Legacy*, New Brunswick, NJ: Transaction.

Prigogine, I. (1967) *Introduction to Thermodynamics of Irreversible Processes*. New York: Interscience Publishers.

Sachs, J.D. and Warner, A.N. (2001) "Natural resources and economic development: The curse of natural resources", *European Economic Review*, 45: 827–38.

Sklair, L. (1999) *The Transnational Capitalist Class*, Oxford: Blackwell.

Tilly, C. (1984) *Big Structures, Large Processes, Huge Comparisons*, New York: Russell Sage Foundation.

Wallerstein, I. (1974) *The Modern World System*, New York: Academic Press.

Wallerstein, I. (1989) *The Modern World-System III: The Second Eve of the Great Expansion of the Capitalist World Economy*, New York: Academic Press.

Watson, A. (1974) "The Arab agricultural revolution and its diffusion 700–1100", *The Journal of Economic History*, 34: 8–35.

Weatherford, J. (2004) *Genghis Khan and the Making of the Modern World*, New York: Crown Publishers.

Wilkinson, D. (1987) "Central civilization", *Comparative Civilization Review*, 17: 31–59.

Wilkinson, D. (2002) "The status of the far Eastern civilization/world-system: Evidence from city data", *The Journal of World-Systems Research*, 8(3): 292–328.

12.3

What is old and what is new?

Considering world-systems in the twenty-first century and beyond

Thomas J. Burns

While its processes have been unfolding for centuries, some aspects of the world-system are new. Particularly since the Industrial Revolution, the world has had unparalleled technological change, wealth accumulation, and population growth and concentration; these have led to unprecedented levels of other problems, particularly environmental degradation (McNeill 2000). Making sense of the world-system in the twenty-first century calls for developments in theory and method that reach beyond our current modes of analysis. This work argues for the centrality of environmental concerns, and offers a set of meta-variables on which a framework for analysis can be based.

Heretofore, environmental considerations have been ignored in many world-systems analyses; yet gaining understanding of these processes necessarily will involve significantly closer attention to environmental concerns. In this essay, I outline families of variables that impact and are impacted by world-systems and ecological factors. While these are analytically distinct, they interact with one another in myriad ways. The list is not exhaustive, but does, I believe, give a sense of some of the most important considerations.

The natural ecology historically has had a profound effect on how societies have developed. Now, particularly with the press of population, technology, and highly skewed distributions of resources, humankind has perhaps more profound effects on the environment than ever before. These are manifested in countless ways, including the rise of greenhouse gas emissions, deforestation, marine pollution, and desertification (for a recent collection of work linking world-systems and the environment, see Jorgenson and Kick 2006).

Although a significant body of work has been amassed over the last few decades examining effects of human social organization on the environment, there is not nearly as much research on effects of environmental degradation on social *outcomes*. As the field matures, it will be important to examine these issues as well.

While people have developed technology throughout history, in the contemporary era, the influence of technology is more profound than ever. We are able to make deeper incursions into the earth than ever in history, and to promulgate technologies more widely than ever before.

There is a temptation to believe technology will provide a fix for current shortages and other environmental constraints. However, history shows that when new technologies are introduced, they typically have unintended consequences, many of which may cause bigger problems than those they were originally developed to address.

Also consider the *Jevons Paradox*—when a new technology is developed that is cheaper or more efficient than what preceded it, there is a tendency to increase usage such that the net benefit to the environment is zero or even negative. This is not to argue against technology; rather it calls for rational planning and attention to the social and environmental effects of what technology is developed.

As literatures on the carrying capacity of animal species have suggested, the possibility of going beyond the ability of the planet to sustain itself and the life dependent on it varies to a large degree with population dynamics. With nearly seven billion and rising, the world has more people than at any time in history.

As societies modernize and become increasingly integrated into the world-system, massive urban agglomerations arise with large numbers of unemployed and cheap labor. Age-ratios of populations tend to co-vary with other social and ecological factors, particularly as they put strains on vital resources.

Institutional fixes in a vacuum are largely ineffective, and sometimes counter-productive. While addressing environmental and other global problems will necessarily involve institutions, multi-faceted approaches involving each of the areas covered in this article are vital.

Each institution has its own modes of operation, and the respective disciplines that examine them have biases that tend to color the research they do. Economics, for example, historically has tended to "externalize" many of the natural resources it takes as givens, such as water, air and soil.

Ways in which societies are organized, especially in terms of local, regional, and global inequality will continue to pose pressing questions, particularly around issues of environmental justice. Additionally, size concerns are important; as the scale of production and consumption increase, ways in which they are *organized* change, as do the effects they have on outcomes.

Human culture supplies networks of meaning that people use to make sense of their surroundings. Implied in cultural knowledge are sets of priorities and ways of seeing. Different cultures place more or less centrality on the natural environment, and many of the strains on the environment can be understood in the context of cultures (for discussion, see Burns 2011).

Cultural lag processes are also crucial. In a world where material conditions change rapidly as a result of technology, world-system processes, and population pressures, the adaptive culture in many ways is not keeping pace. This leads to a mismatch of social and environmental problems on the one hand, and the cultural knowledge and wisdom to address them adequately on the other.

Social sciences have developed methods that are only partially up to the task of making sense of the world-system in the twenty-first century and beyond. It will be important to move beyond linear modeling and consider that systems may have multiple equilibria, non-linear processes, higher order interactions and indirect effects, which look and behave differently within their cultural and ecological contexts than they do when analyzed and de-contextualized.

As society moves into the twenty-first century, it will be increasingly vital to consider environmental constraints as central to any analysis of world-systems processes. We continue to underestimate environmental centrality at the peril of the planet and everyone on it.

References

Burns, T.J. (2011). *This Side of Eden: Culture and the Natural Environment in the Modern/Post-Modern World.* Amherst, NY: Cambria.

Jorgenson, A.K. and E.L. Kick (eds) (2006). *Globalization and the Environment.* Leiden: Brill.

McNeill, J.R. (2000). *Something New under the Sun: An Environmental History of the Twentieth-Century World.* New York: Norton.

12.4

Glad moon rising

A world-systems perspective on the world in space

Marilyn Dudley-Flores and Thomas Gangale

The US's petroleum expenditures far exceed those of its civil space program, estimated by:

- The costs of oil extraction in places barely accessible to current technology (i.e., deep sea drilling)
- The construction of a world-dominating empire

The first expenditure was depicted by the recent Gulf of Mexico spill. Chalmers Johnson has discussed the second expenditure. Among imperial costs is the American empire's military space arsenal, devoted to targeting hostile actors and assets, and killing foreign satellites that counter American interests (2004).

After "flags and footprints" missions to the Moon, the division between American military and civil space endeavors became distinct in function and in cost. The civil space effort functions to explore science and to expand the human ecology off the Earth. The military space effort functions to dominate the world. The latter's true costs are buried in black budgets that exceed by an order of magnitude the miniscule American civil space budget that yearly approximates the annual funding for the US's National Park Service. Despite massive funding, gaining the high ground by space militarization cannot get *humans* to the high ground. The US's military space systems are unmanned.

There is no infrastructure in space today that would see humans permanently living and working there. The Nixon Administration downmoded and de-coupled a twinned civil space project that would have ensured that goal. A reliable, frequently launched space transportation system and a large livable beachhead in space became the Space Shuttle and the International Space Station (ISS). The original project would eventually have deployed space-based energy systems for use on Earth. Solar-to-microwave electrical generation technology alone could transform capital and the technological means of production.

Moreover, the United States has retired its remaining Space Shuttle fleet and is withdrawing from the ISS. It has scrapped development of Constellation, the return-to-the-Moon program. These are signs that its military space aims are more important. Why should the United States occupy any other world when it can dominate this one?

Fortunately, there is a glad moon on the rise. The civil "Space Race" model has given way to a transnational approach beneficial to the global collective. Space production enabled advanced globalization that has allowed routine transnational transactions. In turn, the thing effected has transformed space production. Space has been "globalized" (Dudley-Flores and Gangale 2007).

First, space-based systems are essential to monitoring and mitigating climate change and other natural and anthropogenic problems that can be imaged from space. The discovery of the ozone hole led to its repair. Following the South Asian Tsunami, scientists realized that existing Global Positioning System (GPS) satellites could detect tsunamis within about a 15-minute lead. Second, space technologies can inform green innovation in ways that Earthbound attempts cannot, in order to better run an increasing clutch of advanced industrial societies. Living and working in the most extreme of environments—space—can achieve sustainable technologies for an advancing world-system of societies (WSS) on an Earth growing ever more extreme. The solar power systems that keep ISS operational are but one example. Third, long-term space projects are necessarily transnational, requiring guaranteed multi-year funding streams that no one nation-state or corporation can sustain.

Most terrestrial green technologies hawked today are "green wash." There is nothing transformative about such huckstered technologies; not in the way the plow transformed agriculture, in the way the computer created the Information Age, and in the surge of technologies that the Apollo space program gave the world. Green innovation, like alternative energy systems, requires getting "out there." The first step is a large, quality-of-life, industrial grade space station on orbit and its attendant infrastructure. Such a testbed also ensures human permanence in space that would eventually lead to a multi-planet economy in the inner solar system, a multi-world-system of societies, and the supplanting of the Westphalian nation-state with the transnational state (Dudley-Flores and Gangale 2009, 2010; Gangale 2009).

On present-day Earth, space-capable societies that participate in the transnational approach are changing the core and periphery of the civil space enterprise (Dudley-Flores and Gangale 2007: 7, 25–26). The partnerships led by the supranational European Space Agency and China, Russia, and India are lofting monitoring and mitigation systems, working on green technologies, and planning human spaceflight ventures. In the meantime, the United States suffers a lack of focus regarding its civil space policy (Hearsey et al 2009). While the United States slides further from the WSS's core owing to declining civil space production and other reasons, its militarist space strategy could see the United States coveting things it views as having high value while a transformation of capital is devaluing those global interests. Such a space strategy may be unable to aid mitigation of the large-scale natural and anthropogenic challenges to the WSS. The United States could be positioning itself for a dependent, peripheral role in an increasingly interdependent WSS where core societies are those that can provide assistance throughout the WSS through their cooperative space resources and systems.

References

Dudley-Flores, Marilyn and Thomas Gangale. 2007. "The Globalization of Space—the Astrosociological Approach", *AIAA Space 2007 Meeting Papers on Disc* [CD-ROM], AIAA-2007-6076, Reston, Virginia.

Dudley-Flores, Marilyn and Thomas Gangale. 2009. "Manufactured on the Moon, Made on Mars—Sustainment for the Earth Beyond the Earth," *AIAA Space 2009 Meeting Papers on Disc* [CD-ROM], AIAA-2009-6428, Reston, Virginia.

Dudley-Flores, Marilyn and Thomas Gangale. 2010. "The Rise of the Transnational State: Space Logistics, Sovereignty, and Diaspora off the Earth," *AIAA Space 2010 Meeting Papers on Disc* [CD-ROM], AIAA-2010-8867, Reston, Virginia.

Gangale, Thomas. 2009. *The Development of Outer Space: Sovereignty and Property Rights in International Space Law*. Santa Barbara, CA: Praeger.

Hearsey, Christopher, Thomas Gangale, and Marilyn Dudley-Flores. 2009. "Critiquing Rationales in Space Policy Proposals: Developing a Methodology for Evaluating Space Policy", *AIAA Space 2009 Meeting Papers on Disc* [CD-ROM], AIAA-2009-6819, Reston, Virginia.

Johnson, Chalmers. 2004. *Militarism, Secrecy, and the End of the Republic*. New York: Metropolitan Books.

12.5

Extraction and the world-system

Paul K. Gellert

Extraction is fundamental to the structure and history of the world-system. As early world-systems analysts explained, extraction provides the basic raw materials for production and consumption, and in the capitalist world-system these raw materials are transformed into commodities with exchange values. Declining terms of trade and persistent unequal exchange renders extractive peripheries underdeveloped and ecologically impoverished. Far from an ephemeral period in the evolution of human societies, extraction in the peripheral regions endures and is thus vital to consideration of possible world-system futures.

A growing literature in sociology, geography, and environmental history examines both historical and contemporary dynamics of resource extraction, beginning with questions of how materials of the earth are commodified (Castree 2003) and how such processes are shaped by the characteristics of the commodities and the socionatural formations in particular geographical regions (Nevins and Peluso 2008). In other words, studies of extraction examine both material processes of transforming matter and economic processes of transforming value. Moreover, scholars increasingly recognize the variability and contingency of extraction, as processes of dispossession, commodification, and accumulation (re)occur in various locations (Harvey 2003).

World-systems research makes several distinctive contributions to our understanding of extraction. First, it recognizes the specificity of commodity extraction, especially as corrective to ideologically modernist optimism about the automaticity of steps from raw materials exporter to developed manufacturing center. Whereas core processes may produce "generative effects" that propel cycles of hegemonic ascent, the extractive processes undergirding them produce "diseconomies of scale" (Bunker and Ciccantell 2005), "inattentively reshape[d] local ecosystems" (Bunker 1984; Friedmann 2000: 502).

However, second, the geographic locations of extractive economies do not necessarily coincide with the core and periphery of the world-system so much as with struggles over benefits of the world-economy (Arrighi 1994; Moore 2003). Extraction of oil and minerals from the periphery is common, yet even within the rising core of the United States, Appalachia was an early extractive periphery (Dunaway 1996) while in recent decades oil and coal extraction have continued to be critical. What distinguishes extractive locations are the ways socionatural configurations are disrupted and altered (Swyngedouw 1999).

Third, extractive sectors shape politics. Although some question the "resource curse" thesis for being over-generalized or reductionist (Rosser 2006), there is significant evidence that resources, especially oil, are associated with authoritarian politics (Ross 2001). Indonesia's authoritarian New Order relied on extraction of multiple resources to support its accumulation strategies and longevity (Gellert 2010). Whether and how such strategies persist in a more democratized era is an open question. Movements of resistance to the commodification and extraction of raw materials (Kaup 2008) provide insights into possible foundations for envisioning—and material obstacles to achieving—reconfigurations.

In the present conjuncture of global economic crisis, it is vital to examine extraction within an intertwined relationship with financialization of the world-economy, as well as in relationship to the neo-liberalization of nature (Bridge 2004; McCarthy and Prudham 2004) and attempted spatio-temporal "fixes" (Harvey 2003). Financialization affects the logic of extraction and resource management, for example, while reconfiguring socionature to create (shareholder) value and accumulation (Gunnoe and Gellert 2011). An unfulfilled agenda of research would integrate the heights of finance with materio-spatial analysis to investigate crises and their resolution (O'Hearn 2005). As Foster (2009) argues, the ecological rift of global capitalism is evident in multiple ecological crises, but the pressing question is whether extraction is pushing human society beyond hegemonic transition toward "epochal change" (Friedmann 2000: 504). Examining "non-Cartesian ontologies" (Moore 2011) of capitalism operating through the contingencies of human and nonhuman nature offer the political space to envision reconfigurations of socionatural relations.

References

Arrighi, G. (1994) *The Long Twentieth Century: Money, Power, and the Origins of Our Times*, London: Verso.

Bridge, G. (2004) "Mapping the Bonanza: Geographies of Mining Investment in an Era of Neoliberal Reform", *Professional Geographer*, 56: 406–21.

Bunker, S. G. (1984) "Modes of Extraction, Unequal Exchange, and the Progressive Underdevelopment of an Extreme Periphery: The Brazilian Amazon, 1600–1980", *American Journal of Sociology*, 89: 1017–64.

Bunker, S. G. and Ciccantell, P. S. (2005) *Globalization and the Race for Resources*, Baltimore, MD: The Johns Hopkins University Press.

Castree, N. (2003) "Commodifying What Nature?", *Progress in Human Geography*, 27: 273–97.

Dunaway, W. (1996) "The Incorporation of Mountain Ecosystems into the Capitalist World-System", *Review: Fernand Braudel Center*, XIX: 355–81.

Foster, J. B. (2009) *The Ecological Revolution*, New York: Monthly Review Press.

Friedmann, H. (2000) "What on Earth is the Modern World-System? Foodgetting and Territory in the Modern Era and Beyond", *Journal of World-Systems Research*, VI: 480–515.

Gellert, P. K. (2010) "Extractive Regimes: Toward a Better Understanding of Indonesian Development", *Rural Sociology*, 75: 28–57.

Gunnoe, A. and Gellert, P. K. (2011) "Financialization, Shareholder Value, and the Transformation of Timberland Ownership in the USA", *Critical Sociology*, 37: 1–24.

Harvey, D. (2003) *The New Imperialism*, Oxford: Oxford University Press.

Kaup, B. Z. (2008) "Negotiating Through Nature: The Resistant Materiality and Materiality of Resistance in Bolivia's Natural Gas Sector", *Geoforum*, 39: 1734–42.

McCarthy, J. and Prudham, W. S. (2004) "Neo-liberal Nature and the Nature of Neo-liberalism", *Geoforum*, 35: 275–83.

Moore, J. W. (2003) "The Modern World-System as Environmental History? Ecology and the Rise of Capitalism", *Theory and Society*, 32: 307–77.

Moore, J. W. (2011) "Ecology, Capital, and the Nature of our Times: Accumulation & Crisis in the Capitalist World-Ecology", *Journal of World-Systems Research*, XVII: 107–46.

Nevins, J. and Peluso, N. L. (eds) (2008) *Taking Southeast Asia to Market: Commodities, Nature, and People in the Neoliberal Age*, Ithaca, NY: Cornell University Press.

O'Hearn, D. (2005) "Cycles of Accumulation, Crisis, Materials and Space". In *Ciccantell*, P. S., Seidman, G. and Smith, D. A. (eds) *Nature, Raw Materials, and Political Economy*, Vol. 10. Amsterdam, the Netherlands: Elsevier JAI.

Ross, M. (2001) "Does Oil Hinder Democracy?", *World Politics*, 53: 325–61.

Rosser, A. (2006) "The Political Economy of the Resource Curse: A Literature Survey". *IDS Working Paper*. Sussex: Institute of Development Studies.

Swyngedouw, E. (1999) "Modernity and Hybridity: Nature, Regeneracionismo, and the Production of the Spanish Waterscape, 1890–1930", *Annals of the Association of American Geographers*, 89: 443–65.

12.6

Geopolitical and socio-ecological constraints to the reproduction of the capitalist world-economy

John L. Gulick

Futurology these days is literally a dismal science (or art). Today's prognosticating often involves projecting future amplitudes of global greenhouse gas emission and global carbon sink removal, and the far trickier business of calibrating the probable effects on climate patterns—and ominously, on human sustenance and mortality itself. For even those who appreciate uncertainties in socio-ecological modeling, or understand that anthropogenic climate change impacts will be unevenly distributed across continents, regions, and classes, the prospect of total catastrophe weighs heavily on the brain.

Because its method is rooted in studying the long-term developmental patterns of historical social formations, world-systems analysis naturally invites futurological speculations. Given looming global biospheric disruption, world-systems forecasters are scarcely immune from the dark mood pervading futurology more generally. Yet the world-systems brand of futurology brings to the table a distinctive conceptual apparatus. World-systems scholars Giovanni Arrighi, Christopher Chase-Dunn and Hiroko Inoue, and Immanuel Wallerstein, respectively, recognize that the global ecological crisis is sharpening into focus during a peculiar moment in the life of the capitalist world-economy, rather than in a temporal vacuum. An *entire range* of structural conditions underpinning the past reproduction of the capitalist world-economy are now eroding, portending the dissolution of this eminently *historical* social system and its eventual replacement by an unknown successor (or successors). While some of these diminishing conditions are indeed "environmental" in character, and all of these conditions dialectically intertwine with the explicitly environmental ones, others are analytically separable. Perhaps most notably, the hegemonic leadership cycle, which has escorted the capitalist world-economy through several rounds of restructuring and renewal, appears to be grinding to a halt. Thus, the activity of world-systems futurologists partaking in "end times" thought nowadays has origins and dynamics that are far less simple than a presently fashionable (but nonetheless justifiable) eco-apocalypticism.

In surveying the leading world-systems perspectives on the structural impediments to the reproduction of the capitalist world-economy, Arrighi's *The Long Twentieth Century* furnishes an indispensable starting point (Arrighi 1994). Here Arrighi makes a series of astute observations. One, with each passing "systemic cycle of accumulation," the capitalist world-economy agglomerates new layers of institutional complexity and acquires a higher degree of institutional interconnectivity (Arrighi 1994: 32–33). Two, concomitant with the increased steering requirements of an increasingly complex and interconnected global capitalist geo-economy, each successive hegemonic power has been equipped with greater territorial scale and organizational resources than its predecessor. Three, each successive hegemonic power's scale and resource endowment has not only been greater than that of its predecessor in absolute terms, but greater *relative* to the (enlarging) spatial breadth and (growing) material depth of the global system in its entirety (Arrighi 1994: 14, 32–33). And four, no singular state possesses (or is ever likely to possess) the enormous concentration of both liquid capital and coercive capacity—or the colossal size—necessary to lead the capitalist world-economy through another systemic cycle on "new and enlarged foundations" as US hegemony subsides. The upshot of these structural obstacles to heretofore "normal" hegemonic transition is the remaking of the capitalist world-economy into something fundamentally different, be it benign or malevolent (Arrighi 1994: 6, 79).

Through its marshalling of concepts such as relative territorial scale, Arrighi's predominantly geopolitical account of the structural barriers to modern world-systemic reproduction provides some blocks for building a bridge toward more socio-ecological accounts, although these building blocks are by themselves inadequate. While their geopolitical accounts are less multi-faceted than Arrighi's, Chase-Dunn, and Inoue and especially Wallerstein take additional measures toward building this bridge. Like Arrighi, Chase-Dunn, and Inoue identify a long-range trend toward spatially bigger core capitalist hegemons, as well as a parallel trend toward more numerous and more effective multilateral governance bodies and transnational social movements (Chase-Dunn and Inoue 2010). However, rather than emphasizing how the upward scaling tendency of successive capitalist hegemons threatens to shatter the essential framework of the modern world-system, Chase-Dunn and Inoue explore how the interactive synergy between social movements and multilateral governance promise to transform this scaling tendency into a substantively rational global democracy. The socio-ecological constraints to world-systemic reproduction (namely the costs of runaway climate change and hydrocarbon fuel shortages) enter the picture primarily as barriers that only a capitalist world-economy reconstituted as a global democracy can surmount.

Wallerstein, on the other hand, apprehends environmental challenges not so much as problems to be solved by a future world order, but moreso as gravediggers of the current one (Wallerstein 2000). As part of a broader assessment of the mounting pressures on world accumulation, Wallerstein contends that capitalist exploitation of nature as if it were a "free gift" (to use Marx's words) can no longer proceed as it has for centuries (Wallerstein 2000: 445–46). Precisely because it has swelled to ecologically unsustainable proportions, the capitalist world-economy will either unravel amid severe resource depletion and eco-systemic degradation, or these problems will inspire a new model of resource conservation and environmental remediation. But such a model would ultimately deprive the accumulation process of the profit margins that are the lifeblood of the capitalist world-economy's very continuation. In a related discussion, Wallerstein argues that the spread of capitalist commodity production will soon meet its geographical limit, entailing another squeeze on surplus value generation as the cheap labor reserve of the world's unproletarianized rural population starts to dry up (Wallerstein 2000: 446–47). In this discussion, as well as in his tracing the unsustainability of the capitalist world-economy to its contemporary ecological footprint, Wallerstein enlists concepts also found in Arrighi's analysis, such as relative size and scalar reach—but in permutations different from Arrighi. Despite these differences, Wallerstein's and Arrighi's respective views on structural limits to the capitalist world-economy's reproduction are far from mutually exclusive, and await a transcendental synthesis.

References

Arrighi, G. (1994) *The Long Twentieth Century*, London and New York: Verso Press.

Chase-Dunn, C. and Inoue, H. (2010) "Accelerating global state formation and global democracy", paper presented at workshop on Present Futures and Future Presents: World State Scenarios for the 21st Century, Skagen, Denmark, June.

Wallerstein, I. (2000) *The Essential Wallerstein*, New York: New Press.

12.7

Energy use and world-systems dynamics

Kirk S. Lawrence

Energy use—the capture, control, and transformation of energy containing resources—has been a fundamental part of the story of evolution in both animate and inanimate systems. Indeed, energy use has propelled the formation and differentiation of societies and is the source of the dual global crises of anthropogenic climate change attributable to the burning of fossil fuels and the conflict and competition over diminishing supplies of those fuels. It is now becoming more apparent—as it had for Malthus—that energy use both enables and constrains societal- and system-level evolution and that the need for an equitable and sustainable relationship with energy will continue to be one of the most important issues of our time.

Studies of these energy-based dynamics, then, are essential to an understanding of intersocietal dynamics informed by world-systems research, and there is a growing body of work toward that end. One broad area considers the role of energy use in the evolution of a hierarchy of countries in the modern world-system. Bruce Podobnik's (2006) work on energy shifts is exemplary in this regard, as he describes how the ability to dominate the control and transformation of primary energy sources can fuel political and economic power and hegemony; coal did this for the British in the nineteenth century and oil for the United States in the twentieth (see also Bunker and Ciccantell 2005). This should be of interest for world-systems analysis, since the world has reached or will soon be reaching the peak of oil production, which will almost certainly lead to hegemonic competition and otherwise destabilize the interstate system as oil is the most widely-used, has the greatest flexibility for use, and has the highest energy intensity of major commercial sources currently utilized (cf. Abramsky 2010; Smil 2008).

The positive relationship between the amount of energy a country uses and its position in the world-system relative to other countries has been demonstrated in numerous studies (cf. Burns et al 1997; Prew 2010; Roberts et al 2003). A similar result is found for the level of militarization, also an indicator of a country's power, where a "treadmill of destruction" exists as the militaries are heavy consumers of fossil fuels that contribute to climate change (Jorgenson et al 2010).

Another area of study includes modernization theory's assumption of an "environmental Kuznets curve" in which environmental cleanliness will improve after a country reaches a certain level of development. This has found little support in tests on recent data (York and Rosa 2003); instead, efficiency gains are often translated into greater energy use as suggested by the "Lauderdale and Jevons' paradoxes" (Foster et al 2010).

The world-systems hierarchy is marked by differences in the type and levels of energy use and emissions. Core countries are fairly differentiated, as some core countries such as the United

Kingdom and Japan have seen their energy use start to level off as they have become more service-based, while others, such as France, rely on nuclear fission for the majority of their energy needs. In the industrializing countries in the semiperiphery, ascent in the world-system can be traced to increases in energy use (and the associated emissions); indeed, this is a key part of China's hegemonic challenge. The semiperiphery has seen energy and emissions increases but at much lower levels than the core and semiperiphery, with correspondingly lower levels of economic development and ecological footprints (cf. Grimes and Kentor 2003; Jorgenson 2009; Lawrence 2009; Roberts and Parks 2007).

References

Abramsky, K. (ed.) (2010) *Sparking a Worldwide Energy Revolution: Social Struggles in the Transition to a Post-Petrol World*, Oakland, CA: AK Press.

Bunker, S.G. and Ciccantell, P.S. (2005) *Globalization and the Race for Resources*, Baltimore, MD: Johns Hopkins University Press.

Burns, T., Davis, B., and Kick, E. (1997) "Position in the World-System and National Emissions of Greenhouse Gases," *Journal of World-Systems Research*, 3(3): 432–66.

Foster, J.B., Clark, B., and York, R. (2010) *The Ecological Rift: Capitalism's War on the Earth*, New York: Monthly Review Press.

Grimes, P. and Kentor, J. (2003) "Exporting the Greenhouse: Foreign Capital Penetration and CO_2 Emissions 1980–96," *Journal of World-Systems Research*, 9(2): 261–75.

Jorgenson, A.K. (2009) "The Sociology of Unequal Exchange in Ecological Context: A Panel Study of Lower Income Countries, 1975–2000," *Sociological Forum*, 24(1): 22–46.

Jorgenson, A.K., Clark, B., and Kentor, J. (2010) "Militarization and the Environment: A Panel Study of Carbon Dioxide Emissions and the Ecological Footprints of Nations, 1970–2000," *Global Environmental Politics*, 10(1): 7–29.

Lawrence, K.S. (2009) "The Thermodynamics of Unequal Exchange: Energy Use, CO_2 Emissions, and GDP in the World-System, 1975–2005," *International Journal of Comparative Sociology*, 50(3–4): 335–59.

Podobnik, B. (2006) *Global Energy Shifts: Fostering Sustainability in a Turbulent Age*, Philadelphia, PA: Temple University Press.

Roberts, J.T. and Parks, B.C. (2007) *A Climate of Injustice: Global Inequality, North-South Politics, and Climate Policy*, Cambridge, MA: MIT Press.

Roberts, J.T., Grimes, P., and Manale, J.L. (2003) "Social Roots of Global Environmental Change: A World-Systems Analysis of Carbon Dioxide Emissions," *Journal of World-Systems Research*, 9(2): 277–315.

Smil, V. (2008) *Energy in Nature and Society: General Energetics of Complex Systems*, Cambridge, MA: MIT Press.

York, R. and Rosa, E.A. (2003) "Key Challenges to Ecological Modernization Theory," *Organization and Environment*, 16(3): 273–88.

13. The environment

13.1

Single and composite sustainability indicators in comparative sociology

Philipp Babcicky

Introduction

Over the past decade, scholars have become increasingly interested in the environmental sustainability of the capitalist world-system. The empirical literature has developed a number of sustainability indicators to measure specific societal and ecological aspects that are seen to determine a country's environmental health and its sustainable development. Certainly, it is a challenging and indeed a controversial exercise to quantify national sustainability, but if no attempts are made, social scientists would not be able to confirm or reject existing theories based on empirical evidence, or, as Böhringer and Jochem (2007: 1) put it, " … an issue that can not be clearly measured will be difficult to improve."

While GDP per capita and other world-system position measures have been developed to manage economies, strategies for sustainable development also require credible, reliable, and legitimate statistical indicators comprising environmental, economic, and social dimensions. Sustainability indicators are seen as crucial tools to address policy issues, and to evaluate how countries are performing in terms of sustainability targets. These measures show how countries in the core differ from non-core countries in specific areas such as CO_2 emissions, and in a broader sense with respect to sustainable development. While compiled indicators measure specific issues, world-system analysis also relies on indicators that reflect essential system processes and inter-relationships, such as material flows, trade patterns, or export-dependency.

System processes and inter-relationships do not only reflect an increasing globalization of the world-economy but also indicate that globalization affects our environment ever more strongly. Figure 13.1.1 is derived from Pezzoli (1997: 573) and describes the main characteristics of the world-system. It shows how the world-system is inherently interconnected with our environmental system. The world-economy is described as *dynamic* (periods of growth, decline, crisis, and restructuring) and *uneven* (costs and benefits are unequally distributed). Moreover, the system is increasingly *international* and *interdependent* (globalized economic activities). As a consequence, global economic activities have also become increasingly *ecologically* interdependent.

One of the consequences of this development is a growing global economy that puts ever more pressure on the ecosystem from which it draws its natural resources. To guide and manage these economy-environment interactions, we initially need to understand how they function. One way

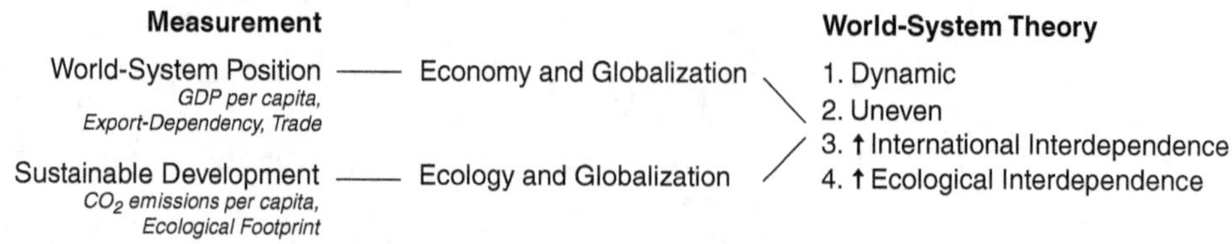

Figure 13.1.1 Measuring economy-ecology interdependence in the world-system

to account for this is to estimate a country's position in the world-system and to relate it to its environmental performance, quantified by sustainability measures. Here, Pezzoli's model has been extended to show how the measurement of the world-system position and sustainable development integrate into the world-system theory (Figure 13.1.1).

Measuring sustainable development is an important component in world-system analysis, and sustainability indicators lie at the heart of our understanding of how core and non-core countries can pursue environmentally sound strategies. Recently, researchers have shown an increased interest in establishing relationships between a country's position in the world-system and its environmental performance (i.e., Pezzoli 1997; Roberts et al 2003). To investigate and to eventually prove such relationships, it is crucial for empirical world-system analysts to choose the most adequate indicators of sustainable development. Therefore, a sound knowledge of the different types of sustainability indicators is essential.

With respect to the growing importance of sustainability measures in world-system analysis, this essay outlines the most widely used environmental and sustainability indicators. It begins by laying out the fundamentals of sustainability measurement, and then goes on to the two different types of sustainability measures: *single* and *composite* indicators. Finally, implications for world-system analysis and sustainability measurement are drawn.

Measuring progress toward sustainability

The meaning of sustainability is far from clear-cut. However, despite differences, there seems to be general agreement on a generic definition as proposed by the World Commission on Environment and Development (WCED 1987: 51): "… sustainable development seeks to meet the needs and aspirations of the present without compromising the ability to meet those of the future." Moreover, it is argued that the term sustainability has "… ideological, and political content as well as ecological and economic content" (Pezzoli 1997: 550). Throughout this chapter, the term sustainable development will, therefore, refer to development that includes environmental, economic, and social dimensions. Thus, a sustainability indicator can be defined as a one-dimensional metric to quantify country-specific information with respect to environmental, economic, and social conditions.

It was only at the end of the twentieth century that the need for environmental indicators was articulated internationally (UN 1992) and when it became clear that economic measures such as GDP per capita can by no means adequately reflect a country's environmental performance. This is one of the reasons why the field of sustainability measurement has grown so rapidly in the last 20 years, and why sophisticated methods have been developed to support environmental theories with quantitative data. Currently, more than 500 sustainability indicators provide researchers and policy-makers with empirical data to bridge socioeconomic and environmental issues (Böhringer and Jochem 2007: 1).

To capture national sustainable development, the environmental literature has derived two different types of sustainability measures: *single* and *composite* indicators. While the former are supposed to measure specific environmental issues such as the emission of greenhouse gases, the latter aim to comprise the environmental, economic, and social dimensions of sustainability. World-system analysts who wish to explore how sustainable development is tied to internal and external conditions are required to pick the most effective type of indicator for their research. The choice of appropriate indicators should be done carefully and, depending on the purpose, analysts can rely on either single or composite indicators.

Single sustainability indicators

The purpose of single sustainability indicators is to track changes in the quality and condition of air, water, land, ecological systems, and their resident flora and fauna. These measures inform analysts about the geographical and temporal trends of environmental conditions and situations. A number of such indicators are used in world-system research to arrive at a better understanding of how the political, social, and economic development of a given place affects its environment. Since GDP per capita is found to be a simple proxy for a country's position in the world-system (Babones 2005), a number of cross-national analyses rely on this measure.

There is a large volume of studies describing the relationship between economic development and sustainability. Economic development has been identified as a major contributing factor with regard to the environmental performance of countries. Grossman and Krueger (1995), for example, use panel data from the Global Environmental Monitoring System to examine the impact of affluence on air pollution (*sulphur dioxide, smoke, and heavy particles*) and the contamination of river basins by *fecals* and *heavy metals*. Torras and Boyce (1998) use a similar approach and further include social dimensions such as literacy and inequality. The pollutant *nitrogen dioxide* is used in a welfare analysis by Welsch (2007), in which he investigates the relationship between well-being and pollution. The economic development of core-countries is often seen to be based on the exploitation (economically and environmentally) of poor countries. This assumption is backed up by Shandra et al (2009) who show how *deforestation* is caused by forestry export flows from peripheral countries to core-countries. Links between a country's economic prosperity and its environmental performance are also made in studies that include other single indicators such as *fertilizer* and *pesticide consumption* (Longo and York 2008) and emission levels of *ozone-depleting substances* (Rosa et al 2004).

With growing concerns about climate change and global warming, a large body of literature has emerged that focuses on the driving forces of greenhouse gas emissions. Much attention has been paid to *carbon dioxide* (CO_2) *emissions* as a single environmental indicator. Since carbon dioxide is considered to be the main anthropogenic greenhouse gas that accounts for 73 percent of the total greenhouse gas emissions (ILO 2009: 99), a number of studies aim to determine the social, political, and economic causes of carbon dioxide emissions. Prew (2010) identifies a strong and positive relationship between the position of a nation in the world-system and levels of carbon dioxide emissions. This suggests that core-countries are still the greatest polluters, although they usually have widespread energy efficient technologies in place, compared to many semiperipheral and peripheral countries. Another world-system related measure is picked up by Ravallion et al (2000) who find that higher inequality—between and within countries—is associated with lower carbon emissions. Different levels of carbon emissions in core and non-core countries are also found to depend on affluence (York 2003; York et al 2003b) and other structural factors such as technology (Dietz and Rosa 1997), population (Rosa et al 2004), or ecological efficiency (Fisher and Freudenburg 2004).

Besides carbon dioxide emissions, the greenhouse gas *methane* (CH_4) has just recently attracted greater attention. Methane is considered to be the second largest contributor to global warming, accounting for 17 percent of all greenhouse gases (ILO 2009: 99). Methane emissions are typically higher in core-countries, since their economic status allows them to consume higher levels of energy (Burns et al 1997; Chase-Dunn 1989). This assumption is backed by Jorgenson (2006) who demonstrates that economic development and foreign direct investment lead to increased methane emissions. However, besides carbon and methane emission rates, other single environmental indicators such as *nitrous oxide*, *chlorofluorocarbons (CFCs)*, and *halons* currently only enjoy minor attention in sustainability research.

One significant weakness of single indicators is that they cover relatively few dimensions of sustainable development, and fail to address sustainability in a broader social, economic, and political sense. It is thus crucial to focus on the balance between the natural and the human environment, and to emphasize that these systems are *multidimensional* and characterized by different environmental, economic, and social dimensions (Mayer et al 2004; Pezzoli 1997). Although single indicators have proven to be highly relevant to word-system analysis, they do not fully meet these requirements, as observed or estimated data usually address only one particular characteristic of the system (Mayer 2008). However, their widespread use in cross-national studies to determine the environmental costs that core-countries impose on the periphery and on the ecosystem in general cannot be denied.

Composite sustainability indicators

When a variety of indicators are compiled into a single figure, this measure shows, at a glance, a "… simplified, coherent, multidimensional view of a system" (Mayer 2008: 279). It is often argued that single indicators cannot adequately capture multidimensional concepts such as sustainable development. Therefore, composite sustainability indicators are based on theoretical frameworks that allow single indicators to be selected, combined, and weighted to better reflect the multi-dimensional nature of sustainability. Today, the use of composite indicators is widespread, and they allow analysts to: (a) monitor and evaluate sustainable development and environmental pressure, (c) anticipate and assess conditions and trends, (c) support policy making, (d) formulate strategies, and (e) determine the structural factors which are most responsible for driving the system (Mayer 2008; Pillarisetti and van den Bergh 2010; Singh et al 2009).

The first impetus for the development of composite environmental indicators was given in Chapter 40 of Agenda 21, at the United Nations Conference on Environment and Development in Rio de Janeiro in 1992. The *Agenda for the 21st Century* emphasized sustainable development and officially called for the development of sustainability indicators:

> Methods for assessing interactions between different parameters (environmental, demographic, social and developmental) are not sufficiently developed or applied. Indicators of sustainable development need to be developed to provide solid bases for decision-making at all levels and to contribute to a self-regulatory sustainability of integrated environment and development systems.
>
> *(UN 1992: 346)*

The proposal called for the necessity to measure and evaluate national development by including ambient, economic, social, ethical, and cultural aspects (Siche et al 2008). Since then, environmental research has developed various kinds of composite indicators to track, evaluate and monitor both environmental sustainability and the level of environmental pressure, and ecological degradation (Pillarisetti and van den Bergh 2010).

Raw national accounting measures such as GDP do not capture the multidimensional nature of sustainable development, and fail to address certain critical dimensions such as the sustainability of production and consumption (van den Bergh 2007). For this reason, Pearce and Atkinson (1993) proposed the *Genuine Savings Indicator (GSI)*. Also known as *Adjusted Net Savings*, the GSI is an economy-based indicator of "weak sustainable development." It defines wealth more broadly than current measures, and aims to represent the value of the net change of assets, including produced assets, natural resources, environmental quality, human resources, and foreign assets. The index is calculated by subtracting the depreciation of human-originated and natural capital from income savings and education expenditures (Mayer 2008). All underlying assets and variables are monetized and, after adding up, a positive indicator value suggests a positive transition toward sustainability, whereas a negative value reflects unsustainable consumption (Atkinson and Hamilton 2002; Hamilton and Dixon 2003; Lawn 2007). While the World Bank interprets this indicator as a measure of a country's development toward sustainability, Pillarisetti (2005) remains critical, and suggests the need to refrain from mixing up natural capital with physical and human capital. Also, Dietz and Neumayer (2007), Lawn (2007), Hueting and Reijnders (2004), and Hamilton and Dixon (2003) all question whether Genuine Savings can actually reflect changes in total wealth per capita, considering issues such as missing data, resources which had been depleted in the past due to direct importation of resources from other countries, and changing consumption patterns.

A further market and economy-based sustainability indicator is the *Index of Sustainable Economic Welfare* which is also known as the *Genuine Progress Indicator* (Böhringer and Jochem 2007: 5). The indicator adjusts net national product for loss of welfare caused by environmental and social issues. The Index of Sustainable Economic Welfare consists of 20 sub-indicators (Singh et al 2009) of which seven reflect a *growth* in welfare (e.g., public expenditures on health and education, net capital growth, net change in international position, adjusted consumer expenditures) and 13 indicators represent a *reduction* of welfare (e.g., costs of commuting, of water, air and noise pollution, of climate change and ozone depletion, loss of biodiversity). In the Genuine Savings Index, all values are monetized and aggregated by direct addition to arrive at a single figure.

One of the most popular ecosystem-based indicators is probably the *Ecological Footprint (EF)*, which was originally introduced by Wackernagel and Rees (1996). The index is published every two years by the WWF in the Living Planet Report (WWF 2006). In short, the EF measures the demand for natural resources, and provides information on whether " … nations are living within or beyond their biological capacity" (Pillarisetti and van den Bergh 2010: 52). This indicator not only accounts for the use of land and water, but also for the ability of the ecological system to absorb the waste that a country generates. The strength of this indicator is that it can be applied at different levels. The sustainability of individuals, cities, countries, and regions, as well as the sustainable development of humanity as a whole, can be measured by comparing the ecological demand against the available supply of natural resources (Siche et al 2008).

One important feature of the Ecological Footprint is that "… resources used for the production of goods and services that are exported are counted in the Ecological Footprint of the country where the goods and services are *ultimately* consumed" (Kitzes 2007: 384). This feature allows world-system analysts to empirically assess how core-countries differ from non-core countries in terms of their natural resource consumption, no matter where the resources have been extracted. The Ecological Footprint is, therefore, an excellent indicator for examining how a country's position in the world-system affects the pressure it puts on the environment.

The Ecological Footprint has undergone a number of refinements in the past, and has become a popular indicator that enjoys widespread application in environmental research. Typical world-system measures such as economic development have been proven to have a significant impact on the Ecological Footprint of a country (Rosa et al 2004; York et al 2003a). Besides economic affluence, the Ecological Footprint has also been linked to other world-system position proxies such as export flows and military expenditure or structural factors like inequality (White 2007), urbanization, or education (Jorgenson 2003, 2005; Jorgenson and Burns 2007; Jorgenson and Clark 2009).

A broader account of sustainability measurement is taken by the *Environmental Sustainability Index (ESI)*, which incorporates environmental, socioeconomic, and institutional dimensions. The ESI is confirmed by many as the most comprehensive approach to the measurement of sustainable development (Whitford and Wong 2009: 191). Like a number of composite indicators (e.g., the Human Development Index), the aggregation of a multidimensional phenomenon into one single measure is often controversial. However, the development of novel composite environmental indicators is still an on-going process, and policy-makers are more than ever dependent on meaningful measures that incorporate not only environmental but also social, political, and economic dimensions. Besides the Ecological Footprint, the ESI (Esty et al 2005) is considered by the scientific community to be a viable tool for the evaluation and measurement of the sustainable development of nations (Jha and Murthy 2003; Morse 2004; Sutton and Costanza 2002).

In contrast to the Ecological Footprint, the ESI aims to inform interested parties about ecological, economic, and social dimensions (Siche et al 2008: 631), thereby providing policy-makers with a more comprehensive picture of sustainability. Nevertheless, the ESI has not escaped criticism and as a result, on-going debates are to be found in both academia and politics (Jha and Murthy 2003; The Ecologist 2001; Morse and Fraser 2005). However, it is crucial to acknowledge the role and widespread usage of the ESI as a policy-making instrument and as a tool for researchers with regard to conducting studies in the field of environmental research. In general, the ESI does not attempt to make aggregate judgments about the sustainability of the entire planet, but to enable and facilitate cross-country comparisons of environmental progress.

The ESI is used in a variety of cross-national studies that aim to explore how economic and social factors determine different levels of sustainability. Whitford and Wong (2009) use the ESI to determine the political and social foundations of environmental sustainability, and to test hypotheses related to economic development, religion, and demographics; an approach that has been criticized by York (2009). Besides studies that aim to explore the relationship between a country's structural position in the world-system (i.e., GDP) and its environmental sustainability (Morse 2008; Park et al 2007), the ESI is also used to link sustainability outcomes with environmental actions (Freymeyer and Johnson 2010), life satisfaction (Bonini 2008), and happiness and quality of life (Zidansek 2007). Further studies include the use of the ESI to gain insights into the relationship between sustainability and political regimes (Bush 2009), education, national culture (Park et al 2007), and corruption (Morse 2008).

Among the difficulties in constructing a composite indicator such as the ESI are methodological issues and the selection of the underlying components (Pillarisetti and van den Bergh 2010; Siche et al 2008). A serious weakness of composite indicators is that they may have a potential bias in favor of their originators and that "… any such attempt inevitably reflects the views of those creating such tools and only produce highly contested notions of 'reality'" (Morse and Fraser 2005: 625).

A potential bias may be due to the fact that countries outside the core do not always have the opportunity to fully participate in the development of such indicators. The consequence of this is that core-countries perform well in sustainability rankings, while periphery and semiperiphery countries rank at the bottom of the list. Indicators such as the ESI make use of a considerable number of variables that reflect social and institutional variables—all characteristics with regard to which core-countries clearly outperform non-core countries. Future composite sustainability indicators should therefore place increased emphasis on a more balanced choice of variables, and consider different weighting schemes to adjust for potential bias.

Implications and conclusion

This essay has given an account of, and the reasons for, the increased use of sustainability indicators in current world-system analysis. It has laid out the most widely used indicators, and explained the characteristics of single and composite measures. The various sustainability indicators allow researchers to investigate how the world-system and the environmental system are related, and furthermore enable policy-makers to design environmentally sound policies in core-countries as well as in countries outside of the core.

Depending on the purpose, world-system analysis draws on either single or composite indicators, although both types have strengths and weaknesses. Single indicators are generally known to be based on clear-cut and unambiguous definitions, to be less subjective and therefore more widely accepted. Their weakness is that they solely reflect one specific dimension of sustainable development (e.g., greenhouse gas emissions, deforestation, water pollution). In contrast, composite indicators reflect multiple aspects of sustainability and usually comprise environmental, economic, and social dimensions. However, it is also argued that this broad approach does sometimes lump together too many aspects into one concept, which results in a lack of clarity. For world-system analysts, it is therefore crucial to be aware of the different qualities of indicators in order to make appropriate choices for their own research.

As the global economy is growing steadily, and large semiperiphery countries such as China, India, or Brazil show remarkable growth rates, we need to focus at the very least on increased energy efficiency and on a reduction of natural resource exploitation. The pressure on the environment might be reduced through technological improvements, government regulations, and smart policies (Roberts et al 2003). Since the core-countries are already the greatest polluters, and non-core countries are "catching up," we need to be sure to pursue a globally-directed effort that involves countries in the core as well as countries outside of the core. In addition, the importance of the semiperiphery must not be ignored. Specific measures such as road and equipment improvement, the use of non-fossil fuel energy sources, conservation, and recycling, should especially be considered for countries in the semiperiphery (Roberts et al 2003: 304). After all, sustainability indicators will show if the efforts to limit environmental impacts will be proven to be successful.

The implication is clear: world-system research requires reliable, credible, and legitimate sustainability indicators to establish links between a country's position in the stratified world-economy and its sustainable development. Only when these links are made and proven to be strong will it become clear which attempts will be required to better manage an increasingly globalized economy-environment interaction.

References

Atkinson, G. and Hamilton, K. (2002) "International trade and the 'ecological balance of payments'", *Resources Policy*, 28: 27–37.

Babones, S. J. (2005) "The country-level income structure of the world-economy", *Journal of World-Systems Research*, 11: 29–55.

Böhringer, C. and Jochem, P. E. P. (2007) "Measuring the immeasurable—a survey of sustainability indices", *Ecological Economics*, 63: 1–8.

Bonini, A. N. (2008) "Cross-national variation in individual life satisfaction: effects of national wealth, human development, and environmental conditions", *Social Indicators Research*, 87: 223–36.

Burns, T., Davis, B. and Kick, E. (1997) "Position in the world-system and national emissions of greenhouse gases", *Journal of World-Systems Research*, 3: 432–66.

Bush, B. J. (2009) "Democracy, autocracy, and hybridity: a cross-national analysis of political regimes, environmental sustainability, and performance", Ph.D. dissertation, Norhtern Arizona University.

Chase-Dunn, C. (1989) *Global formation: structures of the world economy*, Lanham, MD: Rowman & Littlefield.

Dietz, S. and Neumayer, E. (2007) "Weak and strong sustainability in the SEEA: concepts and measurement", *Ecological Economics*, 61: 617–26.

Dietz, T. and Rosa, E. A. (1997) "Effects of population and affluence on CO_2 emissions", *Proceedings of the National Academy of Sciences of the United States of America*, 94: 175–79.

The Ecologist (2001) "Keeping score", *The Ecologist*, 31: 44–47.

Esty, D. C., Levy, M., Srebotnjak, T. and de Sherbinin, A. (2005) "2005 Environmental Sustainability Index: benchmarking national environmental stewardship", New Haven, CT: Yale Center for Environmental Law & Policy.

Fisher, D. R. and Freudenburg, W. R. (2004) "Postindustrialization and environmental quality: an empirical analysis of the environmental state", *Social Forces*, 83: 157–88.

Freymeyer, R. H. and Johnson, B. E. (2010) "A cross-cultural investigation of factors influencing environmental actions", *Sociological Spectrum: Mid-South Sociological Association*, 30: 184–95.

Grossman, G. M. and Krueger, A. B. (1995) "Economic growth and the environment", *The Quarterly Journal of Economics*, 110: 353–77.

Hamilton, K. and Dixon, J. A. (2003) "Measuring the wealth of nations", *Environmental Monitoring and Assessment*, 86: 75–89.

Hueting, R. and Reijnders, L. (2004) "Broad sustainability contra sustainability: the proper construction of sustainability indicators", *Ecological Economics*, 50: 249–60.

ILO (2009) "World of work report 2009: The global jobs crisis and beyond", Geneva: International Labour Organization.

Jha, R. and Murthy, K. V. B. (2003) "A critique of the Environmental Sustainability Index", Working Paper, Australian National University and The University of Delhi.

Jorgenson, A. K. (2003) "Consumption and environmental degradation: a cross-national analysis of ecological footprint", *Social Problems*, 50: 374–94.

Jorgenson, A. K. (2005) "Unpacking international power and the Ecological Footprints of nations: a quantitative cross-national study", *Sociological Perspectives*, 48: 383–402.

Jorgenson, A. K. (2006) "Global warming and the neglected greenhouse gas: a cross-national study of the social causes of methane emissions intensity", *Social Forces*, 84: 1791, 1793.

Jorgenson, A. K. and Burns, T. J. (2007) "The political-economic causes of change in the Ecological Footprints of nations, 1991–2001: A quantitative investigation", *Social Science Research*, 36: 834–53.

Jorgenson, A. K. and Clark, B. (2009) "The economy, military, and ecologically unequal exchange relationships in comparative perspective: A panel study of the Ecological Footprints of nations, 1975–2000", *Social Problems*, 56: 621–46.

Kitzes, J. (2007) "Current methods for calculating national Ecological Footprint accounts", *Science for Environment & Sustainable Society*, 4: 1–9.

Lawn, P. (2007) "A stock-take of green national accounting initiatives", *Social Indicators Research*, 80: 427–60.

Longo, S. and York, R. (2008) "Agricultural exports and the environment: A cross-national study of fertilizer and pesticide consumption", *Rural Sociology*, 73: 82–104.

Mayer, A. L. (2008) "Strengths and weaknesses of common sustainability indices for multidimensional systems", *Environment International*, 34: 277–91.

Mayer, A. L., Thurston, H. W. and Pawlowski, C. W. (2004) "The multidisciplinary influence of common sustainability indices", *Frontiers in Ecology and the Environment*, 2: 419–91.

Morse, S. (2004) *Indices and indicators in development: an unhealthy obsession with numbers?*, Trowbridge: Cromwell Press.

Morse, S. (2008) "On the use of headline indices to link environmental quality and income at the level of the nation state", *Applied Geography*, 28: 77–95.

Morse, S. and Fraser, E. D. G. (2005) "Making 'dirty' nations look clean? The nation state and the problem of selecting and weighting indices as tools for measuring progress towards sustainability", *Geoforum*, 36: 625–40.

Park, H., Russell, C. and Lee, J. (2007) "National culture and environmental sustainability: a cross-national analysis", *Journal of Economics and Finance*, 31: 104–21.

Pearce, D. W. and Atkinson, G. D. (1993) "Capital theory and the measurement of sustainable development: an indicator of 'weak' sustainability", *Ecological Economics*, 8: 103–08.

Pezzoli, K. (1997) "Sustainable development: a transdisciplinary overview of the literature", *Journal of Environmental Planning and Management*, 40: 549–74.

Pillarisetti, J. and van den Bergh, J. (2010) "Sustainable nations: what do aggregate indexes tell us?", *Environment, Development and Sustainability*, 12: 49–62.

Pillarisetti, J. R. (2005) "The world bank's 'Genuine Savings' measure and sustainability", *Ecological Economics*, 55: 599–609.

Prew, P. (2010) "World-economy centrality and carbon dioxide emissions: a new look at the position in the capitalist world-system and environmental pollution", *Journal of World-Systems Research*, 16: 162–91.

Ravallion, M., Heil, M. and Jalan, J. (2000) "Carbon emissions and income inequality", *Oxford Economic Papers-New Series*, 52: 651–69.

Roberts, J. T., Grimes, P. E. and Manale, J. L. (2003) "Social roots of global environmental change: a world-systems analysis of carbon dioxide emissions", *Journal of World-Systems Research*, 9: 277–315.

Rosa, E. A., York, R. and Dietz, T. (2004) "Tracking the anthropogenic drivers of ecological impacts", *Ambio*, 33: 509–12.

Shandra, J. M., Leckband, C. and London, B. (2009) "Ecologically unequal exchange and deforestation: a cross-national analysis of forestry export flows", *Organization & Environment*, 22: 293–310.

Siche, J. R., Agostinho, F., Ortega, E. and Romeiro, A. (2008) "Sustainability of nations by indices: comparative study between Environmental Sustainability Index, Ecological Footprint and the Emergy Performance Indices", *Ecological Economics*, 66: 628–37.

Singh, R. K., Murty, H. R., Gupta, S. K. and Dikshit, A. K. (2009) "An overview of sustainability assessment methodologies", *Ecological Indicators*, 9: 189–212.

Sutton, P. C. and Costanza, R. (2002) "Global estimates of market and non-market values derived from nighttime satellite imagery, land cover, and ecosystem service valuation", *Ecological Economics*, 41: 509–27.

Torras, M. and Boyce, J. K. (1998) "Income, inequality, and pollution: a reassessment of the Environmental Kuznets Curve", *Ecological Economics*, 25: 147–60.

UN (1992) "Conference on environment & development: Agenda 21", Rio de Janeiro: United Nations.

van den Bergh, J. C. J. M. (2007) "Abolishing GDP", *Tinbergen Institute discussion paper*, Amsterdam: Vrije Universiteit Amsterdam and Tinbergen Institute.

Wackernagel, M. and Rees, W. (1996) *Our Ecological Footprint: reducing human impact on the earth*, Gabriola Island: New Society Publishers.

WCED (1987) *Our common future (Brundtland Report)*, Oslo: UN World Commission on Environment and Development.

Welsch, H. (2007) "Environmental welfare analysis: a life satisfaction approach", *Ecological Economics*, 62: 544–51.

White, T. J. (2007) "Sharing resources: the global distribution of the Ecological Footprint", *Ecological Economics*, 64: 402–10.

Whitford, A. B. and Wong, K. (2009) "Political and social foundations for environmental sustainability", *Political Research Quarterly*, 62: 190–204.

WWF (2006) *Living planet report 2006*, Gland: WWF, Zoological Society of London and Global Footprint Network.

York, R. (2003) "A rift in modernity? Assessing the anthropogenic sources of global climate change with the STIRPAT model", *International Journal of Sociology and Social Policy*, 23: 31–51.

York, R. (2009) "The challenges of measuring environmental sustainability: comment on 'political and social foundations for environmental sustainability'", *Political Research Quarterly*, 62: 205–08.

York, R., Rosa, E. A. and Dietz, T. (2003a) "Footprints on the earth: the environmental consequences of modernity", *American Sociological Review*, 68: 279–300.

York, R., Rosa, E. A. and Dietz, T. (2003b) "STIRPAT, IPAT and ImPACT: analytic tools for unpacking the driving forces of environmental impacts", *Ecological Economics*, 46: 351–65.

Zidansek, A. (2007) "Sustainable development and happiness in nations", *Energy*, 32: 891–97.

13.2

Forests, food and freshwater

A review of world-systems research and environmental impact

Rebecca Clausen and Stefano B. Longo

World-systems analysis provides a theoretical foundation for researching the underlying social causes of environmental impacts. This theoretical approach has offered a wealth of insight into the historical and current relationships between human societies and their environment. Drawing originally from the work of Wallerstein (1974) and Braudel (1972, 1973), the world-systems framework provides a global political economic perspective necessary for understanding the degradation and depletion of crucial environmental systems and resources. This contribution has been significant in the examination of the environmental and social dynamics of deforestation, the developments in the global agri-food system, and water pollution and consumption. This chapter will review the vibrant world-systems literature that relates to these environmental impacts.

Deforestation

The extraction of natural resources from the periphery has long been a part of capitalist development in the core. The global inequality of environmental impacts is especially evident in patterns of deforestation. As a global trend, forest loss disproportionately occurs in the less developed countries. Paradoxically, the highest forest product consumption occurs in the developed nations. While this dynamic is related to the fact that core nations have already undergone extensive deforestation, there are other important global processes that contribute. World-systems theory is a crucial tool for understanding how the unequal ecological impacts of deforestation are related to international inequality in production, consumption, trade, and capital accumulation.

Bunker (1984, 1985) first applied world-systems analysis to understand the driving forces of tropical deforestation. Bunker's research examined the deforestation of the Brazilian Amazon to describe how the modes of resource extraction in the extreme periphery lead to further underdevelopment of societies. He explained that while the processing of most natural resources creates additional value, extreme peripheries such as the Amazon tend to export raw forest products and the creation and realization of profit and benefit occurs in the core (Bunker 1984). This is not a sovereign choice; rather, it is one dictated by the economic ties to the world capitalist system. The global organization of production that favors raw material exports often degrades the natural resources of a nation, undermines future development, and concentrates land-holdings and capital into commercial interests. Bunker's (1984) research inspired world-system theorists to "consider the effects of the exploitation of labor and the exploitation of entire ecosystems as separate but complementary phenomena, both of which affect the development of particular regions" (1053).

Early world-systems analysis of deforestation offered a critical refinement to the claim that population alone drives forest decline. Rudel (1989) tested the Malthusian claim that population growth is a primary driver of deforestation. This research demonstrated that population does have an effect on rates of forest loss; however, in some instances capital investment interacts with local population growth to accelerate deforestation. As one of the first empirical investigations of deforestation, this study prompted further inquiry into socioeconomic processes within the world-system that perpetuate environmental impact.

Burns et al (1994) specified the patterns of deforestation by providing exploratory investigation of the demographic, cultural, and economic factors that may lead to forest decline. The authors claimed that social dynamics leading to deforestation operate differently across the structural settings of the core, semiperiphery, and periphery nations within the world-system. Their results attested to this by finding that deforestation had been the most severe in semiperiphery nations during the latter half of the twentieth century. This was due in part to rural encroachment in these regions, combined with access to advanced tree-cutting machinery and low forestry regulation. Kick et al (1996) continued this line of inquiry, confirming that rates of deforestation tend to vary markedly by a country's position in the world-system. The attention to world-system position as causal indicator of forest loss was a key development in this area of research.

Chew (2001) used deforestation measures over a broad sweep of history as an indicator of civilizations' ecological degradation (3000 BCE–CE 2000). He claimed that for 5000 years, societies have used wood to meet the expansionary dynamics of societies connected through trade linkages, facilitating the process of accumulation in world history. His historical survey concludes, "Trees, therefore, have fueled the socioeconomic transformation of every societal system, and for some have provided the means to become the core centers of the world-system" (2001: 4). Sparked by this early comparative historical work (Bunker 1984; Chew 2001; Rudel 1989), world-systems scholars began to advance empirical, cross-national studies of ecologically unequal exchange.

Within the past decade, world-system analysis of deforestation has produced more sophisticated, theoretically driven models. In 2003, the *Journal of World-Systems Research* sponsored a special issue titled "Globalization and The Environment," edited by Edward Kick and Andrew Jorgenson. This issue signaled the recognition of the important contributions that world-systems research was making to the study of environmental problems in general. In this issue, Burns et al (2003) make a convincing argument that global hierarchies must contextualize world-systems research of deforestation. The authors include boreal, temperate, and tropical forests in their study, and find that social factors most closely associated with deforestation vary by world-systems position.

World-systems analysis of deforestation has had a significant role in the academic debate concerning modernization and the environment. Early research into this area suggested that modernization processes, such as economic growth and urbanization, may have more impact on deforestation than world-systems processes of debt, trade, and foreign capital penetration. Ehrhardt-Martinez et al (2002) expanded this argument further to conclude that rates of deforestation may decrease as ecological modernization advances, with environmental impact being gradually alleviated by continued economic prosperity and global economic integration. These quantitative studies challenged the claims that deforestation was related to world-systems factors and sparked healthy debate and refinement within the field.

Numerous world-systems studies have made convincing arguments contesting the propositions of the ecological modernization perspective. Roberts and Grimes (1997) argue that if there is a point at which continued economic growth begins to decrease forest loss, it is probably quite high and will not be reached by most countries across the world. Numerous other scholars have established robust, empirically driven research programs investigating the relationships between world-systems processes (trade, debt, foreign investment) and patterns of deforestation. Jorgenson (2006a) proposed a structural theory of unequal ecological exchange, integrating theoretical perspectives of world-systems theory and environmental sociology. Unequal ecological exchange suggests, "developed countries with higher levels of resource consumption externalize their consumption-based environmental costs to less developed countries, which increase levels of environmental degradation within the latter" (Jorgenson 2006a: 691). The unequal trade relations are facilitated by the strategies of local elites attracting investors and through the direct control of transnational corporations.

Jorgenson (2006a) confirmed the theory of unequal ecological exchange by analyzing deforestation trends based on a new methodological indicator—the "weighted index of vertical trade." By operationalizing unequal ecological exchange, the weighted index of vertical trade allowed Jorgenson to model environmental impact (i.e., deforestation rates) based on the relative extent to which exports are sent to more developed countries, and the characteristics of those receiving nations. This methodological development has allowed other scholars to investigate the relationships between commodity-specific exports and rates of deforestation (e.g., Austin 2010).

Recent studies on the structural drivers of deforestation attempt to specify previous findings concerning deforestation, population, and economic integration. Jorgenson et al (2007) advanced understandings of the relationship between population and deforestation by disaggregating the effects of total population into the particular characteristics of rural and urban population change. The authors found that growth in rural population increases deforestation, and they relate these changes to larger processes of export-oriented development. Similarly, Jorgenson (2008) contextualized previous work on economic integration by finding that less-developed countries with higher levels of foreign investment exhibit great rates of deforestation. Further, Jorgenson (2010) explains how the foreign investment dependence allows developed countries to treat less developed countries as "supply depots" to satisfy their unsustainable levels of natural resource consumption.

Shandra et al (2008) provide further insight into the relationship between deforestation and global hierarchies by focusing attention on national debt. The authors extend prior work by considering the impact of structural adjustment lending to poor nations, and find that the structural adjustment policy reforms required of the indebted nations (increase exports, liberalize trade, cut government spending) are positively associated with forest loss. This finding has timely and relevant implications as social scientists begin to analyze policy proposals such as "debt-for-nature" swaps. Findings that confirm the negative impacts of structural globalization on forests add further support to the theory of unequal ecological exchange, and contribute to the growing body of "ecostructural" analysis within world-systems research (Jorgenson et al 2007).

In addition to researching the root causes of forest degradation, world-systems theorists also investigate the global mechanisms that may reduce deforestation. These studies add a layer of complexity to demonstrate the contradictory nature of globalization processes in general. As discussed, in many instances, global integration leads to inequalities of environmental impact. World polity hypothesis, however, states that institutions may play important roles in the global processes of world-systems dynamics by mediating environmental decision-making and sparking local social movements to demand forest protection. These institutions include international non-governmental organizations (INGOs), inter-governmental organizations (IGOs), and treaties.

Shandra and colleagues have dedicated much of their research to understanding which aspects of world polity theory may reduce deforestation. Shandra (2007a, 2007b) found that nations that are more embedded in global political relationships of all types (INGOs, IGOs, treaties) have less deforestation. These findings support the world polity theory that "behaviors of nations are shaped by pressures that can be put on them through their ties to a wide array of international organizations (Shandra 2007a: 22). Additionally, Shandra (2007b) found that INGOs are able to be more effective in reducing deforestation in nations with higher levels of democracy. In contrast, repression within a nation interacts with various world-systems measures, increasing deforestation in nations with higher indicators of repression (Shandra 2007c). Shandra and colleagues provide important contextualization of how global integration may invite or deter political institutions to redefine national policy in the service of environmental sustainability.

Agri-food systems

The role of food production and consumption has been another important area of research for world-systems scholars. The research has grown out of a number of historical analyses on the political economy of food production and food systems. Inspired by Wallerstein's seminal works (Wallerstein 1974, 1980), these analyses examine the global division of labor and the global dynamics of power represented in the establishment of the modern world-system and within the production and consumption of agri-food products.

In short, the unequal division of labor that emerged in the modern world-system is reflected in the global division of wealth and power. A close historical examination of food production and consumption opens a window into the social organization of the world-economy. While an analysis of agri-food concerns was initially marginalized in world-systems research by a focus on the expansion of industrialization and factory production, by the 1980s, analyses investigating the inherent power of food (McMichael 2000) and agri-food concerns began to provide essential insight into social organization and modern institutions of political economic and military power.

Mintz's classic work *Sweetness and Power* (Mintz 1985) provided an analysis of the fundamental role of sugar production in making the modern world. Mintz (1985) took on a deep historical study of a single product and unlocked the myriad relationships between the expansion of modern capitalist regimes with culture, colonialism, exploitation of humans, and land in the periphery. The work provides insight into the fundamental nature of agri-food production and consumption processes and the modern organization of world culture, economic processes, and military power.

Studies such as this were concurrent with a wealth of research on the interaction between food production and consumption and the development of the modern world-system. During the 1980s and 1990s, the growth of neoliberal policies and their impacts on the global food systems sparked new research programs in agriculture and food systems (Buttel 2001; Buttel et al 1990). The significant changes that restructured the food system after World War II prompted deep analyses by scholars (McMichael 1994). The sociology of agriculture led by scholars such as Harriet Friedman, William Friedland, Philip McMichael, and others, established numerous historical and systems analyses that drew on neo-Marxist political economy and world-systems research to examine agri-food systems, state-policies, globalization, development and, broadly, social life.

Friedland (1984, 2004) pioneered work in commodity systems and commodity chains, which established a new direction in agri-food studies. This approach presented the networks of production and trade in relation to organizations, labor, the application of science and technology, and distribution systems, resulting in innovative research on globalization and food. A commodity chain analysis looks at the interwoven chains of production within the world-system, rather than discrete nations, as points of production and consumption. That is, they are not contained within political boundaries and should not be analyzed as if they are (Gereffi and Korzeniewicz 1994). As a result, this approach originated a methodology that focused on the production system and commodity production within a global system.

Commodity chain research on food items (e.g., coffee [Talbot 2002, 2004], salmon [Phyne and Mansilla 2003], grain flour [Pelizzon 1994], produce [Goldfrank 1994], or shrimp [Skladany and Harris 1995]) can highlight the social relationships within the commodity production process, as well as between numerous chains, and the interactions within the broader dynamic of the modern world-system. Since the onset of global capitalism, core nations have directed agri-food production. Thus commodity chain research has revealed the hierarchical structure of global agri-food production, consumption, and trade (Bonanno et al 1994; Gereffi and Korzeniewicz 1994; McMichael 1995). Further, these analyses have brought to light the increasing role of transnational corporations and concentration of ownership in the agri-food sector, intensification of operations, and a reducing role of the state in recent decades (Heffernan and Constance 1994; Skladany and Harris 1995; Stringer and Le Heron 2008).

Complimenting the work in commodity chains, the concept of "food regimes" emerged during a similar period (Friedmann 1993; Friedmann and McMichael 1989; McMichael 2009; McMichael and Buttel 1990). This research focused on the changing power structures and development strategies that dominated the world economic order with the growth of capitalist social institutions. Placing agri-food production in a world historical perspective, the conceptual strength of food regime research underlines the role of global power structures in dictating the form and direction of agri-food production and consumption. Further, this analytical perspective has revealed an important crisis and contradiction within the existing global food order and provided a necessary critique to the modern development paradigm advanced in the post-World War II era (Friedmann 1993; McMichael 1994, 2000).

The historical work on food regimes developed a critical analysis of international food systems through examining the convergence of state policy and geo-politics of food and food aid, researching its role within the larger global political economic and military arrangements (McMichael 2009). This approach "links international relations of food production and consumption to forms of accumulation broadly distinguishing periods of capitalist transformation" (Friedmann and McMichael 1989: 95).

In 1993, the Political Economy of the World-Systems (PEWS) conference in Ithaca, New York, focused on food and agriculture systems in the world-economy. Both the commodity chain analysis and the food regime approach influenced a great deal of work in world-systems studies, making a central contribution to the conference (McMichael 1995). The expansion of transnational corporate control of the global agri-food system, increasing power of industrial agriculture and the real and potential consequences for the periphery has been clearly documented by developing analyses informed by a world-systems perspective. A more recent concern associated with this transformation in agri-food production addresses the variety of environmental and health problems, such as the increasing use of synthetic chemicals, soil erosion, climate change, and deforestation, among other issues.

Moore (2000, 2003) has expanded on these concerns using a world-systems perspective along with the metabolic rift approach, hence including a necessary ecological critique. Integrating the works of Wallerstein and Marx, Moore (2000) uses the concept of "systemic cycles of agro-ecological transformations" to develop a conception of the growth of the capitalist world-system with both political economic and ecological transformations. Thus, according to this approach, the modern ecological crisis is rooted in the transition to capitalism during the long sixteenth century, and the social relations of production that took shape. This work and others in the world-systems perspective emphasize the historical significance of the dialectic between nature and capital over the *longue duree.* In doing so, these studies have produced important contributions to our understanding of food production and ecology in the crisis of feudalism and the emergence of capitalism (Moore 2003; Myrdal 2007).

Along with the influential theoretical and historical analyses, world-systems scholars have made important contributions to the study of food and agriculture systems by developing empirical studies, principally through employing cross-national quantitative analyses. For example, these studies have examined the interrelationships between agriculture production, global trade, foreign direct investment, and ecological degradation.

The uses of fertilizers and pesticides have long been known to impact human health and contribute to ecological disruptions. Thus, word-systems researchers have examined the relationships between foreign investment and fertilizer and pesticide consumption (Jorgenson 2007; Jorgenson and Kyukendall 2008). These studies find a positive relationship between the infusion of foreign capital and environmentally harmful practices in periphery nations (Jorgenson 2007; Jorgenson and Kyukendall 2008). Other researchers have found a positive relationship between agricultural export intensity and the consumption of fertilizers and pesticides within a nation (Longo and York 2008). Contrary to modernization theories on the environment that expect expanding international trade and finance to have ecological benefits for nations, these empirical studies found that in general, greater integration into the dynamics of the modern world-system was related to environmentally damaging agricultural practices. This research compliments historical studies by world-systems researchers regarding the increasing use of harmful chemicals, and transfer from the core into the periphery (Frey 1998, 2003).

Freshwater pollution

Access to water is a fundamental necessity for all societies. Water, as a primary source of life, an avenue of transportation, and an input for agriculture production, has shaped and been shaped by the organization of human social life. Scholars working in the world-systems tradition have contributed to documenting the long-term causes and consequences of water resource degradation, and it has been well understood that water is a crucial resource that interacts with a variety of human and environmental concerns (Chew 2001).

Recent cross-national empirical work influenced by the world-systems tradition has begun to unravel the complex role of water use and the effects of water pollution. In a series of studies on macro-structural drivers and consequences of water pollution, world-systems scholars have found important relationships between global political economic processes, water pollution, and infant mortality (Burns et al 2003; Jorgenson and Burns 2004). Specifically, nations that have high levels of export commodity concentration also tend to have high levels of water pollution, and there is a significant relationship between higher levels of organic water pollution and infant mortality rates. This research program examines the ways that nations of the global South have been disproportionately burdened with the impacts of water pollution due to their position in the global economic order (Jorgenson 2006b). As a result, pollution of this essential resource can have devastating consequences for the most vulnerable people in the poorest nations of the world.

Other researchers have begun to examine the ways that socio-structural processes impact water withdrawals, combining a world-systems view with the human ecology tradition in environmental sociology (Longo and York 2009). However, this area of research is still in its preliminary stages. As water scarcity and water pollution continue to impact societies around the world, it is clear that world-systems scholars need to do more to understand the ways in which history, the world-economy, and inequality affect use and access to this precious resource.

Conclusion

The vast literature exploring world-systems analysis of environmental impact highlights two primary themes—theoretical specification and empirical sophistication. World-systems scholars have fine-tuned original formulations of dependency theory and unequal trade to speak to the specific characteristics of resource extraction and environmental impact. Early comparative case studies have been expanded to now include robust models of cross-national comparison using advanced quantitative methods. Taken together, the theoretical and empirical accomplishments in this field have led to three critical intellectual advances in understanding social drivers of global environmental change.

First, world-systems research has provided mechanisms to describe how environmental resources have influenced the historical development of unequal hierarchies of power between nations. World-systems scholars demonstrate the important historical role of forests (wood products) and agriculture (food) in creating trade relationships and wealth accumulation in the early stages of capitalism. Up until recently, sociologists have often overlooked the role of the environment in socio-historical development. World-systems research of environmental impact fills that gap. Second, world-systems analysis of forests, food, and freshwater allows for crucial specification of contemporary core-periphery processes such as foreign investment, debt, and trade. For instance, scholars demonstrate how increased foreign investment corresponds with increased agricultural pesticide use, increased levels of debt are associated with increased rates of deforestation, and increased export commodity concentration correspond with increased levels of water pollution. Specifying the relations of these processes has proven invaluable in advancing the larger goal of integrating social and natural sciences in the service of sustainability. Finally, world-systems analysis has provided new conceptual tools such as *commodity chain analysis* and *unequal ecological exchange* to demonstrate how modernization and economic growth place undue burdens on the world's most vulnerable people. These conceptual developments have led world-systems scholars to be at the forefront of academic debates surrounding ecological modernization theory, providing clarity and rigor to a critical discussion.

The accomplishments above provide a strong foundation upon which to expand in new directions. World-systems theory could be enriched by a closer relationship with the realities of ecosystems, locating world-systems in a natural world with material and energetic limits (Abel 2007). The new explorations of thought within world-systems theory will continue to represent innovative hypotheses, suggesting that the biophysical earth system and the historical world-system ultimately form a single inseparable whole.

References

Abel, T. (2007) "World Systems as Complex Human Systems" in A. Hornberg and C. Crumley (eds) *The World System and the Earth System: Global Socioenvironmental Change and Sustainability since the Neolithic*, Walnut Creek, CA: Left Coast Press.

Austin, K. (2010) "Soybean Exports and Deforestation From a World-Systems Perspective: A Cross-National Investigation of Comparative Disadvantage," *Sociological Quarterly*, 51: 511–36.

Braudel, F. (1972) *The Mediterranean and the Mediterranean World in the Age of Philip II*. New York: Harper & Row.

Braudel, F. (1973) *Capitalism and Material Life, 1400–1800*. New York: Harper & Row.

Bonanno, A., Busch, L., Friedland, W., Gouveia, L., and Mingione, E. (1994) *From Columbus to ConAgra: The Globalization of Agriculture and Food*. Lawrence: University of Kansas Press.

Bunker, S. G. (1984) "Modes of Extraction, Unequal Exchange, and the Progressive Underdevelopment of an Extreme Periphery: The Brazilian Amazon, 1600–1980," *American Journal of Sociology*, 89: 1017–64.

Bunker, S. G. (1985) *Underdeveloping the Amazon: Extraction, Unequal Exchange, and the Failure of the Modern State*. Urbana: University of Illinois Press.

Burns, T., Kick, E., and Murray, D (1994) "Demography, Development and Deforestation in a World-System Perspective," *International Journal of Comparative Sociology*, 35: 221–39.

Burns, T., Kentor, J., and Jorgenson, A (2003) "Trade Dependence, Pollution and Infant Mortality in Less Developed Countries: A Study of World-System Influences on National Outcomes" in W. Dunaway (ed.)

Emerging Issues in the 21St Century World System, Volume 1: Crisis and Resistance in the 21St Century World System, Westport, CT: Praeger.

Buttel, F. (2001) "Some Reflections on Late Twentieth Century Agrarian Political Economy", *Sociologia Ruralis*, 41: 165–81.

Buttel, F., Larson, O., and Gillespie, G. (1990) *The Sociology of Agriculture*, New York: Greenwood Press.

Chew, S. (2001) *World Ecological Degradation: Accumulation, Urbanization, and Deforestation 3000 B.C.–A.D. 2000*. New York: Rowman & Littlefield.

Friedmann, H (1993) "The Political Economy of Food: A Global Crisis", *New Left Review*, 197: 29–57.

Friedmann, H. and McMichael, P. (1989) "Agriculture and the State System: The Rise and Decline of National Agricultures, 1870 to the Present", *Sociologia Ruralis*, 29: 93–117.

Frey, R. (1998) "The Export of Hazardous Industries to the Peripheral Zones of the World-System", *Journal of Developing Societies*, 14: 66–81.

Frey, R. (2003) "The Transfer of Core-Based Hazardous Production Processes to the Export Processing Zones of the Periphery: The Maquiladora Centers of Morthern Mexico", *JWSR Editorial Policy*, 9: 318.

Friedland, W. (1984) "Commodity Systems Analysis: An Approach to the Sociology of Agriculture" in H. Schwarzweller (ed.) *Research in Rural Sociology and Development: A Research Annual*, London: JAI Press.

Friedland, W. (2004) "Agrifood Globalization and Commodity Systems", *International Journal of Sociology of Agriculture and Food*, 12: 17–28.

Friedmann, H. (1993) "The Political Economy of Food: A Global Crisis", *New Left Review*, 197: 29–57.

Gereffi, G. and Korzeniewicz, M. (1994) *Commodity Chains and Global Capitalism*, Westport, CT: Praeger Publishers.

Goldfrank, W. (1994) "Fresh Demand: The Consumption of Chilean Produce in the United States" in G. Gereffi and M. Korzeniewicz (eds) *Commodity Chains and Global Capitalism*, Westport, CT: Greenwood Press.

Heffernan, W. and Constance, D. (1994) "TNCs and the Globalization of Food" in A. Bonanno, L. Busch, W. Friedland, L. Gouveia, and E. Mingione (eds) *From Columbus to ConAgra: The Globalization of Agriculture and Food*, Lawrence: University of Kansas Press.

Kick, E., Burns, T., Davis, B., Murray, D., and Murray, D. (1996) "Impacts of Domestic Population Dynamics and Foreign Wood Trade on Deforestation: A World-Wystem Perspective", *Journal of Developing Societies*, 12(1): 68–87.

Jorgenson, A. K. (2006a) "The Transnational Organization of Production and Environmental Degradation: A Cross-National Study of the Effects of Foreign Capital Penetration on Water Pollution Intensity, 1980–95", *Social Science Quarterly*, 87: 711–30.

Jorgenson, A. K. (2006b) "Unequal Exchange and Environmental Degradation: A Theoretical Proposition and Cross-National Study of Deforestation", *Rural Sociology*, 71: 108–25.

Jorgenson, A. K. (2007) "Foreign Direct Investment and Pesticide Intensity in Less-Developed Countries: A Quantitative Investigation", *Society and Natural Resources*, 20: 73–83.

Jorgenson, A. K. (2008) "Structural Integration and the Trees: An Analysis of Deforestation in Less Developed Countries, 1990–2005", *Sociological Quarterly*, 49: 503–27.

Jorgenson, A. K. (2010) "World-Economic Integration, Supply Depots, and Environmental Degradation: A Study of Ecologically Unequal Exchange, Foreign Investment Dependence, and Deforestation in Less Developed Countries", *Critical Sociology*, 36: 453–77.

Jorgenson, A. K. and Burns, T. (2004) "Globalization, the Environment, and Infant Mortality: A Cross National Study", *Humboldt Journal of Social Relations*, 28: 7–52.

Jorgenson, A. K., Dick, C., and Mahutga, M. (2007) "Foreign Investment Dependence and the Environment: An Ecostrctural Approach", *Social Problems*, 54: 371–94.

Jorgenson, A. K. and Kyukendall, K. (2008) "Globalization, Foreign Investment Dependence and Agriculture Production: Pesticide and Fertilizer Use in Less-Developed Countries, 1990–2000", *Social Forces*, 87: 529–60.

Longo, S. B. and York, R. (2008) "Agricultural Exports and the Environment: A Cross-National Study of Fertilizer and Pesticide Consumption", *Rural Sociology*, 73: 82–104.

Longo, S. B. and York, R. (2009) "Structural Influences on Water Withdrawals: An Exploratory Macro-Comparative Analysis", *Human Ecology Review*, 16: 75–83.

McMichael, P. (1994) *The Global Restructuring of Agro-Food Systems*, Ithaca, NY: Cornell University Press.

McMichael, P. (1995) *Food and Agrarian Orders in the World-Economy*, Westport, CT: Greenwood Press.

McMichael, P. (2000) "Global Food Politics" in F. Magdoff, J. Foster and F. Buttel (eds) *Hungry for Profit: The Agribusiness Threat to Farmers, Food, and Environment*, New York: Monthly Review.
McMichael, P. (2000) "The Power of Food", *Agriculture and Human Values*, 17: 21–33.
McMichael, P. (2009) "A Food Regime Genealogy", *Journal of Peasant Studies*, 36: 139–69.
McMichael, P. and Buttel, F. (1990) "New Directions in the Political Economy of Agriculture", *Sociological Perspectives*, 33: 89–109.
Mintz, S. (1985) *Sweetness and Power: The Place of Sugar in Modern History*, New York: Viking.
Moore, J. (2000) "Environmental Crises and the Metabolic Rift in World-Historical Perspective", *Organization and Environment*, 13: 123–57.
Moore, J. (2003) "The Modern World-System as Environmental History? Ecology and the Rise of Capitalism", *Theory and Society*, 32: 307–77.
Myrdal, J. (2007) "Food War and Crisis: The Seventeenth Century Swedish Empire" in A. Hornborg, J. McNeil and J. Martinez-Alier (eds) *Rethinking Environmental History: World-System, History and Global Environmental Change*, Lanham, MD: AltaMira Press.
Pelizzon, S. (1994) "The Grain Flour Commodity Chain, 1590–1790" in G. Gereffi and M. Korzeniewicz (eds) *Commodity Chains and Global Capitalism*, Westport, CT: Greenwood Press.
Phyne, J. and Mansilla, J. (2003) "Forging Linkages in the Commodity Chain: The Case of the Chilian Salmon Farming Industry, 1987–2001", *Sociologia Ruralis*, 43: 108–26.
Rudel, T. (1989) *Tropical Forests: Regional Paths of Destruction and Regeneration in the Late Twentieth Century*, New York: Columbia University Press.
Shandra, J. (2007a) "The World Polity and Deforestation: A Cross-National Analysis", *International Journal of Comparative Sociology*, 48: 5–28.
Shandra, J. (2007b) "Non-governmental Organizations and Deforestation: Good, Bad, or Irrelevant?", *Social Science Quarterly*, 88: 665–89.
Shandra, J. (2007c) "Economic Dependency, Repression, and Deforestation: A Cross-National Analysis", *Sociological Inquiry*, 77: 543–71.
Shandra, J., Shor, E., Maynar, G., and London, B. (2008) "Debt, Structural Adjustment andDeforestation: A Cross-National Analysis", *Journal of World-Systems Research*, 14(1): 1–71.
Skladany, M. and Harris, C. (1995) "On Global Pond: Interantional Development and Commodity Chains in the Shrimp Industry" in P. McMichael (ed.) *Food and Agrarian Orders in the World-Economy*, Westport, CT: Praeger Publishers.
Stringer, C. and Le Heron, R. (2008) *Agri-Food Commodity Chains and Globalising Networks*, Burlington, VT: Ashgate Pub Co.
Talbot, J. (2002) "Tropical Commodity Chains, Forward Integration Strategies and International Inequality: Coffee, Cocoa and Tea", *Review of International Political Economy*, 9: 701–34.
Talbot, J. (2004) *Grounds for Agreement: The Political Economy of the Coffee Commodity Chain*, Lanham, MD: Rowman & Littlefield Pub Inc.
Wallerstein, I. (1974) *The Modern World System I: Capitalist Agriculture and the Origins of the European World Economy in the Sixteenth Century*, New York: Academic Press.
Wallerstein, I. (1980) *The Modern World System II: Mercantalism and the Consilidation of the European World Economy, 1600–173*, New York: Academic Press.

13.3

The sociology of ecologically unequal exchange in comparative perspective

Andrew K. Jorgenson and James Rice

The world is "full" and "unequal" (Andersson 2005; see also Babones 2009; Chase-Dunn 1998; Milanovic 2007). This is scarcely a novel assertion and yet the implications remain fundamentally under-appreciated. With 6.8 billion people and counting and the consumption of energy and other natural resources at historically unprecedented levels, human societies are pushing up against the limits of global ecological systems; that the world remains vastly unequal in the wake of such pressures foreshadows the challenges facing the development prospects of the poorer countries—and it is an indication of the links between power, disparate economic development, and patterns of global environmental change.

As environmental changes have become more extensive, intensive, and unequal, world-systems scholars have responded by focusing upon the systemic processes that underlie the widespread disruption of ecological systems (e.g., Burns et al 2006; Frey 1995; Goldfrank et al 1999; Grimes and Kentor 2003; Jorgenson 2003; Lawrence 2009; Rice 2007a; Roberts and Grimes 2002). If capital accumulation is the driving force of world-system dynamics, as Wallerstein (1974) observes, then the energy and other natural resources transformed by labor into a multitude of commodities are deserving of analytical attention, as all economic production is predicated upon ecological additions and withdrawals; moreover, in an increasingly full and unequal world the externalization or imposition of the unpaid, unaccounted reparations of ecological degradation and change, a predictable strategy for enhancing capital accumulation (Wallerstein 1999), only becomes more consequential. The ecological implications of unequal exchange within a stratified world-economy have always been significant but in a full and unequal world, such implications reverberate much more strongly through comparatively disadvantaged societies.

To more fully conceptualize global environmental change it is imperative to recognize the continual appropriation by human societies of energy and other natural resources and the externalization of waste products; the environment, in turn, underpins socio-organizational complexity through available stocks of natural resource endowments and sink capacity services (Dunlap and Rosa 2000). It is also crucial to document variable capital investment tied to the production and extraction of natural resources and systemic patterns of trade—not simply measured in monetary terms but accounting for the biophysical "flows" between countries occupying distinct positions within an international division of labor. Human societies are embedded not only within local, regional, and global ecological systems but interwoven within systemic patterns of exchange of energy and other natural resources (Podobnik 2002, 2006). To the degree that such exchange is expressed in the overconsumption and under-consumption of natural resources, both affluence *and* poverty are in part predicated upon the prevailing ecological flows within the modern world-system (Hornborg and Jorgenson 2010; Hornborg et al 2007; Jorgenson et al 2010).

Ecologically unequal exchange

Recognition of the substantive, structured ecological relations between countries is increasingly articulated through the theory of "ecologically unequal exchange"; a perspective that describes the unequal material exchange relations and consequent ecological interdependencies within the capitalist world-economy, all of which are fundamentally tied to wide disparities in socio-economic development and power (e.g., Andersson and Lindroth 2001; Hornborg 1998a, 1998b, 2009; Jorgenson 2006, 2009; Jorgenson and Clark 2009, 2011; Rice 2007b, 2009; Roberts and Parks 2007, 2009). Unequal exchange can be broadly defined as the assertion of asymmetrical power relationships between more-developed countries and less-developed countries wherein the former gain disproportionate advantages at the expense of the latter through patterns of trade as well as other structural relationships. The assertion of unequal exchange relations diverges from neoclassical economic thought by inquiring into the historical power relations shaping present comparative advantages, rather than taking present comparative advantage as a given (O'Brien and Williams 2004).

In turn, ecologically unequal exchange refers to the environmentally damaging withdrawal of energy and other natural resource assets from and the externalization of environmentally damaging production and disposal activities within less-developed countries. It constitutes the obtainment of natural capital or stocks of natural resources that yield important goods and services and the usurpation of sink-capacity or waste assimilation properties of ecological systems in a manner enlarging the domestic carrying capacity of more powerful developed countries to the detriment of developing countries. It is therefore focused upon the manner and degree to which less-developed countries tend to fulfill a role in the world-system as a tap for the raw materials and sink for the waste products of industrialized countries, thereby underwriting the disproportionate production-consumption-accumulation processes of the latter.

Ecologically unequal exchange is increasingly recognized as a factor, in addition to labor exploitation, underlying the socioeconomic and environmental disparities between developed countries and less-developed countries. Capital accumulation, in turn, is fundamentally rooted in the alteration of ecological systems and the exploitation of labor (Bunker 1985; Clark and Foster 2009; Foster and Clark 2004; Hornborg 2001). It shapes both the social relations of production and the structure and integrity of ecological systems in an increasingly global manner.

The unequal geographical distribution of energy and other natural resources suggests that trade among nations is a necessity and can contribute to more efficient and productive utilization overall. And yet, in the modern world-economy many export-oriented less-developed countries remain mired in poverty, having failed to exhibit the vertical and horizontal economic diversification and growth that should follow temporally from specialization in their differential comparative advantages (Mahutga 2006). A conundrum, moreover, underlies the juxtaposition between those countries exhibiting the greatest consumption of natural resources relative to those characterized by the greatest degradation or loss of natural resource assets: that is, nations with the highest levels of natural resource consumption, principally the most industrialized countries, are typically characterized by the lowest domestic levels of environmental degradation.

In turn, the most intense natural resource degradation processes frequently beset the poorest countries in the world; those exhibiting minimal natural resource consumption demand. This inconsistency is referred to as the "consumption/environmental degradation paradox" (Jorgenson 2003).

Bunker (1984, 1985) and Bunker and Ciccantell (2005) have crafted a body of comparative-historical work that is particularly responsive to the ecologically unequal connections forged through world-system processes. They illustrate the crucial role reliable access to cheap natural resources has played in fueling the rise of hegemonic powers within a given historical era. They note that orthodox theories of development have insufficiently recognized the fundamental differences between the internal dynamics and logic of accumulation of extractive and productive economies. It is not extraction of natural resources and energy, per se, promoting ecologically unequal exchange but the socio-organizational consequences this tends to produce between and within exporting and importing regions. The historical interactions between modes of extraction and production create path-dependent dynamics shaping the historical development trajectories of differentially situated countries.

Ecological unequal exchange, therefore, is contingent upon differential cross-national social organization and accelerated production-consumption-accumulation linkages in the industrialized countries—facilitating the ability of state and private capital interests to determine global demand for natural resources (Bunker 1985; Hornborg 2001). The capacity to control demand ensures core interests engage in the substantive decisions regarding global export activity and subjects less-developed countries to ever-changing market demands (Bunker 1985; Bunker and Ciccantell 2005). Local populations, social organization, infrastructure, and ecosystems within peripheral extractive regions are often disrupted in the face of malleable core needs. Extractive regions failing to conform to core interests, in turn, are likely to be subject to declining terms of trade or abandoned entirely in lieu of natural resource exports originating elsewhere. Differential cross-national social power, in turn, is based upon historically contingent exchange relationships forged through the ability to control asymmetrical flows of environmental resources and risks (Hornborg 2001).

Quantitative approaches to testing the theory

The use of quantitative comparative methods to assess key assertions of ecologically unequal exchange theory is quite challenging. As implied by the theory, the "vertical flow" of exports from lesser-developed nations to relatively more-developed nations is a structural mechanism through which ecologically unequal consequences are born and maintained. On the flipside, it also matters where human-caused waste is generated, and where it is ultimately sent for disposal.

Over the past few years, a number of macro-sociologists have attempted to create and employ appropriate measures in comparative international studies to test propositions derived from ecologically unequal exchange theory. Early on, Jorgenson (2004) designed a measure—referred to as "weighted export flows"—that quantifies the relative extent to which a nation's exports are sent to more-developed countries. This weighted index, which involves the use of relational data (export flows between sending and receiving nations) and attributional data (levels of development of receiving nations) in its makeup, includes all primary sector and secondary sector exports. These data were initially used in a study of deforestation from 1990 to 2000, which concluded that the vertical flow of all exports contributes to forest degradation in less-developed countries, net of multiple demographic and political-economic factors, including levels of exports and classic trade dependence measures (Jorgenson 2006). In a related study, Jorgenson and Rice (2005) employ the same weighted index in a cross-sectional analysis of the per capita ecological footprints of less-developed countries. Results indicate that the vertical flow of exports suppresses the consumption-based environmental demand of less-developed countries, many of which consume natural resources well below globally sustainable thresholds. Shandra et al (2009) use Jorgenson's (2006) weighted export flows measure as one of multiple predictors in an analysis of industrial organic water pollution in developing nations. Net of other political-economic and human-ecological factors, they find that less-developed countries with relatively greater levels of exports sent to more economically developed nations exhibit higher per capita levels of industrial organic water pollutants emitted per day in the year 2000.

Rice (2007b) conducts a cross-sectional analysis of the per capita ecological footprints of nations for a sample of both developed countries and less-developed countries. To test key tenets of ecologically unequal exchange theory, Rice employs a measure that identifies the percentage of a given country's exports that are sent to core nations. While such a variable differs from the weighted export flows measures used by others, the general logic of the variable and its suitability for testing hypotheses is quite consistent with the former. Most notably, Rice (2007b) creates interactions between his trade measure and dummy variables for nations in differing income quartiles based on the World Bank's country income classifications (i.e., low, lower-middle, upper-middle, and high income). Results of cross-sectional models that include the interactions suggest that low- and lower middle-income countries characterized by a greater proportion of exports to the core countries exhibit suppressed per capita footprints relative to nations that are upper middle-income and high income. These results are consistent with propositions of the theory, and highlight non-trivial differences in such structural relationships for nations at varying levels of development.

Combined, the above studies suggest that the resource consumption/environmental degradation paradox is to some extent two sides of the same coin in the context of ecologically unequal relationships between more-developed and less-developed countries. However, they are limited in particular ways. For example, Jorgenson's (2006) deforestation analysis lacks temporal depth, and the use of a measure for all exports is problematic, since research on forest degradation often emphasizes the relevance of trade in primary sector goods (e.g., Rudel 2005). Jorgenson and Rice (2005) conduct a cross-sectional analysis, which is indeed limited, yet the use of the weighted export flows measure for all commodity types is less problematic in their study, since the dependent variable is the comprehensive per capita ecological footprint. The same cross-sectional limitations apply to Rice (2007b) as well.

Partly in an effort to address some of the weaknesses of prior empirical work, in analyses of deforestation in less-developed countries from 1990–2005, Jorgenson, Dick, and Austin (2010) employ a weighted flows measure for only primary sector goods. Findings reveal a strong association between forest degradation and the vertical flow of primary sector exports (see also Austin 2010; Jorgenson 2010). Likewise, Shandra et al (2009) employ a weighted exports flow measure for primary sector commodities in an analysis of threatened mammals in less-developed countries in 2005. Their export measure is similar to Rice's (2007) in that it quantifies the percentage of a sending nation's primary sector exports that go to OECD countries, the latter of which are mostly core nations. Consistent with the theory of ecologically unequal exchange, the results of their negative binomial regression model estimates indicate that levels of threatened mammals in poor nations are associated with flows of primary sector exports to rich nations.

To help resolve the temporal limitations of prior cross-sectional studies, Jorgenson (2009) employs more rigorous methods in random effects panel analyses of the vertical flow of exports and the per capita ecological footprints of less-developed countries from 1975 to 2000. The results indicate a negative association between the overall consumption-based environmental demands per person and the flow of total exports to relatively more-developed nations, and the association increased in magnitude during the entire 25-year period, suggesting that these relationships have become more ecologically unequal through time. Jorgenson et al (2009) continue in this general tradition, and employ a weighted export flows measure for only primary sector commodities in panel analyses of deforestation and a refined primary sector-oriented ecological footprint—known as the cropland, grazing land, and timber footprint (e.g., "CGT footprint")—for a sample of less-developed countries from 1970 to 2000. Like the preceding studies, their export flows measure is weighted by the levels of economic development of receiving countries. Jorgenson et al (2009) estimate both fixed and random effects models for both outcomes, which allow for more rigorous hypothesis testing. Consistent with prior research, they find that the vertical flow of primary sector exports contributes to increases in forest degradation and concomitant suppression of the CGT footprint, and the results hold, net of various controls and across both types of panel model estimations.

In perhaps the most thorough cross-national analyses in the ecologically unequal exchange tradition to date, Jorgenson and Clark (2009) integrate the tradition with two contemporary theories in environmental sociology: treadmill of production and treadmill of destruction. Treadmill of production theory focuses on how an economic system driven by endless growth, on an ever larger scale, generates widespread ecological degradation (Gould et al 2008). Treadmill of destruction theory suggests that the military has its own expansionary dynamics, which involve significant environmental and ecological costs (Hooks and Smith 2005). Jorgenson and Clark (2009) argue that the ecologically unequal exchange perspective intersects with both treadmill orientations. The treadmill of production propels the world-economy toward constant expansion, demanding more and more resources to meet its insatiable appetite, especially in the articulated consumer markets of developed countries. Similarly, in the interests of national security, technological innovation, political power, and geopolitical influence, the treadmill of destruction facilitates the increased consumption of resources by the nations' militaries and their supporting sectors. What is more, increased military strength enhances access to the natural resources and sink capacity of less powerful, underdeveloped nations.

Jorgenson and Clark (2009) argue that the populations of more-developed *and* militarily powerful countries are positioned advantageously in the contemporary world-economy, and thus more likely to secure and maintain favorable terms of trade allowing for greater access to the natural resources and sink capacity of bioproductive areas within less-developed countries. These advantageous positions facilitate the externalization of environmental costs of resource extraction and consumption to less-developed countries, and help create conditions where more-developed countries and those with more powerful militaries are able to over-utilize global "environmental space." The misappropriation of environmental space suppresses resource consumption opportunities for many less-developed countries. Given the structure and acceleration of both the treadmill of destruction and treadmill of production, it is quite likely that the consequences of these processes for less-developed countries are more pronounced than for developed countries, and possibly increase through time.

To test their arguments and assess the extent to which these perspectives intersect in meaningful and empirically valid ways, Jorgenson and Clark (2009) create and employ two export flows measures, one of which is weighted by the levels of economic development of receiving countries, while the other is weighted by the military expenditures per soldier of receiving countries. Military expenditures per solider has become a commonly used measure of military power (e.g., Clark et al 2010; Kentor and Kick 2008). The two export flows measures are treated as predictors in panel analyses of the ecological footprints of nations from 1975 to 2000. Most notably, the results of their panel model estimates indicate that countries with relatively higher levels of exports sent to economically developed and militarily powerful nations experience suppressed consumption levels, and these effects—that are independent of one another—are especially pronounced and increasingly so for less-developed countries, many of which consume resources well below globally sustainable thresholds. In other words, both forms of structural relationships between nations have become increasingly unequal in ecological contexts.

The entire collection of empirical works discussed thus far focus on how the structure of international trade in ecologically unequal contexts contributes to direct forms of environmental degradation in less-developed countries as well as the suppression of resource consumption for domestic populations, often well below globally sustainable limits. The latter also contributes to the well-being of these populations, which underscores the complexity of resource use and human health associations (e.g., Rice 2008). At the same time, however, it is argued that ecologically unequal exchange relationships are likely to contribute to forms of waste in less-developed countries, and such waste is the result of the off-shoring of environmentally intensive manufacturing or the externalizing of consumption-based environmental costs associated with manufactured goods for the articulated markets in more-developed countries. The focus here is on the treatment of less-developed countries as sinks for waste. While these structural relationships and their potential uneven environmental consequences are also central to the theory of ecologically unequal exchange theory, unlike the approaches discussed earlier, comparative international research on these topics are much more limited. Two notable exceptions in sociology are Roberts and Parks (2007) and Stretesky and Lynch (2009), both of which focus on growth in carbon dioxide emissions in less-developed countries.

Roberts and Parks (2007) estimate cross-sectional models of total emissions, per capita emissions, emissions per unit of production, and cumulative emissions per capita for a large sample of nations. The first three outcomes are measured in 1999, while the fourth measure is a cumulative score for the 1950 to 1999 period. Net of multiple controls, they find that nations with a greater reliance on the export of manufactured goods have higher levels of all four types of emissions. Even though Roberts and Parks (2007) do not include relational measures of trade, their results are consistent with the arguments of ecologically unequal exchange theory.

Stretesky and Lynch (2009) take a moderately different approach that includes relational measures. In their cross-national panel analyses for the 1984 to 2004 period, they assess the extent to which a reliance on exports to the United States relative to other receiving nations contributes to growth in per capita carbon dioxide emissions. Results of fixed effects model estimates indicate that relatively higher levels of reliance on exports to the United States contribute to growth in per capita emissions. While this study highlights the importance in focusing on specific trade relationships and their potential ecologically unequal consequences, the authors do not situate their research in an unequal exchange theoretical framework. Nonetheless, Stretesky and Lynch make notable methodological and substantive contributions to this overall body of literature.

Conclusion

The theory of ecologically unequal exchange has become central in academic discourse about society/nature relationships in international contexts. As we highlighted in the first section of this chapter, the perspective has deep analytical roots in world-systems analysis. Comparative-historical work illustrates the long-term systemic processes that create and maintain conditions permitting such exchanges to occur between nations. Thrusts of quantitative research have developed and employed measurements to test certain propositions of the theory, particularly those that concern the environmental consequences of the vertical flow of exports from less-powerful, less-developed nations to the more powerful and developed nations of the Global

North. It seems likely that research will continue to refine such approaches, and apply them to additional questions. There is a particular need for scholars to design measures that allow for testing propositions in comparative international analyses of the other side of the ecologically unequal exchange coin: the extent to which the flow of hazardous materials and other forms of waste from the Global North to the Global South impacts the environment and well-being of human populations in the latter. Further, in line with Jorgenson and Clark (2009), future research on ecologically unequal exchange relationships needs to consider forms of power other than economic, and as Rice (2008) suggests, forthcoming work in this tradition would do well to pay closer attention to the human well-being consequences of such international and global dynamics.

References

Andersson, Jan Otto. 2005. "Our Full, Unequal World: Ecological Footprints and International Trade." *Inroads: The Canadian Journal of Opinion*, 17: 48–57.

Andersson, Jan Otto and Mattias Lindroth. 2001. "Ecologically Unsustainable Trade." *Ecological Economics*, 37: 113–22.

Austin, Kelly. 2010. "The Hamburger Connection as Ecologically Unequal Exchange: A Cross-National Investigation of Beef Exports and Deforestation in Less-Developed Countries." *Rural Sociology*, 75: 270–99.

Babones, Salvatore. 2009. *The International Structure of Income: Its Implications for Economic Growth*. Saarbrücken, Germany: VDM Publishers.

Bunker, Stephen G. 1984 "Modes of Extraction, Unequal Exchange, and the Progressive Underdevelopment of an Extreme Periphery: The Brazilian Amazon, 1600–1980." *American Journal of Sociology*, 89: 1017–64.

Bunker, Stephen G. 1985. *Underdeveloping the Amazon: Extraction, Unequal Exchange, and the Failure of the Modern State*. Urbana: University of Illinois Press.

Bunker, Stephen G. and Paul S. Ciccantell. 2005. *Globalization and the Race for Resources*. Baltimore, MD: The Johns Hopkins University Press.

Burns, Tom, Edward Kick, and Byron Davis. 2006. "A Quantitative Cross-National Study of Deforestation in the Late 20th Century: A Case of Recursive Exploitation." pp. 37–60 in Andrew Jorgenson and Edward Kick (eds) *Globalization and the Environment*. Leiden, The Netherlands: Brill Academic Press.

Chase-Dunn, Christopher. 1998. *Global Formation: Structures of the World-Economy*. Lanham, MD: Rowman & Littlefield.

Clark, Brett and John Bellamy Foster. 2009. "Ecological Imperialism and the Global Metabolic Rift: Unequal Exchange and the Guano/Nitrates Trade." *International Journal of Comparative Sociology*, 50(3–4): 311–34.

Clark, Brett, Andrew K. Jorgenson, and Jeffrey Kentor. 2010. "Militarization and Energy Consumption: A Test of Treadmill of Destruction Theory in Comparative Perspective." *International Journal of Sociology*, 40: 23–43.

Dunlap, Riley E. and Eugene A. Rosa. 2000. "Environmental Sociology." pp. 800–813 in Edgar F. Borgatta and Rhonda J.V. Montgomery (eds) *Encyclopedia of Sociology*, Vol. 2, 2nd edn., New York: Macmillan.

Foster, John Bellamy and Brett Clark. 2004. "Ecological Imperialism: The Curse of Capitalism." *Socialist Register 2004*. London: Merlin Press.

Frey, R. Scott. 1995. "The International Traffic in Pesticides." *Technological Forecasting and Social Change*, 50: 151–69.

Goldfrank, Walter, David Goodman, and Andrew Szasz (eds). 1999. *Ecology and the World-System*. Westport, CT: Greenwood Press.

Grimes, Peter and Jeffrey Kentor. 2003. "Exporting the Greenhouse: Foreign Capital Penetration and CO_2 Emissions 1980–96." *Journal of World-Systems Research*, 9: 261–75.

Gould, Kenneth, David Pellow, and Allan Schnaiberg. 2008. *The Treadmill of Production: Injustice & Unsustainability in the Global Economy*. Boulder, CO: Paradigm Publishers.

Hornborg, Alf. 1998a. "Ecosystems and World Systems: Accumulation as an Ecological Process." *Journal of World-Systems Research*, 4(2): 169–77.

Hornborg, Alf. 1998b. "Towards an Ecological Theory of Unequal Exchange: Articulating World System Theory and Ecological Economics." *Ecological Economics*, 25: 127–36.

Hornborg, Alf. 2001. *The Power of the Machine: Global Inequalities of Economy, Technology, and Environment*. New York: AltaMira Press.

Hornborg, Alf. 2009. "Zero-Sum World: Challenges in Conceptualizing Environmental Load Displacement and Ecologically Unequal Exchange in the World-System." *International Journal of Comparative Sociology*, 50: 237–62.

Hornborg, Alf and Andrew K. Jorgenson. 2010 (eds). *International Trade and Environmental Justice: Toward a Global Political Ecology*. New York: Nova Science Publishers.

Hornborg, Alf, J.R. McNeill, and Joan Martinez-Alier (eds). 2007. *Rethinking Environmental History: World-System History and Global Environmental Change*. Lanham, MD: AltaMira Press.

Hooks, Gregory and Chad Smith. 2005. "Treadmills of Production and Destruction: Threats to the Environment Posed by Militarism." *Organization & Environment*, 18(1): 19–37.

Jorgenson, Andrew K. 2003. "Consumption and Environmental Degradation: A Cross-National Analysis of the Ecological Footprint." *Social Problems*, 50(3): 374–94.

Jorgenson, Andrew K. 2004. *Export Partner Dependence and Environmental Degradation, 1965–2000*. PhD Dissertation, Department of Sociology, University of California, Riverside.

Jorgenson, Andrew K. 2006. "Unequal Ecological Exchange and Environmental Degradation: A Theoretical Proposition and Cross-National Study of Deforestation, 1990–2000." *Rural Sociology*, 71: 685–712.

Jorgenson, Andrew K. 2009. "The Sociology of Unequal Exchange in Ecological Context: A Panel Study of Lower-Income Countries, 1975–2000." *Sociological Forum*, 24: 22–46.

Jorgenson, Andrew K. 2010. "World-Economic Integration, Supply Depots, and Environmental Degradation: A Study of Ecologically Unequal Exchange, Foreign Investment Dependence, and Deforestation in Less-Developed Countries." *Critical Sociology*, 36: 453–77.

Jorgenson, Andrew K., Kelly Austin, and Christopher Dick. 2009. "Ecologically Unequal Exchange and the Resource Consumption/Environmental Degradation Paradox: A Panel Study of Less-Developed Countries, 1970–2000." *International Journal of Comparative Sociology*, 50: 263–84.

Jorgenson, Andrew K. and Brett Clark. 2009. "The Economy, Military, and Ecologically Unequal Relationships in Comparative Perspective: A Panel Study of the Ecological Footprints of Nations, 1975–2000." *Social Problems*, 56: 621–46.

Jorgenson, Andrew K. and Brett Clark. 2011. "Societies Consuming Nature: A Panel Study of the Ecological Footprints of Nations, 1960–2003." *Social Science Research*, 40: 226–44.

Jorgenson, Andrew K., Christopher Dick, and Kelly Austin. 2010. "The Vertical Flow of Primary Sector Exports and Deforestation in Less-Developed Countries: A Test of Ecologically Unequal Exchange Theory." *Society and Natural Resources*, 23: 888–97.

Jorgenson, Andrew K. and James Rice. 2005. "Structural Dynamics of International Trade and Material Consumption: A Cross-National Study of the Ecological Footprint of Less-Developed Countries." *Journal of World-Systems Research*, 11: 57–77.

Jorgenson, Andrew K., James Rice, and Brett Clark. 2010. "Cities, Slums, and Energy Consumption in Less-Developed Countries, 1990–2005." *Organization & Environment*, 23: 189–204.

Kentor, Jeffrey and Edward Kick. 2008. "Bringing the Military Back In: Military Expenditures and Economic Growth 1990 to 2003." *Journal of World-Systems Research*, 14: 142–72.

Lawrence, Kirk. 2009. "The Thermodynamics of Unequal Exchange: Energy Use, CO_2 Emissions, and GDP in the World-System, 1975–2005." *International Journal of Comparative Sociology*, 50: 335–60.

Mahutga, Matthew. 2006. "The Persistence of Structural Inequality? A Network Analysis of International Trade, 1965–2000." *Social Forces*, 84: 1863–89.

Milanovic, Branko. 2007. *Worlds Apart: Measuring International and Global Inequality*. Princeton, NJ: Princeton University Press.

O'Brien, Robert and Marc Williams. 2004. *Global Political Economy: Evolution and Dynamics*. New York: Palgrave Macmillan.

Podobnik, Bruce. 2002. "Global Energy Inequalities: Exploring the Long-Term Implications." *Journal of World-Systems Research*, 8(2): 252–74.

Podobnik, Bruce. 2006. *Global Energy Shifts: Fostering Sustainability in a Turbulent Age*. Philadelphia, PA: Temple University Press.

Rice, James. 2007a. "Ecological Unequal Exchange: Consumption, Equity, and Unsustainable Structural Relationships within the Global Economy." *International Journal of Comparative Sociology*, 48: 43–72.

Rice, James. 2007b. "Ecological Unequal Exchange: International Trade and Uneven Utilization of Environmental Space in the World System." *Social Forces*, 85: 1369–92.

Rice, James. 2008. "Material Consumption and Social Well-Being within the Periphery of the World Economy: An Ecological Analysis of Maternal Mortality." *Social Science Research*, 37: 1292–1309.

Rice, James. 2009. "The Transnational Organization of Production and Uneven Environmental Degradation and Change in the World Economy." *International Journal of Comparative Sociology*, 50: 215–36.

Roberts, J. Timmons and Peter Grimes. 2002. "World-System Theory and the Environment: Toward a New Synthesis." pp. 167–96 in Riley Dunlap, Frederick Buttel, Peter Dickens, and August Gijswijt (eds) *Sociological Theory and the Environment: Classical Foundations, Contemporary Insights*, Lanham, MD: Rowman & Littlefield.

Roberts, J. Timmons and Bradley Parks. 2007. *A Climate of Injustice: Global Inequality, North-South Politics, and Climate Policy*. Cambridge, MA: MIT Press.

Roberts, J. Timmons and Bradley Parks. 2009. Ecologically Unequal Exchange, Ecological Debt, and Climate Justice: The History and Implications of Three Related Ideas for a New Social Movement." *International Journal of Comparative Sociology*, 50: 385–409.

Rudel, Tom. 2005. *Tropical Forests: Regional Paths of Destruction and Regeneration in the Late Twentieth Century*. New York: Columbia University Press.

Shandra, John, Christopher Leckband, Laura McKinney, and Bruce London. 2009. "Ecologically Unequal Exchange, World Polity, and Biodiversity Loss: A Cross-National Analysis of Threatened Mammals." *International Journal of Comparative Sociology*, 50: 285–310.

Shandra, John., Eran Shor, and Bruce London. 2009. "World Polity, Unequal Exchange, and Organic Water Pollution: A Cross-National Analysis of Less Developed Nations." *Human Ecology Review*, 16: 51–64.

Stretesky, Paul and Michael Lynch. 2009. "A Cross-National Study of the Association Between Per Capita Carbon Dioxide Emissions and Exports to the United States." *Social Science Research*, 38: 239–50.

Wallerstein, Immanuel. 1974. *The Modern World System I: Capitalist Agriculture and the Origins of the European World-Economy in the Sixteenth Century*. New York: Academic Press.

Wallerstein, Immanuel. 1999. *The End of the World As We Know It: Social Science for the Twenty-First Century*. Minneapolis: University of Minnesota Press.

13.4

The displacement of hazardous products, production processes, and wastes in the world-system

R. Scott Frey

Centrality in the world-system allows core countries to impose their anti-wealth on the periphery. Transnational corporations (TNCs), for instance, export some of the core's hazardous products, production processes, and wastes to the peripheral zones of the world-system. Since few peripheral countries have the ability to adequately assess and manage the risks associated with such hazards, the transfer of hazards to the periphery increases the health, safety, and environmental risks facing many peripheral countries. TNC export practices are therefore contributing to the displacement of health, safety, and environmental risks (Frey 1998, 2006a, b). Most current discussions of the risk globalization problem have failed to situate the problem firmly within a global, historical frame. This paper begins such a discourse by contextualizing the dispersion of hazardous products, industrial production processes, and wastes in the world-system frame.

Political and economic forces operating within and between countries promote the transfer of core-based hazards to the periphery. In efforts to create markets and reduce costs, core-based TNCs have moved hazardous products, production processes, and wastes to locations in many peripheral countries. Many peripheral states have enacted trade policies that encourage TNCs to transfer their hazards. In turn, various international organizations (World Bank, IMF, and the WTO) have implemented policies that promote TNC export practices and policies of peripheral states.

Scientific and public concern with the risks associated with hazardous products, production processes, and wastes emerged as an important issue in the core countries during the 1970s and has continued (Frey 2006a, b). These concerns gave rise to increased health, safety, and environmental regulations that reduced markets and increased production costs, forcing TNCs to export products, production facilities, and wastes to the periphery in an effort to increase profits.

Faced with debt and poverty, as well as a world-system marginalizing them, many peripheral states have allowed TNCs to market hazardous products (including asbestos, cigarettes, and pesticides; Frey 2006a). Peripheral states have also implemented export-oriented policies to attract industry from the core. Many peripheral countries are willing to accept dangerous manufacturing processes (Frey 2006b) and the recycling of dangerous wastes such as ocean-going vessels, e-waste, and automobile batteries (Buerk 2006; Frey 1998; Smith et al 2006).

Hazardous products, production processes, and wastes can damage the environment and adversely affect human health and safety through occupational exposure and environmental dispersion of hazardous wastes and substances in the environment, or large-scale failures such as explosions and fires (Auyero and Swistum 2009; Frey 2006b). Numerous undesirable social and economic consequences are also associated with the three hazard types, including staggering economic costs and an inequitable distribution of costs and benefits within and between groups.

Peripheral countries are vulnerable to the risks posed by the hazards because they have young, poorly trained, uninformed, and unhealthy populations. Many other problems exist: limited public awareness of the risks associated with the hazards, weak labor unions, unresponsive state agencies, and inadequate risk assessment and management capabilities. Organized public health and environmental activism is limited because those impacted have little time for such activity and there are limited channels for public participation. Trade liberalization and the general processes of globalization have compounded the problem by increasing some of the risks mentioned and reducing the state's right and ability to regulate the domestic market, the environment, and the health and safety of workers. The problem is further compounded by the fact that hazardous industries and wastes are located in rapidly growing cities faced with many health and environmental risks and inadequate infrastructure and services in terms of health care, housing, water, electricity, sewage and drainage, and garbage collection and disposal (see Auyero and Swistun 2009; Frey 1998, 2006a, b; Frey et al 2007). In the words of Herman Daly (1996: 28): "throughput" is greater than the regeneration and absorptive capacities of these cities.

Efforts to reduce the adverse consequences of the displacement problem have taken many forms: national regulatory efforts, bilateral and multilateral trade agreements, trade treaties such as NAFTA and side agreements, and calls for a World Environment Organization. Such efforts to "globalize responsibility" are problematic because of weak enforcement at the national and supranational levels, resulting from limited resources, capital mobility, and the neoliberal project that frames regulation as a trade barrier (Frey 2006b).

Others have called for the "renationalization of capital" (Daly 1996: 145–62) or the dismantling of what Gould et al (2008) call the international "treadmill of production." Implementation of such measures seems unrealistic given the structural constraints posed by the current world-system (Frey 2006b).

Transnational networks of NGOs may be one of the most effective means for reducing the risks associated with the displacement of specific hazards to the periphery (Frey 2006b; Pellow 2007). Stopping the core's displacement of anti-wealth to the periphery or the appropriation of the periphery's carrying capacity is another matter, because the appropriation process is embedded in the structure of the world-system.

References

Auyero, J. and Swistun, D. (2009) *Flammable*, New York: Oxford University Press.

Buerk, R. (2006) *Breaking Ships*, New York: Penguin.

Daly, H. (1996) *Beyond Growth*, Boston, MA: Beacon Press.

Frey, R.S. (1998) "The hazardous waste stream in the world-system", in P. Ciccantell and S. Bunker (eds) *Space and Transport in the World-System*, Westport, CT: Greenwood Press.

Frey, R.S. (2006a) "The international traffic in asbestos", *Nature, Society, and Thought*, 19: 173–80.

Frey, R.S. (2006b) "The flow of hazardous exports in the world-system", in A. Jorgenson and E. Kick (eds) *Globalization and the Environment*, Leiden, The Netherlands: Brill Press.

Frey, R.S., McCormick, S. and Rosa, E. (2007) "The sociology of risk", in C. Bryant and D. Peck (eds) *The Handbook of 21st Century Sociology*, Thousand Oaks, CA: Sage.
Gould, K., Pellow, D. and Schnaiberg, A. (2008) *The Treadmill of Production*, Boulder, CO: Paradigm.
Pellow, D. (2007) *Resisting Global Toxins*, Cambridge, MA: MIT Press.
Smith, T., Sonnenfeld, D. and Pellow, D. (eds) (2006) *Challenging the Chip*, Philadelphia, PA: Temple University Press.

13.5

Interacting landscapes

Toward a truly global environmental history

Alf Hornborg

Although rich in empirical detail, studies in environmental history often strike world-system analysts as theoretically underdeveloped. They generally do not address the fact that landscape changes in core areas have been recursively interconnected with those in peripheral areas. Although several recent books claim to deal with global environmental history (e.g., Hughes 2001; McNeill 2000; Radkau 2008), they are rarely "global" in this sense. They tend to offer a series of national and local case studies, focusing more on the environmental records of individual nations and groups than on the global historical processes and material flows that have generated their problems as well as their options. In terms of Pomeranz' (2000: 25) useful distinction, most of these global narratives treat different regions in terms of "comparisons" rather than "connections."

The main theoretical conclusion of much of this work is the recurrent observation that technologies designed to solve one kind of problem will ironically tend to generate even more severe problems of another kind, often for other groups of people. This conclusion should instill a powerful antidote against the pervasive belief in technological solutions, but the underlying message is generally that there is something inevitable and politically innocent about the course of global environmental history. Considering that several of these authors use words such as "global" and "world" in the titles of their books, it is remarkable that so few of them really consider the world as a *system*, in which environmental transformations in two geographically distant countries or regions may be closely intertwined in terms of causal connections.

An exception is Kenneth Pomeranz (2000), who joins André Gunder Frank (1998) in showing how circumstances that specifically favored capital accumulation in Europe were not internal to Europe, but the result of external global conjunctures, such as its access to silver and plantations in the New World and slaves from Africa, and the Asian demand for silver. These circumstances made it possible for Europe to transcend the land constraints that it shared with East Asia by using labor and capital to exchange a growing volume of manufactured exports for a growing volume of land-intensive products, mainly from the Americas. Pomeranz calculates the "ecological relief" (also known as "environmental load displacement") gained by Britain in 1830 through imports of cotton, sugar, and timber from overseas at between 25 and 30 million "ghost acres," that is, much more than the nation's total arable land (cf. Hornborg 2006). European landscapes—whether devoted to crops, pastures, forestry, mining, or urban-industrial infrastructure—have for centuries been organically connected with landscapes (plantations, clear cuts, mines, etc.) on other continents. The global environmental impacts of these interconnected and asymmetric material flows, geared to capital accumulation in core areas, are today being assessed in terms of "ecological footprints."

If a world-system perspective is crucial for understanding the local details of environmental history, ecology is thus no less fundamental for understanding world-system history. To trace the metabolic flows of world-systems requires a basic familiarity with their biophysical aspects such as the use of energy and land, the displacement of entropy (such as CO_2 emissions), and ecologically unequal exchange (Jorgenson and Clark 2009). It also requires a radical rethinking of our everyday conception of *technology* as an accumulation of material infrastructure, or capital, that should be understood more as the product of a global zero-sum game than as a cornucopia.

The metabolism of world-systems invariably relies on the unequal exchange of either energy or embodied land between core and periphery. Up until the Industrial Revolution, energy and land were one and the same, converging in the production of food for human labor and fodder for draught animals, but for two centuries now, the age of fossil fuels has kept land requirements and energy requirements distinct from each other. During this period, ecologically unequal exchange has not always involved net transfers of energy, nor has it always involved net transfers of embodied land, but it has always involved net transfers of *one* of these resources.

As we are currently contemplating the fact that peak oil and climate change may prompt us to turn to agrofuels, we are in fact imagining a future where land requirements and energy requirements will once again coincide. Once again, it seems, it will be possible to calculate the costs of transport distances in terms of eco-productive space. If we shall once again see competition over scarce land for food, fodder, fibers, and fuel, we may rest assured that the realities of ecologically unequal exchange and environmental load displacement will be recognized as very tangible conditions of human existence and as indispensable tools for thinking about global environmental history.

In such a future, ecologically unequal exchange will again involve concerns with *both* energy *and* embodied land. In terms of economic theory for understanding the course of history, this would amount to the bankruptcy of both Ricardian and Marxian concepts of labor value in favor of a cosmology more akin to pre-industrial Physiocracy. The global energy shifts over the next few decades may lead to the conclusion that much of what we have come to know as modern or industrial technology is feasible only when it requires less land area than the same work conducted by humans and draft animals.

References

Frank, A.G. 1998. *ReOrient: Global Economy in the Asian Age*. Berkeley: University of California Press.

Hornborg, A. 2006. "Footprints in the Cotton Fields: The Industrial Revolution as Time-Space Appropriation and Environmental Load Displacement". *Ecological Economics*, 59(1): 74–81.

Hughes, J.D. 2001. *An Environmental History of the World: Humankind's Changing Role in the Community of Life*. London: Routledge.

Jorgenson, A.K. and B. Clark, (eds) 2009. "Special Issue: Ecologically Unequal Exchange in Comparative Perspective". *International Journal of Comparative Sociology*, 50(3–4): 211–409.

McNeill, J.R. 2000. *Something New Under the Sun: An Environmental History of the Twentieth-Century World*. New York: W. W. Norton.

Pomeranz, K. 2000. *The Great Divergence: China, Europe, and the Making of the Modern World Economy*. Princeton, NJ: Princeton University Press.

Radkau, J. 2008. *Nature and Power: A Global History of the Environment*. Cambridge: Cambridge University Press.

13.6

The environmental impacts of foreign direct investment in less-developed countries

Andrew K. Jorgenson and Jessie Winitzky

During recent decades, many less-developed countries experienced a deepening of foreign debt, which resulted in austerity measures developed by global governance institutions. These austerity measures, such as structural adjustment programs, encourage the governments of indebted countries to create more favorable conditions for foreign investors and transnational corporations. At the heart of the austerity measures and concomitant actions of many indebted nation-states is the assumption that attracting foreign capital is a means of stimulating economic development to assist in debt repayment while increasing the overall well-being for domestic populations. Further, it is often suggested (e.g., OECD 1999) that the longer term economic benefits of foreign direct investment (FDI) will likely outweigh any short-term environmental harms and human well-being costs, and attracting FDI might eventually lead to more environmentally friendly forms of extraction and production in different sectors through technology transfers and "spillover effects." In other words, investing transnational firms will implement and diffuse environmentally friendly technologies among their counterparts in host countries. Often referred to as the "pollution halo hypothesis," this proposition is commonplace in economics (e.g., Cole et al 2008). However, empirical support for the pollution halo hypothesis is lacking, especially in macro-comparative contexts (e.g., Perkins and Neumayer 2009).

Partly in an effort to attract FDI and transnational enterprises, many less-developed countries implement relaxed labor laws and tax reductions as well as exemptions to environmental regulations designed to protect the natural environment from economic activities (Leonard 1988). In many cases, the real or perceived threat of capital flight could be viewed as an additional incentive for less-developed countries to offer regulatory concessions to foreign capital. Further, prior research shows that many less-developed countries are less likely than developed countries to ratify international environmental treaties, many of which deal explicitly with extractive and productive activities that are of direct relevance for transnational corporations (Roberts and Parks 2007). Previous research demonstrates that international debt and the implementation of structural adjustment programs in less-developed countries do contribute to increases in inward FDI (Shandra et al 2003). At least partly resulting from these unfolding political-economic processes, the relative presence of FDI stocks for all economic sectors combined within less-developed countries increased substantially during recent decades (Chase-Dunn and Jorgenson 2007).

With the above structural conditions in mind, and contrary to the pollution halo hypothesis, scholars influenced by world-systems perspectives argue and empirically demonstrate that a large proportion of secondary sector FDI in less-developed countries finances highly polluting and environmentally unfriendly manufacturing processes and facilities, much of which are outsourced from developed countries (e.g., Grimes and Kentor 2003; Jorgenson et al 2007). More specifically, comparative international research in this tradition indicates that within less-developed countries secondary sector, FDI positively affects growth in carbon dioxide emissions (e.g., Grimes and Kentor 2003; Jorgenson 2009a; York 2008) as well as other greenhouse gases (Dick and Jorgenson 2010; Jorgenson 2006), and industrial organic water pollution (Jorgenson 2009b). These results hold regardless of whether the emissions or pollution outcomes are measured by scale, intensity, or per unit of production. In a similar vein, research suggests that FDI in the primary sector commonly finances agricultural activities, forestry operations, and extractive enterprises that contribute to the degradation of forested areas in less-developed countries (Jorgenson 2008; see also Bunker and Ciccantell 2005). Research in this tradition also links primary sector FDI in less-developed countries to growth in the use of synthetic pesticides and fertilizers in agricultural production (Jorgenson and Kuykendall 2008; see also Frey 1995).

Overall, the empirical literature generally supports the world-systems influenced propositions that FDI in the secondary and primary sectors contribute to various environmental problems in less-developed, investment-dependent nations. Considering the globalization and outsourcing of the service industries, future research needs to examine the potential environmental impacts of tertiary sector FDI in less-developed countries as well.

References

Bunker, Stephen and Paul Ciccantell. 2005. *Globalization and the Race for Resources*. Baltimore, MD: The Johns Hopkins University Press.

Chase-Dunn, Christopher and Andrew Jorgenson. 2007. "Trajectories of Trade and Investment Globalization." pp.165–84 in Ino Rossi (ed.) *Frontiers of Globalization Research*, New York City: Springer.

Cole, Matthew, Robert Elliott, and Eric Strobl. 2008. "The Environmental Performance of Firms: The Role of Foreign Ownership, Training, and Experience." *Ecological Economics*, 65: 538–46.

Dick, Christopher and Andrew Jorgenson. 2010. "Sectoral Foreign Investment and Nitrous Oxide Emissions." *Society and Natural Resources*, 23: 71–82.

Frey, R. Scott. 1995. "The International Traffic in Pesticides." *Technological Forecasting and Social Change* 50: 151–69.

Grimes, Peter and Jeffrey Kentor. 2003. "Exporting the Greenhouse: Foreign Capital Penetration and CO_2 Emissions 1980–96." *Journal of World-Systems Research*, 9: 261–75.

Jorgenson, Andrew. 2006. "Global Warming and the Neglected Greenhouse Gas: A Cross-National Study of Methane Emissions Intensity, 1995." *Social Forces*, 84: 1777–96.

Jorgenson, Andrew. 2008. "Structural Integration and the Trees." *Sociological Quarterly*, 49: 503–27.

Jorgenson, Andrew. 2009a. "The Transnational Organization of Production, the Scale of Degradation, and Ecoefficiency." *Human Ecology Review*, 16: 64–74.

Jorgenson, Andrew. 2009b. "Foreign Direct Investment and the Environment, the Mitigating Influence of Institutional and Civil Society Factors, and Relationships between Industrial Pollution and Human Health: A Panel Study of Less-Developed Countries." *Organization & Environment*, 22: 135–57.

Jorgenson, Andrew, Christopher Dick, and Matthew Mahutga. 2007. "Foreign Investment Dependence and the Environment: An Ecostructural Approach." *Social Problems*, 54: 371–94.

Jorgenson, Andrew and Kennon Kuykendall. 2008. "Globalization, Foreign Investment Dependence, and Agriculture Production." *Social Forces* 87: 529–60.

Leonard, Jeffrey. 1988. *Pollution and the Struggle for the World Product*. Cambridge, MA: Harvard University Press.

Organisation for Economic Co-Operation and Development (OECD). 1999. *Foreign Direct Investment and the Environment*. OECD Publications.

Perkins, Richard, and Eric Neumayer. 2009. "Transnational Linkages and the Spillover of Environment-Efficiency into Developing Countries." *Global Environmental Change*, 19: 375–83.

Roberts, Timmons and Bradley Parks. 2007. *A Climate of Injustice*. Cambridge, MA: MIT Press.

Shandra, John, Robert Ross, and Bruce London. 2003. "Global Capitalism and the Flow of Foreign Direct Investment to Non-Core Nations, 1980–96." *International Journal of Comparative Sociology*, 44: 199–238.

York, Richard. 2008. "De-Carbonization in the Former Soviet Republics, 1992–2000." *Social Problems*, 55: 370–90.

Part VI
Society

14. Individuals and families

14.1

The centrality of the household to the modern world-system

Wilma A. Dunaway

World-systems foundational thinkers emphasized the centrality of sexism, women's work, and households to modern capitalism. Soon after the publication of Wallerstein's (1974) ground-breaking book that signaled the emergence of this new paradigm, the Fernand Braudel Center at Binghamton University structured ongoing research teams (Research Working Group 1978) that emphasized women's unpaid labor as a subsidy to capitalism. Under the tutelage of Immanuel Wallerstein (1981, 1998) and Terence Hopkins (1985; Hopkins and Wallerstein 1982), a team of faculty and graduate students conceptualized the household as the smallest organizational unit of the modern world-system (Wallerstein and Martin 1979). Their theoretical formulations integrated ground-breaking ideas about *unpaid household labor* that were formulated by radical feminists at Germany's University of Bielefeld (von Verhlof 1985), as well as emerging work about the *informal sector* (Portes 1983). This early conceptual work was published in issues of the Braudel Center journal (*Review* 1979, 1980, 1981, 1982, 1983, 1985, 1986, now available through JSTOR library database), in Wallerstein (1983), and in Smith et al (1984, 1992). While my own accumulated work is informed by this foundational work, I have revised and extended it in several ways.

The nexus of unpaid and paid labors

There is widespread consensus among international development agencies (United Nations 1999), among radical feminists (Mies 1986; von Verhlof 1983), and among world-system analysts that the vast majority of the world's workers have never been fully proletarianized. The "partiality of wage labor" is an historical hallmark of the modern world-system, and most of the world's households have never been dependent solely on the accumulation of resources and income through the wage mechanism (Wallerstein 1995: 1). Indeed, capitalists have always exploited semi-subsistence producers "who are available as cheap laborers" (von Verhlof 1983: 338) because their households reproduce and provision the labor force through nonwaged livelihoods and resources (Smith and Wallerstein 1992). Rather than being relics from precapitalism, these nonwaged labor forms "are the product of constraints specific to the capitalist world-economy," and they are "no less 'capitalist' than the labor in the most advanced robotized plant" (von Verhlof 1983: 321). However, these workers "are only marginally or partially proletarianized as, over their life cycle, they derive the bulk of the means of subsistence ... from outside the wage economy" (Arrighi and Saul 1968: 149). Because complete proletarianization is too costly for capitalists, new processes of nonwaged labor have been created far more rapidly than workers have been integrated into waged labor (Tabak and Crichlow 2000).

At the same time that capitalism tends toward proletarianization, the system offsets that creation of waged laborers through the process of *housewifization*. Even though household labor is a structural necessity of capitalism, capitalists obscure the economic value of these contributions to capital accumulation (von Verhlof 1983). "The extremely different conditions of work of the free wage laborer and the housewife constitute the two poles of a continuum of capitalist conditions of work relations and of production." Therefore, the parasitic relationship between wage labor and housewife production constitutes "the basis of all capitalist relations of production" (Smith et al 1984: 149, 266). The housewife did not exist in her present form prior to capitalism because precapitalist peoples were tied to larger communities that pooled survival and reproduction resources. When an area is incorporated into the world-system, capitalists effect policies that lead to the elimination of such alternative livelihood arrangements (Dunaway 2011; Smith and Wallerstein 1992). "The integration of women … into a world-system of capital accumulation has not and will not transform them into free wage-labourers. It is precisely this fact—their not being free wage-labourers, but housewives—which makes capital accumulation possible" (Mies 1981: 500).

Historically, proletarianization has integrated only a small percentage of the world's population (ILO 2007). Consequently, the typical capitalist worker has been "the marginalized, housewifized, unfree labourer, most of them women" (Mies 1986: 116). By externalizing the maintenance and reproduction of households outside the range of capitalist costs of production, housewifization justifies low wages and renders economically invisible a large portion of world labor. Whether for consumption or market, women's labors can "be bought at a much cheaper price" because they lie outside formal capitalist waged sectors (Mies 1986: 116) and because they are socially viewed as "household provisioning" without economic value (Dunaway 2012). On the one hand, housewifization has been a functional requirement of capitalism to such an extent that household-based work—not wage labor—has been predominant in the modern world-system. On the other hand, women's labors comprise "the great 'Other' of the market" because capitalism lays a "veil of invisibility" over their household-based work (Blumberg 1979: 448). Indeed, "the secret of all capitalist life" is that unwaged labor supplies much of the surplus that capitalists accumulate (Dalla Costa and Dalla Costa 1999: 154).

How do capitalists extract surplus from households?

I am challenging the western feminist claim (e.g., Matthaei 1982) that the household and the capitalist enterprise are separate spheres. In that dichotomization, the household is designated as the unpaid female sphere of domestic activity while paid work is externalized into the male-dominated capitalist realm. Woman's work in the home is described as "a premarket form of labor" in which the household never engages in income-earning activities that are traded publicly. This conceptualization ignores all the income-earning work that is done by women in households. Because of the parasitic nexus between waged labor and housewifization, the vast majority of the world's workers are marginalized into *semiproletarian* households from which capitalists can more easily drain hidden surpluses (Wallerstein 1995). In reality, the semiproletarian household is a locus of production in which members simultaneously produce use-value and exchange-value (Dunaway 2012; Smith et al 1984: 131–45). In addition to reproduction, maintenance, and socialization of the labor force (Mies 1986), households serve another essential function for capitalism. They represent millions of structural units that enable capitalists to conceal their systematic extraction of surpluses from workers and from women. Because capitalists extract surpluses from households, consumers never pay the full cost of any commodity. If capitalists compensated households for all their hidden subsidies, externalized costs, and unpaid labor, prices would be driven so high that most commodities would not be competitive in the world-economy. At the macrostructural level, the commodity chain is the global mechanism through which surpluses are extracted from households. A commodity chain is more than a long string of spatial points at which mechanical processes occur to generate a marketable product. Because it consists of successive layers of unequal exchanges, a commodity chain is an interconnected network of nodes at which capitalists directly and indirectly extract surpluses from households. Indeed, every exchange within a commodity chain is unequal, for there is a polarized distribution of the means of production (including natural resources) not only between nodes but also within every single node. Consequently, a commodity chain structures multiple opportunities for capitalists to extract surpluses from households (Dunaway 2001, 2011, 2012).

The first level of hidden inputs into commodity chains occurs through women's biological ability to reproduce and to sustain new laborers. Only that which has value in the marketplace was assigned to the formal economy while the work necessary to produce and maintain households was redefined to be "non-work." Labor that earned money in the capitalist workplace or marketplace came to be defined as "productive." Concomitantly, labor inside the household was devalued with the myth that it generated no surplus that could be appropriated (Mies 1986). "It is the wage form," contends Seccombe (1974: 29), "that obscures domestic labour's relation to capital." Even though it is not priced in the marketplace, housework has economic value, and its unwaged character makes it highly profitable. The housewife's unpaid work is "embodied in the waged labor, and it is a direct input into production. It is not necessary for the housewife to sell her labor to the capitalist for her labor to generate economic surpluses. Traded first to the husband for partial subsistence, it then existed in the husband's labor as an element of subsistence made available to capital for free" (Boydston 1986: 21–22). For this reason, non-proletarianized labor is essential to capitalism, for the highest profits result when nonwaged forms of labor subsidize the capitalist production process. In semiproletarian households, wages "can be reduced below the level of household reproduction because the household supplements this income with its other income-generating activities" (Wallerstein 1995: 5–6).

The second level of hidden subsidies occurs through unpaid household labor that is contributed to home-based capitalist commodity production. Because of their combined subsidization of capitalism through unpaid reproductive and/or market-oriented labors, households have been the "pillar of accumulation" throughout the history of the world-system (Smith et al 1984: 131–45). For instance, a majority of unpaid workers in family-based enterprises or farms are females (United Nations 2003). Wives and children are expected to contribute labor as assistants to the market-oriented production of the husband, but that "family labor" is neither valued economically nor acknowledged socially (Boydston 1986: 9). Capitalists do not only extract surpluses through unpaid household labor, for paid work that is household-based is also devalued and rendered invisible. Thus, the third level of hidden subsidies occurs when households and women are under-paid for their nonwaged inputs into a capitalist production process. Routinely, peripheral households generate market commodities or informal sector inputs into the export production process, but these labors are priced at below-market levels (Mies 1981). In agricultural households, women's livelihoods are often publicly devalued and stigmatized, even when they are as economically valuable as male-dominated export crops (Dunaway 2008, 2011). A narrow emphasis upon those waged and nonwaged laborers who are involved directly in commodity production can ignore three types of hidden laborer inputs. There can be direct and indirect flows into the production process from household provision pools, from the informal economy, and from illegal sectors (Dunaway 2011, 2012). The informal sector provides cheaper goods to workers, permitting capitalists to pay lower wages. Moreover, capitalists routinely lower production costs by integrating informal goods and services into their commodity chains (Tabak and Crichlow 2000). Nonwaged household workers supply foodstuffs, raw materials, and other inputs that provision the capitalist production process (Dunaway 2008, 2011). Moreover, women and children receive below-market prices for household-based inputs into commodity chains, including the collection of ecological resources and the retrieval of recyclable garbage (Tabak and Crichlow 2000; United Nations 2003).

The fourth level of hidden subsidies occurs when workers engage in home-based production of goods and services that are integrated into commodity chains. Such putting out systems and cottage industries (termed *casualization* since the 1990s) have been an enduring feature of capitalism since the sixteenth century (Littlefield and Reynolds 1990), and workers in these production systems have always been disproportionately female (Tabak and Crichlow 2000). In the Global South, emerging industries integrate nonwaged workers directly into their commodity chains through employment of home-based contract workers who simultaneously subsidize capitalists through their unpaid household inputs and their under-paid outputs (Beneria and Roldan 1987). For example, many contemporary Asian women "have been housewifized" to subsidize textile commodity chains through their integration into low-paying piece-rate contracts (Mies 1986). Through these exploitative mechanisms, capitalists externalize much of the cost of production to households by paying a level of remuneration that falls far below the wages paid to factory workers. In addition, households assume the costs for provisioning workers, for the integration of unpaid children, for equipment, for electricity and support goods and services. Moreover, these households are legally accountable for any ecological damage caused by the production process. Despite the significance of their labor inputs into commodity chains, homeworkers are rendered economically invisible because they are neither counted in national statistics nor as part of the waged labor force (Waring 1988).

The fifth level of hidden subsidies occurs because capitalist commodity chains structure networks in which consumer and laborer households at higher nodes benefit from the exploitation of households at lower nodes. Through the commodity chain that exports shrimp from the Philippines, for example, capitalism exploits women in a poor country as nonwaged producers of commodities in order to export cheap food to the core. In this context, the low wages, malnutrition, and degraded ecosystems of Philippine subsistence fishing households keep global prices low, permitting the distant consumer to avoid the actual costs of production. While the distant fisher wife and her children go lacking in essential protein and iron, the Japanese housewife feeds her offspring an abundance of hidden Philippine household sacrifices for which she does not pay (Dunaway and Macabuac 2007).

How capitalists externalize costs to households

In addition to these hidden semiproletarian subsidies, capitalists maximize profits by externalizing costs of production to households and to the ecosystems that provision them (Dunaway 2001, 2012). Routinely, capitalists externalize most of the real costs of commodity production, and these "unseen and unpaid bills ... are to be found at every node/link of every commodity chain" (Wallerstein 1995). Households absorb production and distribution costs from commodity chains at six levels. First, capitalism externalizes to mothers the costs associated with the bearing and raising of successive generations of laborers. Despite its dependency upon this natural female contribution, however, capitalism has externalized child rearing outside the realm of the economic (Mies et al 1988: 29). Moreover, the household is the site in which women undertake unpaid labor for those members who are waged laborers. To generate family survival requirements, women engage in "shadow work" outside formal capitalist structures (von Verhlof 1983). Since the household is the foundation from which members are rendered capable of selling their labor to capitalists, the capabilities of households to absorb externalized costs are preconditions for waged labor. Perhaps the most expensive externalized costs are health care-giving and medical treatment for workers injured on capitalist jobs. "Hence, the housewife and her labour are not outside of surplus value production, but constitute the very foundation upon which this process can get started. The housewife and her labour are, in other words, the basis of the process of capital accumulation" (Mies 1986: 31).

The second category of externalized costs consists of threats to household survival. Routinely, capitalism generates chronic scarcity of resources that are needed to supply the basic survival needs of households. Capitalism "has always been unsustainable since it has assumed, from the start and continues to assume, extermination and hunger for an increasingly large part of humanity" (Dalla Costa and Dalla Costa 1999: 17). Even though capitalism cannot grow without the profits that accrue from exploitation of the world's households, the modern world-system perpetuates poverty and inequality by concentrating assets and resources into a very few hands. Capitalists also expropriate and degrade land and natural resources that are essential to household provisioning. Through super-exploitation of semiproletarian labor, capitalists appropriate "so many of the fruits of the workers' labor that the workers cannot maintain themselves or reproduce their labor power" (Frank 1981: 87). In addition, capitalist enterprises target a small minority of Global South workers while "preventing the majority from entering the occupational niches that export-oriented economic policies foster." As a result, "workers are leaving waged employment and self-employment *at earlier moments* in their lives, and they are becoming disposable earlier and younger than in the past" (de la Rocha 2001: 92, 88). Moreover, capitalist commodity chains replace household crafts with imports and capture a high proportion of local consumer goods for export production. In every historical epoch, capitalism has endangered or monopolized the ecological resources that households need for survival. The modern world-system has inflicted on the Global South steadily declining biodiversity and fresh water, global warming, rising sea levels and desertification. Capitalist agriculture has degraded so much of the world's soil that the percentage of arable land has declined to critical levels (Goldfrank et al 1999). Since agricultural production has been increasingly integrated into world commodity chains, Global South food self-sufficiency has declined sharply. Peripheral countries now import two-thirds or more of the food needed for local consumption, raising the prices of basic survival needs above levels that the poor can afford (McMichael 1994). Since 1900, the proportion of hungry and malnourished people in the world has risen, and per-capita calorie intake has steadily declined in the Global South (United Nations 1999).

The third category of externalized costs results when households are forced to restructure their boundaries and internal dynamics in reaction to capitalist disruption of societies. To confront economic crises or rapid innovation, the household has repeatedly been "redefined and reshaped as part of the pulling and tugging that constitutes accumulation on a world scale" (McGuire et al 1986). Historically, households have repeatedly been forced to alter their composition in reaction to growing immiseration caused by capitalism (Korzeniewicz and Smith 1996). Such household restructuring occurs in reaction to capitalist incorporation of a geographical area, to offset the effects of the widening and deepening of capitalism in an area, or to adjust to the economic cycles of the world-economy (Smith and Wallerstein 1992). In order to force the emergence of households oriented around the reproduction of wage laborers, capitalist states implement policies that eliminate the communal and inter-household networking systems of precapitalist communities (Dunaway 2008; Smith and Wallerstein 1992). Even when capitalist states have initiated policies that were intended to speed up proletarianization, the vast majority of the laborers have been marginalized from waged occupations (Arrighi and Saul 1968; Dunaway 2011).

The fourth category of externalized costs results from the household disruption associated with the export of surplus laborers. In every historical era, capitalism has destabilized households by removing members for labor migrations (Dunaway 2008, 2012). In order to centralize capitalist labor forces in cities, capitalists have historically depopulated rural areas, decomposed households, and weakened their provisioning capacity (Goldfrank et al 1999). Women's reproductive work has been restructured internationally to redistribute core household "care" work onto peripheral women, as evidenced by the current transnational migration of domestic servants, the emergence of an international baby market, and the booming international sex tourism industry (Dalla Costa and Dalla Costa 1999). To permit migrants to be employed in distant countries, contemporary "transnational families" absorb the shortfalls caused by absent members. Through the "international transfer of caregiving," migrant domestic workers shift their reproductive labor from their peripheral children to affluent women in richer countries. In turn, the migrant women must transfer to poorer females in their home countries their child care and unpaid household labor (Parrenas 2005).

Because capitalists need to stimulate the emergence of "households of a shape and form that will create an optimal market for wage-goods" (Wallerstein et al 1982: 441), the fifth category of externalized costs results from unpaid consumption work. Through the work she does to locate and purchase, to assemble and to utilize, or to make commodities edible, the consumer-housewife does unpaid labor "in order to lower the costs for the realization of capital" (Mies 1986: 126). As the Global South becomes increasingly dependent on food imports, women must increase their daily work hours to locate scarce goods at affordable prices, to transport them home, and to prepare them for consumption. At the same time, the household also absorbs rising prices for foods that were previously obtained through household provisioning or informal sector activities (von Verhlof 1983). Moreover, reallocation of household labor time toward the pursuit of capitalist commodities leads to the decline of handicraft production, causing loss of livelihood for thousands of households (Dunaway 2011, 2012; Dunaway and Macabuac 2007; Tabak and Crichlow 2000).

The greater risks that are suffered by women and girls represent the sixth category of externalized costs. Because the inequalities are so stark, it is easy to be fooled into thinking that all peripheral men, women, and children experience the same degrees of immiseration. However, the world-system has structured "a modern form of patriarchal relations, in which women experience a social reality very different from their brothers in capital or labor" (O'Connor 1994: 108–09). Consequently, peripheral men and women do not experience the same degrees of exclusion and poverty. If we are to capture the workings of the household, therefore, we must recognize that there are two classes of people among the semiproletarians: the double exploited (women) and those who are both exploited/exploiters (men). Indeed, males are able to mitigate their own exploitation "by appropriating the labor-power/sexuality and bodies of women" (von Verhlof 1980: 40–41). Even though females contribute more work hours than males to household survival, women control very little wealth or land. Moreover, 70 percent of the world's illiterate adults are women, and there is a rising worldwide incidence of female-headed households and feminization of poverty (United Nations 1999, 2003). Women are disproportionately endangered by the ecological degradation that accompanies capitalist development, and they are the household members who must contribute labor to care for those made ill by environmental risks or resource depletion (Dunaway and Macabuac 2007). Water scarcity, desertification, deforestation, land degradation, and coastal pollution pose inequitable hardships for women. Malnutrition is the most fundamental act of environmental sexism that is inflicted by the capitalist world-system upon women and girls. Half of all peripheral children die before age ten, and females are disproportionately represented among those deaths (United Nations 1999, 2003). Capitalism also externalizes to females the nutritional battering of children associated with high fertility rates that accompany malnutrition and illiteracy in about one-third of all peripheral countries (Ward 1985).

Because the modern world-system has institutionalized the cultural devaluation of the work of women and girls (Wallerstein et al 1982), integration into capitalist commodity chains brings destructive economic results for women. Historically and currently, women have been targeted for the dirtiest, most back-breaking aspects of the capitalist production process while higher-skilled, higher-paying artisan jobs have been reserved for males (United Nations 2003). In the face of capitalist expansion, peripheral women lose artisan jobs to imports and see their local markets subsumed by commercialized agriculture (Dunaway 2008, 2011; Mies et al 1988, Pickering 2000). To keep production costs low, capitalists are breaking the bodies of peripheral girls and young women at an alarming rate. By eliminating safety equipment and sanitary working conditions, corporations externalize to females the health costs of industrial injuries and disabilities, work-related diseases, and the higher incidence of birth defects and mother mortality due to exposure to chemicals and industrial waste. In the face of grossly inadequate medical systems, domestic violence increases when capitalist enterprises target female workers and exclude males (Mies 1986; United Nations 2003).

Future research directions

Think of the mythological image of Atlas carrying the planet on his back. Then visualize Global South households carrying world capitalism through their visible and invisible inputs and sacrifices. Originating in foundational world-systems thinking (Wallerstein 1974), the *commodity chain* concept offers a tool through which we can undertake future research about the centrality of households to the modern world-system. Commodity chains incorporate households and communities into complex networks of unequal exchange that insure the polarized division of surplus among core, semiperiphery, and periphery (Wallerstein 1983). Because they maximize profits by merging numerous forms of paid and unpaid labor and resources, commodity chains are the mechanisms through which capitalists drain surpluses from the Global South in the forms of payment of below-subsistence wages, absorption of hidden subsidies from households, and externalization of production costs to local communities. In short, a household does not have to earn capitalist wages or contract piece-rate homework to be integrated into commodity chains, but those other mechanisms of surplus drain are almost always overlooked by scholars.

Bringing households and gender into commodity chains is ground-breaking research that has yet to be done, and I have previously posed five research questions that need to be pursued (Dunaway 2001). Researchers are not facing a void as they undertake this effort. Dunaway (2008, 2011, 2012) more fully conceptualizes the ideas presented briefly here. Previous groundwork has been laid to quantify gendered work inequalities and market value of unpaid household labor (cf. Antonopoulos and Hirway 2010), and global institutions now pressure Global South governments to develop accounting systems that merge unpaid household labor into GNP (UN 2003). Some case studies (e.g., Ramamurthy 2004) offer feminist analyses of commodity chains. Numerous published studies (e.g., Ferguson 2007) provide guidance about the kinds of nonwaged labors and resources that need to be integrated into commodity chain analysis (cf. Dunaway 2012). In addition, several feminists have explored the externalities associated with labor subcontracting (Salzinger 2003), as well as labor resistance within commodity chains (Hale and Wills 2006).

References

Antonopoulos, R. and I. Hirway (2010) *Unpaid Work and the Economy: Gender, Time Use and Poverty in developing Countries*, New York: Praeger.

Arrighi, G. and Saul, J. (1968) "Socialism and economic development in tropical Africa", *Journal of Modern African Studies*, 2: 141–69.

Beneria, L. and Roldan, M. (1987) *The Crossroads of Class and Gender: Industrial Homework, Subcontracting, and Household Dynamics in Mexico City*, Chicago, IL: University of Chicago.

Blumberg, R. L. (1979) "Rural women in development: veil of invisibility, world of work", *International Journal of Intercultural Relations*, 3: 447–72.

Boydston, J. (1986) "To earn her daily bread: housework and antebellum working-class subsistence", *Radical History Review*, 35: 7–25.

Dalla Costa, M. and Dalla Costa, G. F. (1999) *Women, Development and Labor of Reproduction*, Trenton, NJ: Africa World Press.

de la Rocha, M. G. (2001) "From the resources of poverty to the poverty of resources?" *Latin American Perspectives*, 28(4): 72–100.

Dunaway, W. A. (2001) "The double register of history: situating the forgotten woman and her household in capitalist commodity chains", *Journal of World-System Research*, 7: 2–31.

Dunaway, W. A. (2008) *Women, Work and Family in the Antebellum Mountain South*, London: Cambridge University Press.

Dunaway, W. A. (2011) "Nonwaged peasants in the modern world-system: African households as dialectical units of capitalist exploitation and indigenous resistance, 1890–1930", *Journal of Philosophical Economics*, 4: 19–57.

Dunaway, W. A. (2012) "The semiproletarian household over the *longue duree* of the world-system", in R. Lee (ed.) *The Longue Duree of the Modern World-System: In Memoriam to Fernand Braudel*, Albany, NY: SUNY Press.

Dunaway, W. A. and Macabuac, M. C. (2007) "The shrimp eat better than we do: Philippine subsistence fishing households sacrificed for the global food chain", *Review*, 30: 313–36.

Ferguson, Lucy (2007) "Reproductive provisioning and everyday life in global political economy," *Working Paper 32, Institute of Political Economy*, University of Manchester.

Frank, A. G. (1981) *Crisis in the Third World*, New York: Holmes & Meier.

Goldfrank, W., Goodman, D. and Szasz, A. (eds) (1999) *Ecology and the World System*, Westport, CT: Greenwood Press.

Hale, A. and Wills, J. (eds) (2006) *Threads of Labour: Garment Industry Supply Chains from a Workers' Perspective*, London: Wiley Blackwell.

Hopkins, T. (1985) "FBC Planning Group Memorandum, 6 December," *Unpublished manuscript, Fernand Braudel Center*, Binghamton University.

Hopkins, T. and Wallerstein, I. (1982) *World-Systems Analysis: Theory and Methodology*, Beverly Hills, CA: Sage.

International Labour Organization (2007) *Globalization and Informal Jobs in Developing Countries*, Geneva: World Trade Organization.

Korzeniewicz, R. and Smith, W. (eds) (1996) *Latin America in the World-Economy*, Westport, CT: Greenwood Press.

Littlefield, A. and Reynolds, L. T. (1990) "The putting out system: transitional form or recurrent feature of capitalist production?", *Social Science Journal*, 27: 359–72.

Matthaei, J. (1982) *An Economic History of Women in America: Women's Work, the Sexual Division of Labor, and the Development of Capitalism*, New York: Schocken Books.

McGuire, R. H., Smith, J. and Martin, W. G. (1986) "Patterns of household structures and the world-economy," *Review*, 10: 75–97.

McMichael, P. (ed.) (1994) *The Global Restructuring of Agro-Food Systems*, Westport, CT: Praeger.

Mies, M. (1981) "Dynamics of sexual division of labour and capital accumulation: women lace workers of Narsapur," *Economic and Political Weekly*, March: 487–500.

Mies, M. (1986) *Patriarchy and Accumulation: Women in the International Division of Labor*, London: Zed Books.

Mies, M., Bennholdt-Thomsen, V. and von Verhlof, C. (1988) *Women: The Last Colony*, London: Zed Books.

O'Connor, M. (ed.) (1994) *Is Capitalism Sustainable? Political Economy and the Politics of Ecology*, New York: Guilford Press.

Parrenas, R. (2005) "Long distance intimacy: class, gender and intergenerational relations between mothers and children in Filipino transnational families," *Global Networks*, 5: 317–36.

Pickering, K. (2000) *Lakota Culture, World Economy*, Lincoln: University of Nebraska Press.

Portes, A. (1983) "The informal sector: definition, controversy, and relation to national development," *Review*, 7: 151–74.

Ramamurthy, P. (2004) "Why is buying a madras cotton shirt a political act? A feminist commodity chain analysis," *Feminist Studies*, 30: 734–69.

Research Working Group (1978) "Households, labor-force formation and production processes in the capitalist world-economy," Unpublished manuscript, Fernand Braudel Center, Binghamton University.

Salzinger, L. (2003) *Genders in Production: Making Workers in Mexico's Global Factories*, Berkeley: University of California Press.

Seccombe, W. (1974) "The housewife and her labour under capitalism," *New Left Review*, 83: 22–41.

Smith, J. and Wallerstein, I. (eds) (1992) *Creating and Transforming Households: The Constraints of the World-Economy*, London: Cambridge University Press.

Smith, J. Wallerstein, I., and Evers, H. (eds) (1984) *Households and the World-Economy*, Beverly Hills, CA: Sage.

Tabak, F. and Crichlow, M. A. (eds) (2000) *Informalization: Process and Structure*, Baltimore, MD: Johns Hopkins University Press.

United Nations (1999) *Human Development Report*, New York: Oxford University Press.

United Nations (2003) *The World's Women 2000: Trends and Statistics*, New York: Oxford University Press.

von Verhlof, C. (1980) "Notes on the relation between sexuality and economy," *Review*, 4: 32–42.

von Verhlof, C. (1983) "Production relations without wage labor and labor division by sex," *Review*, 7: 315–59.

von Verhlof, C. (1985) *Why peasants and housewives do not disappear in the capitalist world-system*, Working Paper No. 68, Sociology Department, University of Bielefeld, Germany.

Wallerstein, I. (1974) *The Modern World-System I: Capitalist Agriculture and the Origins of the European World-Economy in the Sixteenth Century*, New York: Academic Press.

Wallerstein, I. (1981) "How to conceive our intellectual task: memorandum to the research working group on households, labor-force formation and the world-economy," Unpublished manuscript, Fernand Braudel Center, Binghamton University.

Wallerstein, I. (1983) *Historical Capitalism*, London: Verso Editions.

Wallerstein, I. (1995) "The modern world-system and evolution," *Journal of World-System Research*, 2: 1–11.

Wallerstein, I. (1998) *Mentoring, Methods and Movements: Colloquium in Honor of Terence K. Hopkins by His Former Students*, Binghamton, NY: Fernand Braudel Center.

Wallerstein, I. and Martin, W. (1979) "Peripheralization of Southern Africa II: Changes in household structure and labor-force formation," *Review*, 3: 193–207.

Wallerstein, I., Martin, W., and Dickinson, T. (1982) "Household structures and production processes: preliminary theses and findings," *Review*, 5: 437–58.

Ward, K. (1985) "The social consequences of the world economy system: the economic status of women and fertility," *Review*, 8: 561–94.

Waring, M. (1988) *If Women Counted: A New Feminist Economics*, San Francisco, CA: Harper & Row.

14.2

International migration in the world-system

Matthew R. Sanderson

Neoclassical economics and push-pull

World-systems analyses of international migration developed in response to studies grounded in neoclassical economics and the "push-pull" approach. The study of international migration was initiated with the publication of Ravenstein's "laws of migration" (1885, 1889), a set of empirical observations later developed by Lee (1966). These explanations of migration identified the factors that compelled ("pushed") people to change locations and the factors that attracted ("pulled") them to specific locations. The push-pull was closely aligned with, and influenced by, neoclassical economic theory. Together, neoclassical economic theory and the "push-pull" approach were the conventional intellectual framework for studying international migration for most of the twentieth century, orienting the classic works of Thomas (1941), Lewis (1954), Ranis and Fei (1961), Todaro (1969), and Thomas (1973).

Neoclassical economic theory takes as its point of departure the assumption that migrants are rational, utility-maximizing individuals who make the decision to move on the basis of cost–benefit considerations, the primary consideration being the expected wage gains from moving into the labor market in a destination country where labor is remunerated at a relatively higher level. At the macro-level, international wage differentials are the key explanation of international migration. International wage differentials are an outcome of differences in the relative supply of and demand for labor: wages will be lower in countries in which the supply of labor exceeds the demand for labor, and higher in countries in which the supply of labor is insufficient to meet demand. Migrants are "pushed" by relatively lower wages in the origin country and "pulled" to relatively higher wages in the destination country. International migration should therefore reduce wage differentials between countries, moving them closer to an equilibrium level that reflects only the costs of moving, and ultimately minimizing much of the economic incentive to move. At the micro-level, migrants act rationally in exploiting differentials in international wage levels.

Although this perspective remains the conventional framework, it is increasingly criticized for its inability to account for empirical regularities. Most importantly, wage differentials have been shown to be only a weak explanation of international migration. Despite widening differences in cross-country wage levels, the propensity to send migrants varies dramatically across countries and the stock of international migrants has remained relatively stable at 3 percent of world population since the 1960s. Moreover, wages across countries have not come close to converging, despite the fact that there are more international migrants today in absolute numbers (215 million) than at any time in modern history (UN 2009).

A world-systems perspective

The limitations of the conventional perspective opened intellectual space for alternative perspectives. In the late 1970s, an alternative perspective emerged that drew upon the burgeoning literature in world-systems analysis to investigate international migration. Inspired by Wallerstein (1974), Amin (1976), and Gunder Frank (1966), scholars working from this perspective varied in their degree of association with world-systems analysis per se, but they shared a common interest in exploring the systemic, structural, and historical context of international migration.

In contrast to the ahistorical conventional model, the world-systems perspective placed international migration in a broader historical context. Three overlapping types of population movements were identified, of which international labor migration is but the latest form (Morawska 1990; Petras 1981; Portes and Walton 1981; Sassen-Koob 1981): forced movements in which labor was mobilized coercively within the periphery through slavery, indentured labor contracts, and peonage; colonizing movements in which labor moved from core to periphery to settle frontier areas; and international labor migration characterized by the relatively "free" movement of labor from the periphery to the core. Collectively, this work argued that international migration was critical to the incorporation of peripheral areas and the consolidation of the capitalist world-system: "The movement of human populations has been an intrinsic component of major processes of structural change throughout history … Thus defined, migration is almost coterminous with history" (Portes and Walton 1981: 21).

In contrast to the neoclassical economics concept of the rational, utility-maximizing migrant, world-systems approaches to international migration focus on the structures that condition and constrain individual action. Migration is part of a system: individuals may indeed migrate on the basis of cost–benefit considerations, but both the costs and benefits of movement are structured by an historical context of unequal exchange in a hierarchical international division of labor. By definition, international migration involves the transgression of national boundaries. But for those working from a world-systems perspective, it is not only a movement across national boundaries; it is more importantly a movement within an integrated political-economic system.

Placing international migration in this broader, world-historical context addresses a key limitation in the conventional, push-pull framework: that cross-country wage differentials are not strong explanations of international migrations. By expanding the scope of inquiry from national to world-scale processes, the world-systems perspective opens up for examination the relationship between trans-national political-economic processes and international migration. This is an important analytical advance because these global processes ultimately create the context for individual-level decision-making: "It is within the context of extensive social and economic penetration of peripheral societies by the institutions of advanced capitalism that individual cost-benefit calculations make sense" (Portes 2007: 77). In this sense: "Migrants can be viewed as stepping or falling into a migratory flow, rather than initiating or constituting such a flow through their individual decisions and actions" (Sassen-Koob 1978: 515).

By focusing almost exclusively on wage differentials, the conventional perspective also de-politicizes the world political-economic context in which migrants make decisions. In doing so, it misses key long-term, structural dynamics that motivate and sustain international migration. Here, the concept of unequal exchange is paramount. Over time, unequal exchange between core and non-core zones produces uneven development across zones in the world-economy. International migration is a consequence of this process: if not for (widening) global income disparities, international migration would not exist. In this sense, both the conventional approach and a world-systems perspective view wage differentials as a necessary cause of international migration. However, wage differentials become an explanandum for world-systems analysis; they are foregrounded and explained, whereas they remain "exogenous" to the conventional approach.

International migration, however, is not only an outcome of unequal exchange; it is also a form of it, reproducing uneven development in the modern world-system. Burawoy (1976) provides one of the earliest discussions. International migration is a labor supply system. In this sense, migrants are a labor force, and like all labor forces, it must be maintained and renewed, or reproduced. What differentiates an international labor supply system from a domestic one, however, is that the process of reproduction (i.e., of maintenance and renewal) occurs across national boundaries, so that different institutions are responsible for organizing, and bearing the costs of, the reproduction of the labor force. This opens up the possibility that the benefits of migrant labor may not accrue to the institutions bearing the costs of reproducing the migrant labor system. For example, in the case of Mexican emigration to the United States, the costs of educating, training, and reproducing the labor force are borne largely by the Mexican state and economy, but the benefits of capital accumulation derived from their application to production processes are reaped mainly by the United States:

> Thus, for Mexican migrants, processes of renewal are organized under the Mexican state in the Mexican economy, and those of maintenance in the United States ... the activities of maintenance and renewal are separated ... In other words, a proportion of the costs of renewal is externalized to an alternate economy and/or state.
>
> *(Burawoy 1976: 1052–53)*

Thus, international migration is a form of unequal exchange, reproducing uneven development: "The significance of migrant labor lies in the separation of the processes of maintenance and renewal, so that renewal takes place where living standards are low and maintenance takes place within easy access of employment" (Burawoy 1976: 1082). By capitalizing on uneven levels of wage remuneration across countries, international migration tends to exacerbate those differences, leaving the origin country with sunk costs associated with education, training, and reproducing labor while enhancing capital accumulation in the destination country: "The very sale of labor power by an underdeveloped country ... to an economically advanced nation serves only to reinforce the relations of economic subjugation and domination" (Burawoy 1976: 1068). Understanding international migration as a global labor supply system means understanding migration as an unequal exchange between nation-states within a hierarchical world-system and not across autonomous, self-contained nation-states. That is, international migration is inherently relational.

The State plays a key role in reproducing the global system of migrant labor because it demarcates the line between migrant and domestic labor (Massey 1999; Zolberg 1978, 1989). Sovereignty means that States can endow upon their citizens legal, political, and civil rights, and protections that they need not afford to non-citizens. Without access to these rights and protections, migrants remain politically and economically vulnerable. Their precarious position is associated with their trans-national status as laborers in the world-economy (Fernandez-Kelly and Massey 2007; Petras 1980, 1981). This vulnerability, rooted in the principle of state sovereignty, creates the political basis for unequal exchange from labor migration in an integrated capitalist world-economy:

> The state organizes the dependence of the productive worker on the reproductive worker, while the economy organizes the dependence of the reproductive worker on the productive worker. The interdependence. ... links the state supplying labor and the state employing

> labor: the former requires revenue and employment for its population, the latter requires labor at low wage rates.
>
> *(Burawoy 1976: 1053)*

The inherently relational nature of international migration, and the role of states in perpetuating the global system of migrant labor, make world-systems approaches better-suited to understand international migration than the conventional perspective: "Nothing invalidates traditional approaches to migration as effectively as border control policies ... Border controls reduce the applicability of standard economic models by impeding the free circulation of labor" (Massey et al 2005: 14).

Research on international migration

World-systems approaches to international migration gained traction in the 1980s and 1990s, with seminal pieces by Alejandro Portes and colleages, and Saskia Sassen. Together, these works constructed a theoretical framework supported by empirical research that explains how world-system dynamics both initiate international migrations and direct them to specific destinations.

International migration is initiated as markets expand into, or penetrate, pre-capitalist and non-capitalist societies, disrupting pre-existing social relations, and mobilizing segments of the population into migration streams, some of which are directed toward domestic urban areas and some of which spills over across national boundaries: "Sustained labor migration requires the penetration of the political and economic institutions of the dominant unit ... into the subordinate one ... (creating) internal imbalances between sectors and institutions in the subordinated unit" (Portes and Walton 1981: 31). As local economies are restructured to fit into the international division of labor, new domestic classes emerge with closer ties to foreign capital, and consumption habits, values, and norms are reoriented toward the core: "Structural imbalances between newer and older elements eventually produce migratory pressures" (Portes and Walton 1981: 32). International migrations do not originate "spontaneously" from individual cost-benefit analyses, but these movements are produced by political-economic processes that imbalance the subordinate society (Portes 1978b, 1978a, 1979). The concept of structural imbalancing is supported with case studies that range from South African manual labor migrations to the emigration of Argentine doctors, providing empirical evidence that "... common forces underlie superficially different movements" (Portes and Walton 1981: 30).

Sassen (1988) extends the concept of structural imbalancing by focusing on the role of foreign direct investments (FDI) by core firms into non-core countries during the late twentieth century. Neoclassical economic theory argues that capital and labor are substitutes: capital inflows from core countries should reduce the need for labor to emigrate from non-core countries. From the world-systems perspective, however, capital flows from core countries do not necessarily inhibit emigration from non-core countries; indeed, they may actually promote emigration. In this respect, FDI in the export agriculture and export manufacturing sectors of non-core countries has three effects that make emigration more likely over time: it mobilizes segments of the population through direct incorporation as wage labor or by indirectly promoting rural-urban migration streams; it disrupts traditional work structures by feminizing labor structures and mobilizing labor; and it generates cultural-ideological linkages between non-core and core countries by westernizing incorporated segments of the labor force (Sassen 1988). Thus, FDI penetration is a manifestation of the broader process of incorporation into the world-system, which results in increased international migration: "In an 'isolated' country, that is one lacking extensive direct foreign investment, emigration would be quite unlikely to emerge as an option" (Sassen 1988: 20).

Incorporation into the world-system, however, does not only entail increased push-type pressures from penetration; it also creates pull-type influences that structure the destination points for international migrants. An important pull influence is economic restructuring in the core (Piore 1979). Here, the value of a global political-economic perspective is particularly apparent, as these two movements are viewed as flip-sides of a single trans-national process: the restructuring of world capital accumulation beginning in the 1960s. Deindustrialization in the core and the restructuring of core economies into service-oriented economies increased demand for both high-wage and low-wage service sector jobs, polarizing occupational and income distributions, and increasing the demand for immigrant labor, particularly in the low-wage service sector. Motivated by the need to sustain profitability in the face of rising wages in the core, corporations invest in production abroad, and this investment ultimately mobilizes segments of non-core populations into migration streams that are directed back toward the core to fill positions in the expanding service sectors of core economies:

> It can be argued that the recent large-scale emigration of capital from the northeastern United States, and the recent, large-scale immigration of workers from the periphery to the northeast, are two processes with a common origin.
>
> *(Sassen-Koob 1981: 77)*

Thus, Sassen extends and refines previous work by establishing an empirical association between three processes hitherto considered as analytically distinct: capital flows from the core to the non-core, emigration from the non-core to the core, and global economic restructuring. This thesis is expanded to include Southeast Asian migration to Japan (Sassen 1993), and is fully elaborated in The Global City (1991), which links the processes of deindustrialization in the core, foreign investment flows into the periphery, and international migration to the formation of "global cities" that serve as command and control centers for geographically dispersed global production networks. Sassen's framework is a push-pull perspective that is significantly modified to account for the global political-economic context. As such, it represents a significant advance in international migration research.

World-systems analysis continues to motivate research on international migration that builds upon the foundational works of Portes and Sassen. This research applies both qualitative and quantitative methodologies and marshals an extensive array of empirical data to extend and refine theoretical frameworks (see e.g., Mielants and Jones 2010). Again, what unites this body of research is an emphasis on identifying how the global political-economic context influences international migration.

Several studies use case studies to explore the relationship between world-system dynamics and international migration. Mac Laughlin (1994) relates Irish emigration to the peripheralization of the country, illustrating the processes through which Ireland was transformed into a "global emigrant nursery" for Great Britain and the United States (Mac Laughlin 1994: 257). Sione (1994) traces the movement of textile workers from Northern Italy to New Jersey in the late nineteenth and early twentieth centuries. She documents how these international migrations were associated with economic restructuring in the European textile industry, which expanded corporate connections to the United States and channeled the subsequent labor migration to the United States.

Migration from the Latin America and the Caribbean has garnered significant attention. Fernandez-Kelly's (1983; Nash and Fernandez-Kelly 1983) seminal work on Mexican migrants, gender, and the world division of labor laid the groundwork for a generation of research on the topic (see in particular Bickham-Mendez 2005; Parrenas 2001). Hamilton and Stoltz Chinchilla (1991, 1996) relate emigration from Central America to foreign capital penetration and structural change in the global political economy. Robinson (1992) places Latino migration to the United States in a world-systems context and more recent work develops this thesis (see in particular Chapters 3–4 of Robinson 2008). Delgado Wise and colleagues (2006; Delgado Wise and Covarrubias 2007, 2008; Delgado Wise and Cypher 2007) examine Mexican migration to the United States in the context of the NAFTA, arguing that economic integration between the two countries is based not only on international trade, but a "labor export-led model" comprised of maquiladoras, "disguised maquilas," and direct labor migration to the United States. Several studies expand the discussion beyond economic factors and consequences. Grosfoguel (1997) illustrates the historical connection between the United States' geopolitical strategies, both symbolic-ideological and military-security, and international migrations from the Caribbean in the Cold War and post-Cold War eras. Subsequent works examine Puerto Rican migration, focusing on incorporation and identity formation in the United States (Grosfoguel 1999, 2003). Nanton (1999) describes the process of identity formation among Caribbean migrants in Great Britain, focusing on the role of the state in defining and "othering," migrants. Sansone (1999) uses the cases of Surinamese and Antillean migration to the Netherlands to explore ethnic identity and allegiance against the common background of Dutch colonialism. More recent studies explore the concept of "transnationalism" among migrants in the world-system (Cervantes-Rodriguez et al 2009; Kyle 2000).

Other studies have applied quantitative techniques to cross-national datasets in order to explain how structural features of the world-economy are related to international migration flows in the late twentieth century. Ricketts' (1987) analysis of 18 Caribbean countries over the period 1970 to 1979 found that FDI increased both legal and illegal migration to the United States. Yang (1998) demonstrated that a composite measure of US trade and FDI increased the prevalence of emigration to the United States from less-developed countries over the period 1984 to 1993. Sanderson and Kentor (2008) found that stocks of foreign investment have a ten-year lagged positive effect on net emigration in LDCs between 1985 and 2000. They argued that this effect of FDI stocks on net emigration is due to the long-term negative effect of foreign capital dominance on economic growth. In a subsequent study, Sanderson and Kentor (2009) decomposed FDI stocks by sector (i.e., primary and secondary), to more explicitly test the FDI-migration hypothesis, and extended the time period to 1970–2000. Their findings indicated that the effects of FDI differed across sectors: primary sector FDI increased emigration, but secondary sector FDI decreased emigration. They also found empirical support for two other migration theories: the migration transition theory, in which emigration rises and then falls as GDP per capita levels increase (Martin and Taylor 1996); and the theory of cumulative causation, which argues that prior levels of emigration generate social networks that promote further emigration (Massey 1990). While most of the previous quantitative research on international migration tries to explain variation in emigration levels, recent work by Sanderson (2010) established the importance of international migration as an explanation of variation in development outcomes across countries, as immigration was found to be associated with lower scores on the Human Development Index in a sample of less-developed countries over a series of 10-year periods.

The promise of world-systems analysis

Over the past 35 years, world-systems approaches have significantly advanced knowledge in the area of international migration. Yet much work remains. In general, the literature on international migration remains limited by a disconnect between theory and empirical research and a lack of comparative research. World-systems analysis is particularly well suited to overcome these two obstacles, but it too suffers from these limitations.

Past research demonstrates a strong theoretical orientation and a relatively close connection with empirical data. Yet the use of data often remains illustrative, and with a few exceptions, world-systems studies do not often assess theoretical frameworks against competing theories: "Although the paradigm yields clear and researchable hypotheses … relatively little has been done to test these propositions systematically, and even less has been done to contrast them with competing theoretical models" (Massey et al 2005: 96).

Future work that assesses the relative explanatory or descriptive power of world-systems analysis would help incorporate world-systems approaches more thoroughly into the broader literature on international migration. This is not an argument to dilute or modify world-systems approaches; it is instead an appeal to strengthen these approaches through more systematic tests against competing explanations and theoretical frameworks.

Similarly, with a few exceptions, comparisons across countries, regions, and zones in the world-economy have not been sufficiently explored. There is a real need for research that identifies causal mechanisms linking international migration to world-system dynamics. Comparative research is crucial in order to move beyond country-specific knowledge:

> … what we have today is mostly an amorphous mass of data on immigration to different countries and a series of concepts whose scope seldom exceeds those of a particular nation-state … Needed are explicitly comparative projects that focus on research topics at a higher level of abstraction than those guiding policy concerns and that employ a common cross-national methodology.
>
> *(Portes 1997: 819)*

World-systems approaches are well suited to comparative research, using both qualitative and quantitative methodologies to elucidate how world-systemic processes are associated with international migration in particular historical contexts. However, country-specific case studies of migration that rely on anecdotal or illustrative data are all too common.

Future work incorporating empirical data to investigate international migrations across national contexts is especially warranted. There are many areas of research that merit further attention, but two are particularly worthwhile. We are only beginning to understand the multifaceted association between world-systemic processes, environmental change, and migration dynamics, yet these relationships will only become more important as climate change accelerates. World-system analysts have much to offer in this area by elaborating how, and when, the global political economy impinges on the environment-migration nexus. Similarly, as the world population urbanizes, it is becoming even more crucial to understand how world-systemic processes structure social change across the global urban hierarchy and within particular sites. There is already an established research agenda, drawing upon world-system analysis, that links transnational capital and migration flows with forms of social polarization in "global cities" (Sassen 2001; Smith and Timberlake 2001), but much work remains in identifying more specifically the role of migration in these processes and on the question of whether migration has similar consequences for social polarization in places less central in the world city-system.

In sum, the past 35 years of world-systems research on international migration has laid a strong foundation for future research, but the promise of world-systems analysis in this area has yet to be fulfilled.

References

Amin, S. (1976) *Unequal development: An essay on the social formations of peripheral capitalism.* New York: Monthly Review Press.

Bickham-Mendez, J. (2005) *From the revolution to the maquiladoras: Gender, labor, and globalization in Nicaragua*. Durham, NC: Duke University Press.

Burawoy, M. (1976) "The functions and reproduction of migrant labor: Comparative material from Southern Africa and the United States", *American Journal of Sociology*, 81: 1050–87.

Cervantes-Rodriguez, A. M., Grosfoguel, R., and Mielants, E. H. (eds) (2009) *Caribbean migration to Western Europe and the United States*. Philadelphia, PA: Temple University Press.

Delgado Wise, R. (2006) "Migration and imperialism: The Mexican workforce in the context of Nafta", *Latin American Perspectives*, 33: 33–45.

Delgado Wise, R. and Cypher, J. M. (2007) "The strategic role of Mexican labor under Nafta: Critical perspectives on current economic integration", *Annals of the American Academy of Political and Social Science*, 610: 119–42.

Delgado Wise, R. and Covarrubias, H. M. (2007) "The reshaping of Mexican labor exports under Nafta: Paradoxes and challenges", *International Migration Review*, 41: 656–79.

Delgado Wise, R. and Covarrubias, H. M. (2008) "Capitalist restructuring, development and labour migration: The Mexico-U.S. case", *Third World Quarterly*, 29: 1359–74.

Fernandez-Kelly, M. P. (1983) *For we are sold, I and my people: Women and industry in Mexico's frontier*. Albany: State University of New York Press.

Fernandez-Kelly, M. P. and Massey, D. S. (2007) "Borders for whom? The role of Nafta in Mexico-U.S. migration", *Annals of the American Academy of Political and Social Science*, 610: 98–118.

Frank, A. G. (1966) *The development of underdevelopment*. New York: Monthly Review Press.

Grosfoguel, R. (1997) "Migration and geopolitics in the Greater Antilles: From the Cold War to the post-Cold War", *Review*, 20: 115–45.

Grosfoguel, R. (1999) "Puerto Rican labor migration to the United States: Modes of incorporation, coloniality, and identities", *Review*, 22: 503–21.

Grosfoguel, R. (2003) *Colonial subjects: Puerto Ricans in a global perspective*. Berkeley: University of California Press.

Hamilton, N. and Chinchilla, N. S. (1991) "Central American migration: A framework for analysis", *Latin American Research Review*, 26: 75–110.

Hamilton, N. and Chinchilla, N. S. (1996) "Global economic restructuring and international migration", *International Migration*, 34: 195–227.

Kyle, D. (2000) *Transnational peasants: Migrations, networks, and ethnicity in Andean Ecuador*. Baltimore, MD: Johns Hopkins University Press.

Lee, E. (1966) "A theory of migration", *Demography*, 3: 47–57.

Lewis, W. A. (1954) "Economic development with unlimited supplies of labour", *Manchester School of Economic and Social Studies*, 22: 139–91.

Mac Laughlin, J. (1994) "Emigration and the peripheralization of Ireland in the global economy", *Review*, 17: 243–73.

Martin, P. and Taylor, J. E. (1996) "The anatomy of a migration hump", pp. 43–62 in J. Edward Taylor (ed.), *Development strategy, employment, and migration: Insights from models*, Paris: Organization for Economic Cooperation and Development.

Massey, D. S. (1990) "Social structure, household strategies, and the cumulative causation of migration", *Population Index*, 56: 3–26.

Massey, D. S. (1999) "International migration at the dawn of the twenty-first century: The role of the state", *Population and Development Review*, 25: 303–22.

Massey, D. S., et al (2005) *Worlds in motion: Understanding international migration at the end of the millennium*. Oxford: Clarendon Press.

Mielants, E. H. and Jones, T. A. (eds) (2010) *Mass migration in the world-system: Past, present and future*. Boulder, CO: Paradigm Press.

Morawska, E. (1990) "The sociology and historiography of immigration", pp. 187–240 in Virginia Yans-McLaughlin (ed.), *Immigration reconsidered: History, sociology, and politics*, New York: Oxford University Press.

Nanton, P. (1999) "Migration dynamics: Great Britain and the Caribbean", *Review*, 22: 449–69.

Nash, J. and Fernandez-Kelly, M. P. (eds) (1983) *Women, men, and the international division of labor*. Albany, NY: State University of New York Press.

Parrenas, R. S. (2001) *Servants of globalization: Women, migration, and domestic work*. Berkeley: University of California Press.

Petras, E. M. (1980) "The role of national boundaries in a cross-national labour market", *International Journal of Urban and Regional Research*, 4: 157–94.

Petras, E. M. (1981) "The global labor market in the modern world-economy", pp. 44–63 in Mary M. Kritz, Charles B. Keely, and Silvano M. Tomasi (eds.), *Global trends in migration: Theory and research on international population movements*, Staten Island, NY: Center for Migration Studies).
Piore, M. (1979) *Birds of passage: Migrant labor in industrial societies*. Cambridge: Cambridge University Press.
Portes, A. (1978a) "Toward a structural analysis of illegal (undocumented) immigration", *International Migration Review*, 12: 469–84.
Portes, A. (1978b) "Migration and underdevelopment", *Politics and Society*, 8: 1–48.
Portes, A. (1979) "Illegal immigration and the international system: Lessons from recent legal Mexican immigrants to the United States", *Social Problems*, 26: 425–38.
Portes, A. (1997) "Immigration theory for a new century: Some problems and opportunities", *International Migration Review*, 31: 799–825.
Portes, A. (2007) "Migration, development, and segmented assimilation: A conceptual review of the evidence", *Annals of the American Academy of Political and Social Science*, 610: 73–97.
Portes, A. and Walton, J. (1981) *Labor, class, and the international system*. New York: Academic Press.
Ranis, G. and Fei, J. C. H. (1961) "A theory of economic development", *American Economic Review*, 51: 533–65.
Ravenstein, E. G. (1885) "The laws of migration", *Journal of the Royal Statistical Society*, 48: 167–235.
Ravenstein, E. G. (1889) "The laws of migration", *Journal of the Royal Statistical Society*, 52: 241–235.
Ricketts, E. (1987) "U.S. Investment and immigration from the Caribbean", *Social Problems*, 34: 374–87.
Robinson, W. I. (1992) "The global economy and the Latino populations in the United States: A world systems approach", *Critical Sociology*, 19: 29–59.
Robinson, W. I. (2008) *Latin America and global capitalism*. Baltimore, MD: Johns Hopkins University Press.
Sanderson, M. R. (2010) "International migration and human development: A cross-national analysis", *Social Indicators Research*, 96: 59–83.
Sanderson, M. R. and Kentor, J. D. (2008) "Foreign direct investment and international migration: A cross-national analysis of less-developed countries, 1985–2000", *International Sociology*, 23: 515–40.
Sanderson, M. R. and Kentor, J. D. (2009) "Globalization, development, and international migration: A cross-national analysis of less-developed countries, 1970–2000", *Social Forces*, 88: 301–26.
Sansone, L. (1999) "Small places, large migrations: Notes on the specificity of the population of Surinamese and Antillean origin in the Netherlands", *Review*, 22: 471–501.
Sassen-Koob, S. (1978) "The international circulation of resources and development: The case of migrant labour", *Development and Change*, 9: 509–45.
Sassen-Koob, S. (1981) "Towards a conceptualization of immigrant labor", *Social Problems*, 29: 65–85.
Sassen, S. (1988) *The mobility of labor and capital: a study in international investment and labor flow*. Cambridge: Cambridge University Press.
Sassen, S. (1993) "Economic internationalization: The new migration in Japan and the United States", *International Migration*, 31: 73–102.
Sassen, S. (2001) *The global city: New York, London, Tokyo*. Princeton, NJ: Princeton University Press.
Smith, D. and Timberlake, M. (2001) "World City Networks and Hierarchies, 1977–97: an Empirical Analysis of Global Air Travel Links," *American Behavioral Scientist*, 44: 1656–78.
Sione, P. (1994) "Patterns of international migrations: Italian silk workers in New Jersey, USA", *Review*, 17: 555–76.
Thomas, B. (1973) *Migration and economic growth: A study of Great Britain and the Atlantic economy*. Cambridge: Cambdrige University Press.
Thomas, D. S. (1941) *Social and economic aspects of Swedish population movements*, 1750–1933. New York: Macmillan.
Todaro, M. P. (1969) "A model of labor migration and urban unemployment in less developed countries", *American Economic Review*, 59: 138–48.
United Nations. (2009) *Trends in the migrant stock: The 2008 revision*. New York: Department of Economic and Social Affairs.
Wallerstein, I. (1974) *The modern world-system: Capitalist agriculture and the origins of the european world-economy in the sixteenth century*. New York: Academic Press.
Yang, P. Q. (1998) "The demand for immigration to the United States", *Population and Environment*, 19: 357–83.
Zolberg, A. R. (1978) "International migration policies in a changing world-system", pp. 241–86 in William H. McNeill and Ruth Adams (eds), *Human migration: patterns and policies*, Bloomington: Indiana University Press.
Zolberg, A. R. (1989) "The next waves: migration theory for a changing world", *International Migration Review*, 23: 403–30.

14.3

The world-system, inequality and violent conflict

Shifting the unit of analysis

Kevin Doran

Scholars have long sought to explain cross-national variation in levels of violent conflict. One area of research, generally focused on civil war prevalence, has highlighted the importance of the opportunity structure for engagement, as well as national economic and political indicators, in generating violence (Sambanis 2004). Adding another level of complexity to this first area of research, a second area has incorporated insights from the world-systems and world-polity perspectives into our understanding of the macro-level sources of violent conflict. In doing so, these scholars highlight the importance of stratification in the world-system in patterning violence (cf. Olzak 2006). This essay will begin by briefly summarizing what this latter area of research suggests the world-systems perspective offers to research on violent conflict, and will then outline the benefits of taking the individual as the unit of analysis in studies of violent conflict.

In examining ethnic conflict, Olzak (2006) provides a detailed discussion of the theoretical implications of integration into the world-system on violent conflict. To simplify her discussion, integration widens the economic gaps both within and between countries. In diffusing global capitalism, economic integration solidifies the dependency of peripheral states on core states, while integration into the world political system promotes the spread of human rights ideologies. This creates a situation in peripheralized states where poverty levels, economic inequality, and political inequality intensify, while the legitimacy of those inequalities is increasingly called into question. This, with the propensity for political instability and repression in peripheralized countries, increases the likelihood of violent political activity. In addition to direct effects then, integration is expected to have indirect effects on violence through inequality, political exclusion, and poor economic performance in peripheral states. This account is especially useful, as it explicitly incorporates global processes with national phenomena, but it raises several issues that suggest a need to examine these questions using data on individuals.

First, it raises the question of how these factors actually influence individuals. What, for instance, does it mean, in terms of individual motivations, to say that increases in GDP per capita are correlated with decreased intensity of violent activity? While group-level analyses move research toward answering these sorts of questions, recent work focusing on Nigeria (Oyefusi 2008) takes an important step toward integrating the theoretical insights from cross-national studies with data on individual propensities for violent activity. Oyefusi (2008) finds that having greater opportunities for profit from rebel movements (e.g., the proximity of large oil reserves) and fewer opportunities to support oneself outside of rebel activity increases the likelihood that Nigerian youths will be willing to join rebel movements, while personal grievance is less important. If we seek to understand which individuals are most at risk to engage in violence and the ways in which global processes generate this risk, research should extend this sort of analysis to examine other regions of interest or cross-national comparisons, as well as to incorporate other theoretically important explanatory variables.

Second, integration into the world-system is expected to influence violence through political and economic inequalities; however, many cross-national analyses of violence fail to find a significant effect of income inequality. The lack of evidence has been suggested to be the result of using national, rather than ethnic or regional, measures of inequality as predictors (Sambanis 2004). This critique of economic inequality measurement highlights the need for a closer examination of inequality's role in violent conflict formation. Specifically, arguments concerning the impact of unequal distributions of material and non-material assets on violence rely on the actor's perception that their level of consumption, relative either to a previous level of consumption or to others in the society, is unacceptable (Gurr 1970). As analyses of individuals allow researchers to actually examine the actor's perception of their social environment, they are particularly useful for examining the impact of inequalities on violent activity in that they alleviate the issue of operationalizing inequality in a way that may not be relevant to the individual actors or may not tap into the correct reference group (e.g., ethnic groups or the individual's past position).

Analyzing the individual's perception of their relative social position, then, would benefit our understanding of violent conflict in several ways. First, researchers could distinguish between the impact of unequal distributions, perceived and real, of particular material and non-material goods (e.g., income and access to effective political channels) on individual propensity for violence. Additionally, it allows for the identification of the relevant reference groups against which comparisons are made. For instance, are perceived inequities between ethnic groups or geographic regions the primary forms of inequality motivating violent action? Furthermore, researchers could examine the assumption that it is primarily those ranked lower in social hierarchies that are mobilized by inequalities. As integration into the world political system is expected to promote the eradication of systematic political and economic inequalities, it may be the case that those who hold politically or economically privileged positions feel threatened and turn to violent tactics to retain their advantaged position (Olzak 2006). Taking the individual as the unit of analysis, then, would not only provide a better understanding of how the process of integration into the word-system impacts individuals' propensities for engagement in violent conflict, but also a more nuanced view of inequality's role in this process. Combined with existing knowledge about the national determinants of violence, studies of this sort would thus provide a more complete understanding of the global and national processes driving violent conflict.

References

Gurr, T. R. (1970) *Why Men Rebel*, Princeton, NJ: Princeton University Press.

Olzak, Susan. (2006) *The Global Dynamics of Racial and Ethnic Mobilization*, Stanford, CA: Stanford University Press.

Oyefusi, Aderoju. (2008) "Oil and the probability of Rebel Participation among Youths in the Niger Delta of Nigeria", *Journal of Peace Research*, 45 (4): 539–555.

Sambanis, Nicholas. (2004) "Poverty and the organization of political violence: A review and some conjectures", In Susan M. Collins and Carol Graham (eds) *Brookings Trade Forum*, pp. 165–222. Washington, DC: Brookings Institution.

14.4

Child marriage in India

An overview

Golam Sarwar Khan

Marriage is regarded as child marriage when girls are married off before attaining puberty. Child marriage in India apparently aligns with pre-capitalist economic mode and indicates a form of commodification of women. It is assumed that child marriage not only adheres to retaining cultural traditions, but has political and economic overtones as well. Given the cost–benefit analyses in the social realm, the patronization of child marriages in India can be interpreted in line with the expansive capitalist development.

Child marriage is observed in the Indian subcontinent, sub-Saharan Africa, Latin America, and in the Caribbean (Levine et al 2009). Although there are instances of child marriage in India, it is not widely practiced (Dutta 2002; Nagi 1993). The custom of child marriage originated in the early medieval ages (Birodkar 1994), when the legal framework of marriage was not developed. At that stage, the powerful despotic monarch and patriarchs acted according to their whims. The prevailing law and order favored the autocratic dictators and fulfilled their wishes only. As against modernity, it is comparable to the feudal authority but concomitant to the political economy of world-systems (Frank 1967; Wallerstein 1976), because instituting child marriage involves economical aspirations.

The practice of child marriage in India is deeply rooted in cultural values and grounded in social structures. Though forbidden, the existence of child marriage is still a reality in India. A recent estimate shows that in Rajasthan, about 80 percent of marriages among girls involve those under the age of fifteen (Gupta 2005: 2). However, in civil societies, child marriage is viewed as a curse because it violates child rights. Despite legal prohibition, child marriage continues in India and elsewhere. It is argued that child marriages are common in sub-Saharan Africa and in rural South Asia for "safety issues," that is, the fear of AIDS/HIV, avoidance of pre-marital sex, and also an extension of the economic security of the parents.

From a male perspective, child marriage in Indian society has a significant role in the continuance of marriage rituals of the dowry-demand system. The dowry system provides economic incentives to the boys' parents and influences child marriages for young girls and boys at a tender age (Nand 1989). Despite laws prohibiting child marriages in India (Lahiry 2004), the pre-fixing and quick realization of dowry demand indirectly violates the law. Further, child marriage is patriarchal in nature because the demand is essentially met for the bridegroom only.

The rigid Hindu caste system contributes to the continuance of child marriages in India. Since the caste identity and hierarchy are based on birth, they do not allow inter-marriages between different castes in order to maintain purity and avoid pollution of the caste hierarchy (Dumont 1972; Srinivas 1969). This caste authority can easily be applied to the children, whereas adults are likely to violate such injunctions (Birodkar 1994) in the wave of post-modernity.

Child marriage is socially and culturally accepted in both rural and urban India. Despite continued resistance by the government and non-government organizations, the promotion of child marriages is observed. The following reasons of promotion are outlined below:

1. Patriarchal value: girls have limited access to opportunities in family and social life and hence should remain subservient to men for life.
2. Safety issues: child marriage emphasizes virginity, fear of pre-marital sex, incidence of sexual violence, elopement, and kidnapping, thereby upholding family prestige.
3. Economic reasons: child marriage alleviates the economic liability of the parents of girls from poverty and unemployment.
4. Caste system: child marriage helps to maintain caste purity against caste pollution. Hindus believe in and strictly value the caste priority and hierarchy.

Further, the political and economic affairs under the caste system tend to corroborate a sense of "deliberate control by political authorities over economic affairs" emphasizing economic well-being in society (Friedman 2002: 196). Child marriage also affects girls' individual experience as social actors. Early marriage negatively affects girls' social networks, decision-making power, and ability to negotiate with partners (Bruce 2007).

The patriarchal structure of child marriage (Sagade 2005) in India (according to ancient scriptures, early marriage for girls is preferable either before puberty or immediately after their first menstruation) exists on the face of the feminist movement along capitalist expansion all over the world. Nonetheless, the existing social custom of child marriage is powerful enough to prolong the economic motif of dowry demand system in India. The system of demanding dowry as well as realization of it in the form of money and material goods should be understood as a prelude to expanding the capitalist system.

References

Birodkar, S. (1994) "Child marriage (Bal Vivaaha)". Available at: http://www.vivaaha.org/child.htm accessed July 05, 2010.

Bruce, J. (2007, September) *Child Marriage in the Context of the HIV Epidemic* (Brief no. 11). Available at: http://www.popcouncil.org/pgy.

Dumont, L. (1972) *Homo Hierarchicus: The Caste System and Its Implications*. New Delhi, India: Oxford University Press.

Dutta, A. (2002) *Child Marriage an Adult Obsession: Bengal 1891–1929*. Kolkata, India: Minerva Associates.

Frank, A.G. (1967) *Capitalism and Underdevelopment in Latin America: Historical Studies of Chile and Brazil*. New York: Monthly Review Press.

Friedman, M. (2002) *Capitalism and Freedom*. Chicago, IL, and London: University of Chicago Press.

Gupta, G. R. (2005, September 14) *Forum on Child Marriage in Developing Countries*. Forum presented at U.S. Department of State.

Lahiry, R. (2004) "Crackdown on child marriages", *The Telegraph* (India), Saturday, December11, 2004.

Levine, R., Lloyd, C. B., Greene, M., and Grown, C. (2009) *Girls Count A Global Investment And Action Agenda: A Girls Count Report On Adolescent Girls*. Washington, DC: The Center for Global Development.

Nagi, B. S. (1993) *Child Marriage in India: A Study of Its Differential Patterns in Rajasthan*. Rajasthan, India: Mittal Publications.

Nand, L. C. (1989) *Women in Delhi Sultanate*. Allahabad, India: Vohra Publishers & Distributors.

Sagade, J. (2005) *Child Marriages in India: Socio-legal and Human Rights Dimensions*. New Delhi, India: Oxford University Press.

Srinivas, M. N. (1969) "The Caste System in India", in A. Beteille (ed.) *Social Inequality*, Harmondsworth: Penguin Books.

Wallerstein, I. (1976) *The Modern World-System: Capitalist Agriculture and the Origins of the European World-Economy in the Sixteenth Century*. New York: Academic Press.

14.5

The migration of reproductive labor from the periphery to the core and semiperiphery under neoliberal globalization

Ligaya Lindio-McGovern

Neoliberal globalization has exacerbated class, gender, and racial inequalities between the North and South and within regions (Lindio-McGovern and Wallimann 2009). One of the conceptual sites to examine this pattern is the migration of reproductive labor from the periphery to the core and semiperiphery. Neolibral globalization is conceived here as the global expansion of capitalism through the promotion of neoliberal policies of deregulation of the market, privatization by dismantling state-owned or state-subsidized enterprises, economic liberalization on the global mobility of capital and goods, and labor flexibilization to create an abundant supply of cheap labor (Lindio-McGovern and Wallimann 2010). Reproductive labor refers to wage labor in households that usually includes the care of children and the elderly members of the household, cooking, cleaning, and other tasks involved in maintaining the household.

Women's participation in the international labor migration has increased, mainly for economic reasons. Their global migratory flows variably range from 50 to 70 to 80 percent (UNESCO 2006). Globalization has accelerated women's migration, as it simultaneously demands the cheap labor of Third World women in the "secondary tiers of the manufacturing and service sectors in advanced capitalist countries." (Parrenas 2000)

The structural adjustment policies of the IMF, the World Bank, and the WTO in the labor-sending countries result in debt crisis and increased unemployment—creating the precondition for overseas migration and labor export (Lindio-McGovern and Wallimann 2007). How is this precondition created? It is an inherent contradiction of capitalist globalization to create surplus labor in its persistent search for cheap labor. Capitalist imperialism creates surplus labor through its logic of "accumulation of power through dispossession" (Harvey 2007). This is clearly illustrated when transnational corporations acquire land in Third World countries and displace peasants and indigenous communities. It is not a mere historical coincidence when neoliberal states violently preclude agrarian reform that will redistribute land to landless peasants. Land redistribution will diminish the creation of surplus labor that can be made to supply cheap labor in transnational corporations' agribusiness that largely employ men and in Free Trade Zones that largely employ women, many from the rural areas. Surplus labor that is not absorbed in this transnational circuit of capital and labor displaced from the local industries destroyed by the dominance of transnational corporations now becomes the potential labor for export—to provide cheap labor for the maintenance of capitalism in the core, in the heartland of empires, and in the New Industrializing Countries. Labor export then becomes a neoliberal instrument to maintain global capitalism, coupled with privatization that opens new spheres for capitalist penetration of the whole world. Men and women are entangled in this circuit, but migrant women's experiences get complicated as they confront oppression and discrimination based on class, gender, race, ethnicity, nationality, and immigrant status.

Labor export has "commercialized migration of domestic workers" (UNESCO 2006), resulting in their commodification as governments and employment agencies profit from the trade. This commodification is even more reinforced when their cheap, flexible, and disposable labor in effect subsidizes the privatization of care work in the core and semiperiphery countries.

Women's migration to domestic work is predominantly from the periphery to the core of the global North, but there is also a trend of regionalization: the flow from poorer countries to richer countries within regions, for example, within East Asia, Southeast Asia (Piper 2003), or southern Africa (Ally 2009). Capitalist globalization is not only reinforcing the inequality between the North and South, but it is also producing inequalities within regions wherein there develops an urban hub where women from poorer countries within regions migrate to work as domestic workers in middle-class or wealthy households. This is exemplified by the migration of Filipino, Indonesian, Sri Lankan, and Nepali women to work as domestic workers in Hong Kong, Taiwan, and Singapore (Lindio-McGovern and Wallimann 2005) and the black African women from the poorer neighboring countries of southern Africa (Mozambique, Zimbabwe and Bosotho) to the urban hub of Johannesburg, South Africa (Ally 2009).

In conclusion, the increasing migration of women to work as low-waged domestic workers from the periphery to the core and semiperiphery of the global capitalist system is indicative of the growing gendered, raced, and classed inequalities between the North and South and within regions under neoliberal regimes. Therefore, it is still appropriate to talk about core, periphery, and semiperiphery in the dynamics of the world-system, as forms of inequality continue to be configured and reconfigured under capitalist globalization.

References

Ally, Shireen 2009. "Regional Division of Reproductive Labor: Southern African Migrant Domestic Workers in Johannesburg", pp. 15–33 in Lindio-McGovern, Ligaya and Wallimann, Isidor (eds), *Globalization and Third World Women: Exploitation, Coping and Resistance*. Surrey: Ashgate Publishing Company.

Harvey, David 2007. *A Brief History of Neoliberalism*. New York: Oxford University Press.

Lindio-McGovern, Ligaya and Wallimann, Isidor 2005. "Alienation and Labor Export in the Context of Globalization: Filipino Migrant Domestic Workers in Taiwan and Hong Kong", *Critical Asian Studies* 36: 2, 217–38.

Lindio-McGovern, Ligaya and Wallimann, Isidor 2007. "Neoliberal Globalization in the Philippines: Its Impact on Filipino Women and Their Forms of Resistance", *Journal of Developing Societies* 23: 1–2, 15–35.

Lindio-McGovern, Ligaya and Wallimann, Isidor (eds), 2009. *Globalization and Third World Women: Exploitation, Coping, and Resistance*. Surrey: Ashgate Publishing Company.

Lindio-McGovern, Ligaya and Wallimann, Isidor 2010. "The Global Structuring of Gender, Race, and Class: Conceptual Sites of Its Dynamics and Resistance in the Philippine Experience", pp. 191–204 in Dello Buono, Richard A. and Fasenfest, David (eds), *Social Change. Resistance and Social Practices*. Leiden and Boston, MA: Brill.

Parrenas, Rachel Salazar 2000. "Migrant Filipina Domestic Workers and the International Division of Reproductive Labor", *Gender and Society* 14: 4, 560–80.

Piper, Nicole 2003. "Feminization of Labor Migration as Violence Against women: International, Regional, and Local Nongovernmental Organization Responses in Asia", *Violence Against Women* 9: 723–45.
United Nations Economic and Social Council (UNESCO), POP/945, April 5, 2006, Department of Public Information, News and Media Division, New York, "The Feminization of Migration, Remittances, Migrants' Rights, Brain Drain among Issues, As Population Commission Concludes Debate", http://www.un.org/News/Press/docs/2006/pop945.doc.htm. Accessed 8/10/2010.

14.6

Impacts of individualism on world-system transformation

Roksolana Suchowerska

Thinking of an impending world-system transformation can be both daunting and exciting—as yet the new world-system holds the power of the unknown over us. We find ourselves debating how factors such as economics, creation of knowledge, culture, values, wealth, and power will influence the transformation to shape a new world-system. Change can be considered both in terms of structural change as well as changes in the behavior of individual agents. In this short essay I address the latter, asking: what are the generational traits of the individuals that will be creating and experiencing systemic change? And does this provide any indication of how a new world-system might emerge?

Contemporary youth in a number of current "core" countries (herein "youth") are researched to participate in electoral politics at a lower rate than their adult counterparts (Dalton 2008; Putnam 2000; Zukin et al 2006). Youth's voting rate is not only lower than adults' but often decreases with each election; youth are less likely to join political parties than their parents were at a young age; and youth are becoming more critical of how democracy operates in contemporary society. Youth's declining engagement in electoral politics can been explained by the simultaneous rise of new politics which prioritises self-actualization, aesthetic needs, global responsibility, and full-scale democratization over classic political issues such as the redistribution of wealth (Zukin et al 2006). New politics have emerged because youth respond to the fragmented and unstable structures of late modernity by constantly managing individualized projects of risk assessment and self-actualization.

Youth are able to manage individualized projects of risk assessment and self-actualization because of the technological advances which have led to the emergence of the network society (Castells 2000). In the contemporary network society, communication has been liberalized by the rise of mobile phones and the Internet (Wellman 2002). Individuals have more choice and agency regarding with whom they communicate and subsequently which values, ideals, and interests they absorb. Time and space are virtually insignificant (Castells 2000). As a consequence of this hypered agency and individualism, youth's political and social identities are qualitatively different to those of preceding generations. Youth understand change and governance in individualistic terms, just as they understand the process of decision-making and empowerment in individualistic terms.

Youth's exploration of new modes of empowerment and political engagement sees them retreating from the idea that the nation-state is the dominant unit of social organization. This is interesting, because individuals that are most involved in exploring new modes of political engagement are likely to pioneer a new world-system. In line with Boswell's thoughts, "it is exactly during the formation of a new world order that revolutions and social movements from below can have the greatest influence on the shape of the world system" (Boswell 2002). Although a new world-system will be defined by new dynamics and currencies, I suggest that agents that have greater access to information and knowledge-creation in the current world-system are likely to have a greater influence on how a new world-system emerges.

There are risks, however, associated with a new world-system emerging out of youth's current individualistic behavior and retreat from electoral politics. One risk may be that retreating from electoral politics may not necessarily diminish the power of the state—it would merely diminish its democratic integrity. Without citizens' political engagement and review, parliaments may choose to abandon a multitude of social support functions including health, housing, and income support whilst increasing protectionist mechanisms such as border security. While youth may engage in the issues they feel passionate about by joining virtual communities and movements, their physical surroundings may remain heavily governed by state policies.

A second consideration should be how access to knowledge-creation and dissemination will affect one's influence over the emergence of a new world-system. Access to knowledge and information is seminal in evaluating one's level of agency and to some degree, power. In the current system, trading patterns are what determine power relations—who exists in the core, and who in the periphery. Advocates of the knowledge economy may argue that the trading of knowledge, data, and ideas through virtual communities will have an impact on currencies in the new world-system. A system hegemon, however masked, is certainly possible within a knowledge economy. However, perhaps the diversity and fragmentation of society that arises from late modernity's uncertainties will pose a barrier to the hegemon having unconditional influence.

References

Boswell, T. (2002) "Hegemonic Decline and Revolution: When the World is up for Grabs" from *World-System Conference*, University of California, Riverside, 3 May 2002. Downloaded from The Institute for Research on World-Systems Website www.irows.ucr.edu/conferences/pews02/pprboswell.doc, accessed 9 January, 2010.

Castells, M. (2000) "Materials for an Exploratory Theory of the Network Society" *British Journal of Sociology*, 51(1): 5–24.

Dalton, R. J. (2008) *The Good Citizen: How a Younger Generation Is Reshaping American Politics*. Washington, DC: CQ Press.

Putnam, R. (2000) *Bowling Alone: The Collapse and Revival of American Community*. New York: Simon & Schuster.

Wellman, B. (2002) "Little Boxes, Glocalization, and Networked Individualism" in M. Tanabe, P. van den Besselaar and T. Ishida (eds) *Digital Cities II: Computational and Sociological Approaches*, from second Kyoto workshop on Digital Cities: October 18–20, 2001, pp. 10–25.

Zukin, C., S. Keeter, M. Andolina, K. Jenkins and M. X. D. Carpini (2006) *A New Engagement?: Political Participation, Civic Life and the Changing American Citizen*. Oxford: Oxford University Press.

15. International and transnational interactions

15.1
Geography and war

Albert J. Bergesen

The general theory of hegemonic succession wars in world-system theory is similar to that of power transition wars in political science/IR. Given uneven global economic development, some states become more dominant than others and they not only rise but decline as well, and importantly, their decline is tied to the rise of others. Eventually a point is reached where rising competitors are so strong, and a declining hegemon so weak, that international order can no longer encase the aspirations of rising, or the defensiveness of falling, and great power war breaks out.

Within the life span of the Wallerstein (1974) world-system, Spain was the first hegemon and with its decline, competition among rising powers emerged over who would follow. It was eventually the British, but to get there they had key wars with Dutch aspirants in the mid-seventeenth century and the French in the Seven Years War (1756–63) and Napoleonic Wars (1803–15). At that point, with rivals vanquished, Britain went on to world-system hegemony in the nineteenth century. By 1914, though, Britain was in decline and a new set of aspirants, led by Germany and the United States, were on the rise. All of them—declining and rising powers—were caught up in another hegemonic succession struggle comprising World War's One and Two (1914–45). Victorious, the United States went on to dominate the second half of the twentieth century. But by the early twenty-first century the United States seems to be clearly in decline while new powers like China and India seem on the rise. Whether a world war over succession to American hegemony comes to pass remains to be seen, but it has long been speculated that the Triple Entente and Triple Alliance of our time will be Washington, Tokyo, and Beijing on one side, and Moscow, Berlin, and Paris on the other, with London/Delhi and perhaps others on the bubble.

But so-called hegemonic transition wars can also be approached through the lens of geography and geopolitics, perspectives that have been out of favor for over a century but which seem increasingly relevant to the twenty-first century polycentric international system. For example, if you re-examine successful world-system hegemons and failed aspirants, you will note that the hegemons have been predominantly sea powers (like Spain, Britain, the United States) while defeated aspirants were predominantly land powers (France and Germany). That hegemons are predominantly sea powers, and the rising, but ultimately failing, aspirants land powers, is an interesting mix of geopolitical and world-systemic dynamics and suggests a rethinking of what is the base and superstructure of the world-system. The base has always been the world-economy, but when oceans and continents co-vary with cycles of hegemony and great power wars, then perhaps the true base of the world-system lies with the deeper geographical substrate, making the world-economy a globological superstructure. This association of hegemony/aspirant powers with water/land could, of course, be spurious, so let's examine this association from a longer historical perspective.

Here we note that the earliest empires were tied to rivers: Pharonic Egypt on the Nile; Mesopotamia between the Tigris and Euphrates; and early Chinese dynasties between the Yellow and Yangtze. Then, in the Western story of power, with a shift toward larger bodies of water, hegemonic centers of power went from the eastern Mediterranean (Athenian Empire) to the western Mediterranean (Carthaginian Empire) to the whole Mediterranean (Roman Empire) and then to the oceanic-based sea power of the Atlantic and partial Pacific (Spanish Empire), to the Atlantic, Pacific, and Indian oceans (British Empire), to all oceans (American Empire). There were, of course, ancient and modern empires that were land based, and some mixed land/sea power, but in general, hegemonic sea power and hegemonic economic power seem to co-vary. There is also, of course, the rise of air and space power, and they will have to be included in future discussions. Note, though, that into the twenty-first century 90 percent of world trade is done by shipping.

If there is a possible causal relation here it probably lies with the fact that water is something of an economic multiplier. From cutting transportation and transaction costs in the import of raw materials and export of finished goods, to providing access to wider markets, and synergies generated by port cities that fruitfully mix cultures, technologies, religions, and virtually anything else that might go into innovation in economic practices, river, sea, and ocean based, or accessed, economies possess a comparative advantage over the solely land based.

There are obvious land-based trade routes (famously the Silk Road), but nonetheless it seems that sea-based economies are more defined by their emphasis upon exchange relations (overseas trade), and land economies by their emphasis upon relations in domestic production (class relations). Behind this observation lie more abstract debates about the relative priority of exchange vs. production, or Smith vs. Marx, which we can approach from a geographical perspective. In general, land constrained consciousness produces thinkers like Karl Marx, who emphasize the role of territorially bounded economic relations, that is, production, or class relations. These sum up to territorially rooted economic wholes, like "modes of production," with exploitative divisions amongst the domestic populace constituting the origin of wealth.

Alternatively, sea constrained consciousness produces theorists like Adam Smith, who place more emphasis upon trade and exchange relations. Here the wealth of nations arises from the acquisition of factors of production and the dispersal of product, for example, exchange relations between different land areas separated by water, that is, overseas trade. Wealth for land theorists, conversely, arises from the more immediate relations of the territorially bound population in a trapped finite political universe. Hence the advance of one sub-division of the territorial whole has to, by definition, come from the wealth of the other sub-division, making the land theorists perspective on social life that of perpetual class warfare. Marx's argument that, "The history of all hitherto existing society is the history of class struggles" is nothing but land theoretically expressing itself as a theory of economic relations. But for sea consciousness expressing itself in theory, wealth arises out of water's presence to bring factors of production in, and transport product out, and so it is out of this indirect relation with people afar (overseas trade/exchange) that produces the wealth of nations.

Water access accelerates economic growth, but growth also occurs with land-based economies, which provides the foundation for their political expansion. The key point here is that economic growth has different political manifestations for sea versus land powers. This is because, given a natural boundary comprising water, like oceans, the transformation of economic power into expansive political power (empire) is constrained where the land meets the sea. In order to politically expand and to move armies into other polities' territories, armies must be placed upon boats and then shipped to their destination.

Land powers, conversely, also grow economically and that growth can be translated into political expansion. However, there is no need to have armies board ships, for they can be directly marched into their neighbor's territory. During the Napoleonic Wars, France invaded German and Russian territory; and then they invaded back. Germany under Hitler invaded France and Russia; and the Russians pushed back and invaded Germany. All of this was possible because these states existed geographically next to each other on the great north European plain. This suggests that for land powers, economic expansion can be more easily translated into direct political expansion at the expense of territorial neighbors, whereas direct territorial expansion is more difficult for sea powers. Nineteenth-century German industrial expansion fueled the growth of the German Empire, which grew through the Franco-Prussian War (1870–71) and the Austro-Prussian War (1866). Although Britain was economically expanding at the same time, there was no invasion of the continent to create a German-like British Empire. There was a British Empire, of course, but it was an overseas empire, as was the empire of Imperial Spain and ancient Athens.

Given the limitation of having to ship armies, sea powers have been primarily expeditionary forces, which are, of necessity, smaller than land-based armies and thereby limited in what opposition they can hope to successfully engage. In effect, given the water-induced constraint of dependency upon expeditionary forces, military engagements shift to what can actually be conquered, and that has involved shifts to the developing world and the acquisition of colonial holdings.

Put more boldly: sea powers, if they develop, which given the economic multiplier of water they inevitably will, have to become colonial powers, and to develop overseas empires. Athens did not conquer and enslave territorial neighbors as did Sparta. But with triremes working hand in hand with trade and exchange power, Athens conquered weaker Aegean island city-states and built a maritime empire. No doubt a Spartan leader said something similar to the Kaiser, who observed that if England were to try to invade Europe, he would send out a local constable and have their army arrested. Athenians, by and large, avoided direct engagements with the Spartan phalanx. Again, the dynamic here is general. Imperial Spain held the Spanish Netherlands and dominated the Holy Roman Empire, but did not expand territorial Spain beyond the Pyrenees. New Spain, though, is in the Americas as an overseas colonial holding facilitated by Spanish galleon sea power. Similarly, hegemonic Britain did not expand by enlarging to include territory in Europe, but rather expanded overseas, and again it was based on a sea power combination of trade and the Royal Navy. Finally, after the continental expansion of the United States (which again, was an act of land power), the power projection of the United States was not based on the incorporation of Canada and Mexico into an even larger United States of North America, but instead took the initial form of overseas colonies and then under full hegemony of bases and periodic interventions. It was, as with Spain and Britain, another naval projection of power over the oceanic commons.

All of this suggests that the land-bound and sea-accessed undergoing economic expansion generate two different patterns of empire. For land-based economies, transforming economic expansion into political expansion results in what could be called empire at neighbors' expense—a greater Sparta, a German Empire, a Third Reich, a Napoleonic Empire, and a Czarist Russia. For expanding sea-based economies, the result is overseas maritime empires—the fifth-century BCE trireme backed Athenian imperium, the sixteenth-century Spanish galleon backed colonial holdings in the Americas, the nineteenth-century Royal Navy backed British Empire in India, Africa, and elsewhere, and twentieth/twenty-first century aircraft carrier backed American imperial influence everywhere there was water.

There is an interesting irony here. It is the weakness of an army's mobility provided by ocean boundaries that forces sea powers to primarily engage weaker opponents, hence to only engage the developing world, hence to only develop colonial empires. If it were not for the national boundary of water, armies could be marched right into neighbor's territories; expansion could be national expansion rather than remaining within the limits of colonial expansion. On the other hand, it is the very ease of marching across one's border to another's heartland, coupled with the absence of significant water access that limits the colonial holdings of land powers.

As these different forms of empire grew, the multiplicative effect of water upon economic growth allowed sea powers to pull ahead of their developing land equivalents. The classic case was Athenian growth after partnering with Sparta to defeat the Persians in 480 BCE, which in turn led to Thucydides famous explanation for the ultimate origin of the Peloponnesian War, "What made the war inevitable was the growth of Athenian power and the fear which this caused Sparta." But more than this needs to be said to fully understand the underlying dynamic of why it is that sea powers repeatedly end up being world-system hegemons but not land powers. After all, if economic hegemony is based upon productive advantage, why can't that be attained by a land power? Neither Marx nor Wallerstein show any preference to sea access as part of their theories of economic development. Part of the answer, I think, lies in the aforementioned role of water for productive advantage, accompanied by the other part which centers on the kinds of political empires water or land engender: the territorially proximate expansion for the landed and the overseas expansion for the sea-accessed.

Over time through their overseas markets, imported raw materials, lower transport costs, and so forth, the water multiplier pulls the sea-accessed ahead of the land-bound. Then, geopolitically, the continued growth of the overseas empire leads to a surrounding of the heartland of the land powers and their empires. As an indicator of the differential in power, note the one-sided nature of complaints. Land powers resent the colonial holdings of sea powers. Napoleon invaded Egypt in 1798 in good part to try to circumvent British ties to India. The Kaiser wanted to build a Berlin-Baghdad railway for the same purpose, and the Great Game between Russia and Britain was about access to and control of India. In contrast, sea powers wait and watch land powers grow their local empires, and only engage in balancing alliances as a means to counter their growing land power (Levy and Thompson 2010; Thompson and Rasler 1994). Interestingly, there seem to be fewer complaints regarding the absence of a land empire by sea powers.

Complaining and rivalries are one thing, but why do these two geographically rooted systems engage in war. First, let's consider the simple Thucydidian account. The sea-accessed grow faster and at some point the land-bounded realize they will never catch up. Why though, that prompts preemptive war on the land power's part is unclear, for the sea-accessed economies power is no threat to the land armies of the land-bound powers. Sparta must have known that no Athenian phalanx could ever defeat them in battle; no land invasion of the Peloponnese by Atticans would ever be successful. And while there is much talk about how Athens hid behind her walls and waited out the Spartan invasion, it seems just as reasonable that the Spartans could have waited out the Athenians who could never have defeated them on land. Why go to war?

Perhaps it is more complicated, as, in fact, neither land nor sea powers can defeat the other on their own terrain. It was Spartan warships purchased with Persian money that finally defeated the Athenians at the end of the Peloponnesian War; it was, then, sea power that defeated sea power rather than the Spartan phalanx, although both could have waited it out. What forces the issue? A second hypothesis would be a variant of Thucydides: the land power as spoiler thesis. From this point of view, the idea of hegemonic transition wars is an inaccurate characterization of what actually goes on, for there is, in fact, no intense competition among competitive national economies over who will be the next global economic hegemon. Land powers are doomed because of their landed status and instead, they play the role of spoilers, and at the point at which the past sea power is about to give way to the up and coming sea power, they make their move. France was never going to succeed Spain economically, so Napoleon moved militarily to try to take over the system. Germany was never going to succeed Britain economically, so the Kaiser and then Hitler moved militarily. In effect, land powers challenge the old hegemon precisely because they cannot replace her, and their actions unwittingly facilitate the rise of a new sea power.

The classic problem, though, still remains: why did Sparta choose to attack? What was the real basis of their fear of the growth of Athenian power, for there was no real existential Athenian threat to Spartan national existence. And if the Kaiser felt that his local police could arrest the British army once it had landed on the continent, there was not much Spartan fear of the Athenian phalanx either.

If the first two accounts do not seem totally satisfactory, let us consider a third: empire spillover. Sea powers continue to add colonies, control trade routes, and dominate the global water commons. In some sense this is no existential threat to land powers. They can continue to expand on the land into adjacent territories. If land and sea powers do not directly threaten each other, there are no grounds for conflict. But land expansion can reach a point where it can threaten sea acquisitions. It was by sea that Britain dominated India but Russian land expansion was trying to approach a British sea holding. Earlier, Napoleonic France had tried to limit Britain in India through land operations in Egypt and thereby access to the Red Sea. Britain responded by engaging Russia in the Great Game of central Asia, and earlier the Royal Navy destroyed Napoleon's fleet at the Battle of the Nile (1798), stranding him in Egypt, and Britain was unhappy about the landed's effort of the Berlin to Baghdad railroad project. The point here is that eventually, land-generated and sea-generated expansions come into contact, generating competition, rivalry and then conflict.

This leads to the following idea: turn Thucydides on his head. That is, it is the hegemon who is fearful of the power of the aspirant; it is Athens who is fearful of Sparta. As with the other models, we begin with both land and sea powers economically expanding and with the water multiplier assumption, we assume the sea power pulls ahead. This is the growth of Athenian power between the defeat of the Persians in 480 and the outbreak of the Peloponnesian War in 431. However, while one is pulling ahead, it is also the case that both are growing, hence developing contiguous land and overseas empires. Therefore, along with the nineteenth-century expansion of the British overseas Empire, there was a similar expansion of the Bismarkian/Wilhelminian German Empire within the land confines of continental Europe.

As noted earlier, colonial expansion requires less military force projection than land wars against other land powers, and therefore in some sense, while the expanse of the sea power's maritime holdings are greater than those of the land power, they are, of necessity, also more weakly held, which makes them, in time, feasible objects to consider acquiring by land power. While this was done in order to obtain access to British India, there are no equivalent machinations by Britain to directly pry territory from the expanding German Empire or Siberian territory from the eastern expansion of the Russian Empire. On the contrary, Britain had tried to accommodate the rise of Germany by giving the Germans colonies in Africa at the Berlin Conference on Africa in 1884.

Geographically, the British Empire was something of a peripheral ring of holdings surrounding the core of the expanding land empires of Russia and Germany. Their growth is clearly the prompt for the famous H.J. Mackinder (1904) idea of the Eurasian heartland as the pivot of history. If we take the Eurasian core as the equivalent of something like the Spartan heartland, then Mackinder is Thucydides sea power fear of Spartan land power. The strategic key for fifth-century Sparta and early twentieth-century Germany was not their colonial holdings, but their land holdings, the German and Spartan Empires themselves. For the British and the Athenians, their strategic key was the ring of imperial/colonial overseas holdings, ports of call and key waterway passages that they controlled. But this British sea power posed no threat to German land power.

There was German resentment over British colonies, but no existential threat. It was the Germans who wanted a rail road to Baghdad, not the British a railway to Berlin, and it was Russians playing the Great Game for influence in central Asia, not the British for influence in Siberia. It was, in reality then, the German/Russian landed core of a geopolitical world-system that was inching toward the weakly held outer rim of British sea powered holdings. It is this sea power anxiety that lies behind Mackinder's famous essay. Therefore, it is the growing fear for the vital interests of the British, rather than the Germans or Russians, as it was for the Athenians rather than the Spartans, that lies behind so-called hegemonic wars, whether in the fifth or twenty-first centuries.

As a working hypothesis, then, sea power generates weak yet wide imperial holdings: the global island ring of colonies. Land power generates the narrower and stronger inner core of land-based empires. The resultant outcome of international politics, empires, great power rivalries, and wars is geographically based. To repeat: land powers can, and do, march directly into their neighbors backyard, and because these are territorially contiguous areas, they can more easily be incorporated into the homeland polity—the growing Czarist, Bismarkian, and Spartan empires. Sea powers cannot do this as easily. Expeditionary forces are required, which are lighter, hence that which can be defeated must be weaker, hence, in general, this tends to be the outer rim of what has been called the developing or colonial world. Britain's overseas empire grew and grew, but with that geo-width came a much more tenuous hold than the landed's domestic empire expansion at the expense of immediate neighboring states. Paul Kennedy's (1989) imperial overstretch is, therefore, a geographic destiny for growing sea powers. Their home base cannot grow because of the water-induced limits on force projection possible vis-à-vis neighbors, whereas the successful land power grows and grows domestically. Napoleonic, Bismarkian, and Czarist Empires are all on the European continent, whereas the maps of the growth of Imperial Athens, Spain, and Britain are all outside of these respective homelands. Such are the constraints and opportunities geography imposes upon human agency and strategy in carrying out domestic and international politics.

This brings us to the geographic foundations of war. The weak ring of sea power holdings is eventually threatened by the growing expansion of the inner core of landed empires. Thucydides is turned on his head. The outer ring, weakly held as it is, is relatively safe from the predations of the land powers until their own geographic destiny pushes them into ever-greater strength and they begin, land-wise, to attempt links to the weaker ring of sea power holdings. At that point the Thucydidian fear is the sea power. And so the preemptive war logic of Thucydides is activated: but now it is the powerful that act out of fear. It was not the fear of the growing power of Athens, or Britain, or Imperial Spain that prodded the land powers to engage in war, but the fear of the expanding Spartan, Napoleonic, and Wilhelmine/Third Reichian Germany that eventually threatened the holdings of the extant hegemonic sea power. Prussia becomes the German Empire; Russia plus Siberia becomes the Russian Empire; Sparta plus conquered neighbors becomes a greater Sparta. The ring of sea power holdings, though, can only yield resentment, not direct threats to national existence. The expanding land power can, if it continues to grow, expand, conquer, make war, and threaten the sea power's more fragile hold on colonies. A sea-homeland is no match for the expanded domestic empire of the land-homeland, and so the Thucydidian logic of preemption emerges. However, the anxious ones are the sea powers, not the land powers.

Hegemonic succession wars, then, are only a succession of one sea power to the next. In the modern world-system it has been Spain to Britain to the United States, and given the growing Chinese navy, perhaps from the United States to China.

References

Kennedy, Paul, 1989. *The Rise and Fall of the Great Powers*. New York: Vintage.
Levy, Jack S. and William R. Thompson. 2010. "Balancing on Land and at Sea: Do States Ally against the Leading Global Power." *International Security*, 35, No. 1: 7–43.
Mackinder, H.J. 1904. "The Geographical Pivot of History". *The Geographical Journal*, 23: 421–37.
Thompson, William R. and Karen Rasler. 1994. *The Great Powers and Global Struggle, 1490–1990*. Lexington: University Press of Kentucky.
Wallerstein, Immanuel. 1974. *The Modern World-System*. New York: Academic Press.

15.2

The global justice movement and the social forum process

Ellen Reese, Ian Breckenridge-Jackson, Edwin Elias, David W. Everson, and James Love

The global justice movement (GJM) is a "movement of movements" that draws together all those who are engaged in sustained and contentious challenges to the current global political economy and all those who seek to democratize it. Narrowly defined, it includes:

> … campaigns of mobilization against global or transnational neoliberalism or its agents, taking place against the policies of international financial institutions or their meetings; against regional economic compacts and summits; and global or regional Social Forums directed against global neoliberalism, like the World Social Forum and the European Social Forum.
>
> *(Hadden and Tarrow 2007: 361)*

More broadly understood, participants in this family of "antisystemic" movements draw attention to various goals that aim to improve and democratize the current world-system, such as ending US militarism, democratizing the United Nations, reducing toxic emissions, defending collective land and public social services, building social cooperatives, and empowering oppressed groups (Arrighi et al 1989).

Since its inception in 2001 in Porto Alegre, Brazil, the World Social Forum (WSF) has has rapidly become the largest international gathering of participants in, and supporters of, movements associated with the GJM. Annual and bi-annual WSF meetings in Brazil, India, and Kenya have attracted hundreds of thousands of participants from around the world. These international meetings have inspired the formation and spread of thematic, regional, national, and local Social Forums across the globe. The Social Forum process has helped activists to connect global and local issues, to coordinate actions, and to expand and strengthen their cross-border and cross-movement networks.

This chapter reviews research on the GJM and the Social Forum process and debates regarding their political significance. We argue that the GJM and the Social Forum process have strengthened efforts to democratize the world-system and encouraged coalition building within an era of globalization and transnational activism. Yet, considerable divisions within the "global left" must be overcome for it to be a truly effective force for global social change.

The global justice movement

The GJM emerged from a long history of transnational activism. Antislavery and feminist radicals in seventeenth-century England and across the Atlantic drew inspiration from egalitarian indigenous American societies. The American and French revolutions of the eighteenth century sparked the Haitian revolution, which contributed to the fall of Napoleonic France, Britain's forthcoming hegemony, and the decolonization of Latin America. In 1840, transnational abolitionists organized the World Antislavery Conference in London. The International Workingmen's Association, organized by Karl Marx and fellow activists in 1864, brought together labor unions as well as socialist, anarchist, and communist groups from across Europe and the United States in class solidarity. Despite its eventual cooptation under Stalin's Soviet Union, the Communist International of 1919 united a highly heterogeneous group of movements representing the disenfranchised on a global scale. Transnational alliances also grew during the waves of decolonization occurring in the 1950s and 1960s, and through the Bandung Conference of 1955, in which Asian and African countries participated to develop policies outside Western and Soviet dominance (Chase-Dunn and Reese 2007; Santiago-Valles 2005).

While certainly not new, the scope and scale of international ties among social activists have risen dramatically over the past few decades along with globalization; activists have increasingly shared information, ideas, and other resources, and coordinated actions across nations (della Porta 2007; Keck and Sikkink 1998; Moghadam 2005; Smith 2004a, b; Staggenborg 2008). Transnational organizing was facilitated by the four fold increase in the number of non-governmental organizations (NGOs) worldwide from 1973 to 1993. Together, these trends fostered the growth of the contemporary GJM, which emerged in response to the global spread of neoliberal policies (Evans 2005).

The GJM, which crosses national boundaries, attracts constituents affiliated with wide-ranging organizations and political groups: unions, direct-action oriented social movement organizations and networks, professionalized NGOs, and transnational advocacy networks. Movement participants also vary in their issue focus and preferred strategies and tactics (della Porta et al 2006; Keck and Sikkink 1998; Reitan 2007; Tarrow 2005). Opposition to neoliberal globalization unifies these disparate forces in the GJM. Refusing to heed Margaret Thatcher's infamous clarion call that "there is no alternative" (TINA), the distinct yet integrated components of the GJM campaign for a more just world-system. The GJM seeks a more equitable distribution of wealth and power and more environmentally sustainable communities. As a counterhegemonic movement, it builds institutions and forms alliances to foster alternative visions and practices of social justice (Evans 2005; Staggenborg 2008).

The increasing importance of economic and cultural exchanges across borders and the proliferation of global governance institutions regulating these exchanges have provided social movements with both powerful incentives and greater opportunities for transnational collaboration (della Porta 2007; Smith 2004a, b; Staggenborg 2008). Neoliberal policies, growing economic inequalities and problems, and a growing awareness of worldwide environmental degradation generated calls for coordinated action. Hardships associated with the lending policies of international financial institutions and the debt crisis led to a series of disruptive actions in the global South known as the "austerity riots" that began in the 1970s, peaked in the 1980s, and continued into the present (Reitan 2007; Walton and Ragin 1990). In the 1980s and 1990s, a series of conferences sponsored by the United Nations, and parallel summits by NGOs, addressed problems associated with neoliberal globalization and the lack of human rights; these meetings fostered the creation of new transnational and North-South alliances, while the slow pace of political change contributed to activists' subsequent radicalization (Moghadam 2005; Smith et al 2007).

The Zapatista revolt in 1994 against the implementation of the North American Free Trade Act in Mexico, and the threats it posed for indigenous peoples' land, resources, livelihoods, and culture, was instrumental in building North-South alliances in the struggle against neoliberal globalization and in linking local struggles to global ones. Subcommandante Marcos and his followers demonstrated the potential of the Internet and the capacity of "encuentros," international gatherings of activists, for bridging cultural divides and building transnational alliances and solidarity, and the power of international pressure and publicity to protect activists against repression (Schulz 1998). The Zapatista revolt revived and drew attention to the struggles of rural indigenous peoples, whose resistance against world-systemic domination predates current struggles by thousands of years (Hall and Fenelon 2009). Furthermore, the Zapatistas' consensual decision-making process and emphasis on local autonomy influenced future GJM organizations and their handling of internal diversity (Wood 2005; Zugman 2005).

Since the mid-1990s, the meetings of international governance institutions such as the World Trade Organization (WTO), the World Bank and the International Monetary Fund (IMF), and the Group of Eight (G8) have constituted the principal focal points for transnational activists to protest. These institutions represent the managerial epicenter of global capitalism and core economic interests, and so their meetings remain a natural target for the GJM. Mass media attention to these meetings and protests brought greater visibility to the GJM (Amin 2009; Pianta 2001). The absence or marginalization of government officials from the global South within these institutions also provided powerful allies for the GJM (Smith 2004a, 2004b; Staggenborg 2008; Tarrow 2005). The "Battle of Seattle" in 1999, marked a new era of high-profile protests against the WTO in the global North, and introduced many US activists to the GJM (Boykoff 2006).

Given the high cost of international travel, the availability of organizational resources has been critical to the spread of the GJM. Surveys of participants at five global protest events show that organizations played important roles in notifying participants, bringing them to the protest, and supporting their travel, especially among non-local participants (Fisher et al 2005). Case studies also document how effectiveness of grassroots transnational social movement organizations depended on access to money and researchers (Batliwala 2002).

Transnational movements, especially those bridging North-South divides, tend to be plagued by geographic and organizational inequalities in activists' access to resources as well as internal democratic deficits. As Bandy and Smith (2005: 232) put it, "the struggle for a democratic and equal form of globalization must overcome the often non-democratic and unequal forces that filter down from the world-system into movements themselves and the civil society they create." Indeed, the North-South divide has been a feature of global political economy for over 500 years and remains the most pressing and unresolved issues in the GJM. Northern nations continue to exert their economic, cultural, and political power to force their interests on Southern nations, often acting "paternalistically" toward them. Northern states are seen as "teachers" to the still "developing" counties of the South, and these patterns are frequently reproduced in the relations among transnational social justice activists. For example, many Northern environmental activists—a major force in the GJM—continue to advocate for "conservation/preservation" that leaves little space for coalition building with impoverished southern nations who seek "sustainable development." A truly "global movement" needs to incorporate the ideas of Southern activists, who can teach important lessons to Northern activists about how to effectively challenge neoliberal globalization (Teivainen 2002). Organizations such as the Peoples' Global Action and transnational feminist and other social movement networks have attempted to mitigate North domination by implementing strategies such as equitable representation in governance structures and decentralization to allow grassroots participation, and by highlighting the ideas and practices of southern activists (Moghadam 2005; Reitan 2007; Staggenborg 2008; Wood 2005; Zugman 2005). Despite such reflexive thinking and practice, the GJM movement still grapples with overcoming the North-South divide.

Ideological tensions also exist regarding goals and visions of democracy, organizational structures, and strategies and tactics. Reformists seek minor changes in nation-states and global institutions, such as the United Nations and the World Bank, and their policies so that they are less repressive and exploitative. Socialists, on the other hand, seek more fundamental transformations in power relations and to abolish capitalism; they often seek to gain control of, and to transform, the state and global political institutions in order to counter the power of global capital, redistribute wealth, and democratize society. Anarchists wish to destroy and eliminate the existing nation-states and global institutions along with capitalism, and replace these institutions with popularly controlled and highly decentralized institutions (Gautney 2010: 112). Religious tensions also exist among the various faith-based activist groups affiliated with the GJM, and between religious and secular activists (de Sousa Santos 2006). Leftists tend to be particularly skeptical of the agendas of NGOs, which rely on corporate and governmental funding. Organizations also vary in their decision-making processes. While anti-authoritarian groups (the "horizontals") implement direct democratic procedures and establish loosely affiliated networks with little hierarchy, other organizations (the "verticals," such as NGOs and unions) rely on more hierarchical and representational forms of democracy (Reiter 2009; Rucht 2011).

Finally, while reformists and NGOs tend to rely on institutionalized tactics such as lobbying, more radical groups employ militant, and sometimes even violent, collective actions to achieve their goals (Gautney 2010). The "amping up" of tactics by the GJM to include destruction of property was partly a reaction to the limited space afforded them by mainstream media outlets as well as frustration with the slow pace of change. The mass media worked diligently to discredit the GJM by highlighting its "violent" nature, as was evident in news coverage of WTO protests in Seattle in 1999 and the World Bank/IMF protests in Washington, DC, in 2000 (Boykoff 2006). Thus, despite the power of mainstream media to disseminate information widely, the GJM frequently relies on alternative media in order to communicate to the public.

Many scholars have emphasized the failures of the traditional left in the twentieth century as an influential catalyst and a defining break from the past of the contemporary GJM. Indeed, participants in the GJM tend to celebrate its internal diversity and seek to avoid the problems of hierarchy and division associated with "vanguardism" and democratic centralism. This has helped to make the movement broad and inclusive and to achieve unity over short-term goals, even as participants disagree over long-term goals and strategies (Amin 2009; della Porta et al 2006; Seibert 2008; Wallerstein 2003).

The world social forum process

The WSF was initially organized by the Brazilian labor movement and the landless peasant movement and their international allies in 2001 and it has most often occurred in Porto Alegre, Brazil, a traditional stronghold of the Workers' Party. The WSF was conceived as a popular alternative to the World Economic Forum in Davos, Switzerland, an annual private gathering where members of the global elite discuss current issues and develop policy ideas. The WSF, which occurs at the same time as the World Economic Forum, draws together all those opposed to neoliberal globalization to meet, share political ideas, experiences, and to make action plans. All types of individuals and organizations were welcome as long as they agree with the WSF "Charter of Principles," do not attend as official representatives of governments or political parties, and do not advocate the use of violence. While participants are divided in terms of priorities and preferred goals and tactics, the Social Forum process has also helped to bridge divides within the GJM and to strengthen various transnational campaigns for social and environmental justice. While workshops taking place at Social Forums are largely self-organized by participants, the International Council and a local Organizing Committee make decisions regarding the event logistics, collective actions, and the content of large plenary meetings. Regional, national, and local forums are likewise organized by planning and organizing committees (Smith et al 2007).

The size of WSF meetings grew dramatically. Allegedly about 20,000 people from 117 countries participated in the first WSF meeting in 2001, while the fourth WSF meeting held in Porto Alegre, Brazil, in 2005, drew a record 155,000 registered participants from 135 countries. The 2007 WSF meeting in Nairobi, Kenya, drew about 74,000 registrants, a remarkably feat given that it was the first WSF meeting to take place in Africa. The 2009 meeting in Belem, Brazil, drew more than 130,000 registered participants. Meanwhile, hundreds of regional, thematic, and local Social Forums have been organized. Although this trend has been more concentrated within Latin America and Western Europe, sub-global Social Forums are now reaching into Africa, Asia, and North America (Rucht 2011; Smith and Smythe 2010).

Social Forum participation and organization has spread unevenly across the globe, affecting who attends and the issues raised. Smith and Smythe (2010) suggest that the spread of forums is contingent upon multi-level networks and the topics addressed. Diffusion follows both an indirect and direct path, while thematic sessions are largely centered upon context and place. Since most participants come from the host country, differences in population density, the strength of organizational infrastructures, and the political opportunities within those countries help to explain why some Social Forums are larger than others (Reese et al 2008; Smith and Smythe 2010).

National political cultures and the shifting composition of local and international organizing committees also influence the issues addressed at Social Forums. According to Brunelle (2009), the history of popular opposition to neoliberalism and strength of socialist parties within Latin America shaped the discussions at the initial WSF meetings in Brazil. At the 2004 WSF in Mumbai, India, Dalit activists (formerly known as the "untouchables"), feminists, other Indian activists, broadened the agenda of the GJM by highlighting issues of casteism, patriarchy, and religious fundamentalism (Hewitt and Karides 2011; Smith 2011). Similarly, issues of racism and the global HIV crisis were brought to the forefront of discussions at the 2007 WSF held in Nairobi, Kenya.

Location and local organizers influence the composition of Social Forum participants. WSF meetings held in Porto Alegre, Brazil, were surprisingly "white" events (Teivainen 2004), while the location of the WSF in Nairobi, Kenya, in 2007 and Mumbai, India, in 2004, helped to increase non-white participation in these meetings. Based on findings from surveys of participants at the WSF meetings in Porto Alegre in 2005, the WSF in Nairobi in 2007, and the US Social Forum in Atlanta in 2007, researchers find "considerable variation in the political views and activities of attendees across these three venues" (Reese et al 2008: 1). At both Porto Alegre and Atlanta, participants were politically homogenous and predominantly leftist. The Porto Alegre forum had strong representation from the local and national Workers' Party, leftist movement organizations, and unions. Participants in Atlanta were far to the left of the political mainstream in the United States and disproportionately non-white, which is largely attributed to organizers' intentional incorporation of left-leaning mass-based community organizations in forum leadership and outreach. The Nairobi forum, however, was largely affiliated with politically moderate NGOs and churches, and participants were more religious and less radical in orientation. Yet across these venues, young adults and the college-educated predominated, as did activists affiliated with two or more social movements (Reese et al 2008). Surveys of participants at the 2005 WSF meeting reveal both North-South convergence and division in terms of preferred goals. Although a majority of participants from all world-system zones favored the abolition and replacement of both capitalism and international financial institutions (IFIs), Southerners were more favorable than Northerners toward simply abolishing IFIs (Chase-Dunn et al 2008).

Despite the successes of the WSF, participants regularly critique its organizational shortcomings leading to structural exclusion, as well as co-optation by political parties and agents of global capitalism. When it comes to organizational shortcomings, Smith (2004a) notes the constant tensions between the demands of organizing large, heterogeneous meetings and maintaining inclusive, decentralized decision making that allows for wide expressions of diversity. While some activists conceive of the *laissez faire*, "open space" format at the WSF for workshops and presentations as democratic and inclusive, other activists are more critical of it. Critics have called for more purposeful interventions to counteract power imbalances between social groups and organizations so that more socially privileged and resource-rich organizations do not dominate, and such views informed the organizing of the US Social Forum (Gautney 2010; Guerrero et al 2009; Juris 2008; Karides and Hewitt 2011). The privileging of formal panels at the WSF featuring famous intellectuals and politicians was also criticized for leaving limited time and space for grassroots dialogue (Gautney 2010; Smith 2004c). Concerns about the participation of moderate NGOs and church organizations, many dependent on Northern donors, were especially heightened during the 2007 WSF in Nairobi and 2004 WSF in Mumbai. Due to such concerns, radical activists have sometimes organized parallel summits to WSF or European Social Forums, drawing greater participation from more radical and grassroots activists (Pommerole and Simeant 2011; Reese et al 2008; Smith 2004c; Smythe and Byrd 2010).

Although some contend that Social Forums are merely "talk shops," these meetings have been instrumental in coordinating mass actions, such as international protests against the US wars in Iraq and Afghanistan and against free trade policies. Local and national campaigns, such as for Dalits' rights, workers' rights, or indigenous rights, have also gained wider support and publicity through involvement in the WSF process. Social movement networks have formed and expanded during Social Forum meetings, and funders meet with and identify international partners (Smith et al 2011). Attendees also take action during Social Forums; they participate in an opening march to highlight their wide-ranging demands for "another world," engage in protests around specific concerns (such as reducing Social Forum entry fees or improving banks' lending and investment practices), sign petitions, and raise money for their organizations (Smith et al 2011). "Social Movement Assemblies" were eventually developed as part of Social Forum process so that participants could make and express support for calls to action (Smith and Doerr 2011).

Conclusion

The GJM and the Social Forum process have brought together a variety of antisystemic movements from around the world through the overarching frame of "global justice" and opposition to neoliberalism. Both of these forces have helped raise the political consciousness of social activists, particularly young activists. The current GJM has largely overcome the problem of vanguardism that plagued the old left by celebrating diversity, incorporating critiques offered by identity-based movements, and through decentralized networks. Yet, the greatest strength of the GJM and the Social Forum process is also their greatest weakness. The "global left" (de Sousa Santos 2006) is still highly fragmented ideologically and organizationally, and remains quite small relative to the global population. While activists are united in what they oppose, they remain divided in their long-term visions of social justice and how to best achieve it. North-South inequalities and divisions also persist, creating tremendous challenges for transnational alliances.

Scholars find commonalities and differences in the spread, organizational development, mobilization efforts, and political tactics of the GJM by country. This is true even within Western Europe (della Porta 2007), although European global justice organizations have generally experienced a boom in membership, partly in reaction to European Union policies. Meanwhile,

US GJM participation has declined after the "Battle in Seattle," partly due to activists' increasing involvement in the anti-war movement and electoral politics (Hadden and Tarrow 2007). Differences in the GJM are also noticeable in the global South. Nanga (2009) notes the weakness of continental networks and the dominance of a small number of NGOs in the African GJM; many NGOs there are tied to Western multinational partners and conservative, United States backed Christian charity organizations that threaten to push back gains made by feminist and lesbian, gay, transgender, bisexual, and queer organizations. However, further growth of Pan-African solidarity could counteract these tendencies (Nanga 2009; Reese et al 2008). In contrast, the GJM has been more robust within Latin America, where impressive transnational campaigns against neoliberalism and socialist parties have flourished, especially among semiperipheral nations such as Venezuela and Brazil (Chase-Dunn et al 2010). Greater research on such cross-national differences in the development of the GJM is needed.

As Staggenborg (2008: 136) concludes, "the extent to which the movement succeeds depends on its ability to develop long-term campaigns with solutions to the problems that global justice activists have identified." Whether or not such long-term campaigns will emerge and be strong enough to effectively challenge the existing world-system remains to be seen. The economic crisis and the hardships associated with it present opportunities for mobilizing the public and expanding the reach of the global left. However, they also present opportunities for conservative forces, as the rise of public sector cutbacks, fascist organizations, right-wing parties, and anti-immigrant sentiment in many countries demonstrates. The appeal of nationalism and protectionism could also increase as economic conditions worsen, exacerbating national and regional divisions among activists, while building and sustaining transnational alliances may become more difficult as funding for such projects becomes more scarce. Countering such forces and trends effectively requires greater unity and concerted mobilization within the GJM, and a concurrent expansion of its reach within the broader public. While the Social Forum process and its movement assemblies has helped to increase the coherence and effectiveness of the GJM, there is still much work to be done.

References

Amin, S. (2009) "For struggles, global and national", in J. Sen and Peter Waterman (eds) *World Social Forum: Challenging Empires*, Montréal: Black Rose Books.

Arrighi, G., Hopkins, T. K. and Wallerstein, I. (1989) *Antisystemic Movements*, London and New York: Verso.

Bandy, J. and Smith, J. (2005) "Factors affecting conflict and cooperation in transnational movement networks", in J. Bandy and J. Smith (eds) *Coalitions Across Borders: Transnational Protest and the Neoliberal Order*, Lanham, MD: Rowman & Littlefield Publishers, Inc.

Batliwala, S. (2002) "Grassroots movements as transnational actors: implications for civil society", *Voluntas: International Journal of Voluntary and Nonprofit Organizations*, 13(4): 393–409.

Boykoff, J. (2006) "Framing dissent: mass-media coverage of the global justice movement", *New Political Science*, 28(2): 201–28.

Brunelle, D. (2009) "Citizen mobilization in the Americas and the birth of the World Social Forum", in J. Sen and P. Waterman (eds) *World Social Forum: Challenging Empires*, Montréal: Black Rose Books.

Chase-Dunn, C., Kaneshiro, M., Love, J., Lawrence, K. and Elias, E. (2010) *Neoliberalism, populist movements and the Pink Tide in Latin America*. http://irows.ucr.edu/papers/irows58/irows58.htm.

Chase-Dunn, C. and Reese, E. (2007) "Global party formation in world historical perspective." pp. 82–120, in K. Sehm-Patomaki and M. Ulvila (eds) *Democratic Politics Globally*, Tampere: NIGFD.

Chase-Dunn, C. and Reese, E. (2007) "Global party formation in world historical perspective", in K. Sehm-Patomaki and M. Ulvila (eds) *Global Political Parties*, London: Zed Books.

Chase-Dunn, C., Reese, E., Herkenrath, M., Giem, R., Gutierrez, E., Kim, L. and Petit, C. (2008) "North-South contradictions and bridges at the World Social Forum", in R. Reuveny and W.R. Thompson (eds) *North and South in the World Political Economy*, West Sussex: Blackwell.

della Porta, D. (ed.) (2007) *The Global Justice Movement: Cross-National and Transnational Perspectives*, Boulder, CO: Paradigm Publishers.

della Porta, D., Andretta, M., Mosca, L., and Reiter, H. (2006) *Globalization from Below: Transnational Activists and Protest Networks*, Minneapolis: University of Minnesota Press.

de Sousa Santos, B. (2006) *The Rise of the Global Left*, London: Zed Press.

Evans, P. (2005) "Counterhegemonic globalization: transnational social movements in the contemporary global political economy", in T. Janoski, R. Alford, A. Hicks, and M. Schwartz (eds) *The Handbook of Political Sociology*, New York: Cambridge University Press.

Fisher, D., Stanley, K., Beerman, D., and Neff, G. (2005) "How do organizations matter? Mobilization and support for participants at five globalization protests", *Social Problems*, 52: 102–12.

Gautney, H. (2010) *Protest and Organization in the Alternative Globalization Era: NGOs, Social Movements, and Political Parties*, New York: Palgrave MacMillan.

Guerrero, M.L., Luu, T.B., and Wiesner, C. (2009) "The road to Atlanta: reflections on the organizing history of the U.S. Social Forum", in J. Sen and P. Waterman (eds) *World Social Forum: Challenging Empires*, Montréal: Black Rose Books.

Hadden, J. and Tarrow, S. (2007) "Spillover or spillout? The global justice movement in the United States after 9/11", *Mobilization: An International Journal*, 12 (4): 359–76.

Hall, T.D. and Fenelon, J.V. (2009) *Indigenous Peoples and Globalization: Resistance and Revitalization*, Boulder, CO: Paradigm Publishers.

Juris, J.S. (2008) "Spaces of intentionality: race, class, and horizontality at the United States Social Forum", *Mobilization*, 13 (4): 353–72.

Karides, M. and Hewitt, L. (2011) "More than a shadow of a difference? Feminist participation in the World Social Forum", in J. Smith, S. Byrd, E. Reese, and E. Smythe (eds) *A Handbook of World Social Forum Activism*, Boulder, CO: Paradigm Publishers.

Keck, M.E. and Sikkink, K. (1998) *Activists Beyond Borders: Advocacy Networks in International Politics*, Ithaca, NY: Cornell University Press.

Moghadam, V. (2005) *Globalizing Women: Transnational Feminist Networks*, Baltimore, MD: Johns Hopkins University Press.

Nanga, J. (2009) "The world social forum in Africa", in J. Sen and P. Waterman (eds) *World Social Forum: Challenging Empires*, Montréal: Black Rose Books.

Pianta, M. (2001) "Parallel summits of global civil society", in H. Anheier, M. Glasius and M. Kaldor (eds) *Global Civil Society*, Oxford: Oxford University Press.

Pommerole, M.E. and Simeant, J. (2011) "African voices and activists at the WSF in Nairobi: the uncertain ways of transnational African activism", in J. Smith, S. Byrd, E. Reese, and E. Smythe (eds) *A Handbook of World Social Forum Activism*, Boulder, CO: Paradigm Publishers.

Reese, E., Chase-Dunn, C., Anantram, K., Coyne, G., Kaneshiro, M., Koda, A.N., Kwon, R., and Saxena, P. (2008) "Research note: surveys of World Social Forum participants show influence of place and base in the global public sphere", *Mobilization*, 13(4): 431–45.

Reitan, R. (2007) *Global Activism*, London and New York: Routledge.

Reiter, H. (2009) "Participatory traditions within the global justice movement", in D. della Porta (ed.) *Democracy in Social Movements*, New York: Palgrave and McMillan.

Rucht, D. (2011) "Social fora as public stage and infrastructure of global justice movements", in J. Smith, S. Byrd, E. Reese, and E. Smythe (eds) *A Handbook of World Social Forum Activism*, Boulder, CO: Paradigm Publishers.

Santiago-Valles, K. (2005) "World historical ties among 'spontaneous' slave rebellions in the Atlantic", *Review*, 28(1): 51–84.

Schulz, M.S. (1998) "Collective action across borders: opportunity structures, network capacities, and communicative praxis in the age of advanced globalization", *Sociological Perspectives*, 41(3): 587–616.

Seibert, T. (2008) "The global justice movement after Heiligendamm", *Socialism and Democracy*, 22(1): 101–13.

Smith, J. (2004a) "Exploring the connections between global integration and political mobilization", *Journal of World-systems Research*, 10(1): 255–85.

Smith, J. (2004b) "Transnational processes and movements", in D.A. Snow, S.A. Soule, and H. Kriesi (eds) *The Blackwell Companion to Social Movements*, Oxford: Blackwell Publishers.

Smith, J. (2004c) "The world social forum and the challenges of global democracy", *Global Networks*, 4(4): 413–21.

Smith, J. (2011) "The Dalit movement meets the Social Forum: a global struggle for social justice and human rights," in J. Smith, S. Byrd, E. Reese, and E. Smythe (eds) *A Handbook of World Social Forum Activism*, Boulder, CO: Paradigm Publishers.

Smith, J., Byrd, S., Reese, E. and Smythe, E. (eds) (2011) *A Handbook of World Social Forum Activism*, Boulder, CO: Paradigm Publishers.

Smith, J. and Doerr, N. (2011) "Democratic innovation in the U.S. and European Social Forums", in J. Smith, S. Byrd, E. Reese, and E. Smythe (eds) *A Handbook of World Social Forum Activism*, Boulder, CO: Paradigm Publishers.

Smith, J., Karides, M., Becker, M., Brunelle, D., Chase-Dunn, C., della Porta, D., Garza, R.I., Juris, J.S., Mosca, L., Reese, E., Smith, P.J., and Vázquez, R. (2007) *Global Democracy and the World Social Forums*, Boulder, CO, and London: Paradigm Publishers.

Smith, J. and Smythe, E. (2010) "(In)fertile ground? Social Forum activism in its regional and local dimension", *Journal of World System Research*, 16(1): 6–28.

Smythe, E. and Byrd, S.C. (2010) "World Social Forum activism in Belem and beyond", *Journal of World-Systems Research*, 16(1): 94–105.

Staggenborg, S. (2008) *Social Movements*, Ontario: Oxford University Press.

Tarrow, S. (2005) *The New Transnational Activism*, Cambridge: Cambridge University Press.

Teivainen, T. (2002) "The World Social Forum and global democratisation: learning from Porto Alegre", *Third World Quarterly*, 23(4): 621–32.

Wallerstein, I. (2003) *The Decline of American Power*, New York: The New Press.

Walton, J. and Ragin, C. (1990) "Global and national sources of political protest: third world responses to the debt crisis", *American Sociological Review*, 55(6): 876–90.

Wood, L. (2005) "Bridging the chasms: the case of People's Global Action", in J. Bandy and J. Smith (eds) *Coalitions Across Borders: Transnational Protest and the Neoliberal Order*, Boulder, CO: Roman & Littlefield.

Zugman, K. (2005) "Autonomy in a poetic voice: Zapatistas and political organizing in Los Angeles", *Latino Studies*, 3(3): 325–46.

15.3

Global civil society or global politics?

Jon Shefner

The use of civil society as an analytic concept has become stunning in its ubiquity. Students of globalization, development, community organizing, political transitions, and social movements have used the concept to explain settings far and wide. In this essay, I focus especially on the literature on international and global civil society. I argue that the concept has been vastly overused, and obscures our understanding of resistance and political contention in the world-system. Civil society especially fails world-systems' and other analyses of global social change and power because in the search for a unifying theoretical concept, its users ignore class, and prove unwilling to address transnational differences in terms of power and resources.

The theoretical genealogy of the concept is available elsewhere (Cohen and Arato 1992; Keane 2006; Seligman 1992; Walzer 1995). Rather than add to this discussion, I demonstrate its wide use and discuss problems with it as an empirical concept within nations. These problems provide entry into the even more significant problems of a concept that suggests global theoretical unity, as opposed to relying on more reliable concepts. As we will see, the use of the civil society label especially serves to obfuscate the class project of neoliberalism.

Who and where is civil society?

Civil society analysts provide a myriad of definitions. Edwards tells us it is "the arena in which people come together to advance the interests they hold in common not for profit or political power, but because they care enough about something to take collective action" (2001: 2). Almost all agree it is independent of, and even opposed to, the state (Habermas 1996; Haynes 2007; Michelettii 1995; Shanks 1995).

It is telling that the concept is operationalized widely enough to include almost all associational life (Bermeo 2000; Harbeson 1994; Norton 1995; Waisman 2006). Nongovernmental organizations are emblematic of civil society, even if there is no consistency in their political role, relationship to the state, or national and international reach (Bazaara 2005; Kissling 2008; Hirata 2002; Ghimire 2005; Sater 2007; Servolo de Medeiros 2005; Veltmeyer 2005).

Social movements and grassroots movements are also members of civil society. These include pro-democracy movements (Baker 2002; Harbeson 1994; Sater 2007); women's rights organizations (Bazaara 2005; Ducke 2007; Gyimah-Boai 1994; Tripp 1994); human rights movements (Levine 2006; Norton 1995); indigenous organizations (Fischer 2009); anti-corruption groups

(Sater 2007); environmental groups and peace groups (Ducke 2007); unions (Servolo de Medeiros 2005); and farmer's organizations (McKeon 2005).

Such wide operationalization betrays definitional problems, as we add both explicitly "political" actors such as political parties (Micheletti 1995), representative political bodies and independent judiciaries (Hicks and Al-Najjar 1996), mass organizations (Muslih 1996), and specialist scientific and academic policy-analysis institutes or think tanks (Hirata 2002; Kissling 2008). The confusion is further reflected by the inclusion of market actors such as professional groups (Gyimah-Boai 1994; Norton 1995; Tripp 1994), business associations (Bellin 1995; Hinnebusch 1996), and trade groups (Hirata 2002).

Religious groups are also consistently considered civil society actors (Haynes 2007; Norton 1995; Shanks 1995; Walzer 1995). Others who have been fitted with the label include middle class groups building on pre-existing traditional kin, clan, ethnic, guild and religious functions (Ibrahim 1996), cooperatives and cultural education groups (Hirata 2002; Levine 2006).

Almost everybody is civil society, so long as they have an organizational life and a constituency to represent, and some form of ill-defined yet contentious relationship with the state. Additionally, civil society is not only everyone but everywhere. Scholars have found civil society throughout the Americas (Avritzer 2006; Fischer 2009; Levine 2006; Salamon 2003; Servolo de Medeiros 2005; Oxhorn 2006; Veltmeyer 2005; Waisman 2006). Northern, Central, and Eastern Europe both currently and in the past provide homes for civil society (Encarnación 2003; Hann and Dunn 1996; Micheletti 1995; Munkler 2006). Civil society even populates the virtual world (Ducke 2007; Hajnal 2002).

Indeed, we find civil society in places where the concept's Eurocentric origins suggest it would be unlikely to appear. Civil society is present in Central Asia (Ruffin and Waugh 1999), Japan, and throughout the developing world (Bernard et al 1998; Haynes 2007; Howell and Pearce 2001). Civil society is emergent or present throughout Africa and the Muslim and Arab worlds (Bratton 1994; Kasfir 1994; McKeon 2005; Roniger 1994; Sajoo 2002).

Some find the concept too burdened by Western theoretical baggage to be useful outside of that tradition (Gellner 1994), or that other societies lack the history of associational life crucial to traditional visions of civil society (Hann 1996). But civil society is tenacious. Despite recognizing the critique of its applicability outside Western contexts, Chatterjee, analyzing India, finds "civil society … is best used to describe those institutions of modern associational life set up by nationalist elites in the era of colonial modernity … These institutions embody the desire of this elite to replicate in its own society the forms as well as the substance of Western modernity" (2001: 174). If the colonial envy is not clear enough, Chatterjee refers to civil society as useful in analyzing "the institutions of modernity as they were meant to be" (2001: 172). Parekh suggests a more indigenous conceptualization, arguing that civil society may take forms that reflect different national origins (2004).

The wide application across space, time, and demonstrable differences suggests that civil society analysis, whether it is of the one-size—fits-all variety or a more culturally flexible one, obscures the actions of real people in organizations that defend or further their own interests defined by their placement on specific societal hierarchies. The combining of so many different kinds of organizations and interest groups as civil society ignores the fact that many of these groups indeed have conflicting, instead of unifying interests. This easily demonstrable critique may be effectively hidden by the anti-state bias that is endemic to civil society analysis. If all oppose the state, it appears, other differences are unimportant. Additionally, such analysis fails to note the differential power of different nationally rooted organizations. Core NGOs have greater resources and reach than their semiperiphery counterparts. Painting them all as civil society ignores this clear power differential. Finally, and perhaps most mystifyingly, despite the recognition that many groups are responding to the ravages of global neoliberalism, civil society analysts fail to engage with class beyond the spur to action. They often recognize the class basis of the neoliberal project, but few of them consider what resistance means in class terms, or how coalitions of organizations with different class bases work together, successfully or otherwise.

This makes the presumption of unity extremely problematic. Strategies, agendas, and resources differ among allied movements and coalesced actors. Indeed, those groups considered civil society fragment with alarming regularity around issues of sectoral or group interests. There may be unity of grievances among these groups, and that may translate into short- or even medium-term alliances. The unity shared by common experience of hardship, however, does not translate into a unified response or a common strategy. Instead, groups harmed by similar political economic dynamics may choose a variety of widely diverging strategies that contradict those of their allies.

Because of these power differences, civil society organizations are also subject to careful strategies of division by the state. States may choose to make accommodations or alliances with certain sectors, while denying the demands of their allies. The choices the state makes about which sectors to ally with, co-opt, or repress further demonstrates the divisions inherent within civil society. The state is much more likely to ally with sectors that are less threatening to its interests, that it can control, or that add to the power base of the state in ways that allow it to achieve various goals.

All of these critiques point to the stratification of civil society. The omissions demonstrated by this review are even more exaggerated as we turn to global civil society analysis.

Obscuring difference in the pursuit of unity

Definitions of global civil society (GSC) are no clearer than their national counterparts. Lipschutz defines GSC as … "self-conscious constructions of networks of knowledge and action, by decentred, local actors, that cross the reified boundaries of space as though they were not there" … due to "… the inability of states to deal with certain social welfare problems … and the crumbling of old forms of political identity, centered on the state, and the growth of new forms of political and social identity" (1992: 390, 392).

Keane writes of "a dynamic nongovernmental system of interconnected socioeconomic institutions that straddle the whole earth, and that have complex effects … an unfinished project that consists of sometimes thick, sometimes thinly stretched networks, pyramids and hub-and-spoke clusters of socioeconomic institutions and actors who organize themselves across borders, with the deliberate aim of drawing the world together in new ways" (2003: 8). Its participants include "athletes, campaigners, musicians, religious believers, managers, aid-workers, teleworkers, medics, scientists, journalists, academics … charities, think tanks, prominent intellectuals, campaigning and lobby groups, citizens' protests … small and large corporate firms, independent media, Internet groups and websites, employers federations, trade unions, international commissions, parallel summits, and sporting organizations" so long as they work across borders and "outside the boundaries of governmental structures" (Keane 2003: 7, 9).

One of the most sophisticated discussions is found in Michael Walzer's work, as he juggles issues of citizenship, social, political, and economic inclusion, and the ongoing importance of states. Yet in his strikingly lucid treatment, the very requirements Walzer stipulates demonstrates a difficulty with the concept, certainly in any global application. "Only a democratic state can create a democratic civil society; only a democratic civil society can sustain a democratic state" (Walzer 1995: 24). Such a comment points out the enormously limited and Western-biased elements of the concept.

The concept of GSC has critics among its advocates, but most nibble around the theoretical edges, finding its promise for global governance yet unarticulated (Baker 2005), unequal to states as interest representatives (Chandler 2005), or insufficiently political in its intersection with neoliberal capitalism (Lipschutz 2005). Heartfield (2005) sees it failing to directly confront neoliberal globalization, acting instead as an expression of radicals trying to find a space of contention that is not class-based.

Many analysts root the emergence of GSC in their belief that the state is diminishing in importance (Baker and Chandler 2005; Lipschutz 1992, 2005). Falk (2005), for example, finds GSC acting in a "post-Westphalian world" in which states have not disappeared but have been joined by a number of different actors that exert influence on a variety of global political issues. Global civil society then affords a place, within "voluntary associations, movements, parties, unions, the individual is able to act publicly" … to … "negotiate, argue, struggle against or agree with each other and with the centers of political and economic authority" (Kaldor 2005: 104).

Those opposing the state out of leftist disappointment in its failure to fulfill popular needs, and those from a rightist market-centric stance share the anti-state position so central to civil society discussion. For the world-systems perspective, of course, the question of the power of the state is better couched as a question of which state is declining, static, or in ascension. Indeed, both classic and recent analyses find certain states remain large, centralized, and powerful (Arrighi 2007; Hopkins et al 1982). What GSC analysts miss is how states remain enormously powerful, even if it is different states and different polices that are ascendant. The Eurocentric bias of GSC analysts may have blinded them to transitions in global power. Additionally, despite the expressed intention of neoliberals to diminish the power of the state in the global economy, the truth is that power has not decreased. It is instead the welfare state function and provisions that have diminished to provide global capital with greater flexibility (Harvey 2005).

The failure to think more precisely about state power and the Eurocentric bias explain what I consider one of the more indefensible elements of GSC and CS analysis—the focus on civility (Keane 2003; Walzer 1995). This focus is more than rhetorical, but is central to many who use the concept, and has disturbing implications for thinking about politics. Clearly, organizations focused on social change pursue a variety of strategies, violent and nonviolent among them, at different times amid different contexts. The idea of civility itself is linked to real protest decisions insofar as context helps define strategy. People pursuing their political interests are sure to question the likely outcomes of different political strategies. But to define a group into or out of political contention by their strategic choices ignores socio-political context and power.

The preoccupation with civility allows analysts to count out those they perceive as the bad guys—uncivil types such as the international movement for white supremacy—or those whose goals we may support but whose strategies create discomfort. Are those that pursue guerilla tactics defined outside of civil society because of their strategic use of violence, yet within it when they practice civil disobedience? Does civility really offer us a useful analytical category to explain different varieties of collective action over the last 35 years?

The protest and interest group action that GSC analysts document are political conflicts, in which people contend for their interests in different ways. In most cases, such contestation is anything but civil, not only because massive inequalities drive protest, but also because they are fiercely defended. To focus on civility is to exclude a large amount of political actions that are typical of political contention in the world-system. What the analysts of GCS are documenting is politics—the exercise, or attempted exercise, of power for defined interests. Sometimes conflicts may be relatively amicably resolved within a democratic political system, an assumption in itself. Often they are not.

Another more important omission by GSC analysts is their focus on theoretical unity as opposed to verifiable divisions, especially regarding power. The recognition of power differences based in class and core-semiperiphery-periphery relations is central to world-systems analysis. This basic theoretical contribution is ignored by GSC analysis of plural conflict. The combination of so many forms of struggle under one theoretical rubric allows its advocates to argue that GCS organizations act in ways that defy exclusive governments and faithfully represent a common societal will, even across national borders.

Because their discussion privileges shared goals and interests, little space remains for any examination of power across class or core-periphery position. Can core-based NGOs exert greater power in comparison to those representing similar social sectors of the Global South? George (2006) documents the victory of "North American and European civil society" over transnational capital in the defeat of the Multilateral Agreement on Investment without reflecting on how such a victory might be related to the representation of a core constituency. Neither do others documenting the actions of NGOs representing different class interests explore the possibility that some successes may be due to the class makeup of the activists. If we examine organized activity representing different class interests across core-periphery distinctions, do we find any different patterns in interaction with and results from various states? The answer from the GSC analysts is that we do not know, because the drive toward theoretical unity presupposes similarity where difference may define the experience. In short, the proponents of GSC fail to understand the different implications of the vast political activity they have documented.

The strongest argument of the advocates of global civil society is that there is indeed something going on. Keane, for example, marshals impressive numbers, whether they are growth of INGOs, world congresses, budgets, or other organizational manifestations that demonstrate the massive explosion of organizations with global interest and reach (2003: 5). So too do Baker and Chandler (2005: 3), as they demonstrate the substantial increase in the "number of non-state actors involved in international policy" whether these be development NGOs, or other GSC actors. And of course it is one of the main missions of the Global Civil Society Yearbooks (Glasius et al, various years) to document such activity.

There has indeed been impressive empirical growth in organizations with global or transnational reach. But its meaning for GSC advocates is almost always couched in romantic language that limits critical perspective. Even among those purporting to advocate for open-eyed discussion, the normative pull of civil society analysis is too strong to resist, as they find within it an actor set on extending community among people across the globe, providing agency to check oppressive structures of economic and political domination, and in this way extending democracy (Baker and Chandler 2005).

It is not their grasp on empirical phenomena that I question, but their analysis of political role, impact, and implications. The disregard for class analysis is especially egregious as they document resistance to neoliberalism. Global civil society advocates may be comfortable with an understanding of neoliberalism as a class project, but they are much less willing to understand how class mediates those responses. For example, opposition to austerity policies articulated by students, workers, and small business owners is likely to find united response when transportation costs are increased, but may fragment when wage raises are at issue. Urban and rural workers may similarly find common cause in resistance to the removal of protective tariffs that increase a flood of costly food imports. But if the state increases wages to export-oriented workers from one industry without making such a reform more generally, solidarity may quickly fail.

Were the romantic lenses stripped, and a careful examination conducted of all such actors, it is likely that we would find that civil society neither resists nor accepts globalization. Instead, certain sectors of society act in ways that demonstrate how globalization affects their interests, and their level of power to do so. Business associations in certain contexts, times and places will respond in certain ways to global trends and policies, yet they are likely to act in very different ways in other global contexts.

Conclusions

Once one brings a world-systems perspective to bear, the GSC assumptions become even more apparent. World-systems allegiance to the concepts of class and power differentiated and defined by region make it even more unlikely that we can somehow think of these enormously varied groups as a unified actor whose global reach somehow erases the different political and economic interests that pertain to class and region.

In sum, the notion of civil society is imprecise empirically, and fails theoretically. World-systems analysis provides an especially useful corrective to demonstrate the errors in imposing too much unity on actors that are better defined by their difference. Globalization processes create a more integrated and interdependent political and economic system, it is true. But that system has become even more unequal. This suggests that independent or voluntary political actors from privileged locales in the global system still possess greater power, and will therefore have greater capacity to participate in transnational action. The increasing gap in political power of global actors suggests that the stratification we find consistently among civil society within nations will similarly be the case among global actors. Transnational work provides means by which many progressive groups work together. But the assumption that there is equality of power in agenda setting, decision making, and mobilization denies the divisions so clearly documented by decades of world-systems informed research.

The concept of civil society, global or otherwise, is not only empirically and theoretically problematic. The assumption of unity is dangerous because it masks the hierarchies that are embedded in the world-system under an umbrella of likeness. Ignoring difference and inequality while chasing a theory of similarity obfuscates our understanding of real struggles. Various social harms and benefits affect different groups differently. In turn, affected groups respond in varied ways—sometimes from ideological positions, and other times as a reflection of material interests. Both ideology and group interests are likely to conflict with those of others at times, and not always in patterned and predictable ways. Organized citizens come together to press their demands due to advantages or disadvantages posed by race, class, gender, locale, demographics, and other divisions that manifest either short-term grievances or longlasting inequalities rooted in the world-system. To resort to civil society analysis is to impose an imaginary theoretical unity above the real interests that people have stemming from the location on the social hierarchies that define us. Whether it is the emphasis on civility over strategy, unity over difference, status over class, or assumptions of global unity over locality-rooted experience, the obsession with global civil society serves us badly.

References

Arrighi, Giovanni. (2007) *Adam Smith in Beijing: Lineages of the 21st Century*. London: Verso.

Amanda Bernard, Henny Helmlick, and Percy Lehning, (eds) (1998) *Civil Society and International Development*. Paris: Development Centre of the Organisation for Economic Co-operation and Development.

Avritzer, Leonardo. (2006) "Civil Society in Latin America in the 21st Century: Between Democratic Deepening, Social Fragmentation, and State Crisis" in R. Feinberg, C.H. Waisman, and L. Zamosc (eds) *Civil Society and Democracy in Latin America.* New York: Palgrave MacMillan.

Baker, Gideon. (2002) *Civil Society and Democratic Theory: Alternative Voices.* London: Routledge.

Baker, Gideon. (2005) "Saying Global Civil Society with Rights" in G. Baker and D. Chandler (eds) *Global Civil Society: Contested Futures.* London: Routledge.

Baker, Gideon and David Chandler. (2005) *Global Civil Society: Contested Futures.* London: Routledge.

Bazaara, Nyangabyaki. (2005) "The Ability of Civil Society Groups to Influence the Debate on the Role of the Market in Rural Asset Building and Redistribution in East Africa" in K.B. Ghimire (ed.) *Civil Society and the Market Question: Dynamics of Rural Development and Popular Mobilization.* New York: Palgrave MacMillan.

Bellin, Eva. (1995) "Civil Society in Formation: Tunisia" in A. Norton (ed.) *Civil Society in the Middle East.* Leiden: Brill.

Bermeo, Nancy and Philip Nord. (2000) *Civil Society before Democracy: Lessons from Nineteenth-Century Europe.* Lanham, MD: Rowman & Littlefield.

Bratton, Michael. (1994) "Civil Society and Political Transitions in Africa" in John Harbeson, Donald Rothchild, and Naomi Chazan (eds) *Civil Society and the State in Africa.* Boulder, CO: Lynne Rienner Publishers.

Chandler, David. (2005) "Constructing Global Civil Society" in G. Baker and D. Chandler (eds) *Global Civil Society: Contested Futures,* London: Routledge.

Chatterjee, Partha. (2001) "On Civil and Political Society in Postcolonial Democracies" in S. Kaviraj and S. Khilnani (eds) *Civil Society: History and Possibilities.* Cambridge: Cambridge University Press.

Cohen, Jean and Andrew Arato. (1992) *Civil Society and Political Theory.* Cambridge, MA: MIT Press.

Ducke, Isa. (2007) *Civil Society and the Internet in Japan.* London: Routledge.

Edwards, Michael. (2001) "Introduction" in Michael Edwards and John Gaventa (eds) *Global Citizen Action.* Boulder, CO: Lynne Rienner Publishers.

Encarnación, Omar G. (2003) *The Myth of Civil Society: Social Capital and Democratic Consolidation in Spain and Brazil.* New York: Palgrave MacMillan.

Falk, Richard. (2005) "The Changing Role of Global Civil Society" in G. Baker and D. Chandler (eds) *Global Civil Society: Contested Futures.* London: Routledge.

Fischer, Edward, (ed.) (2009) *Indigenous Peoples, Civil Society and the Neoliberal State in Latin America.* New York: Bergahn Books.

Gellner, Ernest. (1994) *Conditions of Liberty: Civil Society and its Rivals.* London: Hamish Hamilton.

George, Susan. (2006) "Transnational Risks: A New Challenge for Global Civil Society" in I.K. Richter, S. Berking, and R. Muller-Schmid (eds) *Building a Transnational Civil Society,* London: Palgrave MacMillan.

Ghimire, K.B. (ed.) (2005) *Civil Society and the Market Question: Dynamics of Rural Development and Popular Mobilization.* New York: Palgrave MacMillan.

Glasius, Marlies, Mary Kaldor, and Helmut Anheier. (Various years) *Global Civil Society (Yearbook).* London: Sage.

Gyimah-Boai, E. (1994) "Associational Life, Civil Soceity, and Democratization in Ghana" in John Harbeson, Donald Rothchild, and Naomi Chazan (eds) *Civil society and the state in Africa.* Boulder, CO: Lynne Rienner Publishers.

Habermas, Jurgen. (1996) *Between Facts and Norms: Contributions to a Discourse Theory of Law and Democracy.* Cambridge: MIT Press.

Hajnal, Peter, (ed.) (2002) *Civil Society in the Information Age.* Aldershot: Ashgate.

Hann, Chris. (1996) "Introduction: Political Society and Civil Anthropology" in C. Hann and E. Dunn (eds) *Civil Society: Challenging Western Models.* London: Routledge.

Hann, Chris and Elizabeth Dunn. (1996) *Civil Society: Challenging Western Models.* Routledge: London.

Harbeson, John W., Donald Rothchild, Naomi Chazan, (eds) (1994) *Civil Society and the State in Africa.* Boulder, CO: Lynne Rienner Publishers.

Harvey, David. (2005) *A Brief History of Neoliberalism.* New York: Oxford.

Haynes, Jeffrey. (2007) *Democracy and Civil Society in the Third World: Politics and New Political Movements.* Cambridge: Polity Press.

Heartfield, James. (2005) "Contextualizing the 'anti-capitalism' movement in global civil society" in G. Baker and D. Chandler (eds) *Global Civil Society: Contested Futures.* London: Routledge.

Hicks, Neil and Ghanim Al-Najjar. (1996) "The Utility of Tradition: Civil Socieiy in Kuwait" in A. Norto (ed.) *Civil Society in the Middle East.* Leiden: Brill.

Hinnebusch, Raymond. (1996) "Staet, Civil Society, and Political Change in Syria" in A. Norton (ed.) *Civil Society in the Middle East*. Leiden: Brill.

Hirata, Keiko. (2002) *Civil Society in Japan: The Growing Role of NGOs in Tokyo's Aid and Development Policy*. New York: Palgrave MacMillan.

Hopkins, T., Immanuel Wallerstein et al (1982) "Patterns of Development of the Modern World-System" in T. Hopkins, I. Wallerstein R. Bach, C. Chase-Dunn, and R. Mukherjee (eds) *World-Systems Analysis: Theory and Methodology*. Beverly Hills, CA: Sage.

Howell, Jude and Jenny Pearce. (2001) *Civil Society & Development: A Critical Exploration*. Boulder, CO: Lynne Rienner Publishers.

Ibrahim, Saad Eddin. (1996) "Civil Society and Prospects for Democratization in the Arab World" in A. Norton (ed.) *Civil Society in the Middle East*. Leiden: Brill.

Kaldor, Mary. (2005) "The Idea of Global Civil Society" in G. Baker and D. Chandler (eds) *Global Civil Society: Contested Futures*. London: Routledge.

Keane, John. (2003) *Global Civil Society?* Cambridge: Cambridge University Press.

Keane, John. (ed.) (2006)*Civil Society: Berlin Perspectives*. New York: Berghahn Books.

Kasfir, Nelson. (1994) "Strategies of Accumulation and Civil Society in Bushenyi, Uganda: How Dairy Farmers Responded to a Weakened State" in John Harbeson, Donald Rothchild and Naomi Chazan (eds) *Civil society and the State in Africa*. Boulder, CO: Lynne Rienner Publishers.

Kissling, Claudia. (2008) *Civil Society and Nuclear Non-Proliferation: How do States Respond?* Aldershot: Ashgate.

Levine, Daniel. (2006) "Civil Society and Political Decay in Venezuela" in R. Feinberg, C.H. Waisman, and L. Zamosc (eds) *Civil Society and Democracy in Latin America*, New York: Palgrave MacMillan.

Lipschutz, Ronnie. (1992) "Reconstructing World Politics: The Emergence of Global Civil Society", *Millenium: Journal of International Studies*, 21 (3): 389–420.

Lipschutz, Ronnie. (2005) "Global Civil Society and Global Governmentality: Resistance, Reform or Resignation?" in G. Baker and D. Chandler (eds) *Global Civil Society: Contested Futures*. London: Routledge.

McKeon, Nora. (2005) "The Farmers' Movement and the Market Question in Senegal" in K.B. Ghimire (ed.) *Civil Society and the Market Question: Dynamics of Rural Development and Popular Mobilization*. New York: Palgrave MacMillan.

Micheletti, Michele. (1995) *Civil Society and State Relations in Sweden*. Aldershot: Avebury.

Munkler, Herfried. (2006) "Civil Society: Desparate Wishful Thinking?" in J. Keane (ed.) *Civil Society: Berlin Perspectives*. New York: Bergahn Books.

Muslih, Muhammed. (1996) "Palestinean Civil Society" in A. Norton (ed.) *Civil Society in the Middle East*. Leiden: Brill.

Norton, Augustus, (ed.) (1995) *Civil Society in the Middle East*. New York: Brill.

Oxhorn, Philip. (2006) "Conceptualizing Civil Soceity from the Bottom Up: A Political Economy Perspective" in R. Feinberg, C.H. Waisman, and L. Zamosc (eds) *Civil Society and Democracy in Latin America*. New York: Palgrave MacMillan.

Parekh, Bhikhu. (2004) "Putting Civil Society in Its Place" in M. Glasius, D. Lewis, and H. Seckinelgin (eds) *Exploring Civil Society*. London: Routledge.

Roniger, Luis. (1994) "Images of Clientelism and Realities of Patronage in Israel" in L. Roniger and Ayse Gunes-Ayata (eds) *Democracy, Clientelism, and Civil Society*. Boulder, CO: Lynne Rienner Publishers.

Ruffin, M. Holt and Daniel Waugh, (eds.) (1999) *Civil Society in Central Asia*. Seattle: University of Washington Press.

Sajoo, Amin, (ed.) (2002). *Civil Society in the Muslim World: Contemporary Perspectives*. London: I.B. Tauris Publishers.

Salamon, Lester. (2003) *The Resilient Sector*. Washington, DC: Brookings Institution Press.

Sater, James N. (2007) *Civil Society and Political Change in Morocco*. London: Routledge.

Seligman, Adam. (1992) *The Idea of Civil Society*. New York: Free Press.

Servolo de Medeiros, Leonilde. (2005) "Social Movements, the State and the Experience of Market-Led Agrarian Reform in Brazil" in K.B. Ghimire (ed.) *Civil Society and the Market Question: Dynamics of Rural Development and Popular Mobilization*. New York: Palgrave MacMillan.

Shanks, Andrew. (1995) *Civil Society, Civil Religion*. Oxford: Blackwell Publishers.

Tripp, Aili Mari. (1994) "Rethinking Civil Society: Gender Implications in Contemporary Tanzania" in John Harbeson, Donald Rothchild, and Naomi Chazan (eds) *Civil Society and the State in Africa*. Boulder, CO: Lynne Rienner Publishers.

Veltmeyer, Henry. (2005) "The Dynamics of Market-led Rural Development in Latin America: the Experience of Mexico, Ecuador, Bolivia, and Peru" in K.B. Ghimire (ed.) *Civil Society and the Market Question: Dynamics of Rural Development and Popular Mobilization*. New York: Palgrave MacMillan.

Waisman, Carlos. (2006) "Autonomy, Self-Regulation, and Democracy: Tocquevillian-Gellnerian Perspectives on Civil Society and the Bifurcated State in Latin America" in R. Feinberg, C.H. Waisman, and L. Zamosc (eds) *Civil Society and Democracy in Latin America*, New York: Palgrave MacMillan.

Walzer, Michael, (ed.) (1995) *Toward a Global Civil Society*. Providence, RI: Bergahn Books.

15.4

Language in the world-system

Gary Coyne

This article makes the case that there is a place within world-systems analysis for research focused on language. The various fields that study language have produced a large amount of material that could be brought into productive dialogue with the world-systems perspective. Given limited space, an attempt will be made to make this case in a single point. Namely, there is a highly significant correlation between a state's strength and the linguistic diversity found within it; the stronger, resource rich, states of the core display less linguistic diversity than the relatively weaker, resource poor, states of the periphery. Specifically, when linguistic diversity is the probability that two randomly selected people from the country would have a different primary language (Lewis 2009) and state strength is tax revenue as percent of GDP (World Bank World Development Indicators 2010), this author finds a highly significant negative relationship between state strength and linguistic diversity in a large, diverse sample of countries for the year 2000.

World-systems analysis can help make sense of the lower degree of linguistic diversity in core states by pointing to the fact that the processes of creating centralized, bureaucratic state apparatuses were generally carried further here. This project created a group of central elites with a developing interest in communication between emerging political centers and the territory increasingly under its control, as well as facilitating the flow of information within the state bureaucracy itself. Historically, this meant that the dialect of the capital region became the language of nation-building. It gained status, and speakers, while other languages and dialects were marginalized or actively discouraged. It is noteworthy that in Europe this process occurred contemporaneously, and in close proximity to, the development of the printing press and nationalism, and it has been argued that this facilitated the process of linguistic standardization (Anderson 1983). Although complete linguistic standardization has not, and likely never will, occur, it was carried further by this conjectural synergy. It is noteworthy that the first dictionaries and written works of grammar for most of the European languages appeared in the sixteenth and seventeenth centuries, just in time for these newly standardized languages to be carried on waves of European colonization. The incorporation of the New World typically resulted in the decimation of indigenous populations, the eradication of their languages, and then a filling of territories with large numbers speakers of European languages. These states were arguably more coherent to the extent that their population enjoyed cultural cohesion through shared traditions and languages. Some of these former colonies also experienced upward mobility in the core/periphery hierarchy over the centuries.

Turning to the weaker states in the periphery, insights from the fields of language studies help to flesh out what world-systems scholars have written. Many peripheral states in Africa and Asia were incorporated into the world-system in the nineteenth century, and for various reasons, some indigenous populations and their languages faired substantially better. These states then, typically, gained independence with a high degree of linguistic- and ethnic-diversity and this has made nation-building and the construction of a capable and democratic state that are widely perceived as legitimate more difficult. Even after formal independence, many former colonies continue to use the language of their colonizer in at least some spheres of life (such as courts, diplomatic channels, or higher education); to the extent that this interferes with cultural, political, and economic self-determination, the relationship between colony and colonizer is reproduced (Mazrui 1998).

In economic terms, it can also be pointed out that wherever linguistic diversity is higher, the coordination of individual economic activity and integration of national markets is likely to be a slower process. Here, insights from language studies fields can shed light on how capitalism effects, and is in turn effected by, underlying social structures. Indeed, linguists have linked the marked decrease in the number of languages over the last four or five centuries to colonialism and capitalist expansion. An analogy is often made to capitalism's impact on the biosphere, and here again a world-systems perspective may be useful in deepening our understanding.

The structure of global language usage is also important for the future of global integration, especially the issue of English as a global *lingua franca*. While the imposition of European languages on the non-core was often a violent process, the more recent spread of English seems to be at least partly driven by its usefulness as a language of wider communication and its links to science and technology (e.g., Fishman et al 1996), and not by imperialism (as in Phillipson 1992). English became an important language for international communication largely because it was the language of two successive hegemons (Britain and the United States). But even within the context of a continuing slow decline of the hegemony of the United States, a wholesale replacement of English as an international language seems unlikely. The sheer growth in numbers of English speakers and the strong efforts of many important non-core states to develop English as a second language (especially China but also India and parts of Africa) means that individuals in disparate parts of the world have a shared language and can communicate with one another. This has important implications for the world-economy and also for world politics. And it is likely to facilitate the further development of widely shared normative elements in the world-system—the globalization of culture.

References

Anderson, B. (1983) *Imagined Communities*, New York: Verso.

Fishman, J. A., Conrad, A. W. and Rubal-Lopez, A. (1996) *Post Imperial English: Status Change in Former British and American Colonies, 1950–1990*, Berlin: Mouton de Gruyter.

Lewis, M. P. (ed.) (2009) *Ethnologue: Languages of the World*, 16th edn., Dallas, TX: SIL International.

Mazrui, A. A. A. A. M. M. (1998) *The Power of Babel: Language & Governance in the African Experience*, Chicago, IL: University of Chicago Press.

Phillipson, R. (1992) *Linguistic Imperialism*, Oxford: Oxford University Press.

World Bank World Development Indicators. (2010) *World Development Indicators*, Washington, DC: World Bank.

15.5

Anti-systemic movements compared

Valentine M. Moghadam

In *Anti-Systemic Movements*, Arrighi et al (1989) argued that since 1968, new resistance movements had emerged that challenged the logic of the capitalist world-system even more centrally than the "old" ones did. Here I consider three such new movements—political Islam, global feminism, and global justice (Moghadam 2009). Each is a transnational social movement, as it connects people across borders around a common agenda and collective identity; mobilizes large numbers of supporters and activists, whether individuals or members of networks, groups, and organizations; and engages in sustained oppositional politics with powerholders. Each is a product of late capitalism even as it is historically rooted, and it targets both states and the global order, in part through cyber-activism.

There are, however, distinct features and differences. Although Islamist movements are varied, their grievances, methods, and goals differ from the non-violent radical democratic or socialist visions and methods of global feminism and the global justice movement. Many Islamist movements seek state power, and, like revolutionary movements, are willing to use violence to achieve this aim. In contrast, the feminist and the global justice movements are disinterested in state power, although they seek wide-ranging institutional and normative changes, and they eschew violence. Islamist movements are less interested in economic redistribution and global social justice than in promoting Sharia law, legislating public morality, and defending the rights of Muslims.

The Islamist movements that burst onto the international scene in the late 1970s were rooted in the eighteenth-, nineteenth-, and early twentieth-century revival movements (e.g., Salafiyeh, Mahdi, Wahhabi) that claimed to be following the path taken by Prophet Muhammad in the seventh century CE. Other inspirations were the writings of Abul Ala Mawdudi (who founded the Jamiat-e Islami in India in 1941), and the Egyptians Hasan al-Banna (who founded the Muslim Brotherhood in 1929) and Seyyid Qutb, who took issue with modernity as it was proceeding in their countries and called for a return to strict implementation of Sharia law. To be sure, not all Islamist movements are jihadist. The "moderates," however, such as the Muslim Brotherhood and the ruling AKP party in Turkey, have no quarrel with capitalism.

Political Islam seeks to recuperate traditional patterns, including patriarchal gender relations, in reaction to the westernizing trends of globalization. Ironically, political Islam was once the beneficiary of US moral and military support, especially during the Afghan conflict of the 1980s and early 1990s. The US-sponsored war in Afghanistan had far-reaching world-historical consequences: the collapse of the Soviet Union and world communism; the expansion of militarized Islamist movements; and the emergence of a unipolar world allowing the United States to conduct military operations in Kuwait, Afghanistan, and Iraq.

Here, hyper- or heroic-masculinities become salient. Given the role of the US military in economic and political expansion, and the foundational narratives of heroic masculinity in Islam, one can imagine a "clash of heroic-masculinities" between the American security state and Islamist networks such as al-Qaeda or the Taliban of Pakistan and Afghanistan. Hyper-masculinity is a central ideological pillar of both neoliberal capitalist globalization and some forms of "resistance"; and it is a causal factor in war as well as in women's oppression. The rampant masculinity of radical Islam, its recourse to violence, patriarchal agenda, and disinterest in mobilizing against capitalism serve to separate it from the progressive agendas of the global feminist and justice movements.

The global justice movement echoes earlier movements of workers, socialists, communists, progressives, and anarchists. Many of the older activists in today's global justice movement were once affiliated with left-wing organizations or solidarity movements; many of the younger activists are involved in labor and economic justice causes; and the writings of Karl Marx are well known to many activists. The movement's loosely organized groups meet annually at the World Social Forum and regional forums to protest the downside of globalization and demand economic and social justice.

The global feminist movement has roots in first-wave feminism, with its focus on suffrage and justice, and second-wave feminism, with its demands for equality and cultural change. Scholars have identified moderate, socialist, and militant strands of both waves and described alliances with socialist and nationalist movements (Jayawardena 1984). Although feminists and leftists have not always agreed on priorities or strategies, there is a longstanding affinity that explains the involvement of feminists in the global justice movement.

The principal mobilizing structure of global feminism is the transnational feminist network, consisting of women from three or more countries engaged in research, lobbying, advocacy, and civil disobedience. Feminist international solidarity and humanitarianism are motivated by concern for women's human rights and objection to neoliberal globalization, militarism, war, and patriarchal fundamentalisms. Feminist humanitarianism differs fundamentally from imperialistic humanitarianism, which justified the 1991 Gulf War, the 1999 bombing of Serbia, and the invasions of Afghanistan and Iraq—all of which TFNs opposed. Transnational feminist activism challenges the logic of the world-system.

Arrighi et al erred in considering all "new social movements" as anti-systemic, when in fact many were either identity-focused or comfortable with the capitalist world-system (notably those of Eastern Europe). In contrast, Boswell and Chase-Dunn (2002) and Chase-Dunn (2002) focused on progressive transformation. Late capitalism has produced identity-focused social movements; others opposed to the system's inequalities; reactive movements; and deadly rebellions, martyrdom operations, and transnational networks of violent extremists. We need to either redefine "anti-systemic" or acknowledge that not all anti-systemic movements are interested in, or help bring about, an alternative, socialist world.

References

Arrighi, G., T. Hopkins and I. Wallerstein (1989) *Anti-Systemic Movements*. London: Verso.

Boswell, T. and C. Chase-Dunn (2000) *The Spiral of Capitalism and Socialism: Toward Global Democracy*. Boulder, CO: Lynne Rienner Publishers.

Chase-Dunn, C. (2002) "Globalization From Below: Toward a Collectively Rational and Democratic Global Commonwealth." *Annals of the American Academy of Political and Social Sciences*, 581 (May): 48–61.

Jayawardena, K. (1984) *Feminism and Nationalism in the Third World*. London: Zed Books.

Moghadam, V. M. (2009) *Globalization and Social Movements: Islamism, Feminism, and the Global Justice Movement*. Lanham, MD: Rowman & Littlefield.

15.6

Stabilization operations and structural instability in the contemporary world-system

Jeremy Simpson

Stabilization operations combine asymmetric low-intensity or insurgent-counter-insurgent conflict with traditional "peace-keeping" and "war-fighting" and with development and state building in hybrid conflicts involving both civil and military, state and non-state actors. Afghanistan and Iraq are present paradigm cases. Stabilization operations are necessitated by the structure of the contemporary world-system and exhibit both continuities and discontinuities with prior states of this system. In simplified form, the world-system divides nation-states into core and periphery, relative to past trajectories of development and present position within the world-economy. This may be extended to the peripheral sub-state level, with peripheral state populations divisible into core or elite and peripheral or non-elite positions (Galtung 1971). The subordination of the periphery, while primarily economic, is maintained by political regimes and enforced when necessary by military intervention. Such intervention is legitimated in terms of a core cultural model of the legitimate form and internal and external relations and functions of the state (Meyer et al 1997), or as "humanitarian" or "stabilizing"; peripheral states are represented as potentially unstable or as "weak" or "failing." The dependency perspective considers military intervention the most salient form of enforcement of dependence or reproduction of under-development, with core dominance dependant on peripheral subordination as a resource base and market (Hahn 2008).

Peripheral subordination is maintained by a convergence of interest in exploitation of peripheral non-elite populations between elite core states and elites within peripheral states (Galtung 1971; Kowaleski and Hoover 2000). The defection of peripheral state-institutional elites from state socioeconomic redistributive functions, either through predation or institutional incapacity, delegitimates the state and destabilizes recently and/or partially established state monopolies of coercion. Sub-national institutions of coercion, varying in their relations to sub-national populations from predation and exploitation to provision of socioeconomic redistribution and governance, may develop in response to state incapacity, evolving into paramilitary insurgencies contesting the state coercive monopoly within the space of the nation-state. Asymmetry of military capability or extreme ideological positions on state illegitimacy may lead to strategies of covert and indiscriminate targeting of state civil and military institutions, or those considered representatives of same, a practice defined as illegitimate, that is, as "terrorism" by states and those liable to be targeted, especially where insurgent targeting criteria may be cultural or symbolic and hence appear arbitrary (Lizardo 2008). Where targeting of state representatives extends beyond the peripheral state to representatives of core states as the perceived agents of peripheral state institutional incapacity or illegitimacy, peripheral state instability threatens the security of core states and may have destabilizing effects at the world-system level (Rennstich 2005). Stabilization operations are the core state and core-dominated international government organization (IGO) response to security instabilities resulting from peripheral state instability, resulting from world-systemic structures of dependence.

Stabilization operations may be seen as a form of global policing continuous with colonial police actions, proxy wars, or counter-insurgent conflicts under prior world-system conditions (Hahn 2008), but discontinuities are also apparent. The trans-nationality of insurgency under technological conditions of globalization allows direct peripheral paramilitary intervention in the core, in contrast to prior core immunity. This technological shift is also manifest in information technology led capital intensiveness of core militaries (Kentor and Kick 2008) and in increasing militarization of the periphery through dispersion of modern assault weapons post-Cold War. The post-Cold War shift from bipolar hegemony to multipolarity and unipolar hegemony, initially seen as signaling the end of geostrategic or realist-oppositional legitimation of core military intervention in the periphery, led to increasing legitimation of intervention in humanitarian, peripheral-developmental, or state building terms (Hahn 2008; Meyer et al 1997). While there has been a return to legitimation of intervention in realist or state-security terms in response to peripheral paramilitary intervention in core states, the necessity for core militaries engaged in counter-insurgency in destabilized states to secure local support and proxies, combined with increasing scrutiny of core military interventions by core populations through globalized media and international non-government organizations (INGOs), has ensured that legitimacy remains a key factor in stabilization operations. Core militaries are increasingly required to demonstrate "legitimacy-in-practice" vis-à-vis peripheral populations and civil actors, such as IGOs and INGOs, through civil-military cooperation. Insurgents are conversely increasingly less dependant on demonstrating their legitimacy vis-à-vis peripheral or core populations, having increasingly turned to exploitation of peripheral economic resources in place of prior core state or peripheral popular support, an exploitation facilitated by technological shifts and the parallel development of a "shadow" world-economy and world-system (Mackinlay 2002).

From the long duration world-system perspective, these discontinuities are balanced by continuity with historical conflicts between core imperial state militaries and peripheral non-state paramilitaries or militarized societies (Wilkinson 1996). However, this perspective, constructing peripheral populations as culturally deficient in capacities for state-formation and culturally predisposed to warfare, has been challenged as a core discourse legitimating past and present core military intervention in peripheral states or quasi-states (Hahn 2008).

References

Galtung, Johan (1971) "A Structural Theory of Imperialism", *Journal of Peace Research*, 8: 81–117.

Hahn, Niels S. C. (2008) "Neo-liberal Imperialism and Pan-African Resistance", *Journal of World Systems Research*, 13 (2): 142 –78.

Kentor, Jeffrey and Kick, Edward (2008) "Bringing the Military Back In: Military Expenditures and Economic Growth 1990 to 2003", *Journal of World Systems Research*, 14 (2): 142 –72.

Kowaleski, David and Hoover, Dean (2000) "International Security in the World-System: A Model of Future Dynamics", *International Studies*, 37 (3): 184 –225.

Lizardo, Omar (2008) "Defining and Theorising Terrorism: A Global Actor Centred Approach", *Journal of World Systems Research*, 14 (2): 91–148.

Mackinlay, John (2002) "Globalisation and Insurgency", Adelphi Paper 352, London: International Institute for Strategic Studies.

Meyer, John W., Boli, John, Thomas. George M. and Ramirez, Francisco, O. (1997) "World Society and the Nation State", *American Journal of Sociology*, 103 (1): 144–81.

Rennstich, Joachim Karl (2005) "Chaos or ReOrder: The Future of Hegemony in a World-System in Upheaval", *Journal of World Systems Research*, 11 (2): 209–38.

Wilkinson, David (1996) "Civilisation, Cores, World Economies and Oikumenes", in Andre Gunder Frank and Barry K. Gills (eds) *The World System: Five Hundred Years or Five Thousand*, London: Routledge.

15.7

Conclusion

World-systems analysis as a knowledge movement

Immanuel Wallerstein

World-systems analysis is more than a perspective; more even than a theory, if it is a theory. It is a knowledge movement, and that is its crucial importance to the further development of the historical social sciences. A knowledge movement is an intellectual social movement. It proposes a reorientation of the ways we organize our understanding of the world. In the case of world-systems analysis, it is based on a rejection of social science categories inherited from the nineteenth century. It proposes to replace these categories with a new historical social science.

Over the millennia, almost any argument or proposition or concept in the historical social sciences has probably been asserted a thousand times. Tracing the history of concepts is an interesting and sometimes valuable exercise in intellectual history. But it is only when a concept or set of concepts is adopted by a large enough minority of persons that it becomes able to affect the ongoing evolution of collective knowledge. When that threshold is reached, one can speak of their being a knowledge movement, which means that there is a group of scholars strong enough numerically and coherent enough organizationally that they can hold their own in the collective debates, and perhaps win the debate over time. Of course, if and when they do win the debate, these concepts constitute a new temporary dominant mode of analysis, subject in its turn to a later challenge by new knowledge movements.

Today's dominant premises of the historical social sciences were established in a period running approximately from 1850 to 1945. These premises were analyzed in the report of the Gulbenkian Commission, which I presided. (Wallerstein et al, *Open the Social Sciences: Report of the Gulbenkian Commission on the Restructuring of the Social Sciences*, Stanford, CA: Stanford University Press, 1996. This report has been translated into 25 languages.) The context within which these premises were adopted was that of the state of the world-system as it was during that period. It was the period of the height of Western domination of the world-system—politically, economically, and culturally. In the thinking of the dominant sector of the world-system, there were radical differences between "the West and the rest."

This context changed after 1945. And the changed global realities presented various dissonances to the organizational model of the historical social sciences in effect as of 1945. The two principal changes in the global realities after 1945 were (1) the assumption by the United States of the role of hegemonic power, and the particular relationship it established with the USSR, and

(2) the considerable strength that the traditional antisystemic movements came to show throughout the world-system in the post-1945 period.

The organizational model of the social sciences based on the radical epistemological difference between the West and the rest led to a sharp disciplinary separation of the mode of studying each. There emerged a clear division of academic labor. Research into the past of western societies was reserved to history. Contemporary western societies became the focus of the trio of nomothetic disciplines—economics studying the market, political science studying the state, and sociology studying the civil society. The study of the non-western world was divided between anthropology studying small so-called "tribal" groups and Oriental studies studying large but said-to-be frozen "high" civilizations. This pattern of study had trouble coping with the new post-1945 realities. This led to a debate about whether and in what ways one might adapt the dominant premises to make them more relevant to these new global realities.

In the period running from about 1945 to 1965/70, there were four different attempts to adapt the dominant premises of world social science to these new global realities. Each attempt seemed to make some plausible adjustments to the model, but each eventually demonstrated its limitations.

The first and probably most important attempt was that of modernization theory. Instead of separating the study of the "civilized" world from that of the rest of the world as distinct epistemological sites, modernization theory attempted to historicize the differences between the two sites. It argued that the "developed" world was not ontologically different from the "underdeveloped" world, but simply temporally ahead of it. The underdeveloped countries could "catch up" with the developed countries by learning from the model of more advanced countries and making certain essential changes in their socio-cultural practices.

The second attempt was that of dependency theory, emerging first out of the core-periphery analyses of ECLA under Raúl Prebisch and then elaborated with a more political emphasis by various Latin American and South Asian scholars. Unlike modernization theory, dependency theory had a different time model. As opposed to the idea that all states had started from the same point but some had moved forward faster than others, dependency theorists emphasized the "development of underdevelopment" (to use Gunder Frank's famous phrase).

What this meant was that from the same starting point, some zones had moved forward to becoming "developed" and others had moved forward in time to becoming "underdeveloped." It followed that the changes that were essential in order to catch up lay not in the socio-cultural arena but in the political and economic arenas. Only in that way could "underdeveloped" countries break out of their inferior position.

The third attempt was that of Marxist revisionism, which took two forms. The first variety was the consequence of the famous Khrushchev speech to the Twentieth Congress of the CPSU in 1956. Like modernization theory, the post-1956 Soviet model historicized the difference, and the proposed way to catch up turned out to be surprisingly similar to that advocated by modernization theory, with, however, one crucial difference. The Soviet version suggested that the "advanced" country/model, the one to emulate, was the USSR and not the United States.

A second and possibly more important variety of Marxist revisionism went in another direction. It was launched by the discussion about the "Asiatic mode of production" that took place primarily in Hungary and some western European countries. The Asiatic mode of production was one of Marx's less happy concepts, and one openly banned by Stalin. Giving renewed credence to this concept had two theoretical consequences. It raised into question the automaticity of the sequence of modes of production that presumably led from primitive Communism to the Communist world of the future. It thereby made possible discussion of the validity of the Enlightenment concept of inevitable, unilinear "progress."

The second consequence related to the discussion of the "national question." If some countries (or societies or social formations), but not all, passed through an Asiatic mode of production (or something equivalent), this meant that there was no longer a single path through which all countries passed. This implied that "Marxist" social analyses of particular parts of the world had to be based on the historical particularities of that part of the world. Classical Marxism was essentially nomothetic. This discussion led one in the direction of an idiographic epistemology. It enabled Marxist analysis to move away from trying to fit non-Western history into a sequence derived from the analysis of European thought and institutions.

The fourth attempt was that based on the Braudelian concept of the *longue durée* and its double emphasis on the central importance of socioeconomic history combined with the minimization of the importance of episodic political history, so-called *histoire événementielle*. This attack on traditional narrative politico-diplomatic historiography achieved great success in significant parts of the world historical community.

The limitations of each of the first three revisions is that they continued to regard states/societies/social formations as autonomous separate entities following parallel autonomous paths, at different paces, toward a more or less inevitable future. This failed to explain the continuing polarization of different zones of the world-system, a polarization that seemed to be widening rather than narrowing. The limitation of the Braudelian path was that its practitioners tended to confine their work to an analysis of the fourteenth to eighteenth centuries, and were largely unready to look either at the present time or at the long itinerary of historical change across the millennia.

What undid the relative successes of all four forms of revisionism was the world-revolution of 1968. To be sure, the primary concern of the students and young people who led the various uprisings that we associate with 1968 was not the structures of knowledge. In their attack on the various structures of authority, they were first of all concerned with what they saw as the nefarious consequences of US hegemony as well as with what many (perhaps most) of them saw as Soviet collusion with the United States. And secondly, they were concerned with the failure of the historic antisystemic movements to carry through on their promised second step in the two-step strategy—first obtain state power, then change the world—adopted by these movements in the late nineteenth century. In effect, they said to these movements: you have achieved (for the most part in the 1950s and 1960s) state power, more or less, but you have definitely not changed the world.

However, as the world revolutionary process went on, more and more of the participants in these uprisings began to feel that the existing modes of organizing knowledge and the categories that were being used were themselves major obstacles to the kinds of transformation they hoped to achieve. They turned their attention to the ways in which the dominant epistemological framework systematically neglected the "forgotten peoples." They began to demand that the institutions of knowledge refocus their attention on historical and sociological realities.

This new thrust—seen both by its advocates and its opponents as a political thrust—created another change in the realities of the world-system and made it possible for knowledge dissidents in all the disciplines to obtain sufficient support such that they could be said to have become knowledge movements.

World-systems analysis as a knowledge movement was born at this time and within this context. What world-systems analysis tried to do was to take features of each of the four revisionist attempts and, by joining them together, construct a tool that would be able to challenge more fundamentally the previously dominant epistemological premises that had fashioned the so-called disciplines—as intellectual arguments, as organizational apparatuses, and as cultural phenomena.

Like any other knowledge movement, world-systems analysis is not constituted by a disciplined army but rather is a collection of persons who, while they share certain key premises, pursue different emphases within this framework. I shall start by outlining what the combination of arguments that I call world-systems analysis means to me. I shall follow this with some discussion of other variants within the general camp of world-systems analysis.

The key element for me in world-systems analysis is the emphasis on the unit of analysis—a world-system rather than a state/society/social formation. The word "world" is not at all synonymous with global or planetary but is simply meant to refer to a relatively large unit (relatively large in terms of area and population) within which there is an axial division of labor. We are talking of "a" world, not "the" world, as Fernand Braudel would phrase it.

The second key element for me is that "world-systems" (like all systems) are not eternal. They have lives. They come into existence; they pursue their historical itineraries within the framework of the rules that define and govern the system; and they eventually move so far from equilibrium that the system enters into terminal structural crisis. The crucial thing here is the argument that therefore all systems are historical as well as systemic.

The emphasis of modernization theory on the historicization of the difference between core and periphery is fundamental. But so is the notion of Prebisch and the *dependistas* that the gap between core and periphery is widening rather than closing—a necessary part of explaining the drift from equilibrium over time.

A third crucial element is the refusal of the ontological separation of the imagined arenas so dear to the old dominant set of premises—the political, the economic, and the socio-cultural. For modernization theorists, just as for those who adhered to the dominant set of premises before 1945, the intellectual autonomy of the three arenas was the primary defining feature of what they called modernity. For world-systems analysis, the three so-called arenas are intrinsically linked. They define each other. None of the three is "primary" and all must be analyzed in their mutual defining of each other. Hence, world-systems analysis is inherently unidisciplinary (as opposed to being multi-, inter-, or trans-disciplinary) in terms of the historical social sciences.

Finally, world-systems analysis refuses the nineteenth-century institutionalization of the concept of the two cultures and stands for the overcoming of this false (and historically quite recent) epistemological divide. The idiographic-nomothetic divide between philosophy and science dates only to the second half of the eighteenth century. With the nineteenth-century invention of the "social sciences" as an in-between category, this divide was incorporated into the social sciences as the divide between idiographic history and the three nomothetic social sciences. World-systems analysis asserts that this epistemological divide between history and the nomothetic social sciences was always false, and is now obsolete.

As world-systems analysis gained strength as a knowledge movement, there were versions more or less within the broad camp, which placed different and/or additional emphases on the research and epistemological agenda.

One such version was that put forward by Chris Chase-Dunn, Thomas Hall, and others. This version argued against limiting practical research efforts to that of the "modern world-system" as a "capitalist world-economy"—one that was located for most of its existence in less than the entire globe. Doing so, it was suggested, tended to leave certain major questions undiscussed. One was the analysis of what was happening in modern times in regions defined as outside the axial division of labor of the capitalist world-economy, as well as the complex processes by which external zones were incorporated into the axial division of labor.

Furthermore, this group worried not only that a practice of devoting research efforts primarily, even exclusively, to the capitalist world-economy led to what might be called spatial exclusions of the analysis. It worried also about what might be called long-term temporal exclusions of the analysis. This group wished to look at two longer-term issues. One was the very long-term historical development of human social interaction. They actively confronted the long-standing issue of historical "evolution"—what "evolved," and whether evolution was teleological.

In addition, however, this group felt that there was valuable knowledge to be unearthed by systematic comparison of different kinds of historical systems, for which the cases would necessarily have to be drawn from analyses of historical systems of all kinds and in all geographical areas over several thousand years. One might call this comparative historical systems analysis.

A second version of comparative historical systems analysis that nonetheless limited itself to the "modern" historical period (ca. 1500 to the present) was that put forward by Giovanni Arrighi, Takeshi Hamashita, and others. Basically, they proposed to compare the evolution of a China-centered trading system with that which developed as a western Europe-centered trading system over the post-1500 period. They looked at the ways in which the structures of the two systems differed—Arrighi arguing that the differences persist to this day—as well as at the increasing linkages between the two systems over the centuries.

The increasing economic and geopolitical importance of China in the world-system since the 1980s increasingly turned the attention of world scholarship to the historic role of China, and led in particular to complaints about the intellectual neglect of China's role by pan-European scholars. In the period since then, a relatively large amount of literature has been produced on China and the world in both East Asian and European languages. This literature is diverse and only some of it can be considered to be within the broad framework of world-systems analysis.

Andre Gunder Frank in his post-1990 writings insisted on the concept that only one world-system ever existed (and he therefore spelled it without the hyphen), and he traced its existence back at least some 5000 years. For Frank, China was always the center of this system (except rather briefly in the nineteenth century and part of the twentieth century). While Frank used many methodological tools derived from world-systems analysis, he attacked other versions (indeed all other versions) as being Eurocentric, and rejected the very concept of capitalism as a variable to include in the analysis.

Others in this group of China-oriented scholars, such as Kenneth Pomeranz, insisted on a re-analysis of the data comparing western Europe and China in the sixteenth to eighteenth centuries, and sought to show that what he called "the great divergence" occurred only as of the nineteenth century. Pomeranz, however, does not seek to place himself within the family of world-systems analysts, even though his concrete analysis in some ways concords with the Arrighi–Hamashita version. In fact, Pomeranz's version reinforces the traditional and mainstream social science view that the key shift in modern times was the "industrial revolution" that was considered to have occurred (at least primarily) in England at the cusp of the nineteenth century.

While this argument among world-systems analysts was going on in the period 1970 –2010, two things happened, altering the character of world-systems analysis as a knowledge movement. The first was the rise, even triumph, of neoliberal globalization in the world-system. The second was the changed attitude toward world-systems analysis by the major disciplinary organizations and textbooks. Let us consider each in turn.

The stagnation of the world-economy beginning in the 1970s (a Kondratieff B-phase) was combined with the undermining of the dominance of centrist liberalism as a result of the world-revolution of 1968. The combination allowed conservative forces to launch a worldwide attempt to reverse all the political, economic, and cultural changes that had occurred in the 1945–70 period. This political campaign was given the deceptive label of neo-liberalism, and was incarnated originally in the political success of Mrs. Thatcher's transformed Conservative Party in the United Kingdom and Ronald Reagan's transformed Republican Party in the United States.

The neo-liberals shifted the analytic framework they applied to the world-system from "developmentalism" (which had prevailed in the 1945–70 period) to something they called globalization. They used this new framework to impose, primarily via the US Treasury and the International Monetary Fund (IMF), a practical program that came to be called the Washington Consensus. It demanded that all countries that were not "developed" institute a program that gave priority to export-oriented growth, while simultaneously opening their border to foreign direct investment, privatizing state enterprises, reducing their welfare programs, and downsizing their bureaucracies. Geopolitically, this political effort was enormously successful worldwide in a period running roughly from the mid-1970s to circa 1995.

Within the historical social sciences, the response to this new world political reality was to make globalization the principal buzzword of research and publication. One of the results was somewhat paradoxically to make world-systems analysis somewhat more academically respectable. Previously, world-systems analysis either was subject to strong denunciation for its alleged errors or was treated by a scornful refusal to acknowledge its scholarly character. Suddenly, world-systems analysis came to be seen, was even acclaimed, as a forerunner of globalization theory, if in a version that was too politically committed. World-systems analysis (usually referred to as world-system theory) came to be included in writings and textbooks as one alternative theoretical view among a list of alternative theoretical views of globalization.

In fact, however, world-systems analysis was not a forerunner of globalization theory but something quite different. World-systems analysis had never sought to be one among a list of alternative theories. It thought of itself as formulating a rejection of the entire framework of mainstream social science. World-systems analysis called for a drastic reshaping of the intellectual framework of the social sciences, calling for a unidisciplinary reorganization. World-systems analysis combined this view of the historical social sciences with a demand to overcome the epistemological division of "two cultures" and the recreation of a singular epistemological framework for all knowledge.

The triumph of the Washington Consensus came under political challenge in the second half of the 1990s, as the neoliberal promises of universal economic betterment turned out to be a mirage. This increasing disillusionment was reinforced by the successive financial crises that have been going on ever since, and which finally led to serious questioning about the viability of the promised return to universal economic "growth."

The degree to which the capitalist world-economy can resume its traditional and repeated returns to normal expansion is a matter of some debate even within the camp of world-systems analysts. If one believes, as I do, that the modern world-system is in structural crisis, is therefore bifurcating, and is in the midst of a transition to some new global system, then one question is what happens to world-systems analysis as a knowledge movement in this process.

The strength of world-systems analysis as a knowledge movement is that it has resisted the temptation to define itself too narrowly and dogmatically, while still not allowing itself to be defined so loosely that anything that seems to deal with questions beyond the space of single nations/societies/social formations is deemed within the family. This has been a difficult organizational project, one however that thus far has worked. Indeed, world-systems analysis as a knowledge movement has been relatively successful in spreading its adherents within all the existing major disciplines of the historical social sciences and spreading its organizational base beyond the United States to other parts of the world—notably, but not only, to Latin America, western Europe, and east Asia.

The question for world-systems analysis as a knowledge movement is whether it can continue to play the organizational game the way it has played it up to now. To the degree that a structural transition occurs, success for world-systems analysis might be measured by its disappearance as a knowledge movement as a result of the radical reorganization of the world of knowledge. It is much too early to tell whether this will indeed happen. But if world-systems analysis ends up as being merely one more theoretical position within the social sciences, it will have failed in what it had hoped to accomplish.

Index

www.ingramcontent.com/pod-product-compliance
Lightning Source LLC
LaVergne TN
LVHW080114011026
849160LV00013B/1099
* 9 7 8 1 1 3 8 1 0 7 1 9 9 *